BOSTON STUDIES IN THE PHILOSOPHY OF SCIENCE

VOLUME XXXII

PSA 1974

SYNTHESE LIBRARY

MONOGRAPHS ON EPISTEMOLOGY,

LOGIC, METHODOLOGY, PHILOSOPHY OF SCIENCE,

SOCIOLOGY OF SCIENCE AND OF KNOWLEDGE,

AND ON THE MATHEMATICAL METHODS OF

SOCIAL AND BEHAVIORAL SCIENCES

Managing Editor:

JAAKKO HINTIKKA, *Academy of Finland and Stanford University*

Editors:

ROBERT S. COHEN, *Boston University*

DONALD DAVIDSON, *Rockefeller University and Princeton University*

GABRIËL NUCHELMANS, *University of Leyden*

WESLEY C. SALMON, *University of Arizona*

VOLUME 101

BOSTON STUDIES IN THE PHILOSOPHY OF SCIENCE

EDITED BY ROBERT S. COHEN AND MARX W. WARTOFSKY

VOLUME XXXII

PSA 1974

PROCEEDINGS OF THE 1974 BIENNIAL MEETING
PHILOSOPHY OF SCIENCE ASSOCIATION

Edited by

R. S. COHEN, C. A. HOOKER,
A. C. MICHALOS AND J. W. VAN EVRA

D. REIDEL PUBLISHING COMPANY

DORDRECHT-HOLLAND / BOSTON-U.S.A.

Library of Congress Cataloging in Publication Data

Philosophy of Science Association.
    Proceedings of the biennial meeting.

    Dordrecht, D. Reidel Pub. Co.
    v. 23 cm. (Boston studies in the philosophy of science)
    (Synthese library)
    1. Science—Philosophy—Congresses.
    I. Series.

Q174.B67   subser.   501   72-624169
                                      rev   MARC-S
ISBN 90-277-0647-6
ISBN 90-277-0648-4 pbk

Published by D. Reidel Publishing Company,
P.O. Box 17, Dordrecht, Holland

Sold and distributed in the U.S.A., Canada, and Mexico
by D. Reidel Publishing Company, Inc.
Lincoln Building, 160 Old Derby Street, Hingham,
Mass. 02043, U.S.A.

All Rights Reserved
Copyright © 1976 by D. Reidel Publishing Company, Dordrecht, Holland
No part of the material protected by this copyright notice may be reproduced or
utilized in any form or by any means, electronic or mechanical,
including photocopying, recording or by any informational storage and
retrieval system, without written permission from the copyright owner

Printed in The Netherlands

PREFACE

For this book, we have selected papers from symposia and contributed sessions at the fourth biennial meeting of the Philosophy of Science Association, held at the University of Notre Dame on November 1-3, 1974.

The meeting was lively and well-attended, and we regret that there was no way to record here the many stimulating discussions after the papers and during the informal hours. We also regret that we had insufficient space for all the contributed papers. Even more, some of the symposia were not available: those on systems and decision theory (C. W. Churchman, P. Suppes, I. Levi), and on the Marxist philosophy of science (M. W. Wartofsky, R. S. Cohen, E. N. Hiebert). Unhappily several individual contributions to other symposia were likewise not available: I. Velikovsky in the session on his own work and the politics of science, D. Finkelstein in the session on quantum logic.

Memorial minutes were read for Alan Ross Anderson (prepared by Nuel Belnap) and for Imre Lakatos (prepared by Paul Feyerabend). They initiate this volume of philosophy of science in the mid-seventies.

> ROBERT S. COHEN, *Boston University*
> C. A. HOOKER, *University of Western Ontario*
> ALEX C. MICHALOS, *University of Guelph*
> JAMES W. VAN EVRA, *University of Waterloo*

TABLE OF CONTENTS

PREFACE v

MEMORIAL MINUTES: ALAN ROSS ANDERSON, IMRE LAKATOS XI

SYMPOSIUM:
THE UNITY OF SCIENCE

ROBERT L. CAUSEY / Unified Theories and Unified Science 3
LAWRENCE SKLAR / Thermodynamics, Statistical Mechanics and the Complexity of Reductions 15
THOMAS NICKLES / Theory Generalization, Problem Reduction and the Unity of Science 33

CONTRIBUTED PAPERS: SESSION I

WILLIAM A. WALLACE / Galileo and Reasoning *Ex Suppositione*: The Methodology of the *Two New Sciences* 79
J. V. STRONG / The *Erkenntnistheoretiker*'s Dilemma: J. B. Stallo's Attack on Atomism in his *Concepts and Theories of Modern Physics* (1881) 105

SYMPOSIUM:
GENETICS, IQ AND EDUCATION

N. J. BLOCK / Fictionalism, Functionalism and Factor Analysis 127
NORMAN DANIELS / IQ, Heritability, and Human Nature 143
KENNETH KAYE / The IQ Controversy and the Philosophy of Education 181

CONTRIBUTED PAPERS: SESSION II

HENRYK SKOLIMOWSKI / Evolutionary Rationality 191

LINDA WESSELS / Laws and Meaning Postulates (in van Fraassen's View of Theories) 215
PAUL FITZGERALD / Meaning in Science and Mathematics 235
PHILIP A. OSTIEN / Observationality and the Comparability of Theories 271

SYMPOSIUM:
SCIENCE EDUCATION AND THE PHILOSOPHY OF SCIENCE

MICHAEL MARTIN / The Relevance of Philosophy of Science for Science Education 293
HUGH G. PETRIE / Metaphorical Models of Mastery: Or, How to Learn to Do the Problems at the End of the Chapter of the Physics Textbook 301
ROBERT PALTER / Philosophy of Science, History of Science, and Science Education 313

CONTRIBUTED PAPERS: SESSION III

RAIMO TUOMELA / Causes and Deductive Explanation 325
LAIRD ADDIS / On Defending the Covering-Law 'Model' 361
Comment:
    CARL G. HEMPEL / Dispositional Explanation and the Covering-Law Model: Response to Laird Addis 369
JAMES H. FETZER / The Likeness of Lawlikeness 377
PETER KIRSCHENMANN / Two Forms of Determinism 393
PETER A. BOWMAN / The Conventionality of Slow-Transport Synchrony 423

SYMPOSIUM:
TECHNOLOGY ASSESSMENT

TOM SETTLE / The Bicentenary of Technology Assessment 437
JOSEPH AGASSI / Assurance and Agnosticism 449

HENRYK SKOLIMOWSKI / Technology Assessment as a Critique of a Civilization 459

SYMPOSIUM:
VELIKOVSKY AND THE POLITICS OF SCIENCE

LYNN E. ROSE / The Domination of Astronomy Over Other Disciplines 469
M. W. FRIEDLANDER / Some Comments on Velikovsky's Methodology 477
ANTOINETTE M. PATERSON / Velikovsky Versus Academic Lag (The Problem of Hypothesis) 487

SYMPOSIUM:
QUANTUM LOGIC

PETER MITTELSTAEDT / Quantum Logic 501
JOHN STACHEL / The 'Logic' of 'Quantum Logic' 515

CONTRIBUTED PAPERS: SESSION IV

I. I. MITROFF / Integrating the Philosophy and the Social Psychology of Science, or a Plague on Two Houses Divided 529
ROBERT M. ANDERSON JR. / The Illusions of Experience 549

SYMPOSIUM:
DEVELOPMENT OF THE PHILOSOPHY OF SCIENCE

GROVER MAXWELL / Some Current Trends in Philosophy of Science: With Special Attention to Confirmation, Theoretical Entities, and Mind-Body 565
ERNAN MCMULLIN / History and Philosophy of Science: A Marriage of Convenience? 585
HILARY PUTNAM / Philosophy of Language and Philosophy of Science 603

SYMPOSIUM:
HISTORY AND PHILOSOPHY OF BIOLOGY

KENNETH F. SCHAFFNER / Reductionism in Biology: Prospects and Problems	613
MICHAEL RUSE / Reduction in Genetics	633
DAVID L. HULL / Informal Aspects of Theory Reduction	653
WILLIAM C. WIMSATT / Reductive Explanation: A Functional Account	671

CONTRIBUTED PAPERS: SESSION V

NANCY DELANEY CARTWRIGHT / How Do We Apply Science?	713
WILLIAM DEMOPOULOS / What is the Logical Interpretation of Quantum Mechanics?	721
INDEX OF NAMES	729

MEMORIAL MINUTES

(1)

ALAN ROSS ANDERSON died at home in Pittsburgh on December 5, 1973, of cancer. It was characteristic of him that throughout the long period of his illness he remained active and cheerful. He is survived by his wife, Carolyn, his mother, and four children.

Alan made a number of significant and typically seminal contributions to mathematical logic, always presenting his ideas in a delightfully informal literary style designed to satisfy both sense and sensibility. Notable are his early work in modal logic, his reduction of deontic logic to modal logic via introduction of a propositional constant for The Bad Thing, and his work, largely in collaboration with me, on relevance logics, issuing in a two-volume work *Entailment: the logic of relevance and necessity*.

With interests always wide-ranging, he collaborated over the years with the social psychologist O. K. Moore on early education of children, autotelic folk-models, and related topics in the theory of social interaction, placing unvarying emphasis on the central theme of his career: the potential application of formal methods. Alan also edited two books outside his specialties: *Minds and Machines*, and (with others) *Philosophic Problems*.

Alan's teaching life was spent at Dartmouth, Yale, and Pittsburgh. Both in and out of the classroom Alan was a spectacularly good teacher with unerring pedagogic judgement and a superb sense of the *a propos*. One of his favorite phrases was 'the spirit of the enterprise'; and no one excelled him at communicating not just results but feels, approaches, styles – and at teaching others the importance of doing likewise.

Alan was deeply involved in service to the profession he loved, always combining seriousness about getting the job done with his unfailing casual good cheer. He served in reviewing and editorial capacities for a number of journals, as one or another officer of a variety of professional organizations, and also contributed unstintingly to his own university, both at Yale and at Pittsburgh.

I do not know anyone who generated as much affection in as many people as Alan. We surely miss him. For those who may wish to remember him, an endowed Alan Ross Anderson Memorial Fund has been established at the University of Pittsburgh.

NUEL D. BELNAP, JR.

(2)

IMRE LAKATOS died at home in London, on February 2, 1974. Lakatos was a fascinating person, an outstanding philosopher, and the best theoretician of science of the past fifty years. He was a rationalist, for he believed that man should use reason in his private life and in his attempt to understand the world. He was an optimist for he thought reason capable of solving the problems generated by the attempt. He had a realistic view of this capability, for he saw that it cannot be completely distilled into a set of abstract rules, into a 'logic', and that it cannot be improved by an abstract comparison of such rules either: if reason is to have a point of attack in our real world with its complex episodes and its hair-raising ideas and institutions, then it must have some *sophistication*, it must not be less complex and less cunning than the theories it is supposed to evaluate; on the other hand, it must *not be too severe* or the attempt to improve science will lead to the elimination of science. We need a form of reason that is neither content with the mere contemplation of science (as are most so-called reconstructions), nor intent on Utopian reform. The demands of reason must be adapted to the historically given material without losing the power to transform this material.

This *practical* side of methodology is only rarely examined by philosophers of science. The usual procedure is to compare abstract rules with other abstract rules, considering also certain simple logical restrictions. This is how Popper defends his rule of falsification: singular statements may entail the *negation* of universal statements, they do not entail universal statements. But the question still remains whether scientific theories in the form in which they are used by scientists and in which they lead them to their surprising discoveries are ever in sufficient harmony with observation to survive the rule. Lakatos has seen that they are not and he has invented a criterion for the comparative evaluation not of ideal statements, but of the logically imperfect, anomaly-ridden and often absurd theories that constitute science. The criterion can be applied to 'pure' mathematics as well. The underlying theory of scientific change combines strict judgement and free decision, it uses rules of thought without overlooking historical accident, it implies a historical view of reason and a rational view of history. It is one of the most important achievements of 20th century philosophy.

But Imre Lakatos was not only a theoretician. He was also a most efficient propagandist. He strengthened the power of his arguments with dazzling oratorical displays in lectures, biting humour in discussions, and charming persuasion in private conversations. Soon he was surrounded by a new kind of intellectual community based on friendship and sceptical interest rather than on the slavish acceptance of an idea. His collaborators were not always rationalists, for he was not at all narrow-minded, and even a Dadaist like myself could derive great pleasure from working with him. This was possible because Imre Lakatos was above all a kind and warmhearted human being, deeply impressed by the growing irrationality and injustice in this world, by the almost unsurmountable power that mediocrity possesses today in almost all fields, the younger generation not excluded, always ready to defend his ideas and his principles, even in trying and dangerous situations, not easily silenced and yet not without insight into the basic absurdity of all human effort. A man like him cannot be grasped by studying his writings alone. And his achievement will survive only in the hands of those who have his freedom, his inventiveness, and his love for life.

PAUL FEYERABEND

SYMPOSIUM

THE UNITY OF SCIENCE

ROBERT L. CAUSEY

UNIFIED THEORIES AND UNIFIED SCIENCE*

ABSTRACT. Discussions of unified science frequently suppose that the various scientific theories should be combined into one unified theory, and it is usually supposed that this should be done by successive reductions of the various theories to some fundamental theory. Yet, there has been little systematic study of the characteristics of unified theories, and little foundational support for the use of reductions as a unifying procedure.
In this paper I: (a) briefly review some of my previous work on microreductions, (b) state some conditions which are necessary in order for a theory to be unified, (c) argue that when certain identities exist between the elements in the domains of two theories, then the only satisfactory way to combine these two theories into one unified theory is by a microreduction, and (d) indicate briefly some further applications and consequences of this work.

I

Scientific research is largely concerned with the discovery of laws about the attributes and behavior of various kinds of things, and with the construction of theories in which some laws are explained by others. It is convenient to think of the various branches of non-historical science as codified in various theories about different domains of things. Unified science would then be achieved, at least in part, if all of these theories were combined into one *unified* theory. This is a familiar ideal, which motivates much of the great interest in the reduction of theories. Yet there has been little systematic study of the characteristics of unified theories. It is usually just taken for granted that one good way to unify two theories is to reduce, or if possible to microreduce, one to the other. It is then frequently supposed that the best (and perhaps the only) way to achieve unified science is by the reduction of all theories to one fundamental theory. This is no arbitrary ideal; yet, it is desirable that it have a foundation in terms of some general conception of what a unified theory should be.

In the following I will: (a) briefly review some previous work on microreductions, (b) state some conditions which are necessary in order for a theory to be unified, (c) argue that when certain identities exist between the elements in the domains of two theories, then the only satisfactory way to combine these two theories into one unified theory is by a micro-

R. S. Cohen et al. (eds.), PSA 1974, 3–13. All Rights Reserved.
Copyright © 1976 by D. Reidel Publishing Company, Dordrecht-Holland.

reduction, and (d) indicate briefly some further applications and consequences of this work.

II

Consider a theory T about the objects in the domain, Dom. T is formulated in a language which describes kinds of objects in Dom, as well as attributes of these objects, where an attribute is a property, relation, or quantity. This language consists of symbols used in set theory and mathematics (the 'logical symbols') plus a set $\mathscr{L}$ of nonlogical predicates (including function symbols). $\mathscr{L}$ is the union of two disjoint subsets, $\mathscr{T}$ and $\mathscr{A}$, where $\mathscr{T}$ is the set of *thing-predicates* and $\mathscr{A}$ is the set of *attribute-predicates*. A thing-predicate is interpreted as a name for a *kind* of element in Dom, and an attribute-predicate is a name for an *attribute* applying to elements of Dom. T may use some new predicates defined in terms of those of $\mathscr{L}$, but no definite descriptions of particulars, kinds, or attributes are used in T.

In general, T will contain two kinds of sentences, *law-sentences*, which represent or state laws, and *identity-sentences* (*identities*) which state identities. These can be *thing-identities* between two thing-predicates or *attribute-identities* between two attribute-predicates. Such identities imply that the identified predicates are co-extensional, but they also imply that this co-extensionality is not a law and not subject to causal explanation. They are thus different from an attribute-correlation, which is a nomological co-extensionality.

A sentence that is not subject to causal explanation is said to be *noncausal*. Analytic sentences and logical truths are necessarily noncausal; thing-identities and attribute-identities are also noncausal, and they can be synthetic. All sentences in T are supposed to be true, and it appears that the only possible kinds of true noncausal sentences are logically true sentences, analytically true sentences (which would follow logically from definitions of defined terms), true thing- and attribute-identities, true assertions of nonidentity between predicates, and logical consequences of sets of true noncausal sentences.[1]

Now it is possible that two law-sentences, L_1, L_2, in $\mathscr{L}$ might state the same law. If they do, they are said to be *nomologically-equivalent*, *n-equivalent*, or *n-eq*. Elsewhere, I have proposed the following criterion for *n*-equivalence:

C: Let L_1, L_2 be law-sentences in $\mathscr{L}$. Then L_1 n-eq L_2 iff there is a set N of true noncausal sentences of $\mathscr{L}$ such that $L_1 \leftrightarrow L_2$ is derivable from N. (Where '$\leftrightarrow$' is the material biconditional.)

The most useful application of C is to the case where L_1 contains occurrences of predicate α, L_2 contains occurrences of predicate β, and L_1 and L_2 are uniform substitution instances of each other under substitution of β for α in L_1, and α for β in L_2, where α and β denote the same kind or attribute. Of course, it is necessary that the denotations of α and β be the same and not merely nomologically co-extensional.

Within T explanations of some laws will be represented by certain explanatory derivations from other, fundamental laws. Corresponding to C, it is also possible to state at least a sufficient condition for two explanatory derivations to represent the same explanation and thus be *explanatorily-equivalent, e-eq*.

Now suppose that we have two such theories, T_1, T_2 in two languages $\mathscr{L}_1$, $\mathscr{L}_2$, and these theories are about the elements in the domains, Dom_1, Dom_2. In order to reduce T_2 to T_1, the laws in T_2 must be understandable, and therefore explainable, solely in terms of the laws in T_1 without the help of any other laws. If, as is frequently suggested, we derive the laws of T_2 from those of T_1 plus nomological co-extensionalities used as connecting principles, then we have not really explained the T_2-laws by the T_1-laws alone, for we have added extra laws to the T_1-laws. I maintain that, in order to reduce T_2 to T_1, each predicate in $\mathscr{L}_2$ must be identified with some predicate in, or defined in, $\mathscr{L}_1$, and that the law-sentences in T_2 must be explainable by law-sentences in T_1 plus the appropriate thing-identities and attribute-identities. Indeed, when these identities exist, each law-sentence in T_2 is n-eq to some law-sentence in T_1 and therefore should be explainable within an adequate T_1.

This has been an extremely condensed review of work previously presented in Causey (1974, 1972a, 1972b). Many details, and essentially all justification, have been omitted here; but the general ideas should be clear. Since the connecting principles are identities, the kinds and attributes and laws of T_2 are identified with kinds and attributes and laws of T_1. We now need some conditions for a unified ontology and for unity of laws.

III

Consider a theory $T = F \cup I \cup D$, where F is the set of *fundamental law-sentences*, I is the set of *identities*, and D is the set of *derivative law-sentences*. Not all deductive-nomological derivations are explanatory, i.e., represent causal explanations, but presumably suitable derivations from the fundamental law-sentences of a good theory are explanatory.

First we say that a derivation of a sentence ϕ from a set of sentences Γ is a *minimal derivation* iff ϕ is not derivable from any proper subset of Γ. Now let Γ be a subset of $F \cup I$, and let L be a law-sentence in T such that no law-sentence n-eq to any member of Γ is derivable from L. Then a minimal derivation of L from Γ is said to be an *explanatory derivation*. When there exists in T an explanatory derivation of L, then L is *explainable in* T. Of course, these are very general conditions. They will be restricted further below.

We now need to impose some structure on the language and domain of T. There is a distinguished subset of predicates of $\mathscr{L}$ which are called *basic*, and which satisfy the condition that no member of this set is identical with a predicate definable in terms of other members of this set, where a simple identity between two single predicates is not considered a definition. Any kind or attribute which is denoted by a basic predicate is also called *basic*. Any other kinds or attributes of T are called *derivative*. For each derivative kind or attribute, there is a predicate defined in terms of basic predicates which denotes that kind or attribute. The nature of the definitions will depend on the type of theory under consideration, but a thing-predicate is not definable only in terms of attribute-predicates, nor an attribute-predicate only in terms of thing-predicates. Derivative kinds or attributes may in addition be named by single predicates in $\mathscr{L}$, but such predicates would be called *derivative*.

Now let $\alpha, \beta_1, \ldots, \beta_n$, be in $\mathscr{L}$, and let ϕ be in $\mathscr{L}$ or be defined in terms of $\beta_1, \ldots, \beta_n$, and suppose that

$$(x_1)\ldots(x_n)(\alpha x_1 \ldots x_n \leftrightarrow \phi x_1 \ldots x_n)$$

is a law-sentence of T. Then α is said to occur as an *isolated correlate* in this law-sentence.

Finally, we say that a predicate α occurs *essentially* in a sentence ϕ iff there is no sentence logically equivalent to ϕ which has no occurrence of α. In the following, the word 'occurrence' will mean essential occurrence.

Using all of this terminology, we can now state several conditions which should be satisfied by any unified theory; indeed, several of these conditions should be satisfied by any good theory using the kind of language I have described.

U_1. Each predicate in $\mathscr{L}$ occurs in some sentence of T.

U_2. The extension of any predicate of $\mathscr{L}$ is (at any time) a subset of Dom, or a set of ordered n-tuples of elements of Dom.

U_3. T is closed under derivability.

U_4. Each law-sentence in D is explainable in T.

U_5. No law-sentence in F is explainable in T.

U_6. Each law-sentence L_1 in F has an occurrence of a predicate of $\mathscr{L}$ which also occurs in another law-sentence L_2 of F such that L_1 is not n-eq L_2. (A somewhat similar condition occurs in Bunge (1967, pp. 394–395).)

U_7. No basic predicate of $\mathscr{L}$ occurs in law-sentences of F only as an isolated correlate.

U_8. All true identities between predicates in, or defined in, $\mathscr{L}$ are included in I.

U_9. If α is an n-place attribute-predicate of $\mathscr{L}$ such that the k-th component of any ordered n-tuple satisfying α is always of a derivative kind, then α is a derivative predicate (and hence names a derivative attribute, which is also denoted by some definition in terms of basic predicates).

Time limitations preclude any detailed justifications of these conditions here, but a few clarifying remarks are in order. First of all, U_2 is a homogeneity condition which prevents reference to extraneous things or attributes outside of T. It is very important and will be discussed further below.

U_4 and U_5 are quite natural. U_5 allows the possibility that a member of F may be derived from other members of $F \cup I$. However, no such derivation is explanatory, and it can be shown that in most derivations of this kind, the law derived will be n-eq to one of the premises.

U_6 and U_7 are not needed for our present purposes, but they are important because they prevent T from having certain isolated items. In addition, U_7 is a kind of limited Occam's razor principle. It does not exclude the possibility of nomological co-extensionalities between basic predicates, but it does exclude some utterly trivial ones.

If U_8 is not satisfied, then some phenomena which are really the same will *appear* to be different in T, for law-sentences and explanatory derivations which are really n-eq and e-eq, respectively, will, according to T, be nonequivalent. Moreover, if a true attribute-identity is not stated, then it is likely to be misinterpreted as a nomological co-extensionality. In such a case T will contain a pseudo-law, and fruitless attempts might be made to explain it.

To see the rationale for U_9, recall that a derivative kind is denoted by a definition, and according to T this kind is exactly what its definition makes it. Thus, if T is to provide a single, coherent characterization of this derivative kind, the attributes of this kind should be determined by its definition in terms of basic predicates and the laws containing these basic predicates. Hence, all of its attributes must be denoted by, or definable in terms of, basic predicates. Furthermore, it should be noted that T is not required to have any derivative kinds, although many theories do. Any derivative kinds which do occur in T are supposed to play a secondary role. It is natural to require that all fundamental laws of T be laws about basic kinds and attributes applying to basic kinds. U_9, together with the other conditions, guarantees that this will be the case.

Altogether, $U_1 - U_9$ go a long way towards requiring that T has unified and interlocking systems of laws, ontology, and intratheoretic explanations.

IV

Now we will consider the role of microreductions in the unification of two theories. In general, if two theories, T_1, T_2, are to be unified, it is necessary to construct some unified theory T which contains all of the laws and identities of T_1 and T_2. However, to keep the discussion within reasonable bounds, we will only consider unifications such that the kinds and attributes of T are those used in either T_1 or T_2. Even with this restriction, unification might be accomplished in various ways; fortunately, we are concerned with one type of situation which is quite familiar in modern science, at least from a somewhat idealized point of view.

Assume that T_1 and T_2 are unified theories. T_1 has both basic and derivative kinds and attributes. The elements of Dom_1 which are of the basic kinds are considered as possible parts, and the elements of Dom_1 which are of the derivative kinds are defined as structures composed of two or more parts. In Causey (1972b) I have discussed such theories in considerable detail. Among other things, they should contain laws stating the conditions under which various kinds of structures do or do not form.

In contrast to T_1, T_2 is a theory which has only basic kinds and basic attributes. Moreover, we suppose that it is the case that each (basic) kind of Dom_2 is identical with some derivative kind in Dom_1, and conversely. In practical situations this is an empirical condition which is not easy to establish, although the development of science shows that many such claims do eventually become accepted. In our idealized picture, we simply assume that the appropriate identities between Dom_2-kinds and Dom_1-kinds are given. It is also recognized that the converse part of this assumption is somewhat of an idealization which will simplify the argument without essential loss of generality.

Under the conditions described, many would immediately assert that the proper way to unify T_2 and T_1 is to microreduce T_2 to T_1. But this assertion is not obvious, and at least one philosopher, Schlesinger (1963, Chapter 2), has argued that it is an unjustifiable prejudice. Therefore, we must examine the possible forms of the unifying theory T.

Suppose first that T is a theory whose basic kinds are exactly those of T_2. Then the basic kinds of Dom_1 are derivative kinds of Dom, and they, and the derivative attributes of T which apply to them, must be denoted by definitions. Except for trivial theories it is unlikely that suitable definitions will exist. Furthermore, if the definitions are obtained, then a law-sentence about uncombined parts must be n-eq to a law-sentence about structures composed of parts. This is an extremely bizarre conclusion.

Moreover, consider the fundamental law-sentences of T_1 reformulated in T. If all of these reformulations are fundamental law-sentences of T, then T is essentially a disguised microreduction formulated in an awkward manner.[2] Hence, T is only of interest if some of these reformulations are derivative in T. Therefore, suppose there is such a reformulation which states, for example, that a certain basic kind in Dom_1 has a certain property. Then this sentence must be derivable in T from premises containing law-sentences about the attributes of wholes. I believe that no such derivation can be considered to represent a causal explanation. Our conception of causality is such that we cannot consider a property of an uncombined thing to be caused by attributes of possible structures of which this thing may or may not be a possible part. (Cf. Oppenheim and Putnam (1958, p. 15) and Schlesinger (1963, p. 51).) The contrary view would lead to the conclusion that real events could in part be caused by the attributes of possible objects, and it would be somewhat like introducing final causes into our explanations.

There is another consideration which, although not conclusive, supports the view that T is unsatisfactory. Recall that T_1 has law-sentences stating the conditions under which various structures do or do not form. In T it is likely that some of these law-sentences will be derivative and will be derived from laws about nonstructural attributes of structures. Such attributes of a structure may help to determine its future persistence; but (except for limiting cases) they cannot be causes of the formation of structures. Therefore, within T the derivations of laws of formation of structures would most likely fail to be explanatory.

Now let us consider the possibility that T is a theory whose basic kinds consist of a proper subset of those of T_2 together with a proper subset of those of T_1. Then some basic kind in Dom is a structured whole composed of certain kinds of parts, which are basic in T_1 and hence not definable in terms of other basic kinds of T_1. But these kinds of parts cannot be basic kinds in T, for then the structured whole would not be basic. Thus, these kinds of parts must be derivative kinds in T, and must be denoted by definitions at least partly in terms of predicates of T which name structured wholes and perhaps also attributes of such wholes. Therefore, we would obtain a theory with laws about some uncombined parts n-eq to laws at least in part about wholes and their attributes, in which case we would encounter the same kinds of problems as in the above case.

Now consider the third and last possibility, namely, a theory T which has exactly the same basic predicates as T_1. Because of U_8, T has identities between each basic thing-predicate of $\mathscr{L}_2$ and its corresponding structural definition in $\mathscr{L}_1$. Then, because of U_9, each basic attribute-predicate of $\mathscr{L}_2$ must be identified with some attribute-predicate defined in terms of basic predicates in $\mathscr{L}_1$. By virtue of condition C plus these identities, the fundamental law-sentences of T_2 are n-eq to T_1-law-sentences about wholes and their attributes. If T_1 is well developed, then it most likely already contains these laws, in which case the fundamental laws of T can be those of T_1. Such a unification fits precisely the conditions which I have argued, in Causey (1972a, 1972b), are necessary and sufficient for a microreduction. If the fundamental laws of T_1 are not entirely adequate, then they can be supplemented with some of those from T_2 reformulated in the language of T_1 by means of the thing- and attribute-identities. In this case T amounts to a microreduction to a slightly strengthened T_1. Constructing a suitable, unifying T is our major goal. It is to be expected that T_1, and perhaps T_2, might undergo some modifications in the process of unification. These modifications will be required by the criteria for unity and for causal explanations.

V

The above discussion shows the superiority of microreductions as a unifying procedure when the appropriate identities exist between the domains of two theories. I shall now try to indicate briefly some further applications of the ideas presented here.

Suppose we wanted to reduce classical chemistry to the atomic-molecular theory, and we found a chemistry text which said that sodium cyanide is poisonous to rats. Our reduction would encounter problems with the attribute, *poisonous to rats*, for this is a relational attribute which refers to something, *rats*, outside of the domain of a purely chemical theory. This attribute violates condition U_2. As a general rule, when attempting a reduction, it is advisable that both theories involved be unified.

Frequently human behavior is described in terms of remote consequences of this behavior, for instance, consequences within a particular social system. Such descriptions may also violate U_2 if they are used ad hoc within a theory of individual psychology. A sophisticated, unified theory of individual psychology should explain why certain social structures form or dissociate under various conditions, but it cannot alone be expected to explain the origins of particular social structures. Moreover, a description of a social structure (with institutions, roles, etc.) may be used as a set of boundary conditions in explaining the specialized behavior of persons within it. But this does not imply that general laws of individual behavior are irreducible to causal interactions, or that boundary conditions are irreducible principles. This point is discussed in Causey (1969) and Hull (1974).

Of course, many reasons have been given for the alleged irreducibility of the behavioral sciences. For instance, recently it has been argued that suitable identities are unobtainable because high-level descriptions of complex systems are highly abstracted from descriptions of their internal mechanisms. For instance, Block and Fodor (1972) maintain that psychological states cannot be expected to correspond with physical states because they believe it likely that creatures with very different physical structures can have the same psychological states. Such arguments have already been criticized at length in Gendron (1971) and in Kalke (1969), and I only wish to emphasize one point. Block and Fodor claim (p. 180) that the identity criteria for psychological states should be determined by the psychological, and perhaps neurological, laws governing these states. This is vague, but it seems reasonable in view of my earlier discussion of criteria for n-equivalence and e-equivalence. Yet, in certain arguments in their article, Block and Fodor are at least sympathetic towards the "...doctrine which holds that the type-identity conditions for psychological states refer only to their relations to inputs, outputs, and one another" (p. 173). This doctrine does not follow from their more vague criterion, and this doctrine can lead to far too many alleged identities of psychological states in different creatures.[3] I wish to emphasize that it is very risky to argue for or against certain attribute-identities without clear and stringent criteria for such identities. This is quite serious because the individuation of kinds and attributes is interdependent with the individuation of laws and explanations.

In my view, which has only been sketched here, identities are necessary for reduction. Moreover, given U_1–U_9 plus some empirical assumptions about the domains of theories, reductions are necessary for the unity of science. These conclusions depend on recognition of the non-causal, as well as the causal, connections involved in scientific theorizing. It is clear that connections of either kind can be quite complex. It also appears possible that unification of the physical and behavioral sciences may require the use of psychological predicates quite different from those we ordinarily use now.[4]

The University of Texas at Austin

NOTES

* An earlier version of this paper was presented at the symposium on the Unity of Science at the Fourth Biennial Meeting of the Philosophy of Science Association at Notre Dame University, November 1–3, 1974. This is a revised draft prepared for publication in this volume.

[1] I have no rigorous proof that these are all true noncausal sentences. In fact, it appears that, in certain types of classification systems, there may be true synthetic noncausal sentences which state certain inclusion relationships between kinds of things. However, the existence of such sentences would not affect the arguments and conclusions in the main text.

[2] This remark assumes that the reformulated fundamental law-sentences of T_1 are sufficient to explain all fundamental law-sentences of T_2. If this is not the case, we still could obtain an awkwardly formulated, disguised *partial* microreduction, and T still would not contain any explanations of fundamental laws about parts in terms of fundamental laws about wholes. This case should be compared with the discussion of the third and last possibility below in the main text.

[3] In conversation at the 1974 PSA meetings, Ned Block informed me that the arguments in their paper are not to be interpreted as an endorsement of this doctrine.

[4] This research has been supported by NSF grant GS-39664.

BIBLIOGRAPHY

Block, N. J. and Fodor, J. A.: 1972, 'What Psychological States Are Not', *The Philosophical Review* **81**, 159–181.

Bunge, M.: 1967, *Scientific Research*, Vol. I, Springer-Verlag, New York.

Causey, R. L.: 1969 'Polanyi on Structure and Reduction', *Synthese* **20**, 230–237.

Causey, R. L.: 1972a, 'Attribute-Identities in Microreductions', *The Journal of Philosophy* **69**, 407–422.

Causey, R. L.: 1972b, 'Uniform Microreductions', *Synthese* **25**, 176–218.

Causey, R. L.: 1974, 'Laws, Identities, and Reduction', forthcoming in the *Proceedings* of the Conference for Formal Methods in the Methodology of Empirical Sciences, Warsaw, Poland.

Gendron, B.: 1971, 'On the Relation of Neurological and Psychological Theories: A Critique of the Hardware Thesis', in R. C. Buck and R. S. Cohen (eds.), *PSA 1970*, D. Reidel Publishing Co., Dordrecht, pp. 483–495.

Hull, D.: 1974, *Philosophy of Biological Science*, Prentice-Hall, Englewood Cliffs, pp. 139–141.

Kalke, W.: 1969, 'What is Wrong With Fodor and Putnam's Functionalism', *Nous* **3**, 83–93.

Oppenheim, P., and Putnam, H.: 1958, 'Unity of Science as a Working Hypothesis', in H. Feigl, M. Scriven, and G. Maxwell (eds.), *Minnesota Studies in the Philosophy of Science*, Vol. II, University of Minnesota Press, Minneapolis, pp. 3–36.

Schlesinger, G.: 1963, *Method in the Physical Sciences*, Humanities Press, New York.

LAWRENCE SKLAR

THERMODYNAMICS, STATISTICAL MECHANICS AND THE COMPLEXITY OF REDUCTIONS*

I

That reductions of theories and unifications of science frequently occur by means of identifications is a widely accepted hypothesis of methodology. Usually we are concerned with the micro-reductions of things, in which an entity is identified with a structured aggregate of smaller constituents. And, it is frequently alleged, we must also take into account further identifications in which the attributes of the macro-object, expressed by predicates in the macro-theory, are identified with attributes of the aggregate of micro-entities differently expressed by predicates of the micro- or reducing theory.

I am fundamentally in strong agreement with this view, but intend here to explore one case of inter-theoretic reduction where some important qualifications of this account of reduction are in order. Philosophical scepticism is frequently encountered as to the possibility of ever making coherent sense of the notion of attribute identification. This scepticism often shows itself in the opposition to the identificatory account of the reduction of the theory of mental states to that of neurological states. Sometimes this scepticism is countered in the following way: It is alleged that there are at least some clear-cut cases of attribute identification in physics, and, arguing *ab esse ad posse*, therefore attribute identification is possible. After all, one hears it said, temperature just *is* mean kinetic energy of molecules.

But is this correct? In general, does the reduction of thermodynamics to statistical mechanics take place, in part, by an identification of the attributes expressed by the predicates of thermodynamics with those expressed by predicates of the reducing statistical theory?

At least the following is, I believe, correct: Gases (*et al.*) are aggregates of molecules. And whatever properties gases have, they are the properties of aggregates of molecules and they are to be accounted for in terms of the properties and relations of the molecules, as characterized by our

R. S. Cohen et al. (eds.), PSA 1974, 15–32. All Rights Reserved.
Copyright © 1976 by D. Reidel Publishing Company, Dordrecht-Holland.

best micro-theory, and by the structuring of the molecules which constitutes them into the macroscopic gas.

Yet I think one must still be cautious in taking the reduction of thermodynamics to statistical mechanics to be constituted by an attribute identification. In one way my reason for scepticism is fairly trivial. The classical thermodynamics of Clausius, Kelvin and Carathéodory is simply an incorrect theory of the world. And not merely incorrect in a simple "numerical" way, in which it is, in some sense, conceptually sound but just a little off in some predicted values. It is a "conceptually incorrect" theory.

I doubt that anyone really needs convincing, at the present stage in the philosophy of science, that reduction often consists in replacement, and that what we call the reduced theory is frequently shown by the very reduction itself to be an incorrect theory, or even one whose very conceptual structure is inapplicable to the world. What needs emphasis in this case, though, is the important way in which thermodynamics, while discarded on the basis of atomism and statistical mechanics as the correct theory of the macroscopic behavior of gases, retains a vital and crucial role in science. Unpacking the details of this residual role would require much effort and much care, but, very crudely, we can assert that while thermodynamics is wrong, it remains our only source for supplying us with those concepts we do wish to apply in our now revised macroscopic theory of matter. While we do not expect to "deduce" or "derive" thermodynamics from statistical mechanics in any simple minded way, our one guide to the correct way in which a macroscopic theory is to be extracted from the atomic constitution of matter, the underlying dynamics of molecules and the statistical assumptions of statistical mechanics is that this new macroscopic theory is to have close conceptual ties to the now refuted older thermodynamic theory.

In some sense (or senses) thermodynamics is correct as a "limiting" theory to a correct description of the macroscopic theory of matter. To this, I imagine, most will agree. But just what kinds of limits are involved, and just how thermodynamics is reached when one "approaches" these limits is a complex matter indeed. *Numbers of molecules* are crucial, for, in many cases, one wants to say that the laws of thermodynamics hold only in the case of the (unrealistic) situation where an infinite number of micro-constituents make up the macro-object. *Crudeness of measurement* is involved, for, it is often alleged, one can get the thermodynamic results from the microscopic and statistical theory only by allowing for the fact that our measurements are limited in their accuracy ("coarse graining" in statistical mechanics) and limited in the range of macroscopically conceivable quantities they measure ("the reduction of variables" which takes place when one goes from the micro- to the macro-picture). *Time* is involved. For, it is frequently claimed, we can get thermodynamics from statistical mechanics only by assuming that the times involved in measurement on the microscopic level are sufficiently long (compared with times of molecular interaction, or of mean free path flight time or of "relaxation" times of various micro processes) and yet sufficiently short (compared to Poincaré recurrence times and the like).

I wish that I could at least outline here *how* thermodynamics is related to statistical mechanics by means of these various "limiting" relations, but that I cannot do. And not merely because of limitations of space, nor of limitations on my part, real as both of these limitations are. The fact is that there is no such thing as a coherent, unified, and generally accepted physical theory of the relation of thermodynamics to statistical mechanics. In fact, there is no such thing as *the* theory of statistical mechanics. Rather there are a number of suggestive, tentative approaches which are related to one another at best by crude argument and by hopes for the future. Trying to give a full conceptual analysis of statistical mechanics would be like trying to write the cultural history of a civilization which may someday come to flourish in what is presently a region of tropical rain-forest only as yet tentatively explored and roughly mapped out by a few intrepid exploratory souls.

In the light of this it is no surprise that Gibbs, while occasionally lapsing into such expressions as the "statistical mechanical *definition* of entropy," entitles his chapter on the relation of thermodynamics to statistical mechanics 'Discussion of Thermodynamic Analogies.'[1] Of course Gibbs is still concerned with the fact that even on its own terms the statistical mechanics of his day seemed empirically defective, in that it failed to take account of the details of molecular interaction and of radiation; and in that it sometimes gave brutally wrong results (the specific heat of gases with compound molecules, for example). It is just conceivable that he was also wary of the criticisms leveled against the Boltzmann theory by the remaining energeticists and anti-atomists.

But writing as late as 1938, after quantum mechanics has removed from the statistical approach its last blatant *empirical* difficulties, and long after anti-atomism had become a dead letter in science, Tolman is still being careful to speak only of the "analogues" in statistical mechanics of entropy, temperature and free energy. He, in fact, introduces his own new symbol to use when thermodynamic terms are on the left and statistical mechanical terms are on the right: $\rightleftharpoons$. Unfortunately he introduces the symbol but never really pauses long enough to define it or even elucidate it very carefully.[2] And in 1956 Münster is still speaking of thermodynamic "analogues."[3]

Is 'analogy' the right word to use in characterizing the relation the thermodynamically postulated properties bear to the statistical properties "associated" or "correlated" with them by statistical mechanics? It is indeed appropriate if one takes it to mean that while there is a close scientific connection between the concepts, mediated, at least in part, by their formal roles in their respective theories, and sufficient to allow an inter-theoretic reduction between these theories, still there is good reason to avoid saying that the property expressed by one concept is identical to that expressed by the concept associated with it.

Use of the term can be misleading, however, if one takes it that one has said all there is to say by invoking the word 'analogy,' or where one is misled by over-abundant use of the term into thinking that there are no crucial differences in the relations between concepts grouped together under that all too loose rubric. This confusion is especially dangerous if one then goes on to too hastily identify the "analogy" which appears in this reduction with those relations between concepts which appear, say, in the reduction of Newtonian mechanics to special relativity or classical mechanics to quantum mechanics. If all these conceptual relations are "analogies," then 'analogy' is being used in a very broad way indeed.

II

Instead of offering a general or systematic thesis about the reductive relationship of thermodynamics to statistical mechanics I will merely offer very brief sketches of three problem areas in the study of this reduction. Each will illuminate one or more general issues which arise when the reduction is studied as a whole. The first problem is the nature of the so-called thermodynamic limit, and its role in connecting the multifariously different statistical ensembles considered in statistical mechanics with the multifariously different but *equivalent* descriptions of a macrosystem possible in thermodynamics. The second problem is the invocation of ergodic theory as the *rationale* for choice of ensemble in statistical mechanics. This will reveal one way (by no means the only way) in which considerations of time limits play a role in reductions. The third problem will be a consideration of a few of the possible "definitions" of entropy in statistical mechanics, and what this possibility of different "definitions" for one and the same thermodynamic quantity tells us about the relations of the two theories.

(a) *The Thermodynamic Limit*

It is well known that most statistical mechanical results are calculated "in the limit as the number of particles goes to infinity." Large numbers of particles and large volumes, density being held constant, are necessary both to allow us to replace various combinatorial expressions by "limiting" analytic expressions and to get rid of "edge effects," effects which we know from macroscopic thermodynamics to be inconsequential for real, finite gases.

But the thermodynamic limit has other crucial roles as well, roles of greater *conceptual* interest. One example of this is the problem of phase transitions. In the few calculable exact models (for example, the Ising model of a two-dimensional lattice gas) the following curious situation arises: If we describe phase transitions in the natural thermodynamic way, in terms of singularities in the analytic behavior of some appropriate thermodynamic functions, then, in these calculable models, it can be shown that phase transitions *never* occur in a finite gas, no matter how large. But they can be shown to occur in a gas of infinite size. Naturally our interpretation of this is that phase transitions *as we have macroscopically described them* don't really occur. Not because phase transitions don't occur, but because our macrodescription of them was wrong.[4]

Even more interesting conceptually is the problem of multiple ensembles and the thermodynamic limit. It is well known that in thermodynamics a gas in equilibrium can be equally well described by any of a number of thermodynamic variables (enthalpy, free energy, Gibbs function, Massieu-Planck functions). In the standard ensemble version of statistical mechanics (what I will call from now on Gibbs statistical mechanics, even though Maxwell and Boltzmann anticipated it) each such thermodynamic function is "represented" by a mean value calculated over a statistical ensemble, i.e. calculated by considering all possible systems of a given kind subject to specified constraints and "distributed" according to a standard probability weighting over the micro-conditions allowed a system of that kind subject to those constraints.

Now the following difficulty arises, however. The thermodynamic functions are related to one another, following from the laws of thermodynamics by Legendre transformations. Within statistical mechanics, at least in the case of microcanonical ensembles, the various means calculated from their appropriate ensembles are also related, since the "partition functions" giving the probability distributions over the ensembles are related to one another by Laplace transformations. For finite systems any *identification* of the thermodynamic quantities with the associated means will lead to a contradiction, since the appropriate thermodynamic and statistical mechanical relationships among the variables will conflict. Physically the origin of the discrepancy is that statistical mechanics, unlike thermodynamics, allows for fluctuations of the properties of individual systems about their most probable states, and brings these fluctuations into account when calculating mean values. Each ensemble, however, "brings statistical fluctuations into play in different ways."[5]

What one tries to show is that when the limit of infinitely large systems is taken, however: (1) various limits of the statistically defined quantities exist; and (2) they are related to one another exactly according to the thermodynamically postulated relationship holding between their associated thermodynamic quantities. Physically this is in accordance with our intuition that when the "thermodynamic limit" of an infinite number of particles is taken the fluctuations which distinguish statistical mechanical from thermodynamic quantities should vanish. While it is intuitively clear that this limiting process should do the job of reconciling the thermodynamic and statistical mechanical inter-relationships among the parameters, finding the necessary assumptions to prove these results and then using them to prove both the existence of the limits and the resulting limiting "equivalence" to thermodynamics is a major mathematical project indeed.[6]

What is the moral of this? Roughly something like the following: The meanings of the predicates which appear in thermodynamics are fixed, *in part*, by the role they play in its laws. Since, according to statistical mechanics, these laws are incorrect there are, properly speaking, no such properties as those putatively expressed by the thermodynamic predicate. Yet there are properties which are closely associated with the predicates of the discarded theory in the following ways: (1) their values give a good approximation, at least, to the values asserted by thermodynamics to be held by the magnitudes we no longer believe in; and (2) their lawlike inter-relationship in the limiting (and unrealistic) case of infinitely large systems can be shown to have the same form as the lawlike inter-relationship postulated by thermodynamics to hold among the thermodynamic magnitudes. Surely part of what is meant when the statistical mechanical quantities are called "analogues" to the thermodynamic properties is that (2) as well as (1) is the case.

(b) *Ergodic Theory and Limits in Time*

As we noted in the last section, in the Gibbs version of statistical mechanics thermodynamic values for systems in equilibrium are associated with mean values of microscopically definable parameters calculated over the appropriate ensemble. But why should mean values calculated over an ensemble equal measured equilibrium values?

One approach to this problem, of dubious relevance and satisfactoriness to be sure, is ergodic theory. Here one hopes to show that the mean values calculated by the ensemble method will, for "almost all" initial states of a system, equal the infinite time average of the same parameter calculated over the "life" of a particular system as it evolves from its initial state.

Now it isn't easy to show that this is ever the case. In fact it has only been in the last few years that "ergodicity" has been demonstrated for any system (the Boltzmann model of "hard spheres is a box").[7] But what is of interest to us here is *why* one would want to show a system ergodic. What is so interesting about infinite time averages?

One approach proceeds by arguing, on the basis of non-equilibrium considerations, that nearly all isolated systems will spend nearly all their time at or around the equilibrium state. Hence infinite time averages are good estimators of equilibrium values, since, the equilibrium values so dominating the class over which the averaging proceeds, the value of the infinite time average will perforce be at least nearly identical to that of the dominating equilibrium values in the states averaged over.

Another approach, which I want to focus on here, is less plausible but of some philosophical interest. It is argued that infinite time averages are "reasonable" correlates for empirically measured values because of the enormous length of time required for any macroscopic measurement as compared with the times involved in processes on the microscopic scale. So, the argument goes, the rationalization of the "identification" of "phase" or "ensemble" averages with macroscopically measured values is to be carried out in two steps. First one uses ergodicity to show that these ensemble averages are equal to infinite time averages (for almost all systems). Then one uses the alleged "might as well be infinite" time scale of measurements to rationalize taking infinite time averages as the right statistical "analogue" for measured equilibrium values.

Now this last approach just isn't too plausible. After all we do measure fluctuations from equilibrium, even on the macroscopic scale, and our measurements are certainly brief enough in time to allow us to measure the values of quantities when systems are not yet in equilibrium. Infinite time averages seem to have little to do with macroscopically measured quantities. But the motive here is more important to my argument than the success of the program.[8]

What is the moral to be drawn from the *motive* behind the ergodic theory, at least in the second version of its use which I have presented? I think it is this: The meanings of the terms of thermodynamics are fixed, *in part*, by the "operational" procedures used for determining the values of the magnitudes expressed by the terms. A satisfactory reduction of thermodynamics to statistical mechanics should not only provide an adequate method for calculating by means of statistical mechanics the correct values of these magnitudes, it should also provide some general, systematic *rationale* for associating the values calculated within statistical mechanics with the results of the measurement procedures used to determine the values of the associated thermodynamic quantities. And this "association" may invoke such things as limiting procedures which, although they deviate from the actual procedures used in the macroscopic measurement process and are hence "unrealistic," at least bear a plausible systematic relationship to them.

(c) *Entropy, "Entropies" and Reduction*

We frequently encounter the following situation in replacement reductions: Where the reduced (and replaced) theory had a single concept, the reducing theory allows the construction of a multiplicity of concepts each sharing some of the aspects of the concept of the now replaced theory but none sharing all of its features. Consider mass in Newtonian mechanics *vis-à-vis* rest mass and total mass-energy in the special theory of relativity.

Not surprisingly this happens in the reduction of thermodynamics to statistical mechanics. One interesting feature of statistical mechanics is that the very facts which allow the success of thermodynamics (for example, large numbers of particles) play a role in generating the multiplicity of concepts associated with a single thermodynamic concept in statistical mechanics. Many statistical functions which differ from one another when numbers are small may very well converge in value when one "goes to the limit of an infinite number of particles."

The more "theoretical" the concept in thermodynamics, and the more the meaning of the concept is fixed solely by its role in the thermodynamic laws, the more likely it is to occur that more than one concept of statistical mechanics can be plausibly taken to be its analogue. Thus one doesn't find a plethora of statistical mechanical "volumes" or "energies," but one does find a very large variety of concepts plausibly linked with thermodynamic *entropy*.

Sometimes this variety can be "projected back" into the reduced theory. Thus the realization that the entropy of a system in statistical mechanics is relativized with respect to the level of description we apply to the microstructure of the gas (we get different entropies if we consider the particles individuable than we do if we work only with generic particles, for example) can be carried back to a new, more sophisticated thermodynamic entropy, thereby resolving a classical puzzle within thermodynamics in the process (Gibbs' paradox).[9] But sometimes the statistical mechanical distinctions cannot be drawn within the thermodynamic framework in any plausible way.

Let us consider a number of "entropies" discussed in statistical mechanics. Let me make it clear immediately that I certainly don't claim even this astonishingly variegated list to be exhaustive.

First there is an entropy considered by Boltzmann in his treatment of the ideal gas. It is a logarithmic measure of the "disorder" of the gas insofar as this is determined only by the one-particle distribution function giving the actual distribution of the molecules in the phase space of single molecules. Boltzmann associated the maximum value of this function with the phenomenological entropy of a gas in equilibrium.

If the gas is not ideal, however, one will not even get the correct measured values of entropy in this way, for this entropy completely neglects the addition to the internal energy due to molecular potential interactions. In some of the more general cases, though, a new entropy defined in a more complex way by means of the two-particle distribution function will give us, when we look at its maximum subject to the imposed constraints on the gas, a much better estimate of the phenomenological entropy of equilibrium.[10] But let us stick to the case of the ideal gas.

Notice several features of this entropy. First, the entropy of a gas defined this way can very well vary even though the constraints on the gas remain fixed. Much of the time, for an isolated gas, the entropy will remain near its maximum value, but fluctuations, less and less frequent the greater the deviation from the maximum, will occur. These entropies are like thermodynamic entropy in that they are empirically determined properties of the individual gas. But they are unlike it in that they just don't follow the behavior nomologically imposed on thermodynamic entropy by the Second Law of thermodynamics. Notice also that in this version of the reduction of thermodynamics to statistical mechanics the "association" is between the thermodynamic quantity and a "most probable" statistical mechanical quantity, for it is the fact that the maximum value distributions are the overwhelmingly most probable which rationalizes their "association" with the equilibrium quantities of thermodynamics.

Then there is the Gibbs fine-grained entropy. This is the logarithmic measure of the "spread-outness" of the Gibbs ensemble as determined by the constraints imposed on the gases in the ensemble and by the probability distribution over the molecular conditions which define the ensemble.[11]

Now in the case of the ideal gas, and in the thermodynamic limit, if the ensemble is such that fluctuation values as computed by it vanish when the limit of the number of molecules in the gas goes to infinity, as certainly happens in the standard Gibbsian ensembles, the Gibbs entropy will agree with that Boltzmann entropy assigned to equilibrium. Basically this is because as the thermodynamic limit is approached no *numerical* difference between mean values computed over all molecular states and most probable values computed over the overwhelmingly most likely distributions can be found. So we could compute equilibrium entropies either by finding the Gibbs entropy for the proper ensemble or by finding the correct Boltzmann entropy.[12]

But conceptually these are quite different sorts of things. In one sense the Gibbs entropy will obey a "Second Law" exactly, unlike the Boltzmann entropy. But one gains this close affiliation of the statistical mechanical with the thermodynamic property only at a price. For once the constraints on a gas are fixed, so is its Gibbs entropy; no matter what the molecular distribution characterizing the actual microstate happens to be. This is because the Gibbs entropy is defined by the ensemble which we are taking the gas to be a member of, not by means of the actual microstate of the gas. Whereas the magnitude of the Boltzmann entropies of the gas are empirically determined by its state, like thermodynamic entropy, the Gibbs entropy is fixed once the constraints on the gas are imposed, no matter what the gas happens to do while subject to these constraints. So if we take a gas in equilibrium and then adiabatically change its constraints the Gibbs entropy of the gas cannot decrease, even if the actual gas does the very improbable thing of having its apparent molecular order increase and its Boltzmann entropy decrease when the constraints are changed.

This does not mean, however, that the Gibbs approach ignores the possibility of spontaneous deviation from equilibrium. Indeed one of the great virtues of the ensemble approach is its ability not only to allow the calculation of equilibrium values but also to provide estimates of the likelihood of fluctuations away from these expected means.

Of course the numerical values of the entropy assigned to an ideal gas in equilibrium by the Boltzmann and Gibbs methods will be the same, at least in the thermodynamic limit. For it is the maximum value of the Boltzmann entropy compatible with the constraints which is associated with the phenomenological equilibrium entropy in the Boltzmann approach, and this value, like the Gibbs entropy depends only upon the imposed constraints. But even here we must proceed with caution. For "most probable" quantities and "mean" quantities are not *conceptually* the same, even when they are numerically equal. Crudely the basic conceptual difference between the two approaches is that the Gibbs approach is to look at all possible micro-states of the gas compatible with the constraints and to take as equilibrium values averages of some statistical mechanical parameter over all these possible micro-states using as a "weight" function the ensemble probability distribution function. The Boltzmann approach, on the other hand, tells one to look at the most likely micro-states of the gas compatible with the constraints and to take as equilibrium values a statistical mechanical property associated with these overwhelmingly probable states.

As though this were not complex enough, we must also consider the Gibbs coarse-grained entropy. While the traditional thermodynamic entropy is properly defined only for gases in equilibrium states (although extendible, with caution, to a limited class of non-equilibrium states),[13] the statistical mechanical entropies are generally well defined for gases not in equilibrium. It is therefore plausible to see if one can account for the approach to equilibrium in terms of a predictable increase over time in the statistical mechanical entropy assigned to a gas as the micro-states evolve.

But the very principle which provides much of the rationale for the use of Gibbs ensembles to compute equilibrium quantities in the first place (conservation over time of phase volume of an ensemble of phase points or Liouville's Theorem) allows us to show that the Gibbs entropy assigned a gas in non-equilibrium will remain constant, even when the constraints are such that the gas can evolve to an equilibrium state.[14] Following a suggestion of Gibbs, a new approach is then tried in which one divides up phase space for the gas as a whole into small regions (much as one divides up phase space of a single molecule in the Boltzmann approach) and one then defines a new coarse-grained ensemble entropy relative to one's partition. One then hopes to show that this coarse-grained entropy, if assigned to a gas in non-equilibrium (characterized by the class of ensemble points for the gas not filling the accessible phase-space uniformly), will increase to a maximum value showing an approach to equilibrium. Naturally for equilibrium states the coarse-grained Gibbs entropy assigned a gas will agree, at least *numerically*, with the fine-grained Gibbs entropy and hence (in the thermodynamic limit) with the maximum possible value of the appropriate Boltzmann entropy.[15]

The approach goes on to try and characterize by coarse-graining the details of the approach to equilibrium, and not just to "prove" that equilibrium will in fact be approached. And further attempts are then made to show how the introduction of coarse-graining can be "rationalized" as a suitable statistical device in terms of the "coarseness" of macroscopic measurements; in the sense that they are unable to discriminate all micro-states from one another but can "fix" the micro-state only to within some degree of precision.[16]

But it is on issues like these that current physical theory can be said to be indecisive, to say the least. For current scientists are at great variance with one another about the applicability of such devices in statistical mechanics and about their rationality. They range from those who adopt coarse-graining as *the* method to approach non-equilibrium problems to those who dismiss the whole procedure as totally irrelevant to a correct foundation for statistical mechanics.[17]

And what moral can we draw from these considerations about entropy? I think it is one of some importance: We know what thermodynamics is, and we know that it is an incorrect theory. We know, up to a point, what statistical mechanics is, and, assuming that we have gone over to quantum statistical mechanics, etc., we believe that it is a correct theory. At least we believe that our theory of the micro-structure of gases is correct, that we have (at least to some approximation) the laws governing their interaction, and that we know how to postulate "reasonable" statistical distribution functions and how to calculate means, etc. with them.

What we don't yet really know, however, is what the full structure of the reduction of thermodynamics to statistical mechanics should look like. We have, for example, a well defined notion of entropy in thermodynamics. And we can define several coherent notions of "entropy" in statistical mechanics. But how these are related – what statistical mechanical entropies are to be "associated" with thermodynamic entropy, in which cases, and why – is still an open question.

But why should it matter how thermodynamics "reduces" to statistical mechanics if the former is a theory known to be incorrect? The answer is that incorrect as the theory may be, it is our only systematic guide to a program for understanding fully the consequences of statistical mechanical assumptions on the macroscopic level. We need to know not only what statistical mechanical entropies we can define, but we must also understand fully how to use them for prediction on the phenomenological level. And that understanding we do not yet have. For it is hard to imagine terms further from immediate "operational definitions" than those constructed in a theory which is microscopic, statistical and maybe even "coarse-grained." To determine *which* macroscopic attribute of matter should be associated with *which* mean or most probable value, *when* and *why* is a major theoretical task.

Two things are clear: We can know what the reduced theory is, have a pretty good idea what the reducing theory looks like, and yet still be puzzled over what the *reduction* amounts to. And just because a theory has been replaced, just because it is now known to be incorrect – conceptually as well as numerically, this does not entail that it no longer has a vital living role to play in science. It may survive as an "incorrect but technologically useful" approximation to the truth. But it may also survive as something much more: as an essential guide to the correct completion of the theory which has taken its place.

But incorrect thermodynamics is. And many of its property attributing expressions simply fail to express any real property. Insofar as gases have properties, and insofar as they are correctly expressible by predicates of statistical mechanics, these properties are not identical with those properties expressed by the conceptually failed predicates of thermodynamics. For those properties don't exist. Of course if one prefers being a Platonist with regard to properties one could instead say that the thermodynamic properties exist but are just never instanced. I prefer my Aristotelian way of talking but I do not believe my substantive points depend upon a metaphysical choice of this kind.

One *could* also say that what we have discovered is not that the thermodynamic property doesn't exist, but that it is very different in nature than we thought. Of course when one thermodynamic property becomes associated through reduction with a multiplicity of properties in the reducing theory we are in the embarrassing position of trying to decide which of the new properties the old one is "identical" to. We could, following a suggestion of Hartry Field's, say that the thermodynamic term 'entropy' was "referentially indeterminate," alleging not that it had no referent, but that it had more than one.[18] I myself find this a more cumbersome way of talking than the one I have chosen, but I believe that the choice is one of a "way of talking" and is, again, non-substantive. However we talk the real task we are set is that of describing the reductive relationship between the old and the new theory, and the relationship between the concepts of the one and those of the other, with true regard to the complexity of the scientific achievement and without gross philosophical over-simplification.

At this point those knowledgable in contemporary physics may, very fairly, raise an objection to my thesis. While I have allowed statistical mechanics the benefit of all its recent progress, I have been talking about thermodynamics as though it had made no progress whatever beyond Clausius (or, better, beyond Gibbs) except for the conceptual "tidying up" of Carathéodory, Landsberg, *et al.* But this is not true.

Philosophers are familiar with the idea that when a "replacement reduction" occurs one thing we might do is go back to the reduced and replaced theory and see if we can exchange it for a newer, conceptually similar, but more correct theory. One then hopes that this new theory will "reduce" to the reducing theory in such a way that we need not consider the reduction a demonstration of the incorrectness of the reduced theory.

Physicists have responded to the statistical mechanical reduction of the older thermodynamics in just this way. They have looked for "generalized" thermodynamical theories which, while evolving from the conceptual structure of the older thermodynamics, take account of the modifications of our phenomenological theory imposed upon us by our knowledge of the statistical micro-theory.

One such approach, for example, initiated by Einstein and carried on by Mandelbrot, Szilard, Tisza, Quay and others is to try and construct a "statistical thermodynamics." In this theory one utilizes the conceptual structure of thermodynamics but supplements it by changing from deterministic to statistical laws. In this theory, for example, equilibrium fluctuations, characterized in terms both thermodynamic and statistical, are permitted to an isolated system, insofar as it is viewed as composed of parts in interaction.[19] Just how far such an approach can be extended to do justice to the full macroscopic, phenomenological situation is presently an open question. For example, just how far can one go beyond a statistical thermodynamical theory of equilibrium to a statistical thermodynamical theory of non-equilibrium and irreversible processes?

Sometimes it is said that we should continue with an attempt to construct newer and better thermodynamical theories "autonomously" of statistical mechanics. In one sense I think this is true. There is good reason to believe that theories which are macroscopic and phenomenological but, for example, statistical, and which are formulated on the basis of their own postulates without much regard for what we know to be the case about the underlying micro-structure, will be rich and useful contributions to physical theory.

But this autonomy can only go so far. Even its most ardent exponents admit that when it comes to calculating the values of parameters describing real gases, and to explaining *why* these parameters have the values they do have, one must resort to the "deeper" theory of the micro-constitution of the gas.

In any case, believing as we do in the atomic micro-constitution of macroscopic matter, at some point we will want to explore the reductive relationship which will hold between the new, improved thermodynamics and our best statistical mechanics. What that relationship will look like I don't know. What I anticipate, however, is that the conceptual analysis of it is not likely to be simple. I also anticipate, and I suppose that this is mere dogmatism on my part, that we will hardly do justice to this future reduction by any move as simple as saying that some property expressed by a predicate of the newer thermodynamics is "just identical" to a property expressed by the predicates of statistical mechanics. This new reduction may be more interesting, deeper and more scientifically important, in the long run, than the historically important replacement reduction of orthodox thermodynamics to statistical mechanics. But I am willing to bet that it will be equally conceptually complex.

Let me make one final philosophical point. The reason we had to discard the thermodynamic notions of temperature, entropy, etc., with the advent of the new statistical micro theory was because thermodynamics said so much about them. The more a theory says about the entities and attributes it postulates, the more risk it runs that in the long-run it will be found to be "not exactly true," and hence that its concepts will be found non-viable as they stand. This is just one more instance of the philosophical truism that to say anything interesting at all is to run the risk of saying something false, and that the more interesting one's assertions are, the greater the risk one runs.

University of Michigan

NOTES

* Presented as part of a colloquium on the Unity of Science at the 1974 meetings of the Philosophy of Science Association. I am grateful to the John Simon Guggenheim Memorial Foundation for their generous support during the period in which this paper was written. I am grateful to the following people for their helpful comments and suggestions on an earlier draft of the paper: Evan Jobe, Phillip Quinn, Ken Friedman, and especially P. M. Quay.

[1] J. Gibbs, *Elementary Principles in Statistical Mechanics*, Dover, New York (1960), chap. XIV. The date of original publication of this work is 1902.

[2] R. Tolman, *The Principles of Statistical Mechanics*, Oxford Univ. Press, Oxford (1938), chap. XIII, 'Statistical Explanation of the Principles of Thermodynamics', esp. sec. 122. See p. 536 for the introduction and use of Tolman's special symbol noted in the text of this article.

[3] A. Münster, *Statistical Thermodynamics*, Vol. I, English edition, Springer-Verlag, Heidelberg (1969), sec. 2.11.

[4] A. Münster, *op. cit.*, chap. IV. See also D. Ruelle, *Statistical Mechanics, Rigorous Results*, Benjamin, New York (1969), chap. 5.

[5] A. Münster, *op. cit.*, p. 213.

[6] A. Münster, *op. cit.*, secs. 2.1–3.2 and 4.1–4.4. See also D. Ruelle, *op. cit.*, chaps. 1–3.

[7] See, for a brief outline of some recent results in ergodic theory, Ya. Sinai, 'Ergodic Theory,' in E. Cohen and W. Thirring, *The Boltzmann Equation*, Springer-Verlag, Wien (1973).

[8] For a non-technical outline of the ergodic approach to 'rationalizing' the use of ensemble averages for equilibrium values and some philosophical commentary thereon see the author's 'Statistical Explanation and Ergodic Theory', *Philosophy of Science* **40** (1973), 194–212. In that article I do not do justice, I now believe, to the first type of rationale mentioned here. I still believe, however, that an additional statistical assumption must be added to the ergodic results to complete their rationalizing program. (An additional assumption over and above the inevitable assumption that 'sets of measure zero in the natural measure don't count.') This additional assumption, I believe, is that the equilibrium values shown to be computable from phase averages, given a proof both of the ergodicity and the mixing property of the underlying micro-dynamics, is the equilibrium encountered in ordinary phenomenological experience.

[9] See H. Grad, 'The Many Faces of Entropy', *Communications on Pure and Applied Mathematics* **14** (1961), 323–354, esp. pp. 324–328. See also E. Jaynes, 'Gibbs vs. Boltzmann Entropies', *American Journal of Physics*, **33** (1965), 391–398, esp. sec. VI, pp. 397–398.

[10] On the failure of the one-particle distribution function to give the correct entropy when

inter-particle forces are taken into account see E. Jaynes, *op. cit.*, pp. 391–394. See also his 'Information Theory and Statistical Mechanics', in K. Ford (ed.), *Brandeis University Summer Institute Lectures in Theoretical Physics*, 1962, vol. 3, 'Statistical Physics', pp. 181–218, esp. sec. 6, 'Entropy and Probability', pp. 212–217. For a discussion of how to move to the entropy defined by means of the two-particle distribution function and the generalization of this process see H. Grad, *op. cit.*, *passim*.

[11] On the definition of the Gibbs entropy and its relation to the Boltzmann see R. Tolman, *op. cit.*, sec. 51, pp. 165–179, esp. (d) on pp. 174–177. There are in fact several different definitions for a Gibbs entropy, all of which "converge in the thermodynamic limit." See J. Gibbs, *op. cit.*, chap. XIV.

[12] See E. Jaynes, 'Gibbs vs. Boltzmann Entropies', sec. V, pp. 395–397.

[13] On the limits of extending thermodynamics entropy to non-equilibrium cases see P. Landsberg, *Thermodynamics*, Interscience, New York (1961), sec. 21, pp. 128–142. See also the work of Truesdell cited in note 19, below for important criticism of the "orthodox" view that thermodynamic quantities are "well defined" only in equilibrium situations.

[14] For a defense of the thesis that by use of the fine-grained entropy one is perfectly able to establish the statistical mechanical "analogue" of the Second Law, see E. Jaynes, 'Gibbs vs. Boltzmann Entropies', sec. IV, 'The Second Law', pp. 394–395. See also his 'Information Theory and Statistical Mechanics', sec. 6, pp. 212–217.

[15] The idea of coarse-graining was initiated by Gibbs. See J. Gibbs, *op. cit.*, chap. XII, pp. 139–149.

[16] An introduction to coarse-graining can be found in N. van Kampen, 'Fundamental Problems in Statistical Mechanics of Irreversible Processes', in E. Cohen (ed.), *Fundamental Problems in Statistical Mechanics*, Vol. I, North-Holland, Amsterdam (1961), pp. 173–202. See also O. Penrose, *Foundations of Statistical Mechanics*, Pergamon, Oxford (1970), Chap. I, 'Basic Assumptions', attempts a rationalization of coarse-graining. See esp. chap. I, sec. 3, 'Observation', where the "coarseness" of macro-observation is used to justify coarse-graining in statistical mechanics.

[17] Not surprisingly, in the light of note 14, Jaynes offers a critique of the relevance of coarse-graining to statistical mechanics. See his 'Gibbs vs. Boltzmann Entropies', p. 392.

[18] Field, Hartry, 'Theory Change and the Indeterminacy of Reference', *Journal of Philosophy* **70**, No. 14, pp. 462–481, esp. p. 466.

[19] For a discussion of statistical thermodynamics see Tisza, L. and Quay, P., 'Statistical Thermodynamics of Equilibrium', *Ann. Phys.* (N.Y.) **25** (1963), 48–90. Reprinted in Tisza, L., *Generalized Thermodynamics*, The M.I.T. Press, Cambridge, Mass., 1966. Citations to the work of Einstein, Mandelbrot and Szilard noted in this paper will be found in the bibliography to this article. See also Tisza, L., 'Thermodynamics in a State of Flux. A Search for New Foundations', in E. Stuart, B. Gal-Or, and A. Brainard, (eds.), *A Critical Review of Thermodynamics*, Mono Book Corp., Baltimore, 1970. See also Truesdell, C., *Rational Thermodynamics*, McGraw-Hill, New York, 1969, and Glansdorff, P. and Prigogine, I., *Thermodynamic Theory of Structure, Stability and Fluctuations*, Wiley-Interscience, London, 1971, for other extensions of the concepts of phenomenological thermodynamics beyond the cases and methods of the traditional theory.

THOMAS NICKLES

THEORY GENERALIZATION, PROBLEM REDUCTION AND THE UNITY OF SCIENCE*

> In spite of the fact that, today, we know positively that classical mechanics fails as a foundation dominating all physics, it still occupies the center of all our thinking in physics.
>
> ALBERT EINSTEIN (1936)[1]

INTRODUCTION

Although doubtlessly aimed at later developments in physics, Einstein's famous remark, if interpreted so as to include classical statistical mechanics, nicely captures the spirit of work on the early quantum theory by men like Bohr and Ehrenfest, who, despite their conviction that the classical theories failed, nevertheless mined their riches to the fullest in the development of the new theory. In this paper I try to exhibit and characterize the patterns of reasoning involved in Ehrenfest's attempts to generalize Planck's early theory of the linear harmonic oscillator, and I then employ the same historical case as a basis for arguing that the reduction of problems to problems is an important phenomenon which cannot be fully understood in terms of the reduction of theories. Before turning to Ehrenfest, let me point out the central relevance of these issues to the problem of the unity of science.

As the opening quotation suggests, I am not at all concerned with the unity of apparently distinct sciences (*e.g.*, biology and chemistry) at a given time. Rather, I deal with some aspects of the unity of a "single" body of theory over time, *viz.*, the early quantum theory, both as to its emergence from classical statistical mechanics and as to the later problems of unity or identity arising from attempts by Ehrenfest, Debye, Bohr, Sommerfeld and others to generalize Planck's original quantum theory in different respects. Nor do I concern myself in this paper with so-called "unity of method" but rather with "doctrinal unity," or what might better be called the doctrinal *continuity* of theories.

It is, of course, Feyerabend and Kuhn who have done most to focus attention on the problem of the temporal unity of science, by their bold denials that there is any such doctrinal or conceptual unity. Notoriously, some of their statements have borne the implication that, *pace* Einstein, quantum theory has no more to do with classical physics than with classical genetics. By now the more extreme claims pronounced in the 1960's have been thoroughly criticized [see *e.g.*, (27), (31), (32), (33)], but serious problems concerning temporal unity or continuity nevertheless remain. One of these is the massive meaning-change problem, which I do not discuss. Another, related problem is the need for better conceptual tools with which to characterize this continuity. I have elsewhere[2] sketched an alternative to the standard conceptions of reduction which I believe better represents the relation of many successor theories to their predecessors.

In this paper I tackle a third problem of temporal unity, the problem of characterizing the reasoning patterns involved in the formation and development of theoretical ideas insofar as this process involves reasoning to new theories from certain features of older, established theories. Both the positivists' familiar discovery-justification distinction and the later Feyerabend-Kuhn emphasis on incommensurability and non-rational (I do not say irrational) theoretical leaps of faith have discouraged work on problems of discovery, heuristics, and theory formation until recently. It is by now widely recognized that even reduction – which used to be considered pretty much a matter of logical relations between completed theory systems – has its heuristic side, a view which finds expression in the (somewhat "negotiable") demand that a new theory reduce to the old as a special case under appropriate conditions.[3] This reducibility demand or constraint represents one very important way in which theory formation may be guided by appeal to an available theory, for it suggests, and perhaps requires, that we seek a new theory which will be, in some respect, a *generalization* of the old theory. Let us explicitly assume, as a working hypothesis, that theory generalization and theory reduction [particularly in the second sense of 'reduction' discussed in (19)] are inverse 'operations' or relations: if theory T_1 is obtained by generalizing T_2 in certain respects, then T_1 reduces, in just those respects, to T_2 as a special case. Now it is obvious that until the "certain respects" or "appropriate conditions" are specified, the demand that a new theory reduce to the old under appropriate conditions cannot function as a *specific* guide or constraint on theory formation. For if the working hypothesis is correct, then *any* generalization of T_2 will reduce to T_2 in some respect or other. The reducibility demand accordingly becomes the demand that an adequate new theory be a generalization of the old theory *in some respect or other*, a guide which is completely unhelpful to one who is setting out deliberately to generalize the old theory in the first place. Unlike the more abstract "methodological correspondence principles" with which it is occasionally confused, Bohr's substantive, repeatedly refined correspondence principle specified the "appropriate conditions" in detail.

It is evident, then (if there was ever any doubt), that the single general constraint that "every new theory reduce to the old as a first approximation," taken alone, furnishes neither a specific methodological directive for the scientist seeking new theories nor a philosophically illuminating "account" of theory formation. I make no pretense of furnishing even the latter here, but the discussion of Ehrenfest's reasoning will illustrate some of the ways in which additional constraints on the form of a new theory may be introduced, constraints which may give real content to the demand that the new theory generalize the old in a particular respect and (once formulated) that it reduce to the old in that respect.

The constraints on a sought-for new theory or law may be of several distinct types and may be justified or rationalized in a variety of different ways. Not all constraints need involve an appeal to previous theories or theoretical results at all (*e.g.*, empirical constraints and pragmatic directives such as "Consider the simplest possibility first" *need* not be tied to previous theory, although they may be). And not all constraints which do involve an appeal to previous theory need be "reductive" constraints. This point will find illustration in our discussion of cases below. Not all reasoning from certain features of an established theory to certain features of a newly developing theory amounts to an attempt to find a new theory which reduces, or which reduces to, the old – an attempt, that is, to generalize the old theory. But much of it does, and the focus will be on theoretical reasoning of this sort.

In Sections I and II, I outline Ehrenfest's specific reasons for attempting to generalize Boltzmann's statistical mechanics and Planck's early quantum theory in the ways that he did and then try to characterize his patterns of reasoning more generally. It is arguable (Section I) that

Ehrenfest generalized Boltzmann's statistical mechanics in two different respects, one result being a general theoretical framework from which Boltzmann's own theory, as well as Planck's, Debye's, and Ehrenfest's quantum theories could be obtained by specification. The other result was Ehrenfest's "quantum theory" itself, a generalization of Planck's theory which, like Planck's, reduced to Boltzmann's classical assumptions in the appropriate limits. In Section II, I contend that Ehrenfest's appeal to (classical) mechanics was quite different in nature from his appeal to (classical) statistical mechanics. His reasoning in this case involved neither an attempt to generalize classical mechanics nor, therefore, an attempt to view the early quantum theory as a special application of such a generalized mechanics.

Problems of reduction and theoretical unity are not confined to the relationships of huge theory complexes: they extend also to the development of ideas within a "single," emerging theoretical framework. In Section III, I argue that the unification of Planck's original quantum theory with later conjectural extensions of it by Debye and by Bohr turned largely on Ehrenfest's reduction of certain *problems* to other, initially distinct problems by means of the adiabatic principle and that this reduction of problems and the resulting theoretical unification cannot be understood adequately in terms of *theory* reduction. I conclude that it is not only fruitful but necessary to discuss reduction and the unity of science in terms of new units such as scientific problems and not only as relationships of large theory complexes. My discussion is indebted to Martin Klein's valuable intellectual biography of Ehrenfest (21) and to his numerous articles on the history of the early quantum theory.

I

Discussions of the old quantum theory in relation to classical physics usually center on Bohr, who applied classical mechanics to quantum systems in stationary states and who elaborated a correspondence between classical radiation theory and quantum state-transition rules. It is well known that Bohr took very seriously the question of the relationship of the new theoretical ideas to the old [see, *e.g.*, the opening paragraphs of (2)], and the correspondence principle was a famous product of that concern. The question was taken equally seriously by Ehrenfest,

THEORY GENERALIZATION

who pursued a line of inquiry very different from Bohr's. Ehrenfest's thought involved a fascinating interplay of classical with new, quantum theoretic ideas and, accordingly, is of great interest from the standpoint of the temporal unity of science. His approach *via* statistical thermodynamics ultimately led to partial generalizations of Boltzmann's statistical mechanics, as already noted, and to the so-called "adiabatic principle," second in importance only to the correspondence principle as an heuristic guide in the search for a more general and more intelligible quantum theory. Although these two aspects of Ehrenfest's work are very intimately related, the emphasis in this section will be on his attempt to generalize statistical mechanics.

Ehrenfest's work was quite independent of Bohr's. In fact, Ehrenfest's initial reaction to Bohr's paper (1) of 1913 was one of despair; for, besides its well known, blatant disregard for classical principles and mechanisms and its high-handed treatment of the stability problem, Ehrenfest surely recognized that Bohr's theory threatened to alter the very nature of the quantum theory – as he, Ehrenfest, conceived it – changing it from an elegant modification of classical statistical mechanics into an atomic theory of matter replete with arbitrary and *ad hoc* assumptions. It was some years before he could adapt himself to Bohr's point of view (21, pp. 279, 286). Yet, like Bohr, Ehrenfest sought a way to generalize Planck's quantum hypothesis for the simple harmonic oscillator to a wider class of physical systems, and, ironically, it was precisely Ehrenfest's own work on adiabatic invariants which helped unite the statistical thermodynamical and the Bohr-Sommerfeld atomic spectra approaches to quantum theory. In the opening remarks of his important paper (9) of 1916 on adiabatic invariants and the quantum theory, Ehrenfest wrote:

In an increasing number of physical problems the foundations of classical mechanics (and electrodynamics) are used together with the quantum hypothesis, which is in contradiction with them. It remains of course desirable to come here to some general point of view from which each time the boundary between the "classic" and the "quantum" region may be drawn.
    Wien's [displacement] law has been found by an application of classic principles.... This law, derived without the use of quanta, stands unshaken amid the quantum theory. This fact is now worth our attention.
    Perhaps something similar holds in more general cases, when harmonic vibrations no longer take place, but more general motions [do]....[4]

Looking back over his work several years later, by which time he, too, had fallen under the spell of Bohr, Ehrenfest echoed these remarks and then added:

> Considered from a present day viewpoint, one was on the track of a *special* type of "pseudo-classical" behavior of a quantum system, and from the analysis of the derivation of the displacement law, one should be able to learn something concerning to what extent one could still find correct results in the midst of the quantum world with the help of *classical mechanics* (electrodynamics) and classical thermodynamics – therefore, even *Boltzmann's statistics*! (10, p. 464).

As the quotations indicate, Ehrenfest's starting point was the remarkable fact that Wien's displacement law (not to be confused with Wien's blackbody *distribution* law), a thoroughly classical result derivable from the second law of thermodynamics and classical electromagnetic theory, remained valid in Planck's quantized treatment of the harmonic oscillator, despite the incompatibility of the quantum hypothesis with these classical theories. Indeed, Wien's law had been one of the principal constraints on Planck's theoretical derivation of the radiation law, a constraint which, in the context of Planck's theory, led directly to Planck's quantum condition.[5] In one common form, $\rho(v, T) = v^3 f(v/T)$, the displacement law related the radiation energy density ρ to a function of v/T, where v is the frequency of the radiation and T the temperature of the radiating object or cavity.

In 1911 Ehrenfest rederived the displacement law in a manner calculated to bring out its connection with the blackbody radiation problem, *viz.*, by Lord Rayleigh's (28) method of considering an infinitely slow, adiabatic compression of radiation in a perfectly reflecting enclosure.[6] Other derivations, such as that employed by Planck, were, in Ehrenfest's view, very indirect and failed to illuminate the problem at hand. We have it in his own words:

> ...Already in 1902 *Lord Rayleigh* had derived a mechanical theorem and applied it to the proof of Boltzmann's radiation law [*i.e.*, the Stefan-Boltzmann law – T.N.]; this permitted all mechanical-electromagnetic elements in the derivation of Wien's displacement law to be brought together in an extraordinarily pregnant way; much more pregnant than in the previous presentations which dealt with light rays and the Doppler principle.... This theorem of Rayleigh helped very essentially to clarify the *place* which so rightly belongs to the *displacement law in Planck's radiation theory*.... Here the role which the adiabatic invariants play in the quantum theory... began to unveil itself. (10, p. 464).

Actually Rayleigh's paper was not couched in the language of adiabatic invariants at all, and it takes some digging to extract the result which Ehrenfest discovered to be the key to a "more pregnant" derivation of the displacement law, *viz.*, the result that the quantity E_v/v (where E_v is the energy of a given normal mode of oscillation of frequency v) remained invariant under the adiabatic change.[7] Remarked Ehrenfest, "The existence of this adiabatic invariant may be considered the root of Wien's law" (9, p. 380 fn.). And Wien's law, as he discovered in 1911 (21, p. 249), was essentially a restatement of the second law of thermodynamics for a system of harmonic oscillators, such as Planckian oscillators, which possess the invariant E_v/v. This result explained *why* Wien's classical law remained valid in, and was so central to, Planck's quantum theory. And by linking Planck's quantum hypothesis – an hitherto puzzling departure from Boltzmann's method (5) – so directly with the second law, Ehrenfest's study of the displacement law could only reinforce his view that Planck's theory of linear harmonic resonators amounted to a modification of Boltzmann's statistical mechanics, a view shared by Einstein and others, including Planck himself – up to a point at least (24, Part IV).

As its title suggests, the 1911 paper, 'Which Features of the Light-quantum Hypothesis Play an Essential Role in Planck's Theory of Heat Radiation?' was a penetrating analysis of Planck's theory in relation to Boltzmann's, and an attempt to reconcile the two. It had been Ehrenfest who first[8] called attention to the fact that although Planck had been forced to resort to Boltzmann's statistical mechanics, his resulting quantum theory of the oscillator violated Boltzmann's basic assumption that equal volumes of phase space are to be assigned equal *a priori* probabilities, an assumption which led straight to the equipartition theorem and to what Ehrenfest later termed the "ultraviolet catastrophe." Therefore, adapting Boltzmann's statistical mechanics to the quantum theory was necessary for the consistency, the intelligibility (see the references in Note 8), and the further articulation of the new theory, which otherwise would have been without a clear statistical thermodynamical basis. (Notice that breaking completely with the powerful Boltzmann approach – a move which strong claims about the incommensurability of theories might lead us to expect – was not a very real option. Planck's quantum theory was partly based on Boltzmann's ideas and techniques. Simply rejecting or ignoring "classical" statistical mechanics would have undermined rather than aided Planck's theory.) And now, in the 1911 paper, Ehrenfest observed that Planck had, in effect, "generalized" Boltzmann's statistical weight function, a generalization which Ehrenfest proved to be both necessary and sufficient to yield the highly confirmed blackbody radiation law (21, p. 251). Consequently, a deliberate generalization of Boltzmann's assumption concerning *a priori* probabilities was essential if a Boltzmannian statistical mechanics was to be compatible with the radiation law, *i.e.*, a statistical mechanics which preserved Boltzmann's relationship between entropy and the "most probable" state distribution. This relationship between thermodynamical quantities and the statistical behavior of molecules provided the statistical basis for the second law of thermodynamics and lay at the heart of Boltzmann's theory. In order to place the latter on a sound, quantum theoretical footing – better, to put the fledgling quantum theory on a sound statistical thermodynamical footing – Ehrenfest considered a completely general weight function and proceeded to determine what restrictions were placed on the form of this function by theory and by experience, particularly by the known features of blackbody radiation.

Ehrenfest's 1911 paper is a classic example of reasoning from a set of explicitly formulated constraints (most of which, in this case, were justified by the empirical success of Planck's radiation law) to the essential features of any adequate new theory (in this case both Planck's quantum theory *and* a revised statistical mechanics compatible with the essential features of Planck's theory). The theorem he extracted from Rayleigh's paper enabled Ehrenfest to establish, largely on theoretical grounds, that for simple harmonic oscillators (electromagnetic normal modes of oscillation) the generalized weight function $\gamma(v, E)$ reduced to a function $G(E/v)$, of a single variable, the adiabatic invariant E/v. Moreover, compatibility with the blackbody radiation law in the low and high temperature limits logically required that the weight function possess a point spectrum in the quantum domain, in place of the classical, continuous spectrum of weights. This generalization of the weight function concept to include point weights was necessary to avoid the ultraviolet catastrophe. And if Planck's radiation law were assumed completely valid for all frequencies, then the weight assignment was purely discrete, the weight function $G(E/v)$ possessing equal, nonzero values only at the points $E/v = 0, 1, 2, \ldots$. That is to say, Ehrenfest could show that Planck's quantum hypothesis followed as a *necessary consequence* of the assumption and the other theoretical and empirical constraints (21, p. 251). For this case at least, Ehrenfest had rigorously proved from the radiation law as a premiss that a direct generalization of Boltzmann's weight assumption was necessary: instead of all regions of phase space receiving equal weights, only quantum allowed regions of phase space were to receive equal weights.[9]

Since what I, following Ehrenfest, am terming a "generalization" of Boltzmann's assumption may appear to be not a generalization but a restriction on the classical theory, a slight digression is necessary to distinguish and relate some of the chief uses of 'generalization,' 'special case,' and 'instance' in the context of intertheoretic relations. It appears that one man's generalization is another man's restriction, for Jammer (14, p. 166) comments that while Bohr considered his quantum theory of the atom a "rational generalization" of classical mechanics, Ehrenfest himself, in light of the restrictive quantum conditions, considered it a restriction. A far more comprehensive discussion of these matters is needed than I can provide here, but keeping track of the few distinctions I shall draw will be taxing enough.

(A) *Scope Generality*

Sometimes a theory or law is said to be more general than another simply in the sense of possessing a larger scope or domain of application, particularly when the larger domain includes the smaller. (Here much more discussion is needed to bring out various things one might mean in speaking of the "scope" of a theory – the set of questions the theory can answer at all, the set of questions it can answer "correctly", etc.) Comparisons of scope need not imply a specific logical or mathematical relationship between the two theories or laws. However, unless a logico-mathematical relationship is implied, we say only that one is *more general* than the other and not that it is a *generalization of* the other, or the other a *special case of* the one. Invocation of a generalization/special case distinction turns on the nature of the logico-mathematical relationship between the two theories or laws. At least three basic kinds of cases must be distinguished here: (B) those in which the special case is logically consistent with its generalization and a logical consequence of that generalization (*i.e.*, with the particular generalization in question: the special case may have more than one generalization); (C) those in which the special case is logically consistent with its generalization but is an instance or specification of the latter rather than a logical consequence of it; (D) those cases in which the special case is inconsistent with its generalization — a limiting case or approximation to it rather than a strict logical consequence of it. I shall discuss briefly each of (B–D) in turn and shall relate each to Ehrenfest's work. A more complete discussion would go on to draw further distinctions within (C) and (D) as well as (A).

(B) *Deductive Generality*

Nagel's deductive account of the reduction of theories and laws falls here, with the qualification that auxiliary premises usually are necessary in order to derive the "special case" theory or law from the more general theory or law. Hence, a special case need not be a *strict* logical consequence of its unsupplemented generalization; however, only when the auxiliary premises are theoretically noncommital (*e.g.*, statements of particular fact) or theoretically uncontroversial are we willing to say that the derived law or theory is (essentially) a logical consequence of, and a special case of, the general principle or theory. If a theory T_2 is a special case of T_1 in this sense, then T_1 will also be more "scope general" than T_2. Of course, a "mini-theory" T_2, much smaller in scope than T_1, may nevertheless incorporate a generalization of one or more specific assumptions of T_1 — a fact which is important in understanding Ehrenfest's reasoning. As he informs us in (6, Section 3), "We generalize Boltzmann's treatment only in *one* essential point: through the introduction of an initially arbitrary 'weight function'" (Ehrenfest's emphasis).

(C) *Specification and Abstraction*

Ehrenfest's 1911 paper and especially his 1914 paper (8) involved the consideration of special cases or instances which are consistent with a schematic general principle but which are specifications of the latter and not logical consequences of it, taken alone. I shall follow Ehrenfest's 1914 terminology in speaking of the relation of the generalization to its instances or "special cases" (his term) as one of "specification" or "specialization" *(Spezialisierung)*. Let us term the inverse relation of the instances to the generalization "abstraction." As in 1911 Ehrenfest began with a completely general ("arbitrary," undetermined) weight function and proceeded to inquire what restrictions were imposed on the form of the function by various theoretical requirements, notably that Boltzmann's relation $S = k \ln W$ between the entropy and the probability of a state remain valid. In fact, Ehrenfest, who reported that he was following Planck and Einstein, asserted that Boltzmann's theorem should be raised to the status of a postulate in the new theory. Ehrenfest showed (again) that this postulate, together with Planck's quantum hypothesis, implied that Boltzmann's constant weight function was just one of many possible weight functions which might be introduced into the theory. Moreover, he established that it was necessary and sufficient for consistency with Boltzmann's entropy-probability relation that an admissible weight function be (a function of) an adiabatic invariant; otherwise, the entropy would change during a reversible, adiabatic process in violation of the second law of thermodynamics.[10] In the 1914 paper this conclusion was no longer restricted to harmonic oscillators (and the particular adiabatic invariant E/ν) but held for any "ideal" system of independent molecules. Thus the 1914 paper removed part of the scope restriction on Ehrenfest's 1911 results, which amounted to a generalization of Boltzmann's statistical mechanics only for systems of harmonic oscillators – a generalized theory of very restricted scope. But the remaining restriction to ideal molecular systems in 1914 meant that Ehrenfest still could generalize Boltzmann's theory in one sense only at the cost of restricting it (at least temporarily) in another. The 1914 result was still a generalized theory of restricted scope. Moreover, generalization was achieved at the cost of making the theory highly schematic, for the only characteristic of the generalized weight function which he could establish conclusively was that it must be a function of an adiabatic invariant.

In the 1914 paper Ehrenfest did succeed in exhibiting the various weight functions previously employed by Boltzmann, by Planck, by Debye, and by himself (the generalized function of the 1911 paper) as "special cases," instances, or specifications of a generalized weight function Γ, defined as some function of the hypervolume enclosed by the energy surface in molecular phase space. Ehrenfest showed that all such weight functions satisfy the requirement of adiabatic invariance but could not establish that all possible weight functions must be instances of Γ. The main point I want to make about this case is that while Boltzmann's and Planck's specific weight assignments (for example) were consistent with the generalized weight function Γ and were specifications of it, their assignments were not logical consequences of the generalized weight assignment. For not only were Planck's and Boltzmann's assignments mutually inconsistent (mutually incompatible items may be instances of the same general schema), but also, more importantly, the generalized weight assignment Γ was not really a definite but unknown assignment at all but only the requirement that the assignment be a function of the phase space volume enclosed and thereby satisfy the invariance condition. To say that a specific weight function is an instance or specification of the generalized weight function is therefore to say only that the given function is *consistent* with the general requirement and *not* that it is a *logical consequence* of the general requirement, taken by itself. In short, 'instance of X' or (somewhat more loosely) 'special case of X' does not necessarily mean 'logical consequence of X'; it may mean only that the particular instance is and must be consistent with the general principle or requirement. Ehrenfest's paper also illustrates that one incompletely specified function, principle, or theory may be a specification of another. For Ehrenfest's generalized weight function $G(E/v)$ of 1911 was defined as some function of the invariant E/v, and it turned out to be a specification of the more general function Γ defined in the 1914 paper. Finally, it is also worth noting that *if* specification is considered a form of reduction (e.g., if we choose to say that one specification of the weight function *reduces* the generalized theory to Planck's theory, while another specification reduces it to Debye's theory, etc.), then apparently it constitutes a different sort of reduction from either of the two basic types – Nagelian deductive reduction and limit/-approximation reductions – discussed in reference (22). I cannot pursue this matter here, but the just mentioned limit/approximation concept of reduction itself provides still another way (or a set of ways) of drawing a generalization/special case distinction.

(D) *Limit/Approximation Generalization*

So far I have discussed only generalization/special case distinctions which presuppose the logical consistency of the general case with the special. But physicists and other writers frequently say that a theory or principle T_1 is more general than a special case T_2 even when T_1 and T_2 are mutually inconsistent. For example, equations of classical mechanics are said to be special cases – limiting cases – of equations of special relativity. Indeed, T_1 may have two or more special cases which are themselves mutually inconsistent. Wien's distribution law is incompatible with the Rayleigh-Jeans law, but both are limiting cases of Planck's distribution law, with which they are, of course, both incompatible. The thrust of these examples is that T_2 may be a special case of T_1 even though T_2 is not derivable from T_1 in the strict, logical sense (even allowing auxiliary premises) but is derivable only by means of limit processes or approximations. Here again, 'special case of X' means something other than 'logical consequence of X'.

The minimal and unrefined set of distinctions I have drawn must suffice to indicate the great variety of intertheoretic relationships that may be covered by vague talk about generalizations and special cases. As a further indication of the sometimes complex relationships underlying such talk, notice that if T_1 reduces to T_2 under an approximation or limit, it does not follow that T_2 as a whole is a special case of T_1 or of smaller scope than T_1, for it may be that all postulates of T_1 reduce to statements of T_2 but there exist postulates of T_2 which have no counterparts in T_1. (And even when T_2 is a genuine special case of T_1, we must take care in imputing a wider scope to T_1; *e.g.*, consider the special relativity-classical mechanics relationship.) The relation of the Planck-Ehrenfest quantum statistical mechanics of the harmonic oscillator to Boltzmann's theory exemplifies a somewhat similar sort of case, in which a mini-theory T_1 (Planck-Ehrenfest) reduces to a more general theory T_2 (Boltzmann) such that only the *scope-restricted* T_2 is a special case of T_1. Thus the classical statistical mechanical treatment of the harmonic oscillator is a special case of the Planck-Ehrenfest treatment, although Boltzmann's theory as a whole is certainly not a special case of the Planck-Ehrenfest theory of the oscillator. The Boltzmann theory is far more general in scope, while in another sense the Planck-Ehrenfest theory generalizes the classical approach.

It may occur that two special cases of a theory or principle in the "specification" sense (C) discussed above are themselves related as generalization to special case in the "limit/approximation" sense. Both Boltzmann's and Planck's weight assignments were specifications of the generalized assignment Γ, yet Planck's quantized weight assignment reduced to Boltzmann's uniform one in the limit as $h \to 0$. Again we see that an item may be a special case or a restriction of a given theory or principle in one sense (only discrete, quantum allowed regions of phase space receive nonzero weights) while constituting a generalization of the *same* theory or principle in a second sense. Ehrenfest was perfectly justified in considering Planck's theory to involve a generalization of Boltzmann's weight assumption, provided that we treat Planck's *constant* as a variable in the theoretical "metalanguage." Once we do this, Boltzmann's weight assignment becomes a special case of Planck's and the corresponding Boltzmannian theory of the harmonic oscillator (not the full Boltzmann theory) becomes a special case of Planck's. Actually, the "scope", the "limit/approximation", and the "specification" senses of 'generalization' were all implicit in Ehrenfest's thought, as when he remarked:

> ...Planck, in his radiation theory, was the first to decide on a general choice of [weight function] which escaped the previously unavoidable equipartition theorem. The resulting "quantum theory" introduced a freer choice of [weight function] in the various fields of statistical physics. In particular, Debye soon advanced a very promising generalization of Planck's assumption (8, p. 347).

Planck's choice of weight function in effect opened up a whole new range of possible choices, although it was Ehrenfest who first attempted to delimit the range of choices by suitably defining a generalized weight function. But Planck's own choice was just one very specific choice, tied to the linear harmonic oscillator, a particular specification of the generalized weight function which reduced to Boltzmann's choice in the limit. In turn, both Debye and Ehrenfest sought specific choices more general in scope than Planck's, choices sufficiently general to apply to a wider class of physical systems, choices which would reduce to Planck's when the energy equation of the system in question reduced to that for the simple harmonic oscillator. For periodic systems more general than the harmonic oscillator, this meant finding a weight function which was a more general adiabatic invariant than E/v. Ehrenfest did find a more general one in a classical theorem of Boltzmann and Clausius, a matter to which I return in my discussion of the adiabatic principle in Section II. The more general adiabatic invariant and the adiabatic principle are in turn the key to understanding Debye's generalization of Planck's theory, as will be shown in Section III.

In what ways, then, did Ehrenfest generalize Boltzmann's theory? First, Ehrenfest's 1911 analysis and further articulation of Planck's theory showed, more clearly than Planck had done, that Planck's theory was a *full generalization* of Boltzmann's theory in the limit/approximation sense, although a generalization of very restricted scope (simple harmonic oscillators). By 'full generalization', I mean a consistent set of postulates which includes the generalized component of the older theory (in this case the generalized weight assignment) and from which, for the domain in question, all the postulates of the older theory may be derived, either deductively or as limiting cases or approximations but *not* merely by specification. (Schematic theories do not count as full generalizations of nonschematic theories.) Ehrenfest showed that the central features of Boltzmann's statistical theory were preserved by Planck's.

Second, Ehrenfest's own generalization of Planck's theory to general periodic systems of one degree of freedom – the 1916 results to be discussed below – also amounted to a generalization of Boltzmann's (original) theory in the same sense that Planck's was, *viz.*, the limit/-approximation sense. Planck had observed in 1906 (24, pp. 154ff.) that his quantum theory passed over into the classical theory in the limit $h \to 0$. For black-body radiation the same held true in the limit of high temperatures and/or low frequencies, a result which was later extended to a wider class of physical systems by Bohr.

Third, in his 1914 paper (particularly), Ehrenfest generalized Boltzmann's theory in the abstraction sense, the result being a highly schematic theory of a less restricted scope (ideal molecular systems) than the theory of harmonic oscillators. Let us call this schematic theory, which possessed a generalized weight function (required only to be a function of an adiabatic invariant) the *schematic Boltzmann theory* for ideal molecular systems. For this domain of systems, Ehrenfest established the schematic Boltzmann theory as the most general form of the theory compatible with Boltzmann's entropy-probability relation. (This statement needs some qualification since Ehrenfest succeeded in showing that the schematic theory contained a full counterpart to the thermodynamical temperature only on the additional supposition that the weight function was of the form Γ. However, this qualification will not affect the present point, since I could just as well restrict the schematic theory to weight functions of this form.) The schematic theory was not a theory of any particular physical system: it contained no specific weight function. For this reason, the schematic theory could not, without further specification, be termed a *quantum* statistical mechanics, for it also encompassed the classical theory. Of course, it would make little sense to speak of a full-scale quantum statistical mechanics prior to the development of Bose-Einstein and Fermi-Dirac statistics.

The schematic theory, and its generalized weight function *cum* invariance requirement in particular, possessed something of the character which Einstein ascribed to the main principles of his special theory of relativity in his 1907 reply to a note of Ehrenfest. Indeed, coming full circle, Einstein drew an analogy to the second law of thermodynamics!

The principle of relativity or, more precisely, the principle of relativity together with the principle of the constancy of the velocity of light, is not to be interpreted as a 'closed system,' not really as a system at all, but rather merely as a heuristic principle which, considered by itself, contains only statements about rigid bodies, clocks and light signals. Anything beyond that that the theory of relativity supplies is in the connections it requires between laws that would otherwise appear to be independent of one another.... Thus we are by no means dealing with a 'system' here, a 'system' in which the individual laws would implicitly be contained and from which they could be obtained just by deduction, but rather only with a principle that allows one to trace certain laws back to others, analogously to the second law of thermodynamics." [Quoted in (19, pp. 515f; 21, pp. 151f; translation Klein's).]

Einstein later called theories of this kind "theories of principle" as contrasted with "constructive theories," that is, theories of matter (19, p. 510).

It is now clear why, in all senses of generalization relevant to this case, Ehrenfest believed that generalizing an old, "classical" theory – Boltzmann's statistical mechanics – represented a theoretical advance for the quantum theory and also why Ehrenfest generalized the old theory in precisely the respects that he did. Ehrenfest correctly believed that Planck's quantum theory could be considered essentially a modification, a generalization, of the classical Boltzmann theory as applied to one very special physical system, the linear oscillator. Planck had left the relationship of the theories very obscure, but Ehrenfest reasoned that clarifying their relationship, by determining the precise nature of the generalization, not only would put Planck's little theory on firmer ground but also at the same time would reveal how to generalize the classical Boltzmann theory so as to render it compatible with the new quantum results. Since the Boltzmann theory was one of great power and scope, one might expect that an updated version would itself constitute the framework for a significant generalization of Planck's quantum theory, a generalization in which one could be sure that the main results of statistical thermodynamics (the second law, the entropy-probability relation, etc.) still remained valid. Ehrenfest succeeded in formulating what I have called "the schematic Boltzmann theory," a generalization of restricted scope which preserved the entropy-probability relation and which was related by specification to its particular instances or applications. The schematic theory provided a framework into which not only Planck's theory could be fitted (thereby clarifying its statistical basis) but also into which sought-for *generalizations* of Planck's theory could be fitted. Moreover, the generalized Boltzmann theory provided *positive heuristic guidance* in the search for such a theory in its requirement that weight functions be adiabatic invariants – a requirement which presaged the adiabatic principle for quantum conditions. As Ehrenfest himself described it, it was a requirement "which, it appears to me, must hold for all future generalizations of Planck's energy-step assumption, connecting and even leading to them" (8, 348). Ehrenfest exploited this heuristic constraint in his search for a generalized *quantum* theory, a search for a theory based on a more general weight function (and more general quantum conditions) than Planck's, a function which could be expected to reduce to Planck's and in turn to Boltzmann's classical weight function in the appropriate limits.

Thus, Ehrenfest's path toward a generalization of Planck's theory involved a curious interplay between the comprehensive old Boltzmann theory and Planck's small new one. As one specific example of this interplay, notice that in the 1911 generalization of Boltzmann's theory, the reappearance of the adiabatic invariant E/v as a restriction on the form of the statistical weight function for the harmonic oscillator could once again be attributed to (or was an expression of) the displacement law. This law, which originally entered Planck's theory by virtue of its being a "classical" constraint on the latter, now returned to play a kind of reverse role, since it was now the classical theory which was being altered to comply with Planck's theory *cum* radiation law as a constraint. The give-and-take between the old and the newly emerging ideas can be expressed in more general terms, as follows.

The new theory developed out of the old but, for various experimental and theoretical reasons [see (17) for details], broke with the old in certain respects. Yet analysis of the nature of the break pointed to the kind of modification needed to update the old theory. Ehrenfest showed that generalizing Boltzmann's simplest possible assumption governing *a priori* probability could save the remainder of the theory (at least for those cases in which he succeeded in statistically rederiving the second law). The kind of generalization required was suggested by close study of Planck's radiation theory, and for the harmonic oscillator case the essential features of the Planckian theory determined the required generalization quite specifically. In other words, although the young theory was rooted in the old, it became a guide to the generalization of the old, a generalization which, in turn, would help deepen and extend or generalize the developing, young quantum theory.

Still more abstractly characterized, the pattern or strategy evident here can be set out in the following steps. (1) an old, established theory is discovered to be incompatible with a new, highly confirmed mini-theory. (2) Although much smaller in scope, the mini-theory is found to generalize a particular postulate or component of the old theory. (3) Further analysis reveals that the new theory (subject to further articulation) constitutes what I earlier termed a "full generalization" of the old theory, but one of very restricted scope – *viz.*, the scope of the mini-theory. (4) Assuming the validity of the new theory, one next attempts to remove all or part of the scope restriction still present on the full generalization of step (3). If completely successful, step (4) results in a full generalization of the old theory which possesses a wider scope than does the mini-theory. Such a result constitutes a full generalization of the mini-theory. However, the only obvious strategy for removing the scope restriction on the full generalization of step (3) may be to generate a highly schematic generalization of the old (or new) theory, one which yields the old theory and the new mini-theory only by specification. If this is done, the resulting schematic theory does not itself constitute the sought-for generalization of the mini-theory, so further work is required. (5) One therefore uses the schematic framework theory of step (4) as a theoretical basis and a heuristic guide in the search for a nonschematic generalization of the mini-theory.

It should be noted that this pattern or strategy is not applicable to all cases in which two theories are discovered to clash. It is less clear whether a mini-theory satisfying (1) could fail to satisfy (2), for in a suitably weak sense, perhaps *every* such mini-theory generalizes the established theory with which it clashes. Even if this is true, however, it may require a good deal of work to establish in precisely what respect the new theory generalizes the old – as it did in Ehrenfest's case.

In principle, the generalization involved in step (4) may be any one of the three types (B)–(D). If it is a generalization of types (B) or (C), *i.e.*, if the mini-theory is either deducible from or derivable in the limit (or under an approximation) from the generalized old theory, then step (5) already has been accomplished – provided, of course, that the generalized theory is adequate. For the latter is itself a definite (nonschematic) theory which has the mini-theory as a special case and thus may be counted as the sought-for generalization of the mini-theory. However, the generalization of step (4) may be only a generalization in the specification sense – a general theory schema or framework which can yield a definite theory, such as the mini-theory and any definite generalizations of it only by specification. Here generalization begins by permitting one component of the old theory to assume its most general form; then one proceeds to find constraints which more specifically determine the form of this assumption. In this case – the Ehrenfest case – step (5) requires a good deal of work beyond that required by step (4). If successful, this work will yield a definite theory which is more general than the mini-theory in senses (B) or (C).

It is noteworthy that Ehrenfest's particular mode of theory formation or generalization depended more on great critical insight into the nature of existing theories than on a highly creative and original theoretical imagination of the kind that generates radically new theoretical structures (*e.g.*, Einstein, Dirac). His method of generalization was based almost exclusively on analysis of the natures and relationships of theories already available. While these features are apparent in the pattern presented above, I want to emphasize that this pattern is only a first attempt to exhibit the general form of one aspect of Ehrenfest's reasoning toward more general theories. This pattern represents a strategy or mode of generalization which might well be successfully employed in other cases, but, obviously, the pattern is not an algorithm and carries no guarantee of success. Indeed, a comparative examination of many other historical cases would be necessary to determine just how historically and philosophically useful it is to set out reasoning patterns or strategies at this high level of abstraction at all, and in particular to determine whether such a pattern is more characteristic of Ehrenfest's reasoning than (say) Planck's, Bohr's, or Einstein's. After all, Einstein's own critical insight into the natures, interrelationships, and flaws of established theories was unsurpassed! To ask how Einstein's reasoning differed from Ehrenfest's raises many intriguing historical and philosophical questions. Any attempt to bring out the differences between Einstein and Ehrenfest or Planck *solely* in terms of their adherance to fairly rigid reasoning strategies or to well defined "approaches" would surely be misguided, for one notable difference between Einstein and the other two theorists was Einstein's extraordinary ability to adapt readily to strange new intellectual environments.

As radical as the new ideas were, it is not at all tempting to consider Planck's work, or even Einstein's and Bohr's, much less Ehrenfest's, a complete break with the past. Although Planck's theory of the oscillator was obviously nonclassical in some respects, it took a good deal of effort to discern precisely what classical assumptions and techniques had been abandoned. The precise boundary between the two bodies of theory was far from obvious. According to Klein (21, p. 283), Ehrenfest had trouble convincing anyone that its relationship to Boltzmann's statistical theory of the second law was problematic at all. And, of course, Planck had not realized, what Einstein and Debye only pointed out later, that his simultaneous use of electromagnetic theory and the quantum hypothesis involved an inconsistency. Einstein himself changed his mind more than once as to the compatibility with Planck's theory of his own 1905 "heuristic viewpoint" concerning mutually independent energy quanta.

Generalization rooted in a critical analysis of available theories was also characteristic of Ehrenfest's development and use of the adiabatic principle as a criterion for selecting quantum conditions – the subject to which I now turn. I believe that even under a very abstract characterization the pattern of reasoning involved in this aspect of Ehrenfest's attempt to generalize Planck's theory is distinct from the general pattern set out above. However, the two lines of reasoning are intimately related in Ehrenfest's case, if they can be separated at all, since his development of the adiabatic principle and his attempt to generalize Boltzmann's theory were virtually opposite sides of the same coin. For Ehrenfest's generalization of Boltzmann's probability postulate was basically that quantum allowed states, *i.e.*, states satisfying the quantum conditions, are to receive equal *a priori* weights. Like the generalized weight assignments, the sought-for generalized quantum conditions must be functions of adiabatic invariants.

II

Ehrenfest's first steps toward the adiabatic principle were recounted at the beginning of Section I: his surprise that the classical Wien displacement law, involving a function of the adiabatic invariant v/T, remained valid in Planck's quantum theory and in fact largely determined Planck's quantum condition for the harmonic oscillator; his discovery, in Rayleigh's paper, that E/v is an adiabatic invariant for any one-dimensional, harmonic oscillator; and the "more pregnant" derivation of the displacement law which this result afforded.

The fact that the classical adiabatic invariance results of Wien, Rayleigh, *etc.*, not only remained valid in the quantum theory but also were so central there – staring one in the face in Planck's quantum condition – convinced Ehrenfest of their great importance. Here was an element which the otherwise radically different classical and quantum theories had in common. Convinced that classical adiabatic invariants remain invariant in the quantum theory, and that the quantum condition of any linear harmonic oscillator, whether material or aetherial, was the adiabatic invariant E/v, Ehrenfest set out to exploit this connection to classical theory in order to extend Planck's quantum theory, in so far as possible, to systems of more than one degree of freedom, to nonharmonic motion, and even to nonperiodic motion (although with the last he was unsuccessful).

Ehrenfest eventually recalled a classical dynamical theorem concerning periodic mechanical systems proved decades before by Boltzmann and by Clausius (7 and 21, pp. 260ff), a theorem which implied the adiabatic invariance of the quantity $2\bar{T}/v$ (where $\bar{T}$ is the kinetic energy of the system, averaged over one period). This result – a dynamical, not a statistical result (20) – held good for a wide class of periodic systems, including nonharmonic vibrations and molecules bounding in a box, and immediately extended to systems of several degrees of freedom. In light of Ehrenfest's generalization of Boltzmann's theory, described in Section I, there was good reason to believe that this classical result remained valid in the quantum theory and could become a powerful tool for extending the theory. Since for one degree of freedom the theorem could be expressed in the form $2\bar{T}/v = \iint dq\, dp$ (where q and p are the generalized coordinate and momentum, respectively), the relevance of the theorem, and of Ehrenfest's study of adiabatic invariance generally, to the quantum conditions as formulated by Sommerfeld and others was evident. One could establish the invariance of a given quantum condition (function) by proving it equivalent to $2\bar{T}/v$.

Having established that the known quantum conditions were adiabatic invariants, Ehrenfest became convinced not only that all periodic and conditionally periodic motions were associated with adiabatic invariants but also that every quantized magnitude was expressible as an adiabatic invariant. The quantum conditions, therefore, simply equated two invariant quantities. To say that quantized magnitudes are adiabatic invariants is just to say that any physical system remains in the same quantum state throughout an adiabatic process. This generalization – that adiabatic processes, as defined by Ehrenfest, always transform quantum allowed motions into allowed motions – is the *adiabatic hypothesis* or *principle*. It may be considered the first introduction into quantum theory of an invariance principle special to the theory and a further development of the requirement of the generalized Boltzmann theory that any admissible weight function be a function of an adiabatic invariant. (Ehrenfest's student, J. M. Burgers (3), later showed that the new formulations of the quantum conditions by Epstein and Schwarzschild in terms of action-angle variables were also adiabatically invariant.)

The adiabatic principle guided the development of quantum theoretic ideas in three ways. First, it meant that an adiabatic process did not disturb the quantum state or what Bohr called the "stationary state" of a system. Since classical mechanics applied to systems in stationary states, the principle meant that classical mechanics could be applied not only to isolated systems or to systems under constant external conditions but also to systems undergoing a sufficiently gradual external influencing (*e.g.*, an increasing external electric field). As Ehrenfest put the point in 1923, with Bohr in mind:

THEORY GENERALIZATION

...According to [the adiabatic] principle *quantum systems react quasi classically under certain conditions, i.e., as if they obeyed classical mechanics*, namely, when one submits them to a sufficiently smooth, patiently cautious external influence – *i.e.*, precisely an adiabatic one; ...on the other hand, they *must* show their quantum "claws"... as soon as the influencing no longer proceeds sufficiently cautiously and patiently.[11]

Second, where quantum conditions already had been conjectured, *e.g.*, by Sommerfeld, the adiabatic principle helped to justify these choices by providing a new criterion which they were proved to meet[12] and by unifying them with respect to this common requirement of adiabatic invariance, thus rendering them more intelligible and less arbitrary. Third, the principle helped in the selection of quantum conditions for previously unquantized systems by requiring that quantized magnitudes be adiabatic invariants. And in those cases in which the system in question was adiabatically deformable into a system for which the quantum conditions were already known, the principle constituted a sufficient as well as a necessary condition of quantizing the physical system.

For instance, any system which was adiabatically transformable into a Planckian oscillator – and this included all periodic systems of one degree of freedom – was subject to the quantum conditions which govern the latter. In this special case, the adiabatic principle licensed a direct extension of Planck's original theory to a wider class of systems, and in general it permitted the reduction of one quantization problem to another[13] – a matter which I shall shortly treat in detail.

First, however, I want to conclude this part of the discussion by indicating the general pattern which seems to underlie this aspect of Ehrenfest's thought; for his reasoning is an excellent example of a second way in which one may attempt to generalize or extend a theory, a way which again essentially involves reference to another theory – typically a predecessor theory – but which does not constitute a direct attempt to generalize the older theory. In outline, the main steps are these.[14] (1) First one notes (a) that one or more theoretical results of the old theory (or system of theories) remain strictly valid in the fledgling new theory; (b) that these results are quite central to the new theory; and (c) that facts (a) and (b) are surprising or noteworthy for the particular results in question and/or in light of the significant differences between the old and new theories (In Ehrenfest's case the result in question was Wien's displacement law.) (2) Then, by critical analysis, one seeks the basis of these results in the old theory in such a way as to maximally illuminate their role in the new (cf. Ehrenfest's rederivation of Wien's law and his use of it as an analytical tool in determining the essential features of Planck's theory). (3) Having isolated the most basic element common to the old and new theories, one next tries to formulate this "classical" core in the most general way possible within the context of the old theory (cf. Ehrenfest's appeal to the classical Boltzmann-Clausius theorem). (4) Finally, one attempts to determine whether (or subject to which conditions) one can correspondingly generalize the new theory [Ehrenfest (7 and 9)].

I believe that this particular route to theoretical generalization should be distinguished from that outlined at the close of Section I. To be sure, the two may appear together in interlocking fashion: in some cases the second pattern or strategy as a whole can be inserted into steps (4) or (5) of the first pattern. (The two patterns were still more tightly intertwined than this in Ehrenfest's case.) However, nothing prevents either pattern from being instanced in isolation from the other. One obvious difference between the patterns is that in the first pattern reasoning begins from the fact that a basic principle of the old theory appears in a *different*, "generalized" form in the new mini-theory, whereas in the second pattern inquiry begins from the remarkable fact that exactly the *same* law remains valid in the two, very different theories. A second difference is that the first pattern turns on an attempt to generalize the older theory while the second pattern does not, or need not. The two patterns therefore involve different kinds of appeals to classical theory. In the first mode of reasoning, the classical theory in question serves as a reductive constraint on the sought-for new theory, since the latter is to be a generalization of the old theory (which thus reduces to it under appropriate conditions). But in the second, the classical theory is not, or need not be, a reductive constraint. Instead, it furnishes one or more classical results around which a plausible generalization of the new theory can be constructed. Unlike his appeal to Boltzmann's statistical mechanics, Ehrenfest's appeal to classical mechanics (the Boltzmann-Clausius theorem) involved no attempt to update or generalize the classical theory as a whole[15] – only an attempt to find the most general expression for that element which the old and new theories apparently had in common (a general adiabatic invariant for periodic systems). Of course, Ehrenfest's appeal to classical mechanics did lead him to a generalization of

Planck's quantum theory and therefore also of Boltzmann's statistical theory.

It is important to notice that neither mode of generalization essentially requires theoretical innovation to the degree that it requires critical analysis and clarification. But since neither mode *precludes* highly innovative theorizing, it would be *naive* to suggest that these two reasoning patterns, as I have abstractly articulated them, enable us to *contrast* "highly innovative" with "merely critical" reasoning in physics. My present point is the far more modest one that these patterns of reasoning or strategies represent two *possible* ways of obtaining more general theories without introducing radical new theoretical ideas. Establishing this point is a major step toward solving the historical problem of how it was possible that Ehrenfest should have played such an important role in the generalization of the quantum theory, given that his *forte* was neither theoretical creativity nor calculation. Klein's (21) argument, in barest outline, is that it also largely explains why it was Ehrenfest rather than someone else who discovered this way of generalizing the quantum theory, given that critical analysis *was* his *forte*, that he had been trained on Boltzmann's statistical mechanics, and that he was a careful student of the new quantum theoretic ideas.

Its relevance to attempts by Debye, Bohr, Sommerfeld, Epstein, Schwarzschild, and others to state quantum conditions for a more general class of systems finally brought Ehrenfest's approach into juxtaposition with the more dominant lines of Bohr and Sommerfeld. In fact it was Bohr who made most constructive use of the adiabatic hypothesis from the year 1918 on, as Ehrenfest himself emphasized in (10, §§ 10, 11). However, Bohr renamed it the "principle of mechanical transformability," which was more in keeping with his view of quantum theory as an atomic theory, as a fledgling mechanics of the atom, than with Ehrenfest's view of the theory as an outgrowth or modification of statistical thermodynamics. It is important to our present concern with the unity of science that Ehrenfest's theory of adiabatic invariants contributed very significantly to the unification of the two main theoretical approaches to the early quantum theory, the statistical thermodynamical approach (represented in different ways by Planck, Einstein, Ehrenfest, and even Debye to a considerable extent) and the atomic-structure-and-spectral-lines approach of Bohr, Sommerfeld and others.

Both Planck's theory of the harmonic oscillator and Bohr's theory of the hydrogen atom could be viewed as special applications of Sommerfeld's generalized quantum conditions; and it was precisely the theory of adiabatic invariants which provided theoretical justification for Sommerfeld's generalized quantum rules, as we have seen. Indeed, had it been published sooner and its importance recognized, Ehrenfest's paper (9) could have had a far more determinative influence on the development of the theory instead of (in large part) providing theoretical justification only after the fact. In (9) Ehrenfest showed that Debye's generalized quantum theory was rigorously derivable from Planck's, in conjunction with the adiabatic invariance results, and that Bohr's and Sommerfeld's quantum conditions satisfied the adiabatic principle. But it is evident from the paper that the adiabatic invariance results might have led one, by a straight-forward and plausible if not entirely rigorous argument, from Planck's quantum hypothesis to the *discovery* and not just a justification of Bohr's and Sommerfeld's quantum conditions for atomic systems of one and two degrees of freedom, respectively. Indeed, employing the adiabatic principle in the manner he later recognized to be problematic, Ehrenfest had quantized the angular momentum of rotating, dumbbell molecules in a 1913 paper submitted prior to the appearance of Bohr's famous paper of 1913.

I now turn to the reduction of problems to problems, which is the key to understanding the justificatory and heuristic roles of the adabatic principle as a tool for extending and unifying the early quantum theory.

III

> I had wanted to show them such a very neat thing – the method for making a continuous transition between two problems that are discretely separated to begin with.
>
> PAUL EHRENFEST[16]

The adiabatic principle is extremely interesting from the standpoint of reduction in that it provided a way of reducing certain physical theoretic problems to others. The reduction of problems to problems, incidentally, is a topic which has been almost totally neglected by students of theoretical reduction and scientific reasoning. Now it might be retorted that there is a good reason for this neglect, to wit, that talking about the reduction of problems to problems is really only a misleading way of talking about the reduction of theories to theories; that problem reductions are accompaniments of, or consequences of, theory reductions and can be fully understood as such. This I wish to deny. I shall argue that the reduction of problems to problems is an important issue in its own right. Although the present discussion must be limited to a single case, I am convinced that the comparative study of various mathematical and physical theoretic techniques for transforming or reducing problems is of great philosophical and historical interest. I recognize that the whole matter of scientific problems and solutions, of questions and answers and their conditions of identity and individuation, demands a far more thoroughgoing analysis than is provided by the present, largely intuitive reflections.

An asymmetric, anharmonic oscillator is not equivalent to a linear, harmonic oscillator either physically or mathematically; yet Ehrenfest showed that one is adiabatically deformable into the other and hence, by the adiabatic principle in the form given it by the Boltzmann-Clausius theorem, that the *problem* of quantizing the asymmetric oscillator reduces to (is equivalent to) the problem of quantizing the linear harmonic oscillator. Mathematically, an adiabatic variation is represented as a limit process in which time variation becomes infinite, since the process must proceed infinitely slowly relative to the natural period of the system [see (35, pp. 110f, 297)]. Actually, there is a sort of double limit process in the case under discussion, since the varied parameters are taken, infinitely slowly, to the "limiting" value of zero.

Ehrenfest, in his 1916 paper, considered an anharmonic oscillator of the general form $\ddot{q} = -(v_0^2 q + a_1 q^2 + a_2 q^3 + \cdots)$, where v_0 is the frequency of the harmonic oscillator $\ddot{q} = -v_0^2 q$. (Despite Ehrenfest's notation, v_0 should be interpreted as the angular frequency $\omega_0 = 2\pi v_0$.) If the parameters a_i were now allowed to approach zero in a reversible, adiabatic manner, the expression for $\ddot{q}$ reduced to that for a simple harmonic oscillator. Because the approach to zero could be accomplished in a reversible, adiabatic manner, the two types of oscillator were adiabatically related, and since the adiabatic principle stated that the quantum conditions for adiabatically related systems are the same, the quantum condition for the harmonic oscillator therefore applied to the asymmetric oscillator also. Given the adiabatic principle, to solve one quantization problem was to solve both.

Because Planck already *had* solved the quantization problem for the harmonic oscillator, it is natural to say that the anharmonic oscillator problem reduced *to* the harmonic oscillator problem, and I will sometimes speak this way for convenience. But actually the reductive relation between the two quantization problems is symmetric, subject to the qualification mentioned in note 17, since the relation of being *reversibly* adiabatically related physical systems is symmetric (*cf.* note 7). Thus we are equally entitled to say that the harmonic oscillator problem reduced to that of the asymmetric oscillator – a point which will later prove important. Accordingly, for the type of case under consideration, it is less misleading to speak of reduction to a common problem, of a *coalescence* or a *unification* of problems, than to speak of one problem reducing *to* another.

The asymmetric oscillator case is historically important in the early quantum theory (as well as the later). In his 1913 'Wolfskehl' paper on the equation of state of solids, Peter Debye (4, esp. §§1–3) conjectured the quantum condition for such an anharmonic oscillator by analogy with Planck's theory. This was the first successful extension of Planck's theory to non-harmonic motion, a generalization which in turn helped guide Sommerfeld to his generalization of Bohr's theory. Debye's "trick" (*Kunstgriff*) yielded gratifying experimental results but had insufficient theoretical justification until Ehrenfest showed that Debye's oscillator was adiabatically deformable into Planck's and hence subject to exactly the same quantum condition. In other words, the adiabatic principle licensed the literal and not merely the analogical generalization of Planck's theory to a wider class of physical systems. To put the point in a different way, Debye's theory constituted a "rational generalization" of Planck's, in a sense to be explained. But Ehrenfest, as Bohr would put it later, showed "that [Debye's quantum] condition forms the *only* rational generalization of Planck's condition" (2, p. 109, my emphasis).

Debye's theory of the anharmonic oscillator may be considered *a* rational generalization of Planck's original theory of the linear oscillator in the sense that it was a generalization which could be, and was, scientifically rationalized, even though the available constraints were insufficient to preclude the possibility of competing rationalizable generalizations. What was the nature and rational basis of Debye's generalization?

In Planck's theory as presented in his 1906 *Vorlesungen* (24) and as developed in his Solvay Congress paper (25), the constant energy curves of an oscillator in q-p space (the "phase space" of the oscillator) which represented the allowed energies of the system ($E = \alpha p^2 + \beta q^2 = nh\nu$), formed a discrete set of ellipses symmetric to the q and p axes. Planck's quantum hypothesis stipulated that successive ellipses be separated by an empty band having an area equal to Planck's constant h. In other words, each ellipse had a total area greater by h than the ellipse just inside it [see the figure on p. 5 of (35), for instance], and the quantum condition could be expressed in the form $E/\nu = nh = \iint dq\, dp$, where the double integral (over one period) gives the area enclosed by the nth energy curve. [Planck did not explicitly state the quantum condition in this integral form in the 1906 edition of (24), but he might well have.]

Now the constant energy curves for Debye's asymmetrical oscillator (for which $E = \alpha p^2 + \beta q^2 + \gamma q^3 + \cdots$) were not ellipses, nor were they symmetric to the p axis. Nevertheless, Debye assumed by analogy with Planck's quantum hypothesis that admissible energies could be represented by a discrete set of curves, each larger by h than the one inside it, *i.e.*, that the quantum condition was $\iint dq\, dp = nh$, exactly as for Planck. Obviously, this was a direct analogical extension of Planck's quantum hypothesis to the asymmetrical oscillator. It was a generalization of Planck in the sense that Planck's theory was simply the special case in which the energy curves were ellipses symmetric about both the p and q axes. If we interpret Debye's as a conjectural *general* theory of periodic oscillations of one degree of freedom, then it reduced (deductively implied) Planck's theory as the subtheory pertaining to harmonic oscillations. (Simply set the parameters a_i, now considered variables of the theory, equal to zero.)

Debye offered the following reasons in support of his innovative step:

(1) His "theory" was a generalization of a successful theory. The fact that Einstein already had successfully extended Planck's theory beyond blackbody radiation to the problem of specific heats strongly indicated "a universal validity of Planck's [quantum formula] for oscillatory modes of any sort."

(2) The generalization was based on a close analogy to Planck's the-

ory and, indeed, the phase space formulation of Planck's quantum condition, $\iint dq\, dp = nh$, cried out for generalization to a wider class of physical systems. (Debye expressed the matter in terms of *a priori* statistical weights and not directly in terms of quantum conditions, however.)

(3) For small vibrations, the motion of an asymmetric oscillator around its point of average displacement (which was not equal to zero, in contrast to a symmetric, harmonic oscillator) was approximately harmonic and so, to an approximation, could be treated by Planck's theory. (*E.g.*, relative to this displaced "center" of motion, the constant energy curves were nearly ellipses, and the magnitude of the mean displacement was proportional to Planck's value for the mean kinetic energy instead of the classical, equipartition value.)

(4) The theory was experimentally testable and appeared to accord well with the available data.

(5) The generalized Debye-Planck quantization rule for the oscillator, $\iint dq\, dp = nh$, was expressed in terms of Planck's constant h rather than energy $h\nu$. (The point here was presumably that the expression $h\nu$ had to be less fundamental because the frequency factor restricted its applicability to periodic systems. Furthermore, even in the case of periodic systems such as the asymmetric oscillator ν depended on the amplitude or energy of oscillation, and Planck's condition $E = nh\nu$ was no longer valid.)

(6) In the present, totally incomplete state of the quantum theory, no other, or at least no better generalization suggested itself. While available constraints did not dictate this particular generalization, they certainly did nothing to discredit it.

Despite the indicated reasonableness of his generalization, however, Debye did not hesitate to characterize his innovative step as "in a certain sense arbitrary" and as based on a "trick." He also stressed the approximative nature of his investigation. Clearly, Debye felt that his generalization lacked an adequate theoretical justification and that it was not a definitive solution of the quantization problem for the anharmonic oscillator. Expressed in my terms, the problem of quantizing the asymmetrical oscillator had not been adequately solved despite the fact that, under a natural interpretation, Debye's theory reduced Planck's, or in any case passed over into Planck's in the limit as the asymmetric parameters went to zero. The only sort of *problem* reduction involved here was

THEORY GENERALIZATION   63

the common variety in which special conditions are imposed which enable us to reduce a general problem *to* one of its special cases. But this sort of special case solution typically neither constitutes a solution to the problem in its full generality nor even shows the *general* problem to be solvable at all. If we confine the discussion to those problem reductions which constitute reductions to a common problem (problem unifications) – cases in which the reduced problem is shown to be essentially equivalent to the problem to which it reduces, so that an appropriately formulated solution to one also will solve the other – then we must conclude that Debye did not succeed in reducing his quantization problem to Plank's, nor did he claim to.

Now Ehrenfest's introduction of the adiabatic principle did achieve a reduction of Debye's problem to Planck's: more accurately, a unification of the two problems. (From the symmetry of the reduction, it follows, strictly speaking, that Planck's problem was no longer a special case of Debye's for the two problems were now shown to be essentially identical.) This achievement was the basis of Bohr's remark, in his famous paper of 1918, 'On the Quantum Theory of Line Spectra', that Ehrenfest "had shown [Debye's quantum rule] to be the *only* rational generalization of Planck's [quantum] condition." What Bohr meant, I think, was that when Ehrenfest's adiabatic principle was introduced as a further constraint on the quantization problem, it sufficed to determine Debye's generalization completely. Debye's step was no longer "arbitrary" in the sense of being underdetermined by theoretical constraints. Since the adiabatic principle required quantum conditions to be adiabatically invariant, and since Debye's analogical extension of Planck's theory amounted to the quantization of the only independent, adiabatically invariant quantity $(2\bar{T}/v)$ which exists for one degree of freedom [as Ehrenfest pointed out in (9, p. 384)], it followed that, in the light of Ehrenfest's principle, Debye's was the only possible rational generalization of Planck's theory. The principle implied that all periodic motions of one degree of freedom were subject to Planck's quantum condition, since all such motions were adiabatically deformable into a Planckian oscillator.[17]

We now have some glimmering of what a powerful tool the principle was and how it might have played an even greater role than it did in the early development of the quantum theory.[18] For as he himself pointed out, Ehrenfest's principle plus Planck's theory logically entailed Debye's theory of the asymmetric oscillator, *i.e.*, together they deductively *reduced* Debye's theory. Curiously, the augmented Planck theory now reduced the Debye theory, whereas previously we noted that the conjectural Debye theory reduced the firmly established theory of Planck. The importance of this peculiar reversal of reductive position will soon become apparent. It was not actually a reversal, since the augmented Debye theory also deductively reduced Planck's theory, but this too will be important below.

With this historical case before us, it is time to ask what has become of my thesis that the reduction of problems is not simply a footnote to the story of theory reduction, a thesis to be defended here chiefly in terms of my particular claim that the reduction of problems by means of the adiabatic principle cannot be fully understood in terms of theory reduction. It may appear that in my own description of this case of problem reduction it plainly boils down to a case of theory reduction. The "problem reduction", it may be alleged, amounts to nothing more than the deductive reduction of Debye's *theory* by the conjunction of Planck's *theory* with the adiabatic principle, or with what Bohr (2, p. 109), E. T. Whittaker (37, p. 122) and others have called the "*theory* of adiabatic invariants" (which includes the Boltzmann-Clausius theorem of Section II and Ehrenfest's principle, *inter alia*). It is theory reduction in the best tradition.

I certainly do not wish to deny that in this case the problem reduction took place in the context of a theory reduction. On the contrary, this Planck-Debye-Ehrenfest case is evidence that Nagel's overly criticized account of theory reduction is indeed applicable to real scientific cases. What I do wish to argue is that problem reductions *may* occur outside the context of theory reductions; and that even in the present case the problem reduction was independent of the theory reduction in the sense that the problem reduction *could* have occurred in the absence of the theory reduction, whereas the theory reduction could not have occurred in the absence of the problem reduction. This is not to say that the theory reduction could have been achieved only by explicit consideration of the problem reduction, but only to say that the existence of the problem reduction is logically entailed by the existence of the theory reduction and not conversely. Thus even in a narrow, logical sense the theory reduction depends on the problem reduction as a necessary condition.

In support of these claims, I point out that in the Planck-Debye-Ehrenfest case the *theory* reduction went through only because Planck's and Debye's theories were in fact available – *i.e.*, because the two oscillator quantization problems already had been solved (although only very conjecturally in Debye's case) – whereas the problem reduction did not depend on this. The adiabatic principle reduced the anharmonic oscillator quantization problem to the harmonic oscillator problem, or rather it combined the initially distinct problems into one, whether or not either problem had been solved. But while the adiabatic principle provided the means to reduce the problems, it was clearly powerless to solve them without being primed by Planck's quantum condition or, more generally, by *some* solution, depending on the case at hand. And without solutions to the quantization problems, there would have been no theories to reduce, since the theories *were* essentially those problem solutions and their consequences.

Now the preceding point must be qualified in light of the fact that most scientific problems can be formulated properly (or at all) only within a theoretical context. In our particular case it would be difficult or impossible to speak of quantization problems apart from Planck's, Debye's, Einstein's, or *some* form of quantum theory. And historically, of course, Ehrenfest discovered his principle through an analysis of Planck's theory. Therefore, my contention is better stated in general terms, as follows: The adiabatic principle provided the means to reduce, unify, or coalesce the quantization problems of all adiabatically related systems into a single problem, *whether or not* the problems actually had been solved for two or more of these systems. If there did not exist at least two distinguishable theories (problem solutions), then obviously there could exist no reduction of theories. As noted above, Ehrenfest or someone else could have reduced the anharmonic to the harmonic oscillator problem even if Debye never had advanced his conjectural "theory". Then the problem reduction would have resulted in an extension (generalization) of Planck's theory apart from any theory reduction of available theories, *i.e.*, the problem reduction would have provided a means of *generating* what was in fact Debye's theory. Historically, Debye's conjectural theory was in fact available, and to that extent the two quantization problems had been solved previously and theory reduction was possible. Nevertheless, we can still say that in a certain sense *it is the fact that the problems were unified that permitted the theory reduction to go through in the first place.* There is certainly a point to saying this is regard to Ehrenfest's reduction of Debye's theory by Planck's, augmented by the adiabatic principle. We found that prior to the invocation of the adiabatic principle, Debye's generalization of Planck's theory reduced the latter but did not constitute a genuine unification of the problems in question. On the other hand, the adiabatic principle, by reducing Debye's problem to Planck's, enabled Debye's *theory* to be reduced *by* Planck's theory (in conjunction with the principle). Thus the theory reduction turned on the invocation of a principle which by itself united (interreduced) the two problems apart from any theory reduction. The result was a significant *unification of the early quantum theory* far above and beyond the unification represented by the reducibility of Planck's theory by Debye's prior to the introduction of the adiabatic results. For only Ehrenfest's reduction showed Debye's theory to be the inexorable working out of Planck's quantum hypothesis and unified the theories as interreducible solutions of essentially the same quantization problem.

My contentions that some problem reductions cannot be fully understood in terms of theory reduction and that discussion of problem reduction is necessary to bring out the full historical and philosophical import of theoretical ideas like the adiabatic principle should not be construed as formal logical criticisms of philosophical accounts of theory reduction. I do not deny that one can demonstrate, without any mention of problem reductions, that Debye's theory is logically derivable from, and hence reduced by, Planck's theory, augmented by the adiabatic principle. (It is not a question of the formal argument being invalid or incomplete unless a premiss concerning problem reduction is added.) But it is by now surely apparent that to restrict the discussion of the reduction in this manner is to lose valuable historical and philosophical insights concerning the reduction (even *qua theory* reduction!) and concerning the role of the adiabatic principle in it. For, to summarize my main points: (a) The adiabatic principle (in the form impressed on it by the Boltzmann-Clausius theorem) unified Planck's and Debye's quantization problems without, by itself, solving either. (b) Thus reduction of problems may occur independently of whether the problems have been solved by a theory and hence independently of particular theory reductions. A theory or principle may unify problems, in the sense in question, without itself constituting a theory reduction. (c) Even when a problem reduction does take place within the context of a theory reduction, as in the Planck-Debye-Ehrenfest case, it may be historically and/or philosophically fruitful to view the theory reduction as dependent on the problem reduction rather than the other way around (or in addition to the other way). This is especially true when, as in the present case, the theory reduction depends essentially on the introduction of a principle or theory which, by itself, unifies the problems in question without solving them. Indeed, our discussion of the Ehrenfest reduction warrants the additional conclusion that (d) consideration of problem reductions can provide a new dimension of understanding of at least some kinds of *theory* reduction. The discussion of scientific problems and of problem reductions may help to distinguish different types of theory reduction and may help to illuminate important distinctions such as Einstein's distinction between "theories of principle" and "constructive theories" (see Section I).

My position may be clarified further by considering an objection to claim (b), a challenge directed at the point where my case may appear to be most convincing. The objection is that, even in the absence of theories to be reduced, claims concerning problem reduction still may be couched in the language of theory reduction, for one need only speak of *possible* theories and possible theory reductions.[19] One reply is that there are cases of problem reductions involving unsolvable problems (or problems unsolvable by means of specified techniques or by theories of a specified kind), in which cases problem reduction claims cannot be understood even in terms of claims concerning possible theory reductions. For in such cases there are no possible theories (problem solutions) of the sort required. For example, the early quantum theory of the atom faced the three-body problem already for quite simple atoms, making it impossible to apply Sommerfeld's general quantum conditions (38, p. 113). Here a reduction or unification of one three-body problem with another might well be possible without either problem having a theoretical solution. Another difficulty of the proposal is that nonexistent but abstractly "possible" theories and "possible" theory reductions furnish no concrete theory structures and reduction mechanisms for analysis; hence, talk of unspecified possible theories and theory reductions is of little use to the historian or philosopher attempting to analyze in detail a particular case of problem reduction. (I am not here denying the utility of *setting out* and discussing theories which might have existed historically but in fact did not.) Finally, the objection does not really touch point (b), which asserts that *actual* problem reductions may occur in the complete absence of *actual*, available theories (problem solutions), perhaps even in the absence of possible theories of a known type. Therefore, actual problem reductions may occur in the absence of actual theory *reductions* between particular, available theories, and perhaps even in the absence of possible theory reductions.

A related form of the objection is that to establish that quantization problem A coalesces with quantization problem B (for example) is just to establish the biconditional

$$(x) \, [x \text{ is a theory or quantum condition for systems of type } A \leftrightarrow x \text{ is a theory or quantum condition for systems of type } B]$$

and hence to establish the interreducibility of the theories in question, at least under a scope restriction to these particular problems. As presently formulated, however, the biconditional is too strong, in light of note 17. More importantly, the objection again misses the point. I do not claim that a treatment of problems and their relationships has no implications concerning problem solutions (theories). Naturally, to show that two problems are essentially identical is to establish that they have essentially the same solutions. So I can accept the above biconditional, with the aforementioned qualification. However, it is important to note two things here. First, the biconditional states only that any adequate solution (theory) of one problem will solve the other problem as well. It does *not* guarantee that any theory which adequately solves one problem will be interreducible or unifiable with all theories which adequately solve the other – only that each adequate theory will solve both problems. It is conceivable that there exist distinct theories adequate to solve both problems but which are not interreducible or even simply reducible, one by the other. To say that two problems are essentially the same (interreducible) is not to say that all their solutions, *i.e.*, all distinct theories which solve the two problems, are interreducible – or that any reductive relations at all hold among them. Hence, the objection breaks down. To put the point in normative terms, a problem unification establishes a constraint or condition of adequacy on theories: that an adequate theory which solves one problem must be capable of solving the other also. (Planck's original quantum condition was insufficiently general to solve Debye's problem.) Such a constraint may be heuristically valuable, a helpful guide in the search for an adequate theory. On the other hand, achieving a problem unification does *not* by itself establish the very different normative claim that any adequate theory which solves one problem must reduce, reduce to, or be interreducible with, each theory which solves the other problem.

Second, it is interesting to note that even if achieving a problem reduction or unification *did* establish nothing more than the interreducibility or unifiability of any adequate theories of those problems, as the objection mistakenly contends, this would not be theory reduction or unification of the sort usually discussed by methodologists. Virtually every account of theory reduction in the literature is intended to apply to pairs of particular, specifiable theories which are already in hand–whereas a problem unification would establish something more general, *viz.*, that *all* adequate theory-solutions of the two problems are interreducible. This would be theory reduction of a different type from that usually discussed, since it would establish a *general* reduction claim which holds for *any* two theories possessing a certain feature (ability to solve the problems), whether or not these theories were known and available, rather than a *particular* reduction claim about two available, specifiable theories. What would be established by such a problem unification obviously could not be adequately expressed by any *particular* reduction statement even if the problem unification did provide the mechanism for effecting a unification of two particular theories (*cf.* the Planck-Debye-Ehrenfest case).

A problem oriented approach to philosophy of science would examine the phenomena of reduction and unification in science from the standpoint of problems and their relationships as well as from the standpoint of completed solutions and *their* relationships. The motto of such an approach might be: "Look at the problems, not just at their solutions!" Such an approach offers a natural way in which to study the heuristic, generative, developmental, dynamic aspects of scientific inquiry. It provides one way of appreciating the role of reductive thinking in the process of theory formation or discovery–and one way of avoiding the mistake of treating reduction as merely a "static" matter of logical relationships between theories already in hand and essentially completed, theories which frequently are no longer at the focus of active research.

The generality of the early quantum theory naturally depended upon its success in quantizing physical systems of all kinds. Since the adiabatic principle enabled one to coalesce the quantization problems of all adiabatically related systems into a single problem, the principle reduced or consolidated the domain of quantization questions. This provided some structural organization to the domain of questions the theory was responsible for answering and in itself extended the domain of application of the theory to those problems which coalesced with problems already solved. This phenomenon could be quite striking, as when the adiabatic principle transformed Planck's *special case* solution of the problem of one-dimensional periodic motion into a *general* solution. The same example also illustrates in microcosm how such problem reductions can greatly unify a developing theory by uniting separate attempts to deal with what originally were distinct problems. In addition, the coalescence or unification of problems can clear up puzzles arising from earlier attempts to deal with the problems separately. A well known puzzle of this sort was why Sommerfeld's (nonrelativistic) introduction of elliptical orbits, even highly eccentric ones, did not yield new energy levels for the Bohr atom – only Bohr's original levels for circular orbits. The puzzle was resolved when physicists realized that each circular orbit was adiabatically related to an entire series of elliptical orbits of the same period but of various degrees of eccentricity. In short, the problem of quantizing an elliptical, Sommerfeld atom was discovered to reduce to, or coalesce with, that of quantizing a circular, Bohr atom.

I certainly do not claim that the kind of problem reduction I have described is the only type, and I hope this paper will stimulate the search for others. One possibility, which I can only mention here, is the common case in which the original problem is "reduced in scope," so to speak, by reducing it to the more restricted problem of determining the explicit form of a certain function. A good example of such a procedure is Ehrenfest's old friend, the proof of Wien's displacement law. The original problem of blackbody radiation, to determine the energy distribution over frequencies, was shown by Kirchhoff to depend on an unknown function of frequency and temperature, $\rho(v, T)$. The problem was now to determine the form of this function explicitly. Wien, on the basis of classical thermodynamics and electromagnetic theory, provided a partial solution – the displacement law, which asserted $\rho(v, T)$ to be v^3 times a function of the variable v/T, and which gave the distribution for any temperature once it was known for a single temperature. Therefore, the original problem was reduced to that of determining the explicit form of a function known to be a function $f(v/T)$ of the single argument v/T.[20] As is well known, the quantum theory began with the failure of classical theory to solve the remaining problem. Curiously, this familiar historical observation presupposes precisely what amazed Ehrenfest: that the classical displacement law and the problem it poses carry over, unchanged, into the theory of quanta, where the problem of determining the correct form of the function was solved by Planck.

Having opened with a quotation from Einstein on the importance of one great body of theory to its eventual successor, I now conclude with a quotation from his friend Ehrenfest on the same general theme:

Another objection may be raised against the whole [project of generalizing Planck's theory by reference to classical results]: there is no sense – it may be argued – in combining a thesis which is derived from the mechanical equations with the antimechanical hypothesis of energy quanta. Answer: Wien's law holds out the hope to us that results which may be derived from classical mechanics and electrodynamics by the consideration of macroscopic-adiabatic processes, will continue to be valid in the future mechanics of energy quanta. (7, p. 341).

University of Illinois at Champaign-Urbana

NOTES

* An abbreviated version of this paper was presented at the 1974 meetings of the Philosophy of Science Association as part of a symposium (with Robert L. Causey and Lawrence Sklar) on the Unity of Science. I am grateful to Professor Dudley Shapere and to John Chapman for helpful advice and to the writings of Martin Klein on the history of the early quantum theory. None of these persons are responsible for errors or inadequacies which remain. I am indebted to the National Science Foundation for research support.

[1] Einstein (12, p. 300). Quoted by Tisza (34, p. 151) and by many other writers.

[2] Although I am no longer entirely satisfied with the formulation of the distinctions therein, see my papers (22) and (23).

[3] See my (22) and (23) and the references therein to Popper, Agassi, and Sklar. At the time those articles were prepared, I was unaware of Post's exciting (26). I am sympathetic with Post's attempt to state and defend a "General Correspondence Principle."

[4] I have slightly altered the translation for smoother reading.

[5] (9, p. 390 fn. 2). To obtain agreement with the displacement law, Planck had to set $E = nh\nu$, that is, $E/\nu = nh$, an adiabatic invariant.

[6] The derivation is sketchily described in § 2 of (6), a derivation which easily leads to the displacement law as a relation of two adiabatic invariants: $E\nu/v = F(\nu/T)$.

[7] In (9, p. 381) Ehrenfest gave the following, more precise characterization of a "reversible adiabatic" process:

Let $q_1 \ldots q_n$ be the coordinates of a system, while the potential energy depends not only on the coordinates q, but also on certain "slowly changing parameters" $a_1, a_2 \ldots$. Suppose the kinetic energy T to be a homogeneous quadratic function of the velocities $q_1 \ldots q_n$, while in its coefficients there occur besides the $q_1, q_2 \ldots$ eventually also the $a_1, a_2 \ldots$. Some original motion $\beta(a)$ can be transformed into a definite other motion $\beta(a')$ by an infinitesimal slow change of the parameters from the values $a_1, a_2 \ldots$ to the values $a'_1, a'_2 \ldots$. This special way of influencing the system may be termed "reversible adiabatic", the motions $\beta(a)$ and $\beta(a')$ *"adiabatically related."*

It seems safe to say that the study of a system's behavior under an adiabatic influencing is the most characteristic positive feature of Ehrenfest's method of approach.

[8] In (5, esp. §11). *Cf.* Jeans (15) and also Ehrenfest (6) and (8).

[9] Ehrenfest recognized that another way of expressing Planck's generalization of Boltzmann's principle was to say that, while Planck retained Boltzmann's assumption that all complexions (*i.e.*, the number of microstates compatible with a given macrostate) are equiprobable, Planck counted complexions differently (or introduced a different kind of complexion, if you will), identifying some that Boltzmann had distinguished. In effect, Planck introduced a different, nonclassical, statistical counting procedure, but none of this was completely clear to Planck or to anyone else for many years In fact, with the possible exception of Einstein, Ehrenfest was the first to recognize (a) that Planck had introduced a new statistical counting procedure; (b) that Planck's quanta lacked the classical independence still possessed by Einstein's energy quanta of 1905 (and in this respect, Planck was, despite himself, more radical than Einstein); and (c) that Einstein's quanta could not possibly satisfy Planck's radiation law. Ehrenfest, who began studying these questions as early as 1905 (5), emphasized all of these points at the end of (6). Of course, the full development of quantum statistics did not take place until the 1920's.

[10] See (8) and Ehrenfest's letter to Joffe, quoted in Klein (21, p. 261). See also (7) and (21, p. 282).

[11] The beginning of this statement is reminiscent of Bohr's correspondence principle.

[12] The justification worked both ways. Proving that all known quantum conditions were adiabatic invariants bolstered the status of the principle. A full treatment of the complex question of the justification of the adiabatic "hypothesis" or "principle" itself is beyond the scope of this paper.

[13] See Ehrenfest (9, pp. 385–6). In many cases an adiabatic deformation can be carried out only virtually or conceptually. Ehrenfest distinguished "natural" from "unnatural" adiabatic variations of a physical system, those deformations which are physically realizable (*e.g.*, by slowly varying the strength of an external electric field or the length of a pendulum string) from those which are only "fictional" – physically impossible to carry out (*e.g.*, altering a central force given in nature). See (9, p. 382).

[14] These remarks are inspired by Klein's observations on Ehrenfest's method in (18) and (21). Klein terms Ehrenfest's "the critical mode of reaching scientific generalizations".

[15] Ehrenfest's retrospective statement of 1923, quoted above, does indicate that the

adiabatic principle permitted classical mechanics to be extended to quantum systems undergoing an adiabatic influencing; however, Bohr's theory still severely restricted the scope of classical mechanics, for not only did state transitions have no mechanical explanation or description, but also Bohr made stationary states mechanically stable by fiat. Nor is it arguable that Ehrenfest should or could have generalized classical mechanics in the way that he generalized statistical mechanics, *viz.*, by generalizing a basic principle of the old theory, thereby producing a schematic quantum *mechanics*. It *is* arguable that proper use of the adiabatic principle in conjunction with Planck's quantum hypothesis could have led Ehrenfest to the Bohr-Sommerfeld quantum conditions for simple atomic systems [see (9)]; and one could go on to draw the comparison that just as Ehrenfest generalized Boltzmann's theory by restricting system trajectories to quantum allowed regions of phase space, so Bohr generalized classical mechanics by restricting classical motions to quantum allowed regions of phase space – to systems in "stationary states." But there is an important disanalogy between the two cases. Planck's and Debye's quantum theories were special applications – specifications – of the generalized Boltzmannian theory outlined by Ehrenfest. Bohr's and Sommerfeld's quantum theories of the atom were not specifications of an available general theory of mechanics. At best one could hope that their theories were signposts pointing the way to a generalization of classical theory as Planck's quantum theory had been such a signpost for Ehrenfest. Planck had altered or departed from Boltzmann's theory, while Bohr applied classical mechanics unaltered under steady state conditions, so it is easy to understand how Ehrenfest might have considered this a mere restriction rather than a generalization of the classical theory (14, 166). Bohr's theory did of course depart from classical mechanics and electromagnetic theory, but to an Ehrenfest these breaks must have seemed too radical and arbitrary to signify generalizations of classical principles, indeed to be considered anything but *ad hoc* restrictions on the scope of classical theories. On the other hand, the limit relationships to classical theory give sense to Bohr's claim that his quantum theory was a "rational generalization" of classical theory. Since Bohr's correspondence principle was so central to his thought, it is not surprising that he attached great importance to generalization in this sense. Given the ambiguities of the terms, not to mention the differences in outlook, it is easy to understand how one man's restriction could have been the other man's generalization.

[16] Quoted by Klein (21, p. 7). This remark was not made in the present problem context but pertains to Ehrenfest's first lecture as a new professor at Leyden in 1912.

[17] In this case the adiabatic principle was, for Ehrenfest, tantamount to the Boltzmann-Clausius theorem. It is important to note that the adiabatic principle could not be applied indiscriminately to quantum conditions, however formulated. The adiabatic principle did *not* license the conclusion that Planck's quantum condition, $E/v = nh = \iint dq\, dp$, was also Debye's quantum condition, for the quantity E/v is an adiabatic invariant only for the simple harmonic oscillator and not for Debye's anharmonic oscillator. The Boltzmann-Clausius theorem provided the more general adiabatic invariant $2\bar{T}/v$ which Ehrenfest proved to be invariant for all periodic systems of one degree of freedom. Since $E = 2\bar{T}$ for the harmonic oscillator, Planck's quantum condition was expressible in the more general form $2\bar{T}/v = nh = \iint dq\, dp$. And since Debye's oscillator was adiabatically related to Planck's, Ehrenfest could now conclude, by the adiabatic principle *cum* Boltzmann-Clausius theorem, that the same quantum condition held for Debye's oscillator. The Boltzmann-Clausius theorem did not by itself supply any quantum conditions, of course, anymore than did the adiabatic principle as I originally formulated it in the text.

[18] On the power of Ehrenfest's approach, see the closing paragraphs of Section II.

[19] For this objection I am indebted to Lawrence Sklar.

[21] Physicists do speak of the reduction of problems in just this way. *E.g.*, Richtmyer *et al.* write: "Thus, by reasoning based on thermodynamics, the problem of blackbody radiation is reduced to the determination of a *single unknown function*, $f(\lambda T)$" (29, p. 118, their emphasis). *Cf.* Rosenfeld (30, p. 153) and Jammer (14, p. 9) and, less explicitly, Einstein (11, 3). I should mention that Wien did not state the displacement law in just the form I have given it in the text, but this does not affect the present point. The reader will recall a similar example from Section I: Ehrenfest (6) employed the adiabatic invariance result he extracted from Rayleigh's paper to show that, for harmonic oscillations, the generalized weight function $\gamma(v, E)$ reduced to a function $G(E/v)$ of a single variable.

BIBLIOGRAPHY

[1] Bohr, N.: 1913, 'On the Constitution of Atoms and Molecules', Part I, reprinted in D. ter Haar (ed.), *The Old Quantum Theory*, Pergamon Press, Oxford, 1967, pp. 132–159.
[2] Bohr, N.: 1918, 'On the Quantum Theory of Line-Spectra', Part I, reprinted in (36), pp. 95–137.
[3] Burgers, J. M.: 1917, 'Die adiabatischen Invarianten bedingt periodischer Systeme', *Annalen der Physik* **52**, 195–202.
[4] Debye, P.: 1914, 'Zustandgleichung und Quantenhypothese mit einem Anhang über Wärmeleitung', in *Vorträge über die Kinetische Theorie der Materie und der Elektrizität*, Teubner, Leipzig, pp. 19–60.
[5] Ehrenfest, P.: 1905, 'Uber die physikalischen Voraussetzungen der Planck'schen Theorie der irreversiblen Strahlungsvorgänge', reprinted in (16), pp. 88–101.
[6] Ehrenfest, P.: 1911, 'Welche Züge der Lichtquantenhypothese spielen in der Theorie der Wärmestrahlung eine wesentliche Rolle?' Reprinted in (16), pp. 185–212.
[7] Ehrenfest, P.: 1913, 'A Mechanical Theorem of Boltzmann and its Relation to the Theory of Quanta', reprinted in (16), pp. 340–346.
[8] Ehrenfest, P.: 1914, 'Zum Boltzmannschen Entropie-Wahrscheinlichkeits-Theorem', reprinted in (16), pp. 347–352.
[9] Ehrenfest, P.: 1916, 'On Adiabatic Changes of a System in Connection with the Quantum Theory', reprinted in (16), pp. 378–399. An abbreviated version, reprinted in (36), appeared in 1917 with the title, 'Adiabatic Invariants and the Theory of Quanta'.
[10] Ehrenfest, P.: 1923, 'Adiabatische Transformationen in der Quantentheorie und ihre Behandlung durch Niels Bohr', reprinted in (16), pp. 463–470.
[11] Einstein, A.: 1905, 'On a Heuristic Viewpoint About the Creation and Conversion of Light', reprinted in D. ter Haar (ed.), *The Old Quantum Theory*, Pergamon Press, Oxford, 1967, pp. 91–107.
[12] Einstein, A.: 1936, 'Physics and Reality', reprinted in *Ideas and Opinions*, Crown Publishers, New York, 1954.
[13] Heilbron, J. L. and Kuhn, T. S.: 1969, 'The Genesis of the Bohr Atom', *Historical Studies in the Physical Sciences*, Vol. I, pp. 211–290.
[14] Jammer, M.: 1966, *The Conceptual Development of Quantum Mechanics*, McGraw-Hill, New York.
[15] Jeans, J. H.: 1905, 'A Comparison between Two Theories of Radiation', *Nature* **72**, 293–4.
[16] Klein, M. J. (ed.): 1959, *Paul Ehrenfest: Collected Scientific Papers*, North Holland Publ. Co., Amsterdam.
[17] Klein, M. J.: 1962, 'Max Planck and the Beginnings of the Quantum Theory', *Archive for History of Exact Sciences* **1**, 459–479.

[18] Klein, M. J.: 1964, 'The Origins of Ehrenfest's Adiabatic Principle', in H. Guerlac (ed.), *Proceedings of the Tenth International Congress on the History of Science*, Hermann, Paris, pp. 801–804.
[19] Klein, M. J.: 1967, 'Thermodynamics in Einstein's Thought', *Science* **157**, 509–516.
[20] Klein, M. J.: 1970, 'Maxwell, His Demon, and the Second Law of Thermodynamics', *American Scientist* **58**, 84–97.
[21] Klein, M. J.: 1970, *Paul Ehrenfest*, Vol. I: *The Making of a Theoretical Physicist*, North-Holland Publ. Co., Amsterdam.
[22] Nickles, T.: 1973, 'Two Concepts of Intertheoretic Reduction', *Journal of Philosophy* **70**, 181–201.
[23] Nickles, T.: 1974, 'Heuristics and Justification in Scientific Research: Comments on Shapere', in F. Suppe (ed.), *The Structure of Scientific Theories*, Univ. of Illinois Press, Urbana, pp. 571–589.
[24] Planck, M.: 1906, *Vorlesungen über die Theorie der Wärmestrahlung*, Johann Barth Verlag, Leipzig. The second edition, 1913, appeared in English translation.
[25] Planck, M.: 1912, 'Rapport sur la Loi du Rayonnement Noir et L'Hypothèse des Quantités Élémentaires D'Action', in P. Langevin and M. De Broglie (eds.), *La Théorie du Rayonnement et les Quanta*, Gauthier-Villars, Paris. (Proceedings of the First Solvay Congress, 1911.)
[26] Post, H. R.: 1971, 'Correspondence, Invariance, and Heuristics', *Studies in the History and Philosophy of Science* **2**, 213–255.
[27] Putnam, H.: 1965, 'How Not to Talk about Meaning', in R. S. Cohen and M. Wartofsky (eds.), *Boston Studies in the Philosophy of Science*, Vol. II, D. Reidel, Dordrecht, pp. 205–222.
[28] Rayleigh (Lord): 1902, 'On the Pressure of Vibrations', *Philosophical Magazine* **3**, 338–346.
[29] Richtmyer, F. K., Kennard, E. H., and Lauritsen, T.: 1955, *Introduction to Modern Physics*, 5th ed., McGraw-Hill, New York.
[30] Rosenfeld, L.: 1936, 'La Première Phase de L'Évolution de la Théorie des Quanta', *Osiris* **2**, 149–196.
[31] Shapere, D.: 1964, 'The Structure of Scientific Revolutions', *Philosophical Review* **73**, 383–394.
[32] Shapere, D.: 1966, 'Meaning and Scientific Change', in R. G. Colodny (ed.), *Mind and Cosmos*, Univ. of Pittsburgh Press, Pittsburgh, pp. 41–85.
[33] Shapere, D.: 1971, 'The Paradigm Concept', *Science* **172**, 706–709.
[34] Tisza, L.: 1963, 'The Conceptual Structure of Physics', *Reviews of Modern Physics* **35**, 151–185.
[35] Tomonaga, S.: 1962, *Quantum Mechanics*, Vol. I, North-Holland Publ. Co., Amsterdam.
[36] van der Waerden, B. L. (ed.): 1968, *Sources of Quantum Mechanics*, Dover Publications, New York.
[37] Whittaker, E. T.: 1960, *A History of Theories of Aether and Electricity*, Vol. II, Harper & Brothers, New York.
[38] Yourgrau, W. and Mandelstam, S.: 1968, *Variational Principles in Dynamics and Quantum Theory*, W. B. Saunders, Philadelphia.

CONTRIBUTED PAPERS

SESSION I

WILLIAM A. WALLACE*

GALILEO AND REASONING *EX SUPPOSITIONE*: THE METHODOLOGY OF THE *TWO NEW SCIENCES*

Galileo has been seen, from the philosophical point of view, alternately as a Platonist whose rationalist insights enabled him to read the book of nature because it was written in 'the language of mathematics,' and as an experimentalist who used the hypothetico-deductive methods of modern science to establish his new results empirically (McTighe, 1967; Settle, 1967; Drake, 1970; Shapere, 1974). Both of these views present difficulties. In this paper I shall make use of recent historical research to argue that neither is correct, that the method utilized by Galileo was neither Platonist nor hypothetico-deductivist, but was basically Aristotelian and Archimedean in character. This method, moreover, was not merely that of classical antiquity, but it had been emended and rejuvenated in the sixteenth century, and then not by Greek humanist Aristotelians or by Latin Averroists but rather by scholastic authors of the Collegio Romano whose own inspiration derived mainly from Thomas Aquinas. Other influences, of course, were present, and these came from other medieval and Renaissance writers, but these need not concern us in what follows.

I. CURRENT ALTERNATIVES

Before coming to my thesis I must first explain my dissatisfaction with the two alternatives that have occupied the attention of historians and philosophers of science up to now. Of the two, the Platonist thesis is the more readily disposed of. This enjoyed considerable popularity owing to writings of Alexandre Koyré (1939, 1968), who criticized the experimental evidence adduced by Galileo for his more important results and argued that the experiments were either not performed at all, being mere "thought experiments," or, if they were performed, that they did not yield the results claimed for them but merely provided the occasion for Galileo's idealizing their results. Koyré's analyses had great appeal for many philosophers, and they were not seriously contested until Thomas Settle (1961) explained how he had duplicated Galileo's inclined-plane

R. S. Cohen et al. (eds.), PSA 1974, 79–104. All Rights Reserved.
Copyright © 1976 by D. Reidel Publishing Company, Dordrecht-Holland.

apparatus, actually performed the experiment himself, and shown that the results were not as poor as Koyré had alleged. Since then, Stillman Drake (1973a) has examined anew Galileo's unpublished manuscripts and discovered evidence of hitherto unknown experiments, James MacLachlan (1973) has actually performed one of Galileo's experiments described by Koyré as only imaginary, and Drake himself (1973b), followed by R. H. Naylor (1974) and others (Shea *et al.*, 1975) have variously analyzed and verified the measurements and calculations reported in the re-discovered manuscripts. The results of all this research show that Galileo was far from being a Platonist or a Pythagorean in his practice of scientific method. He was a prolific experimenter and, within the limits of the apparatus and facilities available to him, tried to place his 'two new sciences' on the strongest empirical footing he could find.

The method he used to do so is not easy to discover, despite the mass of materials now available for analysis. The simplest expedient would be to attribute to Galileo the hypothetico-deductive method generally accepted among philosophers of science as typical of modern scientific reasoning. In this view, Galileo would begin with certain hypotheses, such as the principle of inertia and the times-squared law of free fall, and from these deduce the type of motion one might expect from heavy bodies projected or falling under given circumstances. The experimental program would then be designed to verify the calculated characteristics, and if these proved in agreement or near agreement, Galileo would be justified in accepting his hypothesis as exactly or very nearly true.

This account, it may be noted, cannot be rejected out of hand as anachronistic, for Galileo was certainly aware of the possibilities of hypothetical reasoning and indeed explicitly made use of a postulate, or hypothesis, in the *Two New Sciences* (Wisan, 1974, pp. 121–122), just as Sir Isaac Newton was to acknowledge an important hypothesis in his *Mathematical Principles of Natural Philosophy*.[1] The problem with the hypothetico-deductive method as it was used in Galileo's time, however, is that this could never lead to true and certain knowledge. It could be productive only of *dialectica*, or opinion, since any attempt to verify the hypothesis – and here we use 'verify' in the strict sense of certifying its truth – must inevitably expose itself to the *fallacia consequentis*, the fallacy of affirming the consequent. Viewed logically, hypothetical reasoning was at the opposite pole from demonstrative or scientific reasoning. This being so, it is extremely difficult to reconcile the use of hypothetico-deductive argument with Galileo's repeated insistence in the *Two New Sciences* that he had actually discovered a 'new science,' one providing demonstrations that apply to natural motions. Galileo, as we know, was prone to speak of *scientia* and *demonstratio*, using these Latin terms or their Italian equivalents with great frequency, and to my knowledge never conferring on them a sense different from that of his peripatetic adversaries.

The difficulty with the hypothetico-deductivist interpretation of Galileo's experiments, then, lies not so much in its modern flavor as in the fact that such a method could never achieve the results claimed by Galileo for the techniques he actually employed. Thus we are left with the prospect of rejecting not only the Platonic interpretation but the hypothetico-deductivist as well, and searching for yet another alternative to describe Galileo's basic methodological stance.

I would like to propose such an alternative based on Galileo's repeated use of the Latin expression *ex suppositione* to describe the line of reasoning whereby he arrived at strict demonstrations on which a *nuova scienza* dealing with local motion could be erected. Now a peculiar thing about the Latin *ex suppositione* is that it translates exactly the Greek *ex hupotheseōs*, but at the end of the sixteenth century it could carry a meaning different from the transliterated *ex hypothesi*, which also enjoyed vogue at that time in Latin writings. Reasoning *ex hypothesi* was hypothetical reasoning in the modern understanding; arguments utilizing such reasoning could not be productive of science in the medieval and Renaissance sense *(scientia)* but were merely dialectical attempts to save the appearances – a typical instance would be the Ptolemaic theories of eccentrics and epicycles. Reasoning *ex suppositione*, on the other hand, while sometimes used to designate dialectical argument, had a more basic understanding in terms of which it could be productive of demonstration in both the natural and the physico-mathematical sciences. My alternative, then, is that Galileo was not basing his *nuova scienza* on an hypothesis, on a mere computational device that would 'save the appearances' of local motion, as a modern-day positivist might interpret it, but was actually making a stronger claim for demonstration *ex suppositione* and thus for achieving a strict science in the classical sense.

II. THE METHODOLOGY OF DEMONSTRATION *Ex Suppositione*

The expression *ex suppositione*, therefore, provides a clue to Galileo's actual methods, and that clue, if pursued historically, leads back to the medieval Latin commentators on Aristotle who first used the term, e.g., Robert Grosseteste and Albert the Great, but more particularly to Albert's disciple, Thomas Aquinas, who gave classical expression to the teaching on demonstration that it entails.[2] Aquinas did this mainly in his commentaries on Aristotle's *Physics* and *Posterior Analytics*, although he also used the expression *ex suppositione* frequently in his other writings.[3] The need for such a methodology arises from the fact that the physical sciences deal with a subject matter that is in the process of continual change, and that always might be otherwise than it is. Now science, for Aquinas as for Aristotle, has to be necessary knowledge through causes. Causes for them are indeed operative in nature, but the necessity of their operation presents a problem, for sometimes they prove defective and do not produce the effect intended. Is it possible, then, to have scientific or necessary knowledge of contingent natural phenomena? Aquinas devoted much thought to this question and finally answered it in the affirmative. Demonstrations in the physical sciences can circumvent the defective operation of efficient causes, he maintained, but they can do this only when they are made *ex suppositione*.

This technique, to schematize Aquinas's account, begins by studying natural processes and noting how they terminate in the majority of cases. For example, in biological generation it can be readily observed that men are normally born with two hands, or that olive plants are usually produced from olive seeds when these are properly nurtured. From this type of generalization, and the examples are Aquinas's, one can never be certain in advance that any particular child will be born with two hands, or that each individual olive seed will produce an olive plant. The reason is that the processes whereby perfect organisms are produced are radically contingent, or, stated otherwise, that natural causes are sometimes impeded from attaining their effects. But if one starts with an effect that is normally attained, he can formulate this as an ideal *suppositio*, and from this reason back to the causes that are able to produce it, whether or not it will ever actually be attained. In other words, one can use his experience with nature to reason *ex suppositione*, i.e., on the supposition of an effect's attainment, to the various antecedent causes that will be required for its production. It is this possibility, and the technique devised to realize it, in Aquinas's view, that permit the physical sciences to be listed among sciences in the strict sense. They can investigate the causes behind natural phenomena, they can know how and why effects have been produced in the past, and they can reason quite apodictically to the requirements for the production of similar effects in the future, even despite the fact that nature and its processes sometimes fail in their *de facto* attainment.

To illustrate this technique in more detail the favored example of medieval commentators is the causal analysis of the lunar eclipse. Such eclipses are not constantly occurring, but when they do occur they are caused by the earth interposing itself between the sun and the moon. So, on the supposition that a lunar eclipse is to occur, the occurrence will require a certain spatial configuration between sun, moon, and the observer on earth. Thus one can have necessary knowledge of such eclipses even though they happen only now and then and are not a strictly necessary or universal phenomena.

A similar contingent occurrence is the production of the rainbow in the atmospheric region of the heavens. The rainbow is more difficult to explain than the lunar eclipse, and this especially for the medieval thinker, since for him the regular movements of the celestial spheres do not guarantee its periodic appearance as they do that of the eclipse. In fact rainbows are only rarely formed in the heavens, and sometimes they are only partially formed; when they are formed, moreover, they come about quite haphazardly – it is usually raining, the sun is shining, and if the observer just happens to glance in a particular direction, lo! he sees a rainbow. These factors notwithstanding, the rainbow can still be the subject of investigation within a science *propter quid*, if one knows how to go about formulating a demonstration in the proper way.

The correct details of this process were worked out by Theodoric of Freiberg, who studied at Paris shortly after Aquinas's death and while there apparently used the latter's lecture notes on the portions of the *Meteorology* that treat of the rainbow (Wallace, 1959, 1974c). Rainbows do not always occur, but they do occur regularly under certain conditions. An observer noting such regularity can rightly expect that it has a cause, and so will be encouraged to discover what that cause might be. If he moves scientifically, according to the then accepted method, he will take as his starting point the more perfect form that nature attains regularly and 'for the most part', and using this as the 'end' or final result, will try to discover the antecedent causes that are required for its realization. The necessity of his reasoning is therefore *ex suppositione*, namely, based on the supposition that a particular result is to be attained by a natural process. *If* rainbows are to occur, they will be formed by rays of light being reflected and refracted in distinctive ways through spherical raindrops successively occupying predetermined positions in the earth's atmosphere with respect to a particular observer. The reasoning, though phrased hypothetically, is nonetheless certain and apodictic; there is no question of probability or verisimilitude in an argument of this type. Such reasoning, of course, does not entail the conclusions that rainbows will always be formed, or that they will necessarily appear as complete arcs across the heavens, or even that a single rainbow need ever again be seen in the future. But if rainbows *are* formed, they will be formed by light rays passing through spherical droplets to the eye of an observer in a predetermined way, and there will be no escaping the causal necessity of the operation by which they are so produced. This process, then, yields scientific knowledge of the rainbow, and indeed it is paradigmatic for the way in which the physical sciences attain truth and certitude in the contingent matters that are the proper subjects of their investigations.

The eclipse and the rainbow are obviously natural phenomena, but their understanding requires a knowledge of geometry in addition to observation of nature, and on this account demonstrations of their properties are sometimes referred to as physico-mathematical so as to distinguish them from those that are merely physical, or natural. A similar type of physico-mathematical demonstration was employed by Archimedes when demonstrating the properties of the balance. We shall have occasion to return to this later when discussing Galileo's use of Archimedes, but for the present it will suffice to note that with the balance, since it is an artifact, the question of the regularity of its occurrence in nature does not arise in the same way as with the eclipse and the rainbow. Thus the suppositional aspect of the regularity of 'weighing phenomena,' to coin a phrase, does not enter into the process of reasoning *ex suppositione* as we have thus far described it. Other suppositions are involved in its case, however, and these have more the character of a mathematical definition, such as the supposition that the cords by which the weights are suspended from the ends of the balance hang parallel to each other. The demonstration, in this case, obtains its validity on the strength of such a mathematical supposition and how it can be reconciled with the physical fact that the cords, if prolonged, must ultimately meet in a common center of gravity. This type of supposition relates to the closeness of fit between a physical case and its mathematical idealization, whereas the former relates to the actual occurrence of phenomena in the order of nature when these have a contingent aspect to them or can be impeded by physical factors. In both types, however, scientific knowledge of properties can be obtained through demonstration *ex suppositione*, and the reasoning is not merely dialectical, or hypothetical, as it would be if reasoning *ex hypothesi* alone were employed.

III. GALILEO'S EARLY NOTEBOOKS

Having made these methodological observations, and promising to return to them later to clarify the formal difference between *ex suppositione* and *ex hypothesi* reasoning, let me now come to the man who is commonly regarded as 'the father of modern science,' Galileo Galilei. My interest in Galileo began some years ago when studying a sixteenth-century figure, Domingo de Soto, who had been singled out by Pierre Duhem (1913, pp. 263–583) as the last of the scholastic precursors of Galileo. Duhem did this on the basis that Soto had anticipated the law of falling bodies in a work published in 1545, some ninety years before Galileo proposed the law in his *Two New Sciences*. I therefore set about the task of studying Soto's mechanics to find out where and how he formulated the so-called "law," and to trace possible lines of communication to Galileo (Wallace, 1968, 1969, 1971). This search led to a mass of early unpublished writings by Galileo, actually hundreds of folios composed in his own hand. Goodly portions of these materials have been edited, some under the title of *Juvenilia*, or youthful writings, and others under a general heading, *De motu*. (The latter notes are usually referred to as the *De motu antiquiora* to distinguish them from the treatise on motion that is discussed at length in Galileo's last work, the *Two New Sciences*; they have been analyzed recently by Fredette, 1972.) These early writings of Galileo are scholastic in style, they are very much concerned with the works of Aristotle, especially the *Posterior Analytics*, the *Physics*, the *De caelo*, and the *De generatione*, and they cite authors extensively – about 150 authors alone being mentioned in the *Juvenilia*. Among these we find the name of Domingo de Soto, and indeed of several authors whom Soto influenced, including a Jesuit who was teaching at the Collegio Romano in Galileo's youth, Benedictus Pererius. Also mentioned is one of Pererius's Jesuit colleagues at the Collegio, Christopher Clavius, and his erudite commentary on the *Sphere* of Sacrobosco. Not mentioned, but apparently studied by Galileo anyway (one of his textbooks survives in Galileo's personal library), was yet another Jesuit who taught at the Collegio at that time, Franciscus Toletus, who had been Soto's favored disciple at Salamanca before going to Rome.

My more recent studies have been concerned with tracking down the sources cited in these early writings, to ascertain the extent to which such sources were actually used by Galileo, and, if possible, to determine the dates of composition of the various tracts that make up the notes. The work is tedious, but thus far it has led to some interesting results. For example, Galileo refers eleven times to a "Caietanus," whose opinions he discusses at considerable length. In the portions of the notebooks that have been edited by Favaro this author is identified as the Paduan *calculator*, Caietanus Thienensis, or Gaetano da Thiene (*Opere* 1, p. 422). This identification turns out to be inaccurate; the author to whom Galileo was referring in most of these citations is the celebrated Thomist and commentator on Aquinas, Tomasso de Vio Caietanus, generally known simply as Cajetan. And not only are Soto and Cajetan mentioned in the notes, but other *Thomistae*, as Galileo calls them, are there as well: Hervaeus Natalis, Capreolus, Soncinas, Nardo, Javelli, and Ferrariensis. In fact, after Aristotle, whose name is cited more than 200 times, Galileo's next favored group is St. Thomas and the Thomists, with a total of 90 citations; then comes Averroës with 65, Simplicius with 31, Philoponus with 29, Plato with only 23, and so on down to Scotus with 11 and Ockham and assorted nominalists with six or fewer citations apiece (Wallace, 1974b; Crombie, 1975).

This, of course, is a most interesting discovery, for if Galileo truly composed these notes, and if he understood the material contained in them, his intellectual formation would be located squarely in an Aristotelian context with decided Thomistic overtones. But previous scholars who have looked over this material, more often than not in cursory fashion, have been unprepared to accept any such result. Favaro, for example, while admitting that the notes are clearly written in Galileo's own hand, refused to accept Galileo's authorship, maintaining that these were all trite scholastic exercises, copied from another source, probably a professor's notes transcribed by Galileo in 1584 while still a student at the University of Pisa (*Opere* 1, pp. 9–13; 9, pp. 273–282). Thus they are his "youthful writings," or *Juvenilia*, not his own work, material for which he had no real interest and indeed failed to comprehend, and so could have exerted no influence on his subsequent writings.

At the outset of my researches I was prepared to go along with this view, but now I suspect that it is quite mistaken. One piece of evidence that counts heavily against it was the discovery in June of 1971 by Crombie (1975, p. 164) that some ten of the hundred folios that make up the so-called *Juvenilia* were actually copied by Galileo, very skilfully, from Clavius's commentary on the *Sphere*.[3a] A detailed comparison of the various editions of this commentary with Galileo's manuscripts indicates strongly that these notes, and other compositions with the same stylistic features and dating from the same period, were composed and organized by Galileo himself. They were not copied from a professor's notes; in fact, in all probability they were not even done while Galileo was a student at Pisa, but while he was a young professor there between 1589 and 1591. It is known that in 1591 Galileo taught at Pisa the "hypotheses [*hipotheses*] of the celestial motions" (Schmitt, 1972, p. 262), and then at Padua a course entitled the *Sphere* in 1593, 1599, and 1603. His lecture notes for the Padua course have survived in five Italian versions, all similar but none of them autographs, and showing that the course was little more than a popular summary of the main points in Clavius's commentary on Sacrobosco. In one of the versions of these notes, moreover, there is reproduced a "Table of Climes According to the Moderns," which is taken verbatim from Clavius's commentary. Indeed, when these Paduan lecture notes are compared with the summary of Clavius contained in the so-called *Juvenilia* written at Pisa, the latter are found to be far more sophisticated and rich in technical detail. It is probably the case, therefore, that the notes labelled *Juvenilia* by Favaro represent Galileo's first attempt at class preparation, and that the course based on them subsequently degenerated with repeated teaching–a phenomenon not unprecedented in the lives of university lecturers.

This dependence on Clavius is not without further interest for possible influences on Galileo's methodology. Clavius was firmly convinced, and repeatedly makes the point in his commentary on the *Sphere*, that astronomy is a true science in the Aristotelian sense, that it is not concerned mainly with "saving the appearances" but rather with determining the motions that actually take place in the heavens, and that it does this by reasoning from effects to their proper causes (Blake *et al.*, 1960, pp. 32–35; Duhem, 1969, pp. 92–96; Harré, 1972, pp. 84–86; Crombie, 1975, p. 166). There is little doubt that the young Galileo heartily subscribed to this methodological conviction of the famous astronomer of the Collegio Romano, which was consistent with the Aristotelian-Thomistic teachings of his fellow Jesuit philosophers at the Collegio. In fact, there is good reason to believe that this strong realist mind-set on the part of the young Galileo was what encouraged him to apply the canons of demonstrative proof to his discoveries and to claim that he had reached true *scientia* in both his middle period, when his over-riding concern was to demonstrate apodictically the truth of the Copernican system, and in his final period, when he made the claims we have seen for the *Two New Sciences*.

IV. GALILEO'S MIDDLE AND LATE PERIODS

With regard to Galileo's middle period, from 1610 to 1632, we must be brief. The period has already been studied in detail by William R. Shea (1972), who has shown abundantly the extent of Galileo's commitment to science as strict demonstration. Even though his revered colleague, Jacopo Mazzoni, had given an instrumentalist interpretation of eccentrics and epicycles, arguing that astronomy was not a strict science in the Aristotelian sense but merely a system of calculation for "saving the appearances," Galileo was firmly convinced of the opposite (Shea, 1972, p. 68; Purnell, 1972). In his letter on sunspots, significantly entitled *History and Demonstrations Concerning Sunspots and Their Phenomena*, Galileo admitted that the Ptolemaic eccentrics, deferents, equants, and epicycles are "assumed by pure astronomers [*posti da i puri astronomi*] in order to facilitate their calculations." But he went on,

> They are not retained as such by philosophical astronomers [*astronomi philosophi*] who, going beyond the requirement that appearances be saved, seek to investigate the true constitution of the universe – the most important and admirable problem that there is.

For such a constitution exists; and it is unique, true, real, and cannot be otherwise, and should on account of its greatness and dignity be considered foremost among the questions of speculative interest. (*Opere* 5, p. 102)

As this text shows, Galileo had no doubts that the structure of the universe is real and knowable, and that knowledge of it is a legitimate goal of scientific endeavor. This is not to claim, of course, that he was successful in attaining such knowledge. My point is essentially methodological: Galileo was in no sense a logical positivist or an instrumentalist; he was a realist, more Aristotelian than the peripatetics of his day, whom he regarded, to use Shea's phrase, as advocating "nothing more than a thinly disguised nominalism" in their own explanations of nature (1972, p. 72).

At the beginning of his middle period Galileo had written in 1610 to Belisario Vinta outlining what his plans would be should he leave Padua and get the appointment as chief philosopher and mathematician to the Grand Duke of Tuscany:

The works which I must bring to conclusion are these. Two books on the system and constitution of the universe – an immense conception full of philosophy, astronomy, and geometry. Three books on local motion – an entirely new science in which no one else, ancient or modern, has discovered any of the most remarkable properties [*sintomi*] that I demonstrate [*che io dimostro*] to exist in both natural and violent movement; hence I may call this a new science and one discovered by me from its first principles. Three books on mechanics, two relating to demonstrations of its principles and foundations and one concerning its problems... (*Opere* 10, pp. 351–352)

After the disastrous failure of the first of these projects, which culminated in his trial and condemnation in 1633, Galileo turned in his final period to the completion of the second project here mentioned, which he brought out in 1638 under the title of *Two New Sciences*. What is most remarkable is that, despite the rebuffs he had received and the rejection of the demonstrations he had offered in the *Two Chief World Systems*, in his final work he still firmly held to the ideal of *scientia* and strict demonstrative proof. Rather than abandon the ideal he was more intent than ever on preserving it, only now he would be more careful than previously to assure that his demonstrations would gain universal acceptance.

In what follows we shall focus attention on only one aspect of Galileo's final attempt to justify his claims, that, namely, of utilizing the technique of demonstration *ex suppositione*. With regard to the expression *ex suppositione* itself, it is noteworthy that Galileo recognized early in his middle period that it could carry two senses, one that is merely hypothetical and equivalent to an argument *ex hypothesi* that merely "saves the appearances," and the other standing for a supposition that is true and actually verified in the order of nature. He made the distinction, in fact, in reply to Cardinal Bellarmine's letter of April 12, 1615, addressed to the Carmelite Foscarini (*Opere* 12, pp. 171–172), in which Bellarmine had commended Foscarini and Galileo for being prudent "in contenting yourselves to speak *ex suppositione* and not absolutely" when presenting the Copernican system, and thus entertaining this as only a mathematical hypothesis, as he believed Copernicus himself had done. In his *Considerazioni circa l'opinione Copernicana* (*Opere* 5, pp. 349–370), written shortly thereafter, Galileo disavowed that this was either his own or Copernicus's intent, although one might gain such an impression on reading the preface to the *De revolutionibus*, which he noted was unsigned and clearly not the work of Copernicus himself (*ibid.*, p. 360). Galileo did not disclaim the *ex suppositione* character of his own arguments, however, but rather distinguished two different meanings of supposition:

Two kinds of suppositions have been made here by astronomers: some are primary and with regard to the absolute truth in nature; others are secondary, and these are posited imaginatively to render account of the appearances in the movements of the stars, which appearances they show are somehow not in agreement with the primary and true suppositions. (*ibid.*, p. 357)

He went on to characterize the first kind as "natural suppositions" that are "established" and "primary and necessary in nature," (*ibid.*, p. 357) and the second kind as "chimerical and fictive,... false in nature, and introduced only for the sake of astronomical computation." (*ibid.*, pp. 358–359). The whole point of the *Considerazioni*, of course, was to advise Bellarmine that Copernicus's (and Galileo's) *suppositiones* are of the first kind and not of the second.

Coming now to the reasoning advanced in the *Two New Sciences*, we find that the Latin expression *ex suppositione* occurs in at least four crucial places where Galileo is explaining his thought, twice in the text itself, and twice in letters wherein he is elaborating, in fuller detail, the methodology behind the discoveries he records in that work. Of the two uses in the text of the *Two New Sciences*, the first occurs in the Latin treatise being explained and discussed on the Third Day, and leads to the definition of naturally accelerated motion and to the demonstration of the property that the distances traversed in free fall will be as the squares of the times of fall. The second use occurs in the Italian dialogue on the Fourth Day, following Galileo's enunciation of the theorem that the path followed by a heavy object which has been projected horizontally will be compounded of a uniform horizontal motion and a natural falling motion, and will therefore be a semiparabola. In the latter context, with which it is more convenient to begin our analysis, Sagredo states:

It cannot be denied that the reasoning is novel, ingenious, and conclusive, being argued *ex suppositione*; that is, by assuming [*supponendo*] that the transverse motion is kept always equable, and that the natural downward motion likewise maintains its tenor of always accelerating according to the squared ratio of the times; also that such motions, or their velocities, in mixing together do not alter, disturb, or impede one another... (*Opere* 8, p. 273)

Having conceded this, Sagredo then goes on to raise various objections to this demonstration based on the actual physical geometry of the universe, and concludes with the telling observation:

All these difficulties make it highly improbable that the results demonstrated [*le cosi dimostrate*] from such an unreliable supposition [*con tali supposizione inconstanti*] can ever be verified in actual experiments. (*ibid.*, p. 274)

At this point Salviati comes quickly to the rescue. Rising in Galileo's defense and speaking in his name, he admits "that the conclusions demonstrated in the abstract are altered in the concrete," and in that sense can be falsified, but that such an objection can even be raised against Archimedes' demonstration of the law of the lever, for this is based on the supposition "that the arm or a balance... lies in a straight line equidistant at all points from the common center of heavy things, and that the cords to which the weights are attached hang parallel to one another." (*ibid.*). One should recall, however, Salviati goes on, that Archimedes based his demonstrations on the supposition that the balance could be regarded as "at infinite distance" from the center of the earth (*ibid.*, p. 275); granted this supposition his results are not falsified but rather drawn with absolute proof [*con assoluta dimostrazione*]. When great distances are involved, moreover, abstraction can even be made from the small errors introduced by this simplifying supposition and the results are still found to apply in practice. Similarly, he continues, when treating of the dynamic cases taken up in Galileo's new science, as opposed to the old Archimedean statics,

> it is not possible to have a firm science [*ferma scienza*] that deals with such properties as heaviness, velocity, and shape, which are variable in infinitely many ways. Hence to deal with such matters scientifically [*scientificamente*] it is necessary to abstract from these. We must find and demonstrate conclusions abstracted from the impediments [*impedimenti*], in order to make use of them in practice under those limitations that experience [*esperienza*] will teach us. (*ibid.*, p. 276)

From these texts of the *Two New Sciences* it can be seen that Galileo's "new science" of local motion was Archimedean in inspiration, but that it aimed to satisfy essentially the same classical requirements for demonstrative rigor and for application to the world of dynamic experience. On the latter point Galileo was well aware that he had to abstract from many more "impediments" than Archimedes had to, particularly the resistance of the medium traversed by falling and projected bodies, but he felt that he had sufficient experimental evidence to be able to do so. What that evidence was has long eluded historians of science, but Stillman Drake's recent re-discovery of folio 116v in BNF MS Galileiana 72 now supplies the missing link. The cases treated *ex suppositione* by Galileo in the Third and Fourth Days of the *Two New Sciences* were investigated by him in experiments he never reported, and found to be very nearly in agreement with what actually occurs in nature. This is why he could write, in the Latin treatise on naturally accelerated motion, at his first mention of the demonstration *ex suppositione* based on the definition of such motion:

> Since nature does employ a certain kind of acceleration for descending heavy things, we decided to look into their properties [*passiones*] so that we might be sure that the definition of accelerated motion which we are about to adduce agrees with the essence [*essentia*] of naturally accelerated motions. And at length, after continual agitation of mind, we are confident that this has been found, chiefly for the very powerful reason that the properties [*symptomatis*] successively demonstrated by us [*a nobis demonstratis*] correspond to, and are seen to be in agreement with, that which physical experiments [*naturalia experimenta*] show forth to the senses. (*Opere* 8, p. 197)

Further clarification of the method Galileo used in this discovery is given by him in two letters, one written to Pierre Calcavy (or, de Carcavi) in Paris on June 5, 1637, while the *Two New Sciences* was still in press, and the other to Giovanni Battista Baliani in Genoa on January 7, 1639, after its publication. The letter to Calcavy is of particular interest because it is an answer to a query from Pierre Fermat, forwarded by Calcavy to Galileo, concerning a passage in the *Two Chief World Systems* wherein Salviati mentions the treatise on motion that was later to appear in the

Two New Sciences, but which he has already seen in manuscript form. Salviati there had explained Galileo's initial reasoning concerning the path that would be described by a heavy body falling from a tower if the earth were rotating in a direction away from the body's path of fall, and had described that path as compounded of two motions, one straight and the other circular, on the analogy of Archimedes' treatment of spiral motion. The path recounted at that time by Salviati, however, was not a semiparabola but a semicircle, and the composition of motions had obviously been made incorrectly, as Fermat was quick to notice. In his reply to Calcavy Galileo retracted the error – tried, in fact, to cover it up as a mere jest and not as a serious account (Shea, 1972, p. 135) – and then went on to explain how he had derived his new parabolic curve, again noting the analogy with Archimedes' method. In his works *On Weights* and *On the Quadrature of the Parabola*, writes Galileo, Archimedes is supposing [*supponendo*], as do all engineers and architects, "that heavy bodies descend along parallel lines," thereby leading us to wonder if he was unaware "that such lines are not equidistant from each other but come together at the common center of gravity." (*Opere* 17, p. 90). Galileo goes on:

From such an obviously false supposition [*falsa supposizione*], if I am not in error, the objections made against me by your friend [Fermat] take their origin, viz, that in getting closer to the center of the earth heavy bodies acquire such force and energy, and vary so much from what we suppose to take place on the surface, admittedly with some slight error, that what we call a horizontal plane finally becomes perpendicular at the center, and lines that in no way depart from the perpendicular degenerate into lines that depart from it completely. I add further, as you and your friend can soon see from my book which is already in the press, that I argue *ex suppositione*, imagining for myself a motion towards a point that departs from rest and goes on accelerating, increasing its velocity with the same ratio as the time increases, and from such a motion I demonstrate conclusively [*io dimostro concludentemente*] many properties [*accidenti*]. I add further that if experience should show that such properties were found to be verified in the motion of heavy bodies descending naturally, we could without error affirm that this is the same motion I defined and supposed; and even if not, my demonstrations, founded on my supposition, lose nothing of their force and conclusiveness; just as nothing prejudices the conclusions demonstrated by Archimedes concerning the spiral that no moving body is found in nature that moves spirally in this way. But in the case of the motion supposed by me [*figurato da me*] it has happened [*e accaduto*] that all the properties [*tutte le passioni*] that I demonstrate are verified in the motion of heavy bodies falling naturally. They are verified, I say, in this way, that howsoever we perform experiments on earth, and at a height and distance that is practical for us, we do not encounter a single observable difference; even though such an observable difference would be great and immense if we could get closer and come much nearer the center. (*ibid.*, pp. 90–91)

Galileo then describes an experiment by which he is able to verify, by sense observation and not by reasoning alone, the conclusion he has just stated.[4]

Turning now to Galileo's letter to Baliani after the appearance of the *Two New Sciences*, we find him repeating there in summary form what he had already written to Calcavy, and in so doing describing his steps in more accurate detail. Galileo states:

> I assume nothing but the definition of the motion of which I wish to treat and whose properties I demonstrate, imitating in this Archimedes in the *Spiral Lines*, where he, having stated what he means by motion in a spiral, that it is composed of two uniform motions, one straight and the other circular, passes immediately to demonstrating its properties. I state that I wish to examine the characteristics associated with the motion of a body that, leaving from the state of rest, goes with a velocity that increases always in the same manner, i.e., the increments of that velocity do not increase by jumps, but uniformly with the increase of time. (*Opere* 18, pp. 11–12)

Proceeding on this basis Galileo notes that he comes "to the first demonstration in which I prove the spaces passed over by such a body to be in the squared ratio of the times..." (*ibid.*, p. 12) After a brief digression, he then resumes his main theme:

> But, returning to my treatise on motion, I argue *ex suppositione* about motion defined in that manner, and hence even though the consequences might not correspond to the properties of the natural motion of falling heavy bodies, it would little matter to me, just as the inability to find in nature any body that moves along a spiral line would take nothing away from Archimedes' demonstration. But in this, I may say, I have been lucky [*io stato... avventurato*]; for the motion of heavy bodies, and the properties thereof, correspond point by point [*puntualmente*] to the properties demonstrated by me of the motion as I defined it. (*ibid.*, pp. 12–13)

Note here Galileo's explicit affirmation of the methodology of *ex suppositione* argumentation, and his further admission that he had actually discoverd, though by a stroke of luck (it was *avventurato* – recall that in the letter to Calcavy it was by accident, *e accaduto*), that both his definition of motion and the properties resulting therefrom correspond point by point to what actually occurs in nature.

v. *Ex Hypothesi* vs. *Ex Suppositione* ARGUMENTATION

Let us now return to the problem of hypothetico-deductive methodology that was presented at the outset, to clarify how this differs from demonstration *ex suppositione*, and how the latter could achieve the results claimed by Galileo whereas the former could not. Both types of reasoning, it should be obvious, can be expressed in conditional form. Modern hypothetico-deductive reasoning takes the form "if p then q," where p formulates an hypothesis that does not pertain to the order of appearances, whereas q states a consequent that pertains to this order and so is empirically verifiable. The sixteenth-century parallel would be reasoning *ex hypothesi*, "if there are eccentrics and epicycles, then the observed planetary motions result." Here, as in the modern theory, the antecedent cannot be verified directly; one must work through the consequent, either by showing that it is not verified in experience and that the antecedent is therefore false, or that it is so verified, in which case the antecedent enjoys some degree of probability or verisimilitude. The latter alternative gives rise to the problems of contemporary confirmation theory, whereas the former is the basis for Karl Popper's insistence on techniques of falsification. Neither alternative, as is universally admitted, is productive of positive scientific knowledge that could not be otherwise, and so neither can produce *scientia* in the classical sense.

Demonstration *ex suppositione* employs conditional reasoning of a different type. It too can be expressed in the form, "if p then q," but here p stands for a result that is attained in nature regularly or for the most part, whereas q states an antecedent cause or condition necessary to produce that result. Unlike *ex hypothesi* reasoning and its modern hypothetico-deductive equivalent, p usually pertains to the order of appearances, for this is what can be observed to take place in nature regularly or for the most part. Again, with regard to p's content, no claim is made for the absolute necessity or universality of such an observational regularity, since there are always impediments in nature that can prevent the realization of any ideal result. The logical consequent, q, on the other hand, standing as it does for antecedent causes or conditions that produce the appearances, need not itself pertain to the order of appearances, at least not initially, although it may subsequently be found to do so, as in Theodoric of Freiberg's explanation of the rainbow and, as we now know, in Galileo's supposition of uniform acceleration. Unlike purely hypothetical reasoning, finally, this mode of argumentation can lead to certain knowledge and to *scientia* in the strict sense. The form of argumentation, "if p, then, if p then q, then q," will be recognized as one of the valid forms of the *modus ponendo ponens* of the conditional syllogism.

It now remains to show that Galileo's reasoning conforms to the latter pattern, and so could justify his claims for strict demonstration that is productive of a "new science" of local motion. The analysis of the rediscovered folio 116v shows that in his unreported experiments with free fall, as opposed to the inclined-plane experiment described in the *Two New Sciences*, Galileo was able to verify in a surprising way various properties of falling motion compounded of a rectilinear inertial component and an accelerated downward component in accordance with the times-squared ratios. The experiments consisted in dropping a ball from different heights to a deflector located at the edge of a table, at which point the ball was given different horizontal velocities depending on the distance of its fall. Apparently Galileo had computed the horizontal distances the ball should travel depending on the velocity imparted to it, and then had actually measured points of impact to verify his calculations. The accuracy of his results is truly remarkable considering the crude apparatus Galileo had to work with, but he was not able to verify them consistently, and particularly could not reconcile the exceptionally good results of the free-fall experiments with those made on the inclined plane. Galileo rightly discerned that the cause of the discrepancy arose from friction and air resistance, the first of which was particularly serious with the inclined plane, and so he took the various properties he had calculated as something that should be verified in the ideal case, but, as the medieval would have it, need be found true only generally and for the most part. On such a supposition it was a simple matter for him to demonstrate mathematically that the only kind of naturally accelerated motion that could produce the result he had observed, more or less, would be one whose velocity increases uniformly with time. Impediments and defects could, of course, prevent the ideal result from being attained, as Galileo realized, but this is true generally in the physical world, and it is in fact the reason why demonstration *ex suppositione* has to be employed when studying natural processes in the first place.

Still there is a difference in the nuances of demonstration *ex suppositione* as this is employed in an Aristotelian natural science such as biology and in the Archimedean type of science such as those that treat of the balance and falling bodies, and this must now be pointed out. In the former case the force of the demonstration is usually carried through efficient causality, and impediments are seen to arise through imperfections in the matter involved or in the deficiencies of agent causes. In the latter case the force of the demonstration is usually carried through formal causality, understanding this in the sense of mathematical form, where the relationships involved are those between quantifiable aspects of the subject under consideration. When these quantitative relationships are realized in the physical world, however, as opposed to the world of pure mathematical forms, they too can be found defective, or be "impeded," either by material conditions or by defects of agent causes. It is noteworthy that Galileo concentrates only on material defects arising from friction and air resistance and that he is not particularly concerned with the deficiency of agent causes in his analysis. The reason for this is that, along with the Aristotelians of his day, he regards nature as the basic efficient cause of falling motion, which is why he always refers to the phenomenon of free fall as "naturally accelerated motion." So he acknowledges:

...we have been led by the hand to the investigation of naturally accelerated motion by consideration of the custom and procedure of nature herself in all her other works, in the performance of which she habitually employs the first, simplest, and easiest means. (*Opere* 8, p. 197)[5]

With the efficient cause thus taken care of, Galileo's main burden of proof can become that of showing that this "simplest means" is to have velocity increase uniformly with the time of fall, and not with the distance of fall, and this he is able to demonstrate mathematically once he is assured experimentally that the distances traversed are really as he had calculated them to be. But he must still be able to account for the fact that even these results will probably never be realized perfectly in the concrete, and so, to take care of the "impediments," as he calls them, he must resort to the technique of demonstration *ex suppositione*.

To make more explicit the methodology here attributed to Galileo, let us note that it combines elements of Archimedean *ex suppositione* reasoning and of Aristotelian *ex suppositione* reasoning in the following way. At the outset, before the experimental confirmation was available to him, Galileo's demonstration could be expressed in the following logical form:

> If p (definition of motion laterally uniform and downwardly accelerated with time), then, if p then q (by mathematical reasoning), then q (properties of semiparabolic path, e.g., distance of fall, of horizontal travel, etc.).

Note that this is a scientific demonstration, not a merely hypothetical argument, even though it is expressed in conditional form; however, it pertains to the "old science," the Archimedean type of mathematics that is ideal even though it is applicable, in some way, to the physical universe. Now Galileo thought that he had advanced beyond this "old science" to a "new science," for he had been lucky enough to obtain experimental confirmation of the properties he had calculated, but not sufficiently complete confirmation to remove all possibility of error. The error, however, he had by this time come to see could be attributed to the *impedimenti*, understanding these not merely in the Aristotelian sense of physical defects that prevent perfect regularities from being observed in nature, but also in the Archimedean sense of the physical characteristics of the universe (such as its spherical geometry) that prevent simplified mathematical ideals from being applied there perfectly.[6] So, modifying the Aristotelian type of argument *ex suppositione*, Galileo went on to the second stage of his new type of demonstration, which may be expressed logically in the same form simply by interchanging the p's and the q's:

> If q (more or less, physico-mathematically), then, if q then p (by mathematical reasoning), then p (physically verified).[7]

Let us note parenthetically that the scholastic Aristotelian of Galileo's day did not customarily argue in the physico-mathematical mode suggested by the above formulation, but rather employed the following type of argument:

> If q (regularly and for the most part, physically), then, if q then p (by reasoning "philosophically" to a physical cause or necessary condition), then p (physically required, but able to be impeded).

Here the "regularly and for the most part" can be said to be approximate in the qualitative sense, but not in the quantitative sense suggested by the words "more or less" in the reconstruction of Galileo's argument. A primary aspect of Galileo's contribution, it would appear, is that he pointed out how one could legitimately make the transit from the qualitative "regularly" understood in a physical way to the quantitative "more or less" understood in a physico-mathematical way. If this be admitted, then other aspects of Galileo's contribution follow. For example, his use of limit concepts (which we have not been able to go into here, but which are discussed by Koertge, 1974) plus his use of precise experimentation and measurement are what made the above transition scientifically acceptable, if not to the conservative Aristotelians of his day at least to those who were willing to follow in his path. Again, apart from the empirical aspect, Galileo showed how mathematical functionality could serve as a valid surrogate for physical causality in manifesting the necessary connection between antecedent and consequent in a physical situation. This explains why he could proclaim, in the celebrated passage in the *Two New Sciences* (*Opere* 8, p. 202), his indifference to the precise physical cause of the acceleration observed in falling motion (Finnochiaro, 1972; Drake, 1974, pp. xxvii–xxix, 158–159; Wallace, 1974d, pp. 229–230, 239–240). This also clarifies the sense in which he was an Archimedean and could rightfully proclaim the power of mathematics as he had employed it in his *nuova scienza*. Yet again, and this may be the most ingenious aspect of his contribution, Galileo was able to show how, through the use of his physico-mathematical techniques, some "unobservables," i.e., the actual mode of velocity increase (e.g., whether uniformly with distance of fall or uniformly with time of fall, p), could actually be certified empirically, through the use of other mathematical relationships (e.g., the distances of horizontal and vertical travel, q) that were "observables" in the sense that they could be verified approximately in the experiments he had contrived. When all of these aspects of the *Two New Sciences* are taken into account, we see why Galileo can still rightfully be hailed as "the father of modern science." Even if he was not a complete innovator (and who is?), he knew at least in a general way the strengths and limitations of the Aristotelian and Archimedean traditions that had preceded him, and he had the genius to wrest from those traditions the combination of ideas that was to prove seminal for the founding of a new era.

To conclude, then, by resuming the theme stated at the outset of this paper, the logic of *ex suppositione* reasoning was already at hand for Galileo, it was part of the intellectual tradition in which he had been formed, and it was capable of producing the scientific results he claimed to have achieved. Since the same cannot be said for hypothetico-deductive method in the modern mode, there is no reason to impose that methodology on "the father of modern science." Rather we should take

Galileo at his word and see him neither as the Platonist nor as the hypothetico-deductivist he has so frequently been labelled, but as one who made his justly famous contribution in the Aristotelian-Archimedean context of demonstration *ex suppositione*.

The Catholic University of America
Washington, D.C.

NOTES

* The research on which this paper is based was supported by the National Science Foundation, whose assistance is gratefully acknowledged. The author also expresses his thanks to Stillman Drake and Thomas McTighe for their helpful comments on an earlier version read at the Philosophy of Science Association biennial meeting at Notre Dame on November 1, 1974.

[1] See Galileo's *Opere*, A. Favaro (ed.), Vol. 8, p. 207, and Newton's *Principia*, Koyré and Cohen (eds.), Vol. 2, p. 586. Henceforth all citations to Galileo's writings will be made to Favaro's edition, to which Drake's recent translation of the *Two New Sciences* (1974) is also keyed. Usually Galileo refers to an hypothesis as a *postulato* or as an *ipotesi* (*Opere* 7, p. 29), although sometimes he uses the terms *supposizione* and *ipotesi* interchangeably (e.g., *Opere* 2, p. 212), as will be explained *infra*.

[2] Some details of the methodology this entails are given in my *Causality and Scientific Explanation* (1972, pp. 71–80, 102, 104, and 143; and 1974, pp. 247, 250, 293, and 354). Grosseteste translates the Greek *ex hupotheseōs* as *ex supposicione* in his commentary on the second book of the *Physics*, and Albert the Great, in his paraphrase of the same, renders it variously as *ex suppositione* and *ex conditione*; the Latin text of Averroës's great commentary on the *Physics*, on the other hand, gives the reading *ex positione*.

[3] See Aquinas's *In lib. II Physicorum*, lect. 15, n. 2; *In lib. I Posteriorum Analyticorum*, lect. 16, n. 6; and *In lib. II Posteriorum Analyticorum*, lect. 7, n. 2, and lect. 9, n. 11; also *Contra Gentiles*, lib. I, c. 81, and lib. II, cc. 25 and 30; *In lib. I Sententiarum*, dist. 2, q. 1, art. 4, ad 3 arg., and *In lib. III Sententiarum*, dist. 20, art. 1, questiuncula 3.

[3a] The date of Professor Crombie's discovery was not known to me at the time I corrected the galley proofs for my article, 'Galileo and the Thomists' (1974b). In a note added at that time (November 20, 1972), I stated that since my submission of the article on December 11, 1971, "a considerable amount of work has been done on Galileo's early writings both by the author and by Professors A. C. Crombie and Adriano Carugo," and that this work confirmed "the thesis only tentatively advanced" in the paper, "namely, that the *Juvenilia* were probably composed by Galileo himself, with little or no direct use of primary sources but with a recognizable dependence on the writings of Pererius and Toletus, and also with some borrowings from Christopher Clavius's commentary on the *Sphaera* of Sacrobosco" (1974b, p. 330, n. 133). In so expressing myself I inadvertently created the impression that Crombie's and Carugo's work was derivative from, or at least consequent on, my own. This impression is erroneous, for, as Crombie has now made clear (1975, pp. 162–163), Carugo had made his discoveries of Galileo's dependence on Pererius and Toletus in 1969, and Crombie himself had detected the copying from Clavius in June of 1971. As for myself, quite independently of Carugo (of whose work I was unaware), I began to suspect Galileo's

use of Pererius and Toletus early in 1971 while working on the "Galileo and the Thomists" paper, the first draft of which I completed on May 20, 1971, and sent a copy to Crombie on July 16, 1971. Owing to the complete collection of Dominican Thomistic authors available to me at the Dominican House of Studies, Washington, D.C., I was able to verify then that there was no evidence of Galileo's "direct copying from any of the Thomistic authors mentioned in this study" (1974b, p. 327), meaning by that the Dominican authors I had there analyzed in detail. I had no access to the actual texts of Pererius or Toletus at that time, however, having based my suspicion of their influence on notes taken from their works in libraries in Salamanca in November of 1966, in conjunction with my studies of Soto. By September of 1971, moreover, I had submitted a proposal to the National Science Foundation requesting funds to travel to Europe to check the actual texts of Pererius, Toletus, and other possible sources mentioned in the so-called *Juvenilia* (which, of course, included Clavius, and this independently of Crombie's discovery). Finally, on March 31, 1972, after I had received the NSF grant, Crombie wrote to me, acknowledged receipt of my letter and paper of July 16 of the previous year, mentioned that there had been quite a bit of overlap between my work and that of himself and Carugo, and informed me that they had shown that the three main sources of the *Juvenilia* were Clavius, Pererius, and Toletus. From my point of view, therefore, they had confirmed the thesis I had only tentatively advanced in the 'Galileo and the Thomists' article. They had done this, indeed, *without* knowledge of that thesis, and thus I wish now to disavow any imputation that their work was derivative on my own (cf. Crombie, 1975, pp. 170 and 304). That our research interests continue to overlap, however, and that we in fact reinforce each others' conclusions, should be apparent to anyone who compares this present paper with Crombie's recent publication, where he likewise discusses demonstration *ex suppositione* (1975, p. 172). Yet it might be noted, for the record, that the present paper was written, revised, and submitted for publication before Crombie's essay reached me on March 1, 1975, then already in printed form. [Note added by the author, together with some emendations acknowledging Crombie's results, on March 5, 1975.]

[4] The experiment is described by Galileo as follows: "Let us hang from two strings that are equally long two heavy bodies, for example two musket balls, and let one of the aforementioned strings be attached at the very highest place one can reach and the other at the lowest, assuming that their length is four or five feet. And let there be two observers, one at the highest place and the other at the lowest, and let them pull aside these balls from the perpendicular position so that they begin their free movement at the same instant of time, and then go on counting their swings, continuing through several hundred counts. They will find that their numbers agree to such an extent that, not merely in hundreds but even in thousands, they will be found not to vary by a single swing, an argument concluding necessarily [*argomento necessariamente concludente*] that their falls take place in equal times. And since such falls in this motion along the arcs of a circle are duplicable on the chords drawn from them, there results on earth all that your friend [Fermat] says should happen on inclined planes that are parallel to each other and equally long, one of which is closer to the center of the earth than the other. They fall, I say, exactly in unison [*assolutissimamente*], despite the fact that both are placed outside the surface of the terrestrial globe. And that this might happen between similar planes, one of which were outside the surface of the earth and the other so far inside as to terminate even at the center of the same, I do not wish at the moment to deny, although I have no reason that absolutely convinces me to admit that the movable object that comes to rest at the center would traverse its space in a time shorter than the other movable object traverses it. But to say more, it is apparent to me that it is not well resolved and clear that a heavy movable object would arrive sooner

at the center of the earth when leaving from the neighborhood of only a single cubit than a similar body that would depart from a distance a thousand times greater. I do not affirm this but propose it as a paradox, through the solution of which perhaps your friend will have found a demonstration that concludes necessarily [*dimostrazione necessariamente concludente*]." – *Opere* 17, pp. 91–92.

[5] See also the text from the same page of the *Two New Sciences* cited *supra*, p. 92.

[6] The question suggests itself at this point whether Galileo actually conflated *impedimenti* that prevent ideal mathematical accuracy from being attained with those that prevent perfect and unfailing regularities from being observed in nature. An affirmative answer would seem indicated on the basis of the way Galileo proceeds in the *Two New Sciences*; similar uses of the term *impedimenti* in his earlier writing have been noted by Noretta Koertge (1974), and these likewise suggest a conflation of the two types of cases. With regard to the *Two New Sciences*, as we have seen, when discussing the law of the lever as applied to the balance, Galileo says that this yields results that are only approximate at finite distances, but reasoning *ex suppositione*, i.e., on the supposition that the balance is at an infinite distance from the center of the earth, the results can be perfectly demonstrated, *con assoluta dimostrazione*. In this context he seems to be regarding the physical geometry of the universe as an "impediment" that prevents a mathematical ideal from being realized much as air resistance will prevent an ideal in nature (i.e., uniformly accelerated motion) from being realized. And in both cases it seems that he is employing *ex suppositione* reasoning to make the transit from a real to an ideal case, i.e., from a real balance to an ideal balance, and from the falling motion actually observed in nature to the ideal motion nature is attempting to realize.

[7] If one were to focus on the downward component of the motion alone, here q would stand for the times-squared law (verified approximately) and p for the definition of motion uniformly accelerated with respect to time (now known to be true *ex suppositione* in the order of nature). Then the inference "if q then p" would be verified mathematically by strict implication (and not merely by material implication), since the differentiation of $s \propto t^2$ yields immediately that $v \propto t$. Note that here the reverse inference, "if p then q," would also be true for these same parameters, since one can obtain, by integrating $v = ds/dt \propto t$ with respect to t, the result that $s \propto t^2$; Galileo, as we know, did not see this immediately. Effectively this means that the inference here really has the force of the bi-conditional or equivalence function, since both "if p then q" and "if q then p" are true, and thus the "if's" can be read as "iff's" ("if and only if") in virtue of the mathematics involved. The same, it goes without saying, could not be said of any inference to the proximate cause of uniform acceleration or of the times-squared law, as Galileo was well aware.

BIBLIOGRAPHY

Blake, R. M., Ducasse, C. J., and Madden, E. H.: 1960, *Theories of Scientific Method: The Renaissance through the Nineteenth Century*, Univ. of Washington Press, Seattle.

Crombie, A. C.: 1975, 'The Sources of Galileo's Early Natural Philosophy', in M. L. Righini Bonelli and W. R. Shea (eds.), *Reason, Experiment, and Mysticism in the Scientific Revolution*, Science History Publications, New York, 1975, pp. 157–175, 303–305.

Drake, S.: 1970, *Galileo Studies: Personality, Tradition, and Revolution*, Univ. of Michigan Press, Ann Arbor.

Drake, S.: 1973a, 'Galileo's Discovery of the Law of Free Fall,' *Scientific American* **228**, pp. 84–92.

Drake, S.: 1973b, 'Galileo's Experimental Confirmation of Horizontal Inertia: Unpublished Manuscripts', (Galileo Gleanings XXII), *Isis* **64**, pp. 290–305.
Drake, S. (ed.): 1974, *Galileo Galilei: Two New Sciences*, including *Centers of Gravity* and *Force of Percussion*, translated with introduction and notes, Univ. of Wisconsin Press, Madison.
Duhem, P.: 1913, *Etudes sur Léonard de Vinci*, Vol. 3, A. Hermann et Fils, Paris.
Duhem, P.: 1969, *To Save the Phenomena. An Essay on the Idea of Physical Theory from Plato to Galileo*, tr. E. Doland and C. Maschler, Univ. of Chicago Press, Chicago and London.
Favaro, A. (ed.): 1890–1909, *Le Opere di Galileo Galilei*, 20 vols. in 21, G. Barbèra, Florence, reprinted 1968.
Finnochiaro, M. A.: 1972, *'Vires acquirit eundo:* The Passage Where Galileo Renounces Space-Acceleration and Causal Investigation', *Physis* **14**, 125–145.
Fredette, R.: 1972, 'Galileo's *De motu antiquiora*', *Physis* **14**, 321–348.
Harré, R.: 1972, *The Philosophies of Science. An Introductory Survey*. Oxford Univ. Press, London, Oxford and New York.
Koertge, N.: 1974, 'Galileo and the Problem of Accidents', paper read to the British Society for the Philosophy of Science, October 7, 1974, being readied for publication.
Koyré, A.: 1939, *Etudes Galiléenes*, Hermann, Paris, reprinted 1966.
Koyré, A.: 1968, *Metaphysics and Measurement*. Essays in Scientific Revolution, Harvard Univ. Press, Cambridge, Mass.
Koyré, A. and Cohen, I. B.: 1972, *Isaac Newton's Philosophiae Naturalis Principia Mathematica*, 3d ed. of 1726 with variant readings, 2 vols., Harvard Univ. Press, Cambridge, Mass.
MacLachlan, J.: 1973, 'A Test of an "Imaginary" Experiment of Galileo's,' *Isis* **64**, 374–379.
McTighe, T. P.: 1967, 'Galileo's "Platonism": A Reconsideration', in E. McMullin (ed.), *Galileo, Man of Science*, Basic Books, Inc., New York and London, 1967, pp. 365–387.
Naylor, R. H.: 1974, 'Galileo and the Problem of Free Fall', *The British Journal for the History of Science* **7**, 105–134.
Purnell, F.: 1972, 'Jacopo Mazzoni and Galileo', *Physis* **14**, 273–294.
Schmitt, C. B.: 1972, 'The Faculty of Arts at Pisa at the Time of Galileo', *Physis* **14**, 243–272.
Settle, T. B.: 1961, 'An Experiment in the History of Science', *Science* **133**, 19–23.
Settle, T. B.: 1967, 'Galileo's Use of Experiment as a Tool of Investigation', in E. McMullin (ed.), *Galileo, Man of Science*, Basic Books, Inc., New York and London, 1967, pp. 315–337.
Shapere, D.: 1974, *Galileo: A Philosophical Study*, Univ. of Chicago Press, Chicago and London.
Shea, W. R.: 1972, *Galileo's Intellectual Revolution*. Middle Period 1610–1632, Science History Publications, New York.
Shea, W. R. and Wolf, N. S.: 1975, 'Stillman Drake and the Archimedean Grandfather of Experimental Science', to appear in *Isis*.
Wallace, W. A.: 1959, *The Scientific Methodology of Theodoric of Freiberg*, Fribourg Univ. Press, Fribourg.
Wallace, W. A.: 1968, 'The Enigma of Domingo de Soto: *Uniformiter difformis* and Falling Bodies in Late Medieval Physics', *Isis* **59**, 384–401.
Wallace, W. A.: 1969, 'The "Calculatores" in Early Sixteenth-Century Physics', *The British Journal for the History of Science* **4**, 221–232.
Wallace, W. A.: 1971, 'Mechanics from Bradwardine to Galileo', *Journal of the History of Ideas* **32**, 15–28.

Wallace, W. A.: 1972, *Causality and Scientific Explanation*, Vol. I. Medieval and Early Classical Science, Univ. of Michigan Press, Ann Arbor.

Wallace, W. A.: 1974a, *Causality and Scientific Explanation*, Vol. II. Classical and Contemporary Science, Univ. of Michigan Press, Ann Arbor.

Wallace, W. A.: 1974b, 'Galileo and the Thomists', *St. Thomas Aquinas Commemorative Studies 1274–1974*, 2 vols., Pontifical Institute of Mediaeval Studies, Toronto, 1974, Vol. 2, pp. 293–330.

Wallace, W. A.: 1974c, 'Theodoric of Freiberg: On the Rainbow', in *A Source Book in Medieval Science*, E. Grant (ed.), Harvard Univ. Press, Cambridge, Mass., 1974, pp. 435–441.

Wallace, W. A.: 1974d, 'Three Classics of Science: Galileo, *Two New Sciences*, etc.', in *The Great Ideas Today 1974*, Encyclopaedia Britannica, Chicago, pp. 211–272.

Wisan, W. L.: 1974, 'The New Science of Motion: A Study of Galileo's *De motu locali*', *Archive for History of Exact Sciences* **13**, 103–306.

JOHN V. STRONG

THE *ERKENNTNISTHEORETIKER'S* DILEMMA: J. B. STALLO'S ATTACK ON ATOMISM IN HIS *CONCEPTS AND THEORIES OF MODERN PHYSICS* (1881)

Notable among the by-products of the late nineteenth-century controversies over the atomic theory of matter was the first full-scale contribution to the philosophy of science written in America, John Bernard Stallo's *The Concepts and Theories of Modern Physics*. The work of a lawyer and jurist largely self-trained in philosophy, its publication in 1881 provoked considerable (and mostly hostile) reaction[1], for Stallo attacked atomism and mechanism at a time when scientists were still divided into vociferously opposing camps over the question of whether or not the ultimate structure of matter *was* atomic – indeed, were in sharp disagreement over even the utility of the hypothetical concept 'atom' in physics and chemistry.[2]

Ever since the early 1800s, when Dalton had transformed the corpuscular hypothesis from a commonplace of the prevailing natural philosophy into a working part of chemistry, the fortunes of atomism had risen and fallen as competing atomic and non-atomic theories were matched against the accumulating experimental evidence. In such circumstances it is not surprising that many scientists took an agnostic position on the subject. A considerable number (like the prestigious Berzelius) held that though the chemist might *speak* of atoms, the term was interchangeable with 'chemical equivalent' or even 'combining volume', since one was dealing with 'mere methods of representation'[3]; even in 1867, Kekulé could write that "the question whether atoms exist or not has but little significance in a chemical point of view."[4] By the early 1880s, when Stallo's *Concepts and Theories* made its appearance, it is clear enough (at least in retrospect) that the tide had begun to run in the atomists' favor, as the Maxwell-Boltzmann theory of gases and the structural account of chemical isomerism found ever wider acceptance. Yet even into the present century, non-atomic theories (particularly of the 'energeticist' variety) were still defended by a small but respectable sector of the scientific community.[5] Thus Stallo's criticisms of the atomic hypothesis, to say nothing of his objections to the more sweeping 'atomo-mechanical' theory that envisioned all interactions among the elementary particles of matter purely in terms of contact forces and a mechanical aether, could not be dismissed out of hand as the work of a crank or visionary, whatever some of his orthodox atomist opponents might try to insinuate.

To place Stallo's views in their context, I shall begin with a brief outline of his career, which is not without its own intrinsic interest, and touch with equal brevity on his intellectual concerns prior to the publication of *Concepts and Theories*. I shall then survey the more important of Stallo's lines of attack on the atomo-mechanical theory in the latter work, arguing that it embodies two quite distinct though closely related aims: first, a *methodological* critique of the various models for the microstructure of matter put forward by the atomists; and secondly, the grounding of this methodology in a comprehensive *epistemology* – or, as Stallo preferred to say, in a 'theory of cognition'. In the final part of my essay I shall try to suggest some of the underlying (and very basic) reasons why, although the first of these two projects is carried out by Stallo with considerable success, the second remains at most the sketch for a philosophical program.

I

That Stallo's *Concepts and Theories* attracted wide attention when it first appeared in 1881 was owing to more than its topicality. Its inclusion in a series of books which at the time already boasted titles by such eminent Victorians as Walter Bagehot, Herbert Spencer, John Draper and T. H. Huxley insured that it would be widely reviewed; and Stallo's clear if occasionally somewhat rhetorical presentations of complex material made the work sufficiently accessible to the educated reader to justify several additional printings on both sides of the Atlantic.[6] After the initial rush of notices, however, it does not seem to have generated much discussion in print until Ernst Mach rediscovered it towards the end of the century.[7] By that date even those whose knew the book seem to have had only the vaguest notions of who the 'J. B. Stallo' of the title page might be.[8]

But the scientific and philosophical community is never the whole world. In American politics, Stallo's had long been a familiar and respected name. The author of *Concepts and Theories* was a talented German-American who had emigrated to the United States as a youth to seek his fame and fortune, and had in due time acquired a measure of both. Born Johann Bernhard Stallo in Oldenburg, in 1823, he had in his early years been well if informally trained in the classics, modern languages and (fatefully) the rudiments of Kantian philosophy. At the age of seventeen he followed an uncle to Cincinnati, and after a period as a teacher there and in New York City, qualified for the Ohio bar in 1849 and embarked upon a successful legal career. He was also active – indeed, one is tempted to say hyperactive – politically, repeatedly switching party allegiance in his devotion to the principles of free trade and the anti-slavery movement. At length, in the 1880s, he remained a Democrat long enough to be awarded the post of minister to Italy under the first Cleveland administration, subsequently retiring to a villa in Florence and dying there, full of years and respectability, in 1900.[9]

Stallo's intellectual career displayed similar variety. As (popularly) with a number of well-known names in philosophy, there are two Stallos. The first was the precocious author of *General Principles of the Philosophy of Nature* (1848). This, as its title suggests, was a treatise in the tradition of *Naturphilosophie*, tracing out in the manner of Hegel, Schelling and Lorenz Oken the 'self-evolutions' of Spirit as they are manifested as 'phases and processes in the sphere of cosmic individuality' – *i.e.*, as the phenomena and laws of nature.[10] Well before the publication of *Concepts and Theories* three decades later, however, a second Stallo had emerged, who repudiated his own earlier views as having been 'written while I was still under the spell of Hegel's ontological reveries ... and still seriously affected with the metaphysical malady which seems to be one of the unavoidable disorders of intellectual infancy', and proclaimed his regret that the *General Principles* had ever seen the light of day.[11] Just what brought about this *volte-face* is unclear, and Stallo apparently preferred that it remain so.[12] The available evidence suggests that even during the period of his 'metaphysical reveries' he had acquired from his laboratory training in physics and chemistry a feeling for the outlook of the working scientist and a respect for the power of empirical methods. As early as 1855, in an essay on 'Materialism', Stallo was exposing the 'metaphysical fictions' which he saw as undergirding the matter-and-motion universe of the prevailing natural philosophy.[13] Indeed, many of the key elements of *Concepts and Theories* are present explicitly in this short piece; already prominent, for instance, are Stallo's central tenet that 'every so-called material thing is only a complex of relations to other things', and his firm rejection of 'materialism' (i.e., Lockean sensationalism, particularly as expounded by John Stuart Mill).[14] Stallo's specific criticisms of the atomic hypothesis were developed in greater detail in articles for *Popular Science Monthly* in 1873 and 1874, and were subsequently incorporated into *Concepts and Theories*.[15]

II

Despite the space devoted in Stallo's *magnum opus* to the analysis of particular scientific theories, however, the author intended that his audience should be left in no doubt as to the primacy of the second, epistemological task. Atomism, he explained in the Preface to the first edition, in reality rests – however much its proponents believed themselves to have escaped from the 'cloudy regions of metaphysical speculation' – on 'latent metaphysical' foundations. Stallo's analysis is itself to be regarded neither as physics nor as metaphysics, but as a 'contribution to the theory of cognition', which would lay bare the true character of those foundations and thus demonstrate the unsoundness of the conceptual edifice that had been erected upon them.[16]

If the reader now skips to the middle of *Concepts and Theories* – specifically, to Chapter IX (entitled 'The Relation of Thoughts to Things – The Formation of Concepts – Metaphysical Theories') – he will find Stallo carrying out just the program described in the Preface. Stallo here defines thought as 'the establishment or recognition of relations between phenomena'[17]; on the strength of this definition he proposes three 'irrefragable truths' as his basic epistemological principles:

(1) Thought deals not with things as they are, or are supposed to be, in themselves, but with our mental representations of them.

(2) Objects are known only through their relations to other objects. They have, and can have, no properties save these relations, or rather, our mental representations of them.

(3) A particular operation of thought never involves the entire

complement of the known or knowable properties of a given object, but only such of them as belong to a definite class of relations.[18]

"All metaphysical or ontological speculation," Stallo continues, "is based upon a disregard of some or all of the truths here set forth."[19] Specifically, the 'latent metaphysical elements' in the atomic hypothesis can be laid bare by showing that its picture of matter – aggregates of discrete, indestructible, impenetrable units, endowed with motion and interacting by elastic collisions – rests upon a series of metaphysical errors, these errors in their turn arising from a failure to observe the requirements (1) through (3) just enumerated. The same is true of the even more reprehensible 'atomo-mechanical' hypothesis defended by writers like P. G. Tait and James Challis, to the effect that bodies interact *solely* through contact forces; that *actio in distans* is meaningless; and that all forms of energy (including potential energy) are ultimately reducible to the kinetic energy of the atoms of matter or of aether. Stallo devotes the subsequent chapters of *Concepts and Theories* to spelling out in detail just how these errors arise.[20]

Unfortunately for the reader, the 'theory of cognition' on which Stallo proposes to base his views (and to which he so off-handedly attaches the definite article) receives no concise exposition in his book, and can be reconstructed only with considerable difficulty and some conjecture. Even sympathetic critics among Stallo's contemporaries – in particular, Hans Kleinpeter – pointed out the problematic status of his 'irrefragable truths'; Stallo had assembled them eclectically from (or perhaps read them into) the writings of psychologists and logicians like Sigwart, Drobisch, Herbart, Wundt and Sir William Hamilton; and though he seemed to regard them as representing a consensus among the philosophers of his day, this – as Kleinpeter makes clear – was simply not the case.[21] Nonetheless, keeping in mind (even if one cannot entirely reconcile) all these conflicting indicators, it is possible to assemble the elements of Stallo's philosophy of science into the following scheme: Science is defined to be "the establishment of definite relations between phenomena clearly and distinctly apprehended."[22] Its object is physical reality, and hence the world of phenomena, since "there is no physical reality which is not phenomenal."[23] But (*pace* the sensationalists) science cannot deal directly with the results of experiment; these must first be transformed into concepts, derived from experience but also subject to conditions precedent to experience.[24] The task of science is therefore to frame hypotheses – "guesses at the ultimate truth suggested by the analogies of experience" – about the causal relations among these conceptualized phenomena, and to test the consequences of those hypotheses against observation.[25] The phenomena in question need not be (indeed in mature scientific theories rarely are) accessible to any but the most sophisticated experimental techniques[26]; and the hypotheses which relate them may well contain fictitious elements, provided the *relata* they bring into agreement are again phenomenal.[27] The whole process becomes illegitimate only when the concepts employed are reified[28], and their logical inter-connections hypostatized into physically existing entities.[29] The natural world in this latter case is then furnished on the assumption "that every concept is the counterpart of a distinct objective reality," with the priority of the more general concepts to the less general mirrored in the world by the "order and genesis of things"[30] – the 'self-unfolding of Being' that had preoccupied Stallo the young *Naturphilosoph* thirty years earlier.[31] On this view, moreover, "things exist independently of, and antecedently to, their relations"[32] (a doctrine which can be traced from Aristotelean entelechy down to the writings of Stallo's contemporary F. H. Bradley); their properties or attributes have a reality distinct from the reality of the subjects of which they are modifications.

Stallo undertakes to support his contention that "the general principles of the atomo-mechanical theory... are substantially identical with the cardinal doctrines of ontological metaphysics"[33], by examining the atomo-mechanical theory in detail and unmasking its true character. The reader is thus prepared for an epistemological rebuttal of the existential claims about the micro-structure of the universe made by the atomo-mechanists. What Stallo actually presents, however, is much more like an essay on the psychology of concept formation in science: an extended series of glosses on Bacon's aphorism (quoted at the beginning of *Concepts and Theories*) to the effect that "the human intellect receives a coloration as a result of observing what occurs in the mechanical sciences... The result is that the intellect thinks that something similar occurs also in the universal nature of things."[34] Our concept, for instance, of hard, indestructible atoms as the basic constituents of matter arises, according to Stallo, from that tendency of mind which "identifies the Strange... with the Familiar";[35] since "the most obtrusive form of matter is the solid," we model our ideas of microscopic atoms on billiard balls and other macroscopic objects. Our natural analogues for the interaction of these unseen entities are drawn from everyday experience, where we see (or seem to see) the direct and instantaneous transfer of motion from one solid object to another. Closer analysis of such events on the macroscopic level reveals that in actuality they involve very complex processes (compression, deformation, the generation of heat); but such is the bewitchment of our intellect by pictures drawn from our naive impressions of what is taking place, that it is to these latter images that we turn when we seek to describe the fundamental make-up of matter. Stallo sums up his argument with a flourish:

What does the demand of the atomo-mechanical theory, to admit no interaction between bodies other than that of impact, imply? Nothing less than this, that the first rudimentary and unreasoned impression of the untutored savage shall stand for ever as the basis of all possible science.[36]

Paralleling and reinforcing this naive constructing of mechanical models, according to Stallo, is the tendency of the human mind to absolutize what is relative: that is, to see what are in reality the eternally shifting relations of a body to all the other bodies in the universe – the body's attributes – as 'substantial' realities in themselves. Reiterating that "there is no form of material existence which is either its own support or its own measure"[37] against the "ontological prejudice, that nothing is physically real which is not absolute"[38], Stallo concludes:

The assumption that all physical reality is in its last elements absolute – that the material universe is an aggregate of absolutely constant units which are in themselves absolutely at rest, but whose motion, however, is measurable in terms of absolute space and absolute time – is obviously the true logical basis of the atomo-mechanical theory.[39]

But this is precisely what Stallo has *not* shown. Had he written 'historical and psychological basis' (instead of 'logical basis'), the conclusion would have been perfectly cogent; as it is, there is a crucial slide from the genetic back to the properly epistemological. This, I think, is neither mere inadvertance nor conscious intellectual sleight-of-hand. Stallo, as we have seen, envisioned his work as based on his 'theory of cognition', generally eschewing the term 'epistemology' (which occurs in only two isolated instances in the entire book).[40] The catch-all character of the former term patches over a conflation in *Concepts and Theories* of two quite different ways of looking at scientific theories. Now, Stallo's genetic arguments about the origin of our scientific 'illusions' (in the Freudian sense) are interesting and plausible; they may well persuade us that our scientific beliefs arise from simplistic mechanical pictures of the world, and make us decently sceptical about the validity of those pictures; but unless we can show independently that the beliefs in question are false (or, alternatively, that no rational basis for them could possibly be forthcoming), the claim of such theories to represent the world has not been inherently enhanced or diminished.[41] Indeed, Stallo's criticisms, as applied to empirically-based matter theory, tend (as Gerd Buchdahl has remarked about many of the purportedly epistemological arguments adduced in the course of the nineteenth-century atomic debates) to "cut right across the conclusions of the scientific enquirer."[42] It is only when they are brought to bear against speculative natural philosophy *à la* Hegel or Fichte (or the early Stallo!), where the attempt is made to ground a science of nature deductively on the analysis of concepts, that the principles enunciated in *Concepts and Theories* as a 'theory of cognition' really do serve an epistemological function, by calling into question the truth-claims of any such *Naturphilosophie*. But such purported 'science' the mature Stallo does not even consider worthy of attack, but mockingly dismisses out of hand.[43]

III

But there is more – fully twice as much more – to Stallo's case against atomism and mechanism than the arguments thus far outlined. The first half of *Concepts and Theories* (which, like the latter portion, could stand independently as a separate essay) is a masterly dissection of the atomo-mechanical hypothesis on quite different 'principles'. Using a variety of physical and methodological 'criteria' and 'canons' for the validity of a scientific theory, Stallo effectively argues that the prevailing physics is unacceptable on logical and observational grounds. It is worth looking at these arguments in some detail; though eclectic in origin, they are deployed with vigor and subtlety, and provide as well a comprehensive view of the conceptual perplexities into which the physics of the 1860s and 1870s had worked itself.

One of these methodological criteria of Stallo's has already been mentioned in passing. This is his 'canon' to the effect that valid hypotheses must be complete or partial identifications between classes of phenomena; in its regulative aspect, it requires that

(1) Every valid hypothesis must be an identification of two terms – the fact to be explained and a fact by which it is explained;

and that

(2) The latter facts [i.e., those making up the *explanans*] must be known to experience.[44]

The first of these requirements rules out *idem per idem* explanations (the 'dormitive potency' account of opium's soporific powers, and the like). Stallo instances the explanation given by the atomic hypothesis for chemical combination, arguing that the empirical laws of change of volume, the 'evolution or involution of energy', and the 'emergence of a wholly new complement of chemical properties' in such reactions, are either left unexplained by atomism, or actually go counter to it – all of which, for Stallo, points up the "ineptitude" of such an hypothesis as a "figurative adumbration of the real nature of chemical processes."[45] Stallo's requirement, then, is tantamount to the demand that the hypothesis be a "simplification of the data of experience, which reduces the number of uncomprehended elements in the phenomenon."[46]

More interesting in its ramifications is Stallo's second requirement, that the *explanans* of the hypothesis – the 'explanatory phenomenon' – must be an observable fact, a 'datum of experience'.[47] Stallo identifies this with Newton's demand in the *Principia* (in the first of the *Regulae philosophandi*) that we admit only *'verae causae'* in natural philosophy; Stallo takes these (in the light of Query 31 of the *Opticks*) to be laws which are summaries of experiential data – 'general principles... from phenomena'.[48] Stallo's language here suggests a full-blown phenomenalist view of science, as do other passages in *Concepts and Theories*[49], and it was doubtless this strain in Stallo's work that scientists like Mach and (later) P. W. Bridgman found sympathetic.[50] On the other hand, there are, as already pointed out, important residual elements of an idealist view of knowledge in Stallo's thought. Moreover, Stallo allows that the identification which the hypothesis makes between phenomena may be 'partial and indirect'. Thus the luminiferous aether, for instance, though inadmissible because it leads to disconfirmed predictions, is *not* ruled out by this methodological canon, since the required identity "lies, not in the *fictitious* element, the aether, but in the *real* element, the *undulation*."[51] Hence

> an hypothesis may involve not only one but several fictitious assumptions, provided they bring into relief, or point to the probability, or at least possibility, of an agreement between phenomena in a particular that is real and observable.[52]

What Stallo does limit is the number of such fictitious elements which can permissibly be introduced in an *ad hoc* fashion to patch up an hypothesis threatened with disconfirming evidence; following the French philosopher and economist Cournot, Stallo asserts that "while the probability of the truth of an hypothesis is in direct ratio to the number of phenomena brought into relation, it is in inverse ratio to the number of such fictions."[53] He applies this criterion to the specific case of the kinetic theory of gases; auxiliary assumptions in the theory, like the inverse-fifth-power force of repulsion between molecules suggested at one time by Maxwell, or Stefan's temperature-dependent molecular volumes, are "purely gratuitous" and "fatal to all claims of simplicity preferred on behalf of the kinetic hypothesis."[54]

This, if it be phenomenalism, is a very weak brand, since it makes no claims as to whether such 'fictions' are ultimately eliminable from science. Stallo's remark, that in the case of the luminiferous aether the identity "consists, not in the *agent*, but in *the law of its action*,"[55] leaves it unsettled whether much more is at stake than an assertion that our theoretical constructs ought to be formed on analogy with objects and types of interactions familiar on the macroscopic level.[56] He takes a similar line in criticizing, once again, the kinetic theory, claiming that the theory's premises are "not only wholly unwarranted by experience, but out of all analogy with it": the long free paths postulated for the atoms of a gas are altogether unlike the behavior Newtonian gravitational theory would predict for their macroscopic counterparts.[57] None of this squares well with Stallo's earlier warnings against the pitfalls of appealing to readily available analogues in constructing models of the physical world; he seems to be in the awkward (to say the least) position of arguing that analogy with ordinary experience is both a primary desideratum for scientific theories and also *prima facie* cause for suspecting their legitimacy. (The quandary might be resolvable had Stallo treated the place of analogical reasoning in his theory of cognition more explicitly; there are hints that it is partly a means whereby an hypothesis is systemically integrated into a larger body of scientific knowledge. One recalls Newton's 'analogy of nature', Whewell's 'consilience of inductions', and Cournot's *'enchaînement des idées'* when Stallo speaks of the privileged status to be accorded to those many physical truths "whose universality is so well established as to afford strong, if not conclusive presumption against the legitimacy of concepts and the reality of alleged phenomena which would invalidate them."[58])

Still another, and even more variegated, set of criteria prefaces Stallo's discussion, in Chapter VII of *Concepts and Theories*, of the properly 'atomic' component of the atomo-mechanical view:

All physical theories properly so called are hypotheses whose eventual recognition as truths depends upon their consistency with themselves, upon their agreement with the canons of logic, upon their congruence with the facts which they serve to connect and explain, upon their conformity with the ascertained order of Nature, upon the extent to which they approve themselves as trustworthy anticipations or previsions of facts verified by subsequent observation or experiment, and finally upon their simplicity, or rather their reducing power. The merits of the atomic theory, too, are to be determined by seeing whether or not it is in accord with itself and with the known laws of Reason and of Nature.[59]

Stallo's use of these methodological rules against the matter-and-motion view is very telling, as a few examples will illustrate. He includes the luminiferous aether as a component of the atomic hypothesis, since a model which took the aether to be composed of discrete particles – a 'gas' permeating all space (and matter) – had replaced the earlier continuous elastic fluid.[60] Stallo argues that not only is a particulate aether not supported by independent lines of corroborating evidence (although it was originally supposed to explain not only the transmission of light but also the secular retardation reportedly observed in the orbits of periodic comets[61]); but that it predicts, contrary to all observation, the dispersion of light *in vacuo*.[62] Also (and Stallo stresses that Maxwell had himself noted this), postulating that such an aetherial gas fills *all* space gravely exacerbates the already troublesome problem of accounting for the specific heats of ordinary, 'material' gases (since the specific heat capacity of the aether would have to be 'added in').[63] Stallo adduces a variety of additional instances, the most detailed being his discussion of the difficulties encountered by Laplace's model for the formation of the solar system [64]; in all of them, prominent (and sometimes exclusive) use is made of observational confirmation and disconfirmation as the measure of a theory's success or failure.

So effective are Stallo's arguments on this level that there is little reason to doubt Josiah Royce's recollection that they shocked the sense of scientific orthodoxy of Stallo's contemporaries.[65] In the event, many of Stallo's analyses were vindicated: arguments similar to his did play a part in finally breaking down the opposition to action at a distance; the specific heats of gases were adequately accounted for only when quantum mechanics replaced the classical kinetic theory; the questions he raised about the conservation of angular momentum in the solar nebula remain relevant to present-day planetary theory. Unfortunately, as their eclectic origins indicate, Stallo's methodological criteria, though compatible with the higher-level 'principles' and 'irrefragable truths' of his theory of cognition, neither follow from nor imply the validity of that latter theory. Stallo draws in the methodological part of *Concepts and Theories* extensively on Mill and Jevons, whose 'sensationalist' views he elsewhere in the book belabors; and even his citations from other, less 'materialistic' writers like Whewell only serve to emphasize how extensive was the common territory shared by all philosophers of the period who took the empirical sciences into serious account. It is true that Stallo created his own synthesis of these criteria (he seems to have been the first writer in English to incorporate Cournot's sophisticated ideas on probability into his own philosophy of science [66]); and he modified the canons of scientific method offered by earlier authors in small but significant ways: for instance, while adopting Whewell's doctrine of the 'consilience of inductions', Stallo was readier than Whewell to allow for the subsequent rejection of even highly consiliated theoretical entities like the luminiferous aether.[67] Nonetheless, with some qualifications to be noted in what follows, the really telling methodological arguments in *Concepts and Theories* 'float free' to such an extent that the first eight chapters of the book where they are found could stand, as already mentioned, as an entirely independent essay. Stallo must have realized this. He is confronted with a dilemma which is of his own making, in the sense that, having rejected the extremes represented on the one hand by a deductive natural science after the fashion of *Naturphilosophie*, and on the other by a 'strict' phenomenalism whose only criterion is that of 'saving the appearances' in a conceptually economical fashion, Stallo is left with a criteriology for legitimate hypotheses which cannot be (or at least is not) validated from his general theory of cognition; and with a theory of cognition which provides only the most rudimentary standards for testing the adequacy of proposed scientific hypotheses. Admittedly there are, among the wide range of 'canons', 'criteria', and 'principles' to be found in *Concepts and Theories*, some which (like the exclusion of *idem per idem* explanations, or the demand for internal consistency) can plausibly be referred to higher-level 'laws of thought'; but they are such as would govern *any* program for understanding the world, and tend to be so universal in scope that (to paraphrase an aphorism of George Polya's) no interesting application of them is possible; it is only their lack of specificity that makes them unproblematic.[68]

There might remain the hope that what cannot be deduced from above, might arise from below – that the methodological apparatus on which Stallo has relied so heavily might be extracted inductively from the empirical sciences themselves. Not directly, to be sure; to demand *only* that a theory 'save the phenomena' would, as we have seen, be for Stallo to open the door again to 'occult powers' and worse.[69] Rather, one might identify, by a careful examination of past and present scientific theories, the methodological criteria according to which science actually carries out its tasks (something which *Concepts and Theories* itself devotes great attention to); and then argue – and the argument is superficially plausible, at least when science is in one of its self-confident epochs – that the empirical and pragmatic success of the scientific enterprise points to the validity of its methods. But to state this argument is to refute it; unless the 'success' of science is tautological (and Stallo was very far from allowing science to *define* itself by its own canons), such reasoning begs all the really important questions. Norms *for* science cannot be derived from descriptions *of* science in any such simplistic fashion. Even were it not the case that the working norms of science are shaped by a great variety of rational and non-rational factors, and change with the course of time, a descriptive account of what science does leaves us little wiser as to why particular procedures and criteria should generate 'successful' science (though we do emerge with a clearer idea of what this success consists in).[70]

IV

Stallo, then, set out in *Concepts and Theories of Modern Physics* to refute the atomo-mechanical hypothesis – a limited, well-defined critical aim. Using a diverse but widely acceptable collection of methodological criteria, he was able to point out serious inconsistencies and inadequacies in the matter-theory of his day, particularly in the full-blown scheme of elementary, indestructible atoms and mechanical contact forces championed by writers like Challis and Tait, and tacitly accepted with varying degrees of conviction by many of Stallo's contemporaries. The larger project of the book – the development of a comprehensive philosophy of science grounded in a general theory of cognition, which would account not only for the genesis of scientific concepts and hypotheses, but from which the canons for their acceptance or rejection might be derived – remained unrealized. That this is as much due to the intractability of the problems with which Stallo tried to deal, as to Stallo's limitations as a philosopher, is strongly suggested by the tortuousness of the paths which those who have tried to push on beyond the regions mapped out in *Concepts and Theories* have been compelled to follow.

Department of History and Philosophy of Science,
University of Pittsburgh
and Boston College

NOTES

I am indebted to Dr. Larry Laudan for his advice and encouragement during the preparation of this paper, and to the participants in the Faculty-Student Colloquium of my Department (where an earlier version was read) for their helpful criticisms.

For works of Stallo frequently cited in the notes and bibliography, the following abbreviations have been used:
  CT *The Concepts and Theories of Modern Physics* (Cambridge, Mass.: The Belknap Press, 1960). [Introduction and notes by P. W. Bridgman; reproduces the text of the third American edition (1888).]
  GP *General Principles of the Philosophy of Nature, With an Outline of Some of Its Recent Developments Among the Germans, Embracing the Philosophical Systems of Schelling and Hegel, and Oken's System of Nature* (Boston: Wm. Crosby and H. P. Nichols, 1848).
  RAB *Reden, Abhandlungen und Briefe* (New York: E. Steiger and Co., 1893).

[1] Henry Adams (1946, p. 377) records in the *Education* that it was met with a "conspiracy of silence." Josiah Royce (1946, p. 10) remembered how the "sense of scientific orthodoxy was shocked" when the book first appeared. The redoubtable P. G. Tait judged it "a curious

work" that displayed "a subtlety which is occasionally almost admirable" (1882, pp. 521–522). It is also cited by such varied writers as Bertrand Russell (1956), Émile Meyerson (1962) and P. W. Bridgman (1927).

[2] For a brief survey, see D. M. Knight (1968, pp. xv–xvii); a more extended discussion is given by Brock and Knight (1967, pp. 1–30).

[3] J. Berzelius, *Lehrbuch der Chemie* (Dresden, 1827), quoted in Buchdahl (1959, p. 125 and note 2).

[4] Buchdahl (1959, pp. 125–126); the quotation from Kekulé is taken from the latter's 'On Some Points of Chemical Philosophy' (1867), reproduced in facsimile in Knight (1968, p. 256), with comments by Knight (*idem*, p. 237). Stallo was thoroughly familiar with the *status quaestionis*, citing Dalton, Dumas, Berthollet, Kekulé, Brodie, *et al.*

[5] For Ostwald's views as late as 1904, see his Faraday lecture of that year, reproduced in Knight (1968, pp. 323–332). And in 1907, Edward Divers, a pupil of the atomist A. W. Williamson, could still write *à propos* of the atomic debates that "chemists remain not much less divided on the subject now than they were then [*i.e.*, in the 1860s]" (Brock and Knight [1967, p. 30]).

[6] Information about the series of volumes in which *CT* first appeared is given in Drake (1959, p. 23), and in Easton (1966, pp. 72–73). For all the lack of critical enthusiasm, the book was reissued several times in English (Stallo added the long 'Introduction and Preface to the Second Edition' in 1884), and there were translations into French (*ca.* 1882), and, most importantly, German (1901). (For the latter, see below, note 50.) Translations into Spanish, Italian and Russian are occasionally mentioned, but appear to be mythical; see Bridgman (1960, pp. vii–viii).

[7] Mach, writing to Karl Pearson in 1897, lamented that he had not known of Stallo's work earlier, and so been reassured that he did not labor alone in his efforts to reform the philosophy of science. See Thiele (1969, p. 537). Kleinpeter (1913, p. 282) has some perceptive remarks on Mach's enthusiasm for Stallo's ideas, and on key differences in their views.

[8] Pearson had written in reply to Mach's original inquiry about Stallo: "I will endeavour to find out who Mr. Stallo is...; I think he may be an *American*. He is not in any academic position in England; nor have I ever heard of any memoir or work of his except the Concepts of Physics." (Thiele [1969, p. 538]; italics in original.)

[9] Stallo's life is treated in some detail in Easton (1966), but the accounts in Drake (1959, 1967) and in Bridgman (1960) are adequate for the non-political side of his career. (Additional bibliographical material can be found in the article on Stallo in the *DNB*, but this is otherwise none too trustworthy.) Other sources largely quote from one another, all drawing ultimately to a privately printed memoir (1902) by Stallo's long-time acquaintance H. A. Rattermann. Of interest in this connection is a brief *curriculum vitae* included in a letter of Stallo to Mach, quoted in part in Mach (1901) and given in full in Thiele (1969, pp. 541–542).

[10] Stallo, *GP, passim*; the flavor can be gotten from the texts reproduced in Easton (1966, pp. 229–278).

[11] *CT*, pp. 5–6.

[12] A personal crisis of some kind seems to have been involved; around the time of the publication of *GP*, Stallo abandoned the Roman Catholicism in which he had been raised and announced himself a Protestant, simultaneously giving up a promising academic career. (At age twenty-five he was "Professor of Analytical Mathematics, Natural Philosophy, and Chemistry" at Fordham [then St. John's] College in New York City.) For details, see Easton (1966, pp. 31–32, 52).

[13] The essay appears in Stallo's *RAB*, and is discussed in Easton (1966, pp. 53–56), and in Wilkinson (1951).

[14] Easton (1966, p. 55) has characterized Stallo's position from the 1860s onwards as an 'experience-based idealism', and stresses even the later Stallo's debt to Hegel, noting that the relationalist view of attributes is found prominently in Hegel's *Phenomenology* as well as in Stallo's discussion of the latter work in *GP*. This is a reasonable enough argument for the genesis of this aspect of Stallo's thought; but the doctrine in question is as much Kantian as Hegelian (and Stallo had read his Kant as well); indeed, in the rather general form in which Stallo embraces it, it represents a philosophical tradition stretching back at least to the medieval nominalists. In this context, Easton makes a good deal of the 'dialectical' character of Stallo's thought (1966, p. 56), but in the passages he cites from *CT* this seems to amount to little more than the *dividere et componere* of Aristotelean logic. On the other hand, the characteristically Hegelian stress on the dialectical nature of the historical process, though applauded by Stallo in an address on Humboldt in 1859 (in *RAB*; compare Easton, *loc. cit.*), is entirely absent in *CT*; and it probably of some significance, even granting the limitations of subject matter indicated in the title of the latter work, that Darwinian evolution is nowhere mentioned in its pages. Hans Kleinpeter, who was thoroughly familiar with the sources Stallo drew upon, saw Stallo's later philosophy as a modified Kantianism; see Kleinpeter (1901, pp. 401–440).

[15] *CT*, p. 10.
[16] *CT*, p. 4.
[17] *CT*, p. 152.
[18] *CT*, pp. 156–157.
[19] *CT*, p. 159.
[20] Chapters XIII and XIV of *CT* are an attack on the transcendental geometry of Riemann, Bolyai *et al.*, and on Riemann's ideas on the nature of number, on precisely the same grounds. Stallo's arguments here seem to have rested on a fundamental misconception as to what these mathematicians (particularly Riemann) were out to do; this part of *CT* will be taken into account in this paper only to the extent that it casts light on Stallo's general theory of cognition. For the issues involved in the debate about non-Euclidean geometries, see Russell (1956).
[21] Kleinpeter (1901); *cf.* Bridgman (1960, p. xix).
[22] *CT*, p. 21.
[23] *CT*, p. 216.
[24] *CT*, pp. 41–44.
[25] *CT*, p. 16.
[26] *CT*, p. 23.
[27] *CT*, p. 136–137.
[28] *CT*, p. 12; *cf.* p. 144, note 25.
[29] *CT*, pp. 21, 160; *cf.* pp. 189–199.
[30] *CT*, pp. 159–160.
[31] See *GP*, pp. 49–186 ('Evolutions').
[32] *CT*, p. 160.
[33] *CT*, p. 14.
[34] *CT*, p. 2. (Bridgman's translation, from the Latin original given by Stallo.)
[35] *CT*, p. 193.
[36] *CT*, p. 198.
[37] *CT*, p. 201.
[38] *CT*, p. 203.
[39] *Ibid.*
[40] And on one of the occasions when it does, it appears in the italics which, as we have

already seen (above, note 8), could serve the Victorians as discreet shudder quotes; cf. CT, p. 15.

[41] In fact, Stallo also argues that the penetrability of matter – for instance, the 'mutual passivity' (in Dalton's phrase) of gases toward one another – is more primary in our experience than the absolute solidity and impenetrability of bodies, and that this also tells against the atomic hypothesis (CT, pp. 115–117)! There is an important ambivalence in Stallo's use of analogy here and elsewhere, as will become apparent; it is sufficient here to note that Stallo is virtually conceding the impossibility of arguing to the truth or falsity of the atomo-mechanists' claims on purely psychological grounds.

[42] Buchdahl (1959, p. 132).

[43] CT, p. 300 and passim; cf. note 20, above. It is ironic that the work of the 'transcendental' geometers was to play such a role in the demise of the very apriorism in natural philosophy which Stallo set himself to oppose.

[44] CT, p. 131.

[45] CT, pp. 125–128.

[46] CT, p. 133.

[47] CT, p. 137.

[48] Ibid.; for the full text see Isaac Newton (1952, pp. 401–402).

[49] For instance, CT, pp. 216ff.

[50] See Mach (1901, pp. xii–xiii); Thiele (1969, pp. 536–537); and Bridgman (1960, pp. xix, xxvff.). Bridgman had also mentioned Stallo in his early *Logic of Modern Physics* (Bridgman [1927, p. v]).

[51] CT, pp. 135–136.

[52] CT, pp. 136–137.

[53] CT, p. 137 and note 19; Stallo here is drawing on Cournot (1861, I, p. 103). Stallo goes on to say that "more accurately" the improbability of the hypothesis "increases geometrically while the series of independent fictions expands arithmetically" (CT, loc. cit.). This quantitative form of the relation is not attributed to Cournot; it seems to echo an assertion of Drobisch, cited by Stallo (CT, p. 54), that the extension of a concept and its intension vary reciprocally in just this manner.

[54] CT, pp. 147–148.

[55] CT, p. 136.

[56] This kind of argument has already been encountered (above, note 41) in connection with Stallo's strictures against the doctrine of the impenetrability of the ultimate constituents of matter.

[57] CT, p. 146.

[58] CT, p. 168. Stallo also cites with approval (CT, p. 127, note 9) Cournot's affirmation of "the harmony which certainly exists in the totality of all things" (Cournot [1861, I, p. 246]). But there is a further qualification (CT, p. 141): agreement with the 'order of Nature' is not an absolute criterion, since the "known laws of nature ... may be [sic!] incomplete inductions from past experience, to be supplemented by the very elements postulated by the hypothesis" – an intriguing glance in the direction of the theory-ladenness of observation and all that it entails.

[59] CT, p. 112.

[60] CT, pp. 120–121.

[61] CT, p. 137.

[62] CT, pp. 120–122. For the problem of cometary orbits, see Whittaker (1960, II, pp. 148–149).

[63] Classical kinetic theory partitions the internal energy of a gas among the degrees of

freedom of its constituent atoms or molecules; for diatomic gases (for instance, hydrogen or oxygen), this leads to calculated values for the specific heat capacity which are appreciably higher than those actually measured in the laboratory. Stallo does not give a reference for Maxwell's discussion of the heat capacity of the aether; he is likely drawing on the latter's article 'Atom' in the 9th edition of the *Encyclopaedia Britannica* (reproduced in Maxwell [1956, II, pp. 456–457]).

[64] *CT*, pp. 290–300.
[65] See above, note 1.
[66] *CT*, pp. 127, 132, 137.
[67] *CT*, p. 137.
[68] A case in point is Stallo's discussion of the role of the *a priori* in science. He asserts, it will be recalled, that "all processes of deductive reasoning involve an ultimate reference to primary constants which are not given in experience" (*CT*, p. 253, note 23). The only instance of such "reference to primary constants" which Stallo presents in the original edition of *Concepts and Theories* is that of the Euclidean straight line "or simple direction" – this in the course of his attack (above, note 20) on non-Euclidean geometries, the least fortunate part of the entire book. In the 'Introduction and Preface' to the revised edition of 1884, he does include a few pages of remarks (*CT*, pp. 42–45) on the role played in the sciences by the concept of causality and the related notions of constancy and continuity; but these observations are general and tentative in character, suggesting that Stallo's reading of Whewell (whose 'Fundamental Ideas' [Whewell (1847, I, pp. 66ff.)] function in much the same way as Stallo's 'primary constants') had roused his suspicions that, when examined closely, such allegedly *a priori* elements might well turn out either to be of such generality that they gave no indication to *which* empirically-based concepts they were meant to apply; or else themselves to be in actuality empirical and *a posteriori*. Similarly, Stallo's positive remarks on the role of analogy suggest an implicitly regulative use of the concept which seems without foundation in his basic theory of knowledge; see especially his discussion of the kinetic theory in Chapter VIII. It is this aspect of *Concepts and Theories* (note especially p. 253, note 23, and *cf.* p. 21) that to my mind most clearly shows the differences in outlook between Mach and Stallo already referred to (above, note 7), however sympathetic they may have found one another's views.
[69] *CT*, p. 134.
[70] For instance, a requirement like predictivity – as distinct from simple well-confirmedness – plays no important role in formal scientific methodology until the nineteenth century, although the enhancement of a theory's plausibility by its ability to make *surprising* predictions in areas outside the context in which it was developed, was noted as early as the 1690s by the Newtonian William Whiston.

BIBLIOGRAPHY

Adams, H.: 1946, *The Education of Henry Adams*, Modern Library, New York.
Bridgman, P. W.: 1927, *The Logic of Modern Physics*, Macmillan, New York.
Bridgman, P. W.: 1960, 'Introduction', *CT*, pp. vii–xxix.
Brock, W. H., and D. Knight: 1967, 'The Atomic Debates', in W. H. Brock (ed.), *The Atomic Debates: Brodie and the Rejection of the Atomic Theory. Three Studies*, Leicester University Press, Leicester, pp. 1–30.
Buchdahl, G.: 1959, 'Sources of Scepticism in Atomic Theory', *Brit. J. Phil. Sci.* **10**, 120–134.

Cournot, A.-A.: 1861, *Traité de l'Enchaînement des Idées Fondamentales dans les Sciences et dans l'Histoire*, L. Hachette, Paris, I, p. 103.
Drake, S.: 1959, 'J. B. Stallo and the Critique of Classical Physics', in H. M. Evans (ed.), *Men and Moments in the History of Science*, University of Washington Press, Seattle, pp. 22–37.
Drake, S.: 1967, 'Stallo, John Bernard', in P. Edwards (ed.), *The Encyclopedia of Philosophy*, Macmillan and the Free Press, New York, 8, pp. 4–6.
Easton, L.: 1966, *Hegel's First American Followers. The Ohio Hegelians: John B. Stallo, Peter Kaufmann, Moncure Conway, and August Willich, with Key Writings*, Ohio University Press, Athens, Ohio.
Kleinpeter, H.: 1901, 'J. B. Stallo als Erkenntniskritiker', *Vierteljahrsschrift für wiss. Phil.* **25**, 401–440.
Kleinpeter, H.: 1913, *Der Phänomenalismus: Eine naturwissenschaftliche Weltanschauung*, J. Barth, Leipzig.
Knight, D. M. (ed.): 1968, *Classical Scientific Papers: Chemistry*, American Elsevier, New York.
Mach, E.: 1901, 'Vorwart' to J. B. Stallo, *Die Begriffe und Theorien der modernen Physik*, J. Barth, Leipzig, pp. III–XIII.
Maxwell, J. C.: 1965, *Scientific Papers*, Dover, New York.
Meyerson, É.: 1962, *Identity and Reality*, Dover, New York. (Original French edition 1907.)
Newton, I.: 1952, *Opticks*, Dover, New York. (Based on the 4th edition, London, 1730.)
Rattermann, H. A.: 1902, *Johann Bernhard Stallo, Deutsch-amerikanischer Philosoph, Jurist und Staatsmann*, Verlag des Verfassers, Cincinnati, Ohio.
Royce, J.: 1946, 'Introduction' (1913) to H. Poincaré, *The Foundations of Science*, The Science Press, Lancaster, Pa.
Russell, B.: 1956, *An Essay on the Foundations of Geometry*, Dover, New York. (Original edition 1897).
Tait, P. G.: 1882, 'Modern Physics', *Nature* **26**, 521–522.
Thiele, J.: 1969, 'Karl Pearson, Ernst Mach, J. B. Stallo', *Isis* **60**, 535–542.
Whewell, W.: 1847, *The Philosophy of the Inductive Sciences*, 2nd ed., John Parker, London.
Whittaker, E.: 1960, *A History of the Theories of Aether and Electricity, Volume Two: The Modern Theories 1900–1926*, Harper, New York.
Wilkinson, G. D.: 1951, 'John B. Stallo's Criticism of Physical Science', unpublished Ph.D. dissertation, Columbia University, New York.

SYMPOSIUM

GENETICS, IQ AND EDUCATION

N. J. BLOCK

FICTIONALISM, FUNCTIONALISM AND FACTOR ANALYSIS*

Two doubtful philosophical doctrines have contributed strongly to the development of IQ tests, and now play a major role in defending IQ tests as measures of intelligence: they are operationism and fictionalism.

The application of operationism in psychometrics is in such doctrines as the following: one can avoid answering the question whether IQ tests measure intelligence simply by defining the word 'intelligence' as what IQ tests measure. Obviously, such a definition does not avoid the issue. For we can now reasonably ask whether people who accept the operational definition use the word 'intelligence' to refer to the same quantity as that referred to by people who reject the operational definition or have never considered it. Many operational definitions of 'intelligence' would be clearly unsatisfactory. Consider, for example, an operational definition which stipulates that a person's intelligence is the number of pounds indicated when the person is placed on a scale. Whether a given measuring device measures what it is supposed to measure is as much an empirical question in psychology as it is in physics; adopting an operational definition cannot make such an empirical question "go away".[1]

Operationism in psychometrics is insidious in that it contributes to psychometricians' playing a double game. On one hand, many psychometricians feel they are scientists who use 'intelligence' as a technical term to denote whatever it is that IQ tests measure, and thus have no responsibility to use the term to refer to what people ordinarily use it to refer to. On the other hand, they want or at least are willing to accept the important social role which depends on it being widely believed that IQ tests measure intelligence in the ordinary sense of the word.

My target here, however, is not operationism, but fictionalism, a more subtle (but equally false and insidious) view which arises in connection with the question whether intelligence is one thing or many. Now the question "One thing or many?" often is sensible with respect to concrete objects. For example, in wondering whether the Boston Strangler was one or many people, one might be wondering whether a number of dif-

*R. S. Cohen et al. (eds.), PSA 1974, 127–141. All Rights Reserved.
Copyright © 1976 by D. Reidel Publishing Company, Dordrecht-Holland.*

ferent people, each working alone, committed the crimes which because of the common style were attributed to one person: the (non-existent) Boston Strangler. But is there *any* sense to the "One or many?" question with respect to abstract objects like intelligence? It seems there is. For example, in the early 19th Century, chemists used terms like "atomic weight" to refer to what they thought was one quantity, but which as Cannizaro pointed out in 1858 were three different quantities: atomic weight, molecular weight, and equivalent weight. What Newton thought was one quantity, mass, turned out (if Einstein was right) to be two quantities, rest mass and relativistic mass.[2]

The psychologists' approach to the question whether intelligence is one thing or many has centered around a set of mathematical techniques called 'factor analysis'. Jensen and many other psychologists believe factor analysis shows intelligence is one thing – a very general intellectual capacity variously called 'general intelligence', 'general ability' or just 'g' – and that IQ tests measure it.[3] A minority view is that factor analysis shows that most of the variance[4] in IQ scores is due to a number of special abilities such as verbal ability, perceptual ability, reasoning ability, and memory. Those who hold this view often think of intelligence as a set of abilities, and they view factor analysis as showing IQ tests measure intelligence so considered.

Why the disagreement? Factor analysis is simply a set of mathematical methods for finding and analyzing patterns in large numbers of correlation coefficients. There are a number of different sorts of methods, some of which have the effect of ascribing as much as is mathematically possible of the variance in a group of tests to a single factor; others have the effect of spreading the variance out among a group of factors (e.g., the 'Primary Mental Abilities'). The dispute goes on because no one has ever offered a good *psychological* reason for preferring one group of techniques to the other. If believers in general intelligence are now in the ascendancy, it is because the 'Primary Mental Abilities' isolated by the latter sort of methods turn out themselves to be correlated somewhat, thereby giving a slight advantage to believers in general intelligence.

I want to focus on an assumption which is common to both sides of this dispute, an assumption which I have called the 'Correlation-Entails-Commonality' Principle. The CEC Principle is this: insofar as two tests of abilities correlate, this correlation is totally due to a common ability measured by both tests. The CEC Principle can be put more precisely as follows:

$$r_{ab} = r_{ac} \cdot r_{bc}$$

where r_{ab} is the correlation[5] of two tests, a and b, scored on the same scale; r_{ac} is the correlation of test a with common ability (considered as if measured by a test scored on the same scale as a and b); and r_{bc} is the correlation of test b with the value of the common ability. r_{xc} is often referred to as the 'loading of test x on the common factor'. Suppose, for example, that $r_{ab} = .7$. The CEC Principle dictates that there is a single ability common to a and b which has correlations with a and b of at least .7. For if test a is a *pure* measure of the common ability, then $r_{ac} = $ a maximal 1.0, and r_{bc} will be at its minimum, .7 (since $.7 \cdot 1 = .7$). On the other hand, if tests a and b are equally good measures of the common ability, then $r_{ac} = r_{bc} = \sqrt{.7} = .84$. Other things equal, factor analysts would adopt the latter alternative, ascribing equal loadings on the common ability to the two tests. In this case, it would be said that the common ability accounts for 70% ($= (.84)^2$) of the variance[6] in each test. So other things equal, the CEC Principle is used to conclude that when two tests, a and b are correlated .7, 70% of the variance in each test is due to their measuring a single common ability.

Now it is clear that the CEC Principle is false. Correlation is a necessary but not a sufficient condition for commonality. Often the best explanation of correlation will *not* be commonality. For example, a tyrant could bring about a correlation between speaking and bowling ability by picking two large groups at random, damaging all the tongues and fingers in one group and giving speaking and bowling leassons to the other. Among the tyrant's subjects, tests of speaking and bowling ability would correlate highly, but probably not because there is some common ability.

Of course sometimes commonality *is* the best explanation of correlation. For example, probably ability to fix cars is correlated with ability to fix tanks; and perhaps this correlation is due to some sort of mechanical ability which is in some sense common to or a part of both tank fixing and car fixing abilities. Perhaps both car and tank fixing ability are mainly a matter of a thorough understanding of the internal combustion engine and other components common to both cars and tanks.

In sum, the best explanation of correlation is sometimes commonality, sometimes something else, sometimes in part commonality and in part something else. What is wrong with the CEC Principle is that it insists that correlation of ability tests is always due *totally* to commonality.

It might be replied that while the CEC Principle is false in general, it is true, or nearly true with respect to some very restricted domains: intellectual abilities, for example. It might be said that commonality is the only plausible explanation of the correlation of intellectual abilities. It is easy to see, however, that there are many possible explanations of why abilities might correlate. For example:

(1) Perhaps many environmental conditions which nurture one ability tend to nurture a number of abilities. For example, the effect of good health, nutrition, educational and cultural background are likely to be spread over a number of abilities. Personality traits like curiosity and drive for achievement may also affect a number of abilities.

(2) Perhaps many environmental conditions which hinder the development of one ability hinder the development of a number of abilities. This is plausible with respect to disease, injury, malnutrition, poor education, lack of encouragement, etc.

(3) One ability may sometimes be a prerequisite for – though not a component of – another ability. This could be natural, in the way ability to hear may be required for learning to speak, or societally imposed, as would be the case if successful completion of auditing training depends in part on adhering to certain codes of behavior and dress.[7]

(4) People who excel in a number of abilities may tend to marry people who also excel in a number of abilities. Such 'assortative mating' has been observed for many traits. If abilities turn out to be heritable, as has often been claimed (though on the basis of poor evidence) they might correlate for genetic reasons.

I have just given a number of possible explanations of correlation of abilities. I do not claim to know these to be explanations of whatever inter-correlations abilities have; what I do claim is that it would be a silly mistake to blindly accept the claim of the CEC Principle that the *only* explanation of correlation of abilities is commonality.

At the risk of being repetitious, I wish to emphasize that I am not denying that correlation of abilities may sometimes be part of a body of evidence for commonality. Suppose, for example, that we knew that there is something which could be called 'reasoning ability' and that we had tests which measure it. Suppose tests of reasoning ability agreed with good tests of say mathematical ability, except in cases where the disagreement could be explained in terms of 'interfering factors'. E.g. a person might do well on the reasoning test and poorly on the mathematics test, but only because of a neurotic fear of mathematics, rather than a lack of mathematical capacity. In such circumstances, one might hypothesize that mathematical ability is just a kind of reasoning ability.

Is there anything left of the CEC Principle? How about this? If two genuine ability tests correlate and if there is no *other* explanation of the correlation, then the correlation is due to a common ability measured by both tests. This is true enough, but it hardly deserves to be called a 'principle'. For the following is also true: if two genuine ability tests correlate and if there is no other explanation of the correlation, then the correlation is due to the presence of environmental conditions which promote the development of both abilities. Both of these trivialities are consequences of the following general truth: if you can rule out all possible explanations of a phenomenon but one, then you can conclude that the remaining explanation is the correct one.

I have been supposing abilities do correlate with one another. But the only quantitative evidence for this is that psychometric tests which are supposed to measure (e.g.) mathematical ability, correlate with psychometric tests which are supposed to measure (e.g.) verbal ability, and these tests correlate with tests which are supposed to measure (e.g.) reasoning ability. Many 'ability' tests correlate substantially with one another.[8] But what if (as G. H. Thomson (1939) suggested) each test measures a *set* of abilities? Then if the sets measured by different tests *overlap*, this would produce correlation among tests. Thus even if the CEC Principle were *true*, correlation of standard 'ability' tests would not be good evidence for general intelligence.

Moreover, to a substantial degree, standard IQ tests fail to measure abilities and thus fail to be in the domain of the CEC Principle. As many test makers and widely accepted psychometrics texts[9] point out, standard test scores are substantially influenced by personality, motivation and temperamental characteristics, such as persistence in solving problems, concentration, and tendency to check back for mistakes. But such characteristics are likely to help on many subtests, thus contributing to the intercorrelation among subtests.

Further, a glance at IQ tests shows that whatever *else* they measure, they certainly measure skills learned in school. E.g., every IQ test has 'numerical abilities' questions which depend partly on mathematical ingenuity, but also partly on mastery of the level of mathematics involved. E.g., adult IQ tests contain high school algebra problems. Cronbach (1970b, p. 470) remarks

> Factor analysts often speak as if they were discovering natural dimensions that reflect the nature of the nervous system. Correlations and factors, however, merely express the way performances covary in the culture from which the sample is drawn. The usual high correlation between verbal and numerical abilities is due in part to the fact that persons who remain in school are trained on both types of content. If a culture were to treat map-making as a fundamental subject, then map-making proficiency, it is suspected, would correlate highly with verbal and numerical attainments.

Given the obvious falsity of the CEC Principle, and the obvious fallacies involved in applying the CEC Principle to standard 'ability' tests, why are the CEC Principle and its applications so widely accepted? The answer is that many psychologists accept a version of a doctrine in the philosophy of science sometimes known as 'fictionalism' which can be used to support the CEC Principle and these applications of it. Fictionalism is the doctrine that theoretical terms in science (e.g. 'electron,' 'gene') should not be construed as actually referring to the sort of 'unobservable' things they purport to refer to, but rather the use of these terms should be construed as a way of speaking designed to facilitate prediction and control. The fallacy fictionalists attribute to realists – taking theoretical terms at face value – is described as 'reifying'. Fictionalists come in two forms: *hard* fictionalists deny that electrons, genes, etc., exist; *soft* fictionalists assert that electrons do exist, but they in effect take it back by claiming either (1) that electrons should be *identified* with what we might naively think of as the effects of electrons: meter readings, cloud chamber tracks and the like, or (2) that statements about electrons should be *analyzed* in terms of what we might naively think of as the effects of electrons.[10] Fictionalism with respect to mental states, events, processes and capacities has traditionally been of the soft variety. Indeed, the latter 'semantic' version of soft fictionalism with respect to mental phenomena, the view that statements about mental phenomena should be analyzed in terms of what we would naively describe as their behavioral effects, is simply familiar logical behaviorism.[11]

Operationism construes talk apparently about 'theoretical' entities as a disguised way of talking about measuring operations. Thus operationism entails fictionalism. The converse is not the case, however; for while operationism demands explicit definability of theoretical terms in terms of 'observables', fictionalism does not.

I have construed fictionalism as neither requiring nor denying the possibility of explicit definition of 'theoretical' terms in terms of 'observables'. However, psychometric fictionalists and operationists sometimes construe fictionalism as denying this possibility. Their construal has the effect of making operationism and fictionalism incompatible. Those of us who see realism as a live alternative to both operationism and fictionalism prefer to emphasize the similarities of operationism and fictionalism rather than their differences, and thus tend to prefer the definition of 'fictionalism' I am using.

Early factor analysts were often realists. Spearman thought of general intelligence as 'amount of mental energy'. But since factor analytic methods yield no psychological results by themselves, and since factor analytic methods do yield results when combined with the CEC Priniple, it is no surprise that fictionalism – which provides the most natural way of supporting the CEC Principle – is now dominant among factor analysts.

It would be a mistake to suppose that fictionalism alone entails the CEC Principle. For example, consider a version of fictionalism which indentifies ability A with the performance of behavior a, b, and c in circumstance 1; behavior d, e, and f in circumstance 2; g, h, i in circumstance 3, and so on. Ability B is identified with the same behaviors in circumstance 1 and 2 (a, b, c and d, e, f, respectively) but different behaviors in circumstances 3, 4, and so on. A proponent of this version of fictionalism could explain correlation of tests of abilities A and B by appeal to the overlap of behaviors in circumstances 1 and 2 and the claim that circumstances 1 and 2 are involved on both tests A and B. Yet he could reasonably maintain that this overlap does not itself constitute a genuine ability. He may maintain, for example, that a specification of a genuine ability requires a specification of behaviors in a larger (perhaps infinite) set of circumstances. The 'overlap' of A and B is behaviorally defined *only* for circumstances 1 and 2 and for this reason might reasonably be denied the status of an ability.

The CEC Principle requires for its defense a particular brand of fictionalism. The version generally used is the identification of psychological states and capacities with sets of intercorrelating behaviors. Ann Anastasi (1938, p. 391) stated this position as follows

> The writer maintains that to speak in terms of such inferential constructs is not the best way to arrive at psychologically meaningful results. Psychology as an experimental science demands that we remain as close as possible to the objectively observable facts and that we define our concepts operationally....
>
> A "factor" isolated by such analyses is simply a statement of the tendency for certain groups of behavior manifestations to vary concomitantly. It does not indicate the presence of any other characteristic or phenomenon beyond or beneath the concrete behavior.

What is probably the most influential statement of the position appears in P. E. Vernon's *The Structure of Human Abilities*. (1961, p. 1, 3) Vernon says

> An ability or factor should be thought of as a class or group of performances....
>
> Psychologists nowadays tend to adopt a more operational or Behavioristic outlook, though rejecting the wilder excesses of J. B. Watson's doctrines. They realize the fruitlessness of mental entities such as faculties, which can never be directly observed nor verified, and prefer to deal with concepts directly derived from measurable activities of human beings. An ability is inferred from the fact that some people carry out certain tasks more rapidly or more correctly than others. Whether it also depends on some power in the mind is a matter which interests metaphysicians, but not scientists. The clue leading to a scientific solution of the impasse is provided by the term overlapping, i.e., by correlation. By means of correlation we can find whether the scores of a group of people on two or more tasks correspond or not, and therefore whether these tasks involve the same or distinctive abilities.

Cronbach's psychometrics textbook makes the position clearer.

> Looking at a collection of scores such as the Wechsler subtests, the psychologist must ask: just how many different abilities are present? *The word "ability" in such a question refers to performances, all of which correlate highly with one another, and which as a group are distinct from (have low correlations with) performances that do not belong to the group.* To take a specific example, Wechsler Vocabulary items call for recall of word meanings, and Wechsler Similarities items call for verbal comparison of concepts. Are these measures of the same ability? Or do some people consistently do well on one but not the other? [The reader should note the alternatives presented here.] For a group of adolescents, we have these data:
>
> > Form-to-form correlation of Vocabulary on same day = 0.90
> > Form-to-form correlation of Similarities on same day = 0.80
> > Correlation of Vocabulary and Similarities = 0.52
>
> The two tests evidently overlap. About 52% of either test can be regarded as representing a shared ability or "common factor".
>
> Twenty percent of the Similarities variance is due to form-to-form variation. This leaves

28% that must be due to some distinct ability that is not tested by Vocabulary. Likewise, 38% of Vocabulary is due to an ability not involved in Similarities. There is a common factor of verbal facility or reasoning, but each test also involves something extra. Hence the two tests do involve distinct abilities.

Factor analysis works along these general lines, starting from correlations. Binet applied such reasoning when he decided that his tests, all having a substantial relation to each other, must reflect a pervasive general intelligence.[12]

What these psychometricians have done is to define the word 'ability' in such a way that if abilities A and B correlate perfectly, it *follows* that A and B are one and the same ability. On their sense of 'ability', the CEC Principle is trivially true. But as is shown by the points made above about alternative explanations of correlation of abilities, *their* sense of 'ability' is not *ours*. Speaking and bowling ability could correlate perfectly without being the *same* ability, and the same is true for reasoning ability, and mathematical ability, for example.

This version of fictionalism with respect to abilities has other consequences which are at least as peculiar as the CEC Principle. The public understands words like 'intelligence' and 'ability' (especially as used in connection with IQ and aptitude tests) as referring to *capacities* to perform or acquire skills to perform, not to performances themselves. Both makers and victims of educational policies regard IQ tests as more or less successful attempts to measure what people *can* do, not just what they *do* do. Standard IQ tests *are* tests of abilities in the absurd sense of performances, but to say this is not to say that they measure the capacities underlying these performances.

Of course, a definition of 'ability' as a set of performances ought not to be taken literally. It obliterates the distinction between exercising an ability and having an ability. If the performances in question are test performances, then someone who does not take the test has no ability. The charitable reader will replace 'performances' with something about dispositions to perform. But what sort of dispositions? There is a range of quite different theories of mentality which can be broadly characterized as 'dispositional'. For example, one dispositional theory (the one the psychometricians quoted above are perhaps best construed as holding) is that abilities are dispositions to perform so as to produce certain scores on certain (properly administered) standardized tests. Another very different sort of dispositional account – amounting to a version of functionalism (see note 11) – construes abilities as dispositions to perform in certain ways in certain circumstances; the specification of the circumstances involves reference to other mental states, e.g. beliefs and desires; and 'disposition' is understood so as to allow the disposition to do x to be a cause of doing x.[13]

It is not obvious what sort of dispositional account psychometricians who support the CEC Principle would hold. If the arguments I have given so far are correct, no adequate dispositional account of abilities will suffice to establish the CEC Principle; for if abilities are construed as capacities – and this is the relevant sense of the term in discussions of intelligence and aptitude – the CEC Principle is obviously false, as I argued above. On the other hand, if 'ability' is defined as a set of performances, then the CEC Principle is true of abilities so construed. Construing abilities as dispositions to produce certain scores on certain standardized tests will perhaps also suffice to justify the CEC Principle. But these senses of 'ability' are irrelevant to many of the social and educational purposes towards which testing of abilities is directed.

The functionalist account of abilities is in a sense dispositional, and it seems an adequate account. But such an account will not serve to justify the CEC Principle; further, on a functionalist account, standard 'ability' tests would not seem to measure abilities very well.

For example, on the functionalist account of abilities (and on any other adequate account) whether a child correctly answers a standard IQ test question like "What is the Koran?" or "What is the meaning of 'espionage' ('achromatic', 'imminent', 'dilatory')?" depends not only on his abilities, but also on whether he knows the answers. But children of equal intellectual capacities can differ in such knowledge because of differences in *access* to information due, e.g., to differences in parental vocabulary and knowledge, presence of books, magazines and newspapers in the home, and intellectual sophistication of teachers and playmates. Further, differences in parental encouragement of intellectual and scholastic activities, differences in role models, and other differences can cause differences in interest in intellectual activities and motivation to do well in school, and these in turn, can cause differences in knowledge. Finally, children with the *same* knowledge and intellectual skills differ in performances which are intended to tap knowledge and skills for reasons which have little to do with intellectual capacity, e.g., anxiety in the testing situation, testwiseness, desire to do well on the test, response to the tester, concentration, persistence, distractability, tendency to guess, tendency to give up on hard questions, tendency to check over answers, and even familiarity with paper and pencil.

In sum, if we understand abilities to be capacities, the CEC Principle and the factor analytic arguments for general intelligence which depend on it are clearly wrong. On the other hand, if by 'ability' we mean a set of performances the CEC Principle is correct, but it is wrong to suppose correlations among 'abilities' so called tell us anything about abilities or general intelligence in any ordinary understanding of the terms.

The claim that IQ tests measure general intelligence or ability in the fictionalist sense of 'ability' amounts to much the same thing as the claim that IQ test items correlate with one another in a way such that most of them correlate substantially with the test as a whole. But the fact that test items intercorrelate in this way is totally uninteresting, because IQ tests have simply been *cooked* to have this property. This is no secret. Terman and Merrill state flatly in the manual to the current Stanford-Binet (1960, p. 33) that "Tests that had low correlation with the total were dropped even though they were satisfactory in other respects." In general, high correlation with the rest of a test is a strong reason for an item to be included in a test, and low correlation is a conclusive reason for dropping an item.

Of course, psychometricians can use words (including 'ability' and 'intelligence') any way they like, so long as they do not mislead anyone. The trouble is that psychometric fictionalists *are* misleading us. They are playing a double game, or at least contributing to one. They use the fictionalist notion of ability to justify claims that the tests measure abilities or general intelligence but then the educational policy recommendations based on their claims often depend on the assumption that the tests measure abilities or intelligence in the ordinary *capacity* sense.

The claim that certain people are low in general intelligence or general ability yields different predictions and hence different educational policies depending on whether general intelligence is understood as a capacity or as a mere set of performances or dispositions to perform on a test. Consider two children with identical performances on tests, one of whom is highly motivated and has had all the advantages of private tutors and fine schools, while the other has had no opportunities, poor education, no reason to believe learning will get him anywhere, and no interest in intellectual things. Taking poor performances as an indication of low capacity may be reasonable in the former case, but not in the latter. It would be a mistake to base educational treatment of the two children on the supposition that their identical test performance shows they have equal capacity.

The double game I speak of is very clear in Jensen's writings. He often commits himself explicitly to fictionalism. For example, he says "We should not reify g as an entity of course, since it is only a hypothetical construct intended to explain covariation among tests. It is a hypothetical source of variance (individual differences) in test scores."[14]

Further, Jensen makes heavy use of the claim that certain current tests measure g to a large extent, a claim which is based squarely on the CEC Principle, and thus depends on fictionalism. Another reflection of Jensen's fictionalism is the line he takes with respect to the fact that the g loading of a test (the correlation of the test with g) depends on the other tests it is factor analyzed with. According to one factor analysis (with one group of tests) a given test may measure pure g, while according to another factor analysis, the test may measure g to a small degree. Jensen says

> We should think of g as a 'source' of individual differences in scores which is common to a number of different tests. As the tests change, the nature of g will also change, and a test which is loaded, say, .50 on g when factor analyzed among one set of tests may have a loading of .20 or .80, or some other value, when factor analyzed among other sets of tests. Also, a test which in one factor analysis, measures only g and nothing else, may show that it measures g and one or more other factors when factor analyzed in connection with a new set of tests. In other words, g gains its meaning from the tests which have it in common.[15]

Taken literally, Jensen commits himself to the view that the degree to which a test measures a person's g depends on what *other* tests the tester gives him and other people! But this seems crazy if g is anything like "a capacity for abstract reasoning and problem solving". If test A actually is a pure measure of abstract reasoning capacity, and I get a certain score on test A, then that score ought to indicate my abstract reasoning capacity regardless of what other tests are given to other people. To be fair to Jensen, he often identifies g not with the 'general factor' common to any arbitrary group of tests, but rather with the 'general factor' common to *the* tests regarded as standard tests of intelligence. But standard IQ tests are heavily constrained by accidental features of the original tests constructed by Alfred Binet around the turn of the century, and by the desire of subsequent test constructors to produce tests which were similar to and correlated highly with the Stanford-Binet, the successor of Binet's tests. One psychometrics text observes that in each version of the Stanford-Binet (1916, 1937, 1960) "continuity... was ensured by retaining in each version only those items that correlated satisfactorily with mental age in the preceding form."[16] Thus 'standard IQ tests' are a pretty arbitrary lot; different initial accidents would have produced different standard tests. But since according to Jensen, the nature of g depends on the set of tests which are factor analyzed together, different standard IQ tests might well have resulted in a 'different' g. But clearly, if g is understood to be a "capacity for abstract reasoning and problem solving," this is nonsense. A person's abstract reasoning capacity is not a function of accidents of test construction.

While much of what Jensen says seems to reflect fictionalism, many other remarks seem to reflect realism. For example:

The term "intelligence" should be reserved for the rather specific meaning I have assigned to it, namely the general factor common to standard tests of intelligence. Any one verbal definition of this factor is really inadequate, but, if we must define it in so many words, it is probably best thought of as a capacity for abstract reasoning and problem solving.
Intelligence fully meets the usual scientific criteria for being regarded as an aspect of *objective reality*, just as much as do *atoms, genes and electromagnetic fields*. Intelligence has indeed been singled out as especially important by the educational and occupational demands prevailing in all industrial societies, but it is nevertheless a *biological reality and not just a figment of social convention*...[17]

So Jensen appears to believe both that intelligence is a fiction and that it is as real as atoms and genes. Perhaps the resolution of the conflict is that Jensen takes a fictional view of *all* 'theoretical entities' including those of biology and physics. Whatever the resolution of the conflict, the effect of Jensen's peculiar views is to allow him to play (obviously not knowingly) a double game. On the one hand, he recommends that education for lower class lower IQ children make use of memory learning rather than of conceptual learning. This recommendation is based squarely on the claim that lower class lower IQ children are deficient not in a fiction but in g or general intelligence, "a capacity for abstract reasoning and problem solving." But the arguments which show IQ tests substantially measure g *depend* on understanding g as a fiction, a set of performances or dispositions to perform on tests, and not a capacity in any ordinary sense. Fictionalism, like operationism, facilitates the illegitimate transfer of the social and educational role of ordinary concepts like intelligence to their peculiar 'scientific' counterparts.

Massachusetts Institute of Technology

NOTES

* I would like to thank the following persons for their comments on earlier versions of this material: Richard Boyd, Susan Carey, L. J. Cronbach, Gerald Dworkin, Jerry Fodor, Paul Horwich, John Loehlin, Tom Nagel, Hilary Putnam, and Tim Scanlon. Earlier versions of much of this material were included in my contribution to Block and Dworkin (1974).

[1] There are good reasons for thinking operational definitions of 'intelligence' as what IQ tests measure are unsatisfactory. See Block and Dworkin (1974a) and Daniels (1975).

[2] Hartry Field has developed the notion of 'partial denotation' to describe the relation which Newton's uses of 'mass' had to rest mass and relativistic mass. See Field (1973).

[3] Psychometricians (including Jensen) often argue that IQ tests can be seen to measure intelligence in virtue of their measuring g. But it is hard to take this argument seriously, since the most Jensen or any other competent psychometrician claims is that g accounts for about 50% of the variance (see note 4) in standard IQ tests. The procedures which ascribe 50% of the variance in IQ to g ascribe much more of the variance in other tests (Raven's Progressive Matrices, for example) to g. Tests more highly loaded on g, however, are not as useful for prediction of school and job success as is IQ.

[4] The variance of a quantity is a measure of the variation in that quantity in a population i.e., the 'spread' of the distribution of the quantity. The variance of x = the average of: $(x - \text{the mean of } x)^2$

[5] The correlation between x and y, $r_{xy} =$

$$\frac{\text{Average }[(x - \text{the mean of } x)(y - \text{the mean of } y)]}{\sqrt{\text{Variance of } x \cdot \text{Variance of } y}}$$

I shall use 'a' to refer both to test a and to the ability the test putatively measures.

For an interesting misuse of the CEC Principle (even by psychometric standards) to attempt to show IQ tests are pure measures of g see Urbach (1974) and my reply (Block, forthcoming).

[6] Assuming linearity. If the correlation between a and c is due totally to the causal effect of c on a, c accounts for a portion of the variance in a equal to $(r_{ac})^2$.

[7] I am indebted to Richard Boyd here.

[8] Though many do not. Guilford (1967) examined over 7000 correlations among (supposedly) intellectual tests and found 24% of them to be close to zero.

[9] Wechsler (1958); Cronbach (1970a), Anastasi (1968) and Tyler (1965).

[10] The latter version of soft fictionalism might be termed 'semantic fictionalism' since it is a view about the meaning of theoretical terms. The former version (electrons should be identified with cloud chamber tracks, etc.) might be termed 'ontological fictionalism'.

[11] See Smart (1963) and Maxwell (1962) for good critical discussions of fictionalism. See Putnam (1965) and Fodor (1968) for excellent critiques of logical behaviorism. My view is that abilities are *functional* states: states characterized in terms of their causal relations to sensations, behavior and *other mental phenomena*. What makes functionalism incompatible

with behaviorism is its causal nature and its characterization of mental phenomena partly in terms of other mental phenomena. For discussions of functionalism and its relation to behaviorism and materialism see Putnam (1967), Fodor (1968) and Block and Fodor (1972).

[12] Cronbach (1970), p. 310–311. Emphasis added.
[13] Because it takes abilities to be causes of behavior, this dispositional version of functionalism is not a fictionalist account.
[14] Jensen (1972), p. 77.
[15] *Ibid.* p. 79.
[16] Anastasi (1968), p. 204.
[17] *Ibid.* p. 88. Emphasis added.

BIBLIOGRAPHY

Anastasi, A.: 1938, 'Faculties Versus Factors: A Reply to Professor Thurstone', *Psychological Bulletin* **35**, 391–395.

Anastasi, A.: 1968, *Psychological Testing*, Macmillan, New York.

Block, N.: forthcoming, 'A Test Case for the Methodology of Scientific Research Programmes'.

Block, N. and Dworkin, G.: 1974a, 'IQ Heritability and Inequality, Part I', *Philosophy and Public Affairs* **3**, 331–409.

Block, N. and Dworkin, G.: 1974b, 'IQ, Heritability and Inequality, Part II', *Philosophy and Public Affairs* **4**, 40–99.

Block, N. and Fodor, J.: 1972, 'What Psychological States are Not', *Philosophical Review* **81**, 159–181.

Cronbach, L. J.: 1970a, *Essentials of Psychological Testing*, Harper and Row, New York.

Cronbach, L. J.: 1970b, 'Test Validation' in R. L. Thorndike (ed.), *Educational Measurement*, Washington D.B., pp. 443–507.

Daniels, N.: 1975, 'IQ, Heritability, and Human Nature', this volume, p. 143.

Field, H.: 1973, 'Theory Change and the Indeterminacy of Reference', *Journal of Philosophy* **70**, 462–487.

Fodor, J. A.: 1968, *Psychological Explanation*, Random House, N.Y.

Guilford, J.: 1967, *The Nature of Human Intelligence*, McGraw Hill, New York.

Jensen, A.: 1972, *Genetics and Education*, Harper and Row, New York.

Maxwell, G.: 1962, 'The Ontological Status of Theoretical Entities', *Minnesota Studies in the Philosophy of Science*, Vol. 3 (ed. by H. Feigl and G. Maxwell), University of Minnesota Press, Minneapolis, pp. 3–28.

Putnam, H.: 1965, 'Brains and Behavior', *Analytical Philosophy*, Vol. II (ed. by R. J. Butler), Blackwell, Oxford, pp. 1–20.

Putnam, H.: 1967, 'Psychological Predicates', in W. Capitan and D. Merrill (eds.), *Art, Mind and Religion*, U. of Pittsburgh Press.

Smart, J. J.: 1963, *Philosophy and Scientific Realism*, London, Routledge.

Thomson, G. H.: 1939, *The Factorial Analysis of Human Ability*, University of London Press, London.

Tyler, L.: 1965, *The Psychology of Human Differences*, Appleton Century Crofts, New York.

Urbach, P.: 1974, 'Progress and Degeneration in the "IQ Debate" (1)', *British Journal for the Philosophy of Science* **25**, 99–135.

Vernon, P. E.: 1961, *The Structure of Human Abilities*, London.

Wechsler, D.: 1958, *The Measurement and Appraisal of Adult Intelligence*, Williams and Wilkins, Baltimore.

NORMAN DANIELS

IQ, HERITABILITY, AND HUMAN NATURE*

I

Scientific revolutions are no everyday affair. So it is of some importance that *Fortune* (Alexander, 1972) and several other major magazines have recently proclaimed that we are in the midst of a major Kuhnian revolution in the social sciences, one that has significant implications for social policy. According to these magazines, 'environmentalist' theories, which assume that equalization of human environments and opportunities will increase equality of achievement between individuals, groups, and races, are in 'crisis'. The crisis exists because the egalitarian reform programs of the 1960's, which relied on such theories and thus constituted tests of them, failed to equalize achievement. In the face of this crisis, *Fortune* suggests, scientists are welcoming evidence from the study of ducks, baboons, and humans which points to "a basic intractability in human nature, a resistance to being guided and molded for improving society" (Alexander, 1972, p. 132). But if human nature is intractable, the argument continues, then social policy should be adjusted to recognize the inherited capacities and differences between individuals, groups and races, rather than continue to insist on unrealistic egalitarian reform programs.

Leaving aside the ducks and baboons, the most famous human nature theory is the one advanced by Arthur Jensen (1969, 1972, 1973) and Richard Herrnstein (1971, 1973). Their IQ Argument, as I shall call it, is a human nature theory for two reasons. First, it attempts to explain existing features of human behavior and social structure mainly in terms of genetically determined biological or psychological traits. Second, it makes predictions about what can and should be done given this "intractability" in human nature. The IQ Argument can be stated as follows (cf. Daniels, 1973, for a similar statement of the argument):

(1) IQ measures intelligence, a central behavioral trait.
(2) IQ is an important causal determinant of success in school and life.
(3) IQ is highly heritable (80%) in the white population.

(4) Mean Black/White (15 point) and social class IQ differences are large.
(5) Therefore, (a) IQ is resistant to change;
  (b) Inequalities in IQ are not eliminable;
  (c) Black/White and social class mean IQ differences are probably genetic in origin;[1]
  (d) Attempts to achieve greater equality in a variety of social contexts, e.g., in education, where IQ is important, are unrealistic.

In Section III I will document Jensen's and Herrnstein's inferences of (a)–(d) from (1)–(4).

Recent literature has attacked various parts of the IQ Argument. Leon Kamin (1974) has convincingly shown that the main data supporting the heritability estimate in premise (3) is methodologically flawed. For example, the famous twin studies are marred by such fatal methodological flaws as: (i) failure to administer uniform tests; (ii) failure to select twins representative of the range of environments in society as a whole; (iii) failure to find twins raised in genuinely different environments, (iv) failure to control for age and sex similarities that affect IQ variance in twins; (v) and failure to control for experimentor bias of the most obvious and gross kinds.[2] Each of (i)–(iv) acted to inflate heritability estimates. Even Jensen (1974) has recently been forced to admit that the Burt data is worthless for theoretical purposes. In short, there is no significant evidence to support (3) and in my view (1973; Cronin, Daniels et al., 1975) heritability of IQ is most likely insignificant, and may not even be measurable, as Layzer (1974) has argued.

Others like Bowles and Gintis (1972–3) and Block and Dworkin (1974) have attacked premise (2), the claim that IQ is causally important to success. The IQ-success correlations can be explained without claiming IQ causes success. Thus, whatever IQ-success correlations exist are either built into the test by the method of standardization, as Cronin, Daniels, et al. (1975), and Block and Dworkin (1974) argue, or they are largely artifactual, as Block and Dworkin suggest, resulting from 'score causation'. That is, IQ scores influence teacher expectations and the assignments of students to 'tracks', creating increased likelihood of greater school achievement. One of Bowles' and Gintis' most telling points is that IQ is a notoriously poor predictor of success at any given kind of job.

This fact seems to undercut the claim that IQ is a causally necessary condition for reaching that job level. Premise (2) seems to be false, on any serious interpretation of the causal claim involved. If these two fairly straightforward empirical premises, (2) and (3), are false, then there is already little reason to believe (5).

In this paper, I want to further reduce support for the human nature conclusion in (5). My argument falls into two parts. In Section II, I want to challenge the remaining, and in my view most important, premise of the IQ Argument, the claim that IQ measures intelligence. This premise is especially important because it is widely believed by both hereditarians and environmentalists alike, with unfortunate consequences for educational theory, as I argue in Section IV. I begin by looking briefly at sample IQ test items to establish what kind of performance is required by the tests. Then I shall argue that the main validation arguments for IQ tests fail to give us any reason to think that tests requiring such performances measure intelligence or thinking capacity. Next, I show that class and race biases of the early test developers steered them into answering the wrong question when they began to investigate human intelligence, and, finally, I suggest a different strategy for the study of intelligence. If my argument is right, then all three main premises of the IQ Argument are false and we have no reason to believe conclusions like those in (5).

In Section III, however, I want to argue for a stronger claim: not only are the premises false, but the IQ Argument is invalid. I want to show that human nature arguments of the type represented by the IQ Argument, that is, those relying on heritability estimates, cannot support strong conclusions such as those in (5). The concept of heritability simply lacks sufficient explanatory and predictive power to provide a basis for such strong conclusions. Heritability estimates fail to provide explanatory power because they fail to inform us of the functional relationship between genotype, environment, and phenotype, and because they fail to tell us about the causal mechanisms involved. These failures at the explanatory level strip the concept of the needed predictive power for the human nature conclusions. If I am right, then we ought to be wary of any human nature theory which rests its deductive weight on estimates of heritability. In short, the IQ Argument, which is advanced as the best established or paradigm case of the revolution in the social sciences not only has false premises, but rests on a concept which lacks the conceptual power to support its human nature conclusions.

II

For the Jensen-Herrnstein human nature theory to get off the ground, IQ must be a valid measure of intelligence. Is there any reason to think that IQ tests measure intelligence, construed as thinking ability or cognitive capacity? It may help to begin by looking at the type or performance required by IQ test items. Perhaps the nature of the test items will reveal why test developers think they are measuring intelligence or thinking ability.

(A) *The Performance Required by IQ Test Items*

Though many of us have taken IQ tests, it often comes as something of a surprise to be reminded what IQ test items are like. Since we are often told that the IQ test measures general intelligence or general thinking ability or general problem solving ability, it is surprising to find that the test items themselves seem only to tap specific bits of knowledge, specific skills, specific values and perhaps motivations. For example, the best subtest predictor of school success on the Stanford-Binet test is a long list of vocabulary words. Included are words like 'lotus', 'mosaic', 'stave', 'bewail', 'ochre', and 'ambergris'. A vocabulary rich in words such as these may indicate that one has well-educated parents who regularly employ such vocabulary or that one has spent many hours reading English literature, not just newspapers or comic books. But why should we interpret possession of such specific bits of knowledge represented by these skills as a sign of intelligence?

Other subtests seem to tap other specific bits of knowledge. For example, on a Stanford-Binet 'Comprehension' subtest given to seven and eight year olds, a child is asked, "What's the thing for you to do when you are on your way to school and notice that you are in danger of being late?" The guidelines say that "only those responses which suggest hurrying are acceptable." Thus correct responses include "Hurry", "Go right ahead to school" and "Take the street bus." Incorrect responses include "Go on to school and tell my teacher why I'm late," "Not stop," "Just keep on going," or "Get a late card." Is a specific attitude toward punctuality a sign of intelligence? A WISC Comprehension Subtest asks, "Why are criminals locked up?" and "Why do we elect (or need to have) senators and congressmen?" Correct answers include "Criminals need to be segregated from society for the protection of society," and "Electing senators makes government responsible to the people." Wrong answers include, "Criminals should be locked up because they get into trouble," and "Senators help control people in the U.S." A Stanford-Binet (1960) "Picture Identification Test" given to very young children consists of paired pictures of a prim-looking woman with WASP features and a slightly unkempt woman with Negroid features. The child is asked "Which is prettier?" and must answer that the WASP woman is. Why think that absorption of culturally standard and biased criteria or prettiness is a sign of intelligence? (See Cronin, Daniels *et al.*, 1975, for a discussion of the value biases of test items.)

There are some subtests which purport to tap the more abstract, problem solving and thinking abilities we are often told IQ tests are about. But even these seemingly 'culture free' subtests, like the digit span tests or puzzle tests, are highly sensitive to cultural and psychological variables. For example, the digit span test is sometimes used by psychologists to test for anxiety. Moreover, puzzles and puzzle-based problems may be more familiar to some children than to others. Even Jensen (1969, p. 100) admits that children allowed to *play* with items similar to IQ test items for a while before taking the tests showed significant IQ boosts over control groups, though it is unclear how much of the IQ data is flawed in this way.

Two points seem to be obvious about these test items and the kind of performance asked for by IQ tests. First, the content of the items gives us little or no *prima facie* reason to think that they are tapping general cognitive abilities or general intelligence rather than specific bits of information, specific skills, values and attitudes. Second, there is plenty of room for class and cultural biases to enter, since children in society probably experience systematically different learning environments with regard to the kinds of knowledge, values and attitudes required for the test.

(B) *The Validation of IQ*

Setting aside the question of possible bias in the selection of items, is there any reason to think that performance levels on tests with these kinds of items measure intelligence? IQ tests are supposed to be ability, not achievement, tests. What does acquisition of the kinds of knowledge, skills, values and attitudes the test items are about have to do with general thinking ability or intelligence? Part of the problem here is that the word 'ability' is systematically ambiguous. It can mean (a) current level of performance; or it can mean (b) capacity to acquire certain levels of performance. If the IQ test measures *only* current performance on an achievement test, and does not measure underlying capacity, then an IQ score is not useful for the purposes to which IQ tests are put. IQ scores are used to explain why an individual exhibits a given level of performance – that is all his capacity or intelligence allows – and are used to predict what achievement levels an individual will reach in school or life. But if IQ tests are supposed to tell us about capacity, then we are owed some explanation or proof that tests measuring current performance levels are also measuring capacity.

There are two main IQ validation arguments in the literature (Daniels, 1973, p. 26). The first such argument assumes that precocity is a measure of capacity. It rests on the view that the more intelligent a child, the quicker he is to assimilate what is in his environment, the various bits of knowledge, skills, values, preferred attitudes, and so on. Therefore, the content of the particular test items is of secondary importance; what counts is how a child stands relative to others of his own age with regard to the total amount of knowledge absorbed. This idea lay behind the early scoring method for these tests in which the Intelligence Quotient was a quotient of mental and chronological ages.

Unfortunately, there are several untested and unlikely assumptions behind this precocity argument. First, the argument assumes that the rate of development of intelligence is a monotonic function of degree of intelligence, that is, the greater the basic intelligence, the faster it develops. Even if this first unlikely assumption were true, the argument also assumes that individual children are raised in randomly, not systematically, different environments. We *know* this assumption is false since the environments of children distinguished by race and social class are not randomly different with regard to the kinds of test items on the IQ test. Finally, the argument also assumes that precocity in performance, even with randomly different backgrounds, is a function of capacity and not of motivation. This assumption we also know to be false.

The main IQ validation argument, however, is based on the correlation of IQ with success in school and life.[3]

This argument can be characterized as follows:

(a) Success is a measure of (correlates ___ with) intelligence.
(b) IQ correlates .5 with success in school or life.[4]
(c) Therefore, IQ is a measure of (correlates ___ with) intelligence.
The argument formalizes presuppositions behind the wisecrack, "If you're so smart, how come you're not rich?"

There are glaring problems with the argument. First, there is no independent measure of intelligence besides IQ, so we have no way of knowing exactly how well intelligence and success are correlated in (a). For the sake of argument, let us suppose that intelligence and success are perfectly correlated (1.0) in (a). Though correlations are not in general transitive, in the case of a 1.0 correlation in premise (a), we can infer that IQ and intelligence are .5 correlated in (c). Unfortunately for the argument, a .5 correlation of IQ and intelligence is *too small* to count as a measure since IQ would then account for at most 25% of the variation in intelligence. Imagine relying on a thermometer whose variations accounted for only 25% of the variation in temperature. So, *assuming* a perfect intelligence-success correlation in (a) forces us to conclude IQ does not measure intelligence.

Let us try a different tack. Since assuming a perfect success-intelligence correlation was incompatible with IQ measuring intelligence, let us assume a more moderate correlation in premise (a). For example, a .5 correlation in (a) would be *compatible* with a perfect IQ-intelligence correlation in (c). Unfortunately, although (a) and (b) are compatible with IQ measuring intelligence, taken as premises of an argument, they give absolutely no support for that conclusions. Since correlations are not transitive, the correlations between success and intelligence in (a) and IQ and success in (b) could both be .5 and the correlation between IQ and intelligence (c) could still be 0.0. One might think the correlation argument would gain plausibility, if not validity, were it possible to show in premise (b) that the *only* relevant thing correlating moderately well with IQ was success. But this is not the case. IQ correlates about as well with parent social class background as it does with future success of the child. Accordingly, it would be equally plausible (but no more valid) to conclude IQ must *measure* social class background. There is a difference between saying what I believe is true, that IQ tests are biased in favor of the upper classes, and saying what is not strictly true, that IQ tests *measure* class background.

The correlation argument is worthless. Maximizing the correlation between success and intelligence, so that (a) and (b) can carry weight as premises of an argument, entails that IQ cannot measure intelligence. On the other hand, *supposing* a correlation between success and intelligence weak enough to allow IQ to measure intelligence makes the premises (a) and (b) too weak to support the conclusion that IQ measures intelligence. In short, either IQ is not a measure of intelligence, or the correlational argument is invalid and we have no reason to believe that IQ measures intelligence.

But if neither IQ validation argument gives us any reason to believe IQ measures intelligence, how are we to explain the moderately good correlation between IQ and success in school and life? I think we should not be unduly impressed with the correlation. In part, the correlation is artifactual,[5] resulting from such phenomena as 'score-causation', and so cannot be used to shore up validation arguments. And, in part, the correlation is built into the test through the procedure for selecting items. Items were selected for IQ tests by Binet because they served to distinguish children whose teachers judged to be 'bright' or 'dull', i.e., *children who were already doing well in the class-biased French school system of 1905.*[6] Good test items were those a good student got right and a poor student got wrong. Items a poor student got right and a good student got wrong would be excluded. The crucial assumption here is that what makes good students do well in teachers' eyes is their intelligence – rather than the many other factors that might contribute toward pleasing a teacher. This assumption systematically ignores the effect of race or class biases in schooling, e.g., in teachers' expectations. Later test developers continued to select and standardize items in essentially the same way: items had to correlate well with the results of earlier IQ tests and were refined by including factors like level of education and social position among the 'intuitive' measures of intelligence with which test items had to correlate.

Through this procedure, IQ tests most likely came to acquire their class and race bias. The procedure made it possible to construct a test which tapped a variety of information, skills, values, and attitudes which are more likely to be found in upper or upper-middle class child rearing environments. Moreover, it was able to do so statistically, without requiring any explicit theory of what test items had to be about. The only requirement of test items was that they would have to sort people in the appropriate way.

That culture or class determined features of child rearing environments may have a lot to do with determining IQ is confirmed by some of the recent IQ boosting intervention studies. One of the most interesting is Phyllis Levenstein's study (1970, 1971; cf., also Bronfenbrenner, 1974) in which selected lower working class mothers and their two years old children were visited twice weekly for thirty minutes, provided with educational toys and books, and coached on how to interact with their children. Children's IQ's showed a sustained rise of about 15 points.

Putting the various parts of my argument together, it seems likely that IQ tests measure the acquisition of specific bits of information, attitudes, values, motivations, and possibly specific cognitive skills which are more likely to be acquired in upper class child rearing environments than in lower ones. In this sense, the IQ test is clearly class and race biased. At the same time there is no reason at all to think the performance the test has anything to do with intelligence, at least if we view intelligence in the way many IQ promotors view it, as a general, underlying cognitive capacity.[7]

(C) *Two Strategies for Investigating Intelligence*

There is a cruel irony in the history of the IQ testing movement. In my own view, the IQ test developers and promoters have failed to prove the validity of IQ tests because they set out to solve the wrong problem and came up with the wrong measuring instrument. I believe they failed because they were committed to two key assumptions about the nature of intelligence, the second of which is most likely false and the first we have no reason to think true. The early test developers believed (i) that intelligence was normally distributed in the human population, and (ii) that it correlated with 'eminence' in society. These assumptions determined their task: find a measuring instrument which matched these expectations about the distribution of intelligence and one would have an intelligence test. Though early test developers were often quite explicit about both assumptions, as we shall see, later psychometricians have generally affirmed only the first. But, because test standardization procedures are as I have described them, the second assumption has been implicitly retained as well.

Both assumptions about the distribution of intelligence were the result of an elitist, and sometimes explicitly racist, ideology about human differences and human social structure. The ideology was shared by almost all of the IQ test developers and promotors. Lewis Terman,

American developer of the Binet test, felt confident to say in 1916, long before there could be any data to back him up:

> Their dullness seems to be racial, or at least inherent in the family stocks from which they come. The fact that one meets this type with such extraordinary frequency among Indians, Mexicans, and Negroes suggests quite forcibly that the whole question of racial differences in mental traits will have to be taken up anew. The writer predicts that when this is done, there will be discovered enormously significant racial differences in intelligence, which cannot be wiped out by any scheme of mental culture. (Terman, 1916, Ch. I).

Similarly, Robert Yerkes (1921, p. 790–91), who made mass mental testing a reality in World War I wrote, "Quite apart from educational status, which is totally unsatisfactory, the Negro soldier is of relatively low grade intelligence. Education along will not place the Negro race on a par with its Caucasian competitors." And James McKeen Cattell (1947, p. 155) remarked, "The main lines are laid down by heredity – a man is born a man and not an ape. A savage brought up in a cultivated society will not only retain his dark skin, but is likely to have also the incoherent mind of his race." Expressed more in class terms than race terms is Henry Goddard's plea,

> Now the fact is, *that workman* may have a ten year intelligence while you have a twenty. To demand for him such a home as you enjoy is as absurd as it would be to insist that every laborer should receive a graduate fellowship. How can there be such a thing as social equality with this wide range of mental capacity? The different levels of intelligence have different interests and require different treatment to make them happy... (Cited in Kamin, 1974, Ch. I).

These views were not just a background ideology which the test developers happened to have. They functioned, in the form of our assumptions (i) and (ii) above, as criteria determining the acceptance of a test as an intelligence test. Early mental testers, like Francis Galton, Cattell, R. M. Bache, and B. R. Stetson, all attempted tests of various information processing abilities, like memory span, quickness of reaction, and other abilities one might have intuitively thought would be related to intelligence. Since these tests failed to produce the appropriate match with assumption (ii), class or race correlations with performance either being absent or going in the opposite direction to their expectations, the test developers threw out the tests, not the assumptions. They concluded they were not, after all, looking at factors which influenced intelligence. Sometimes the tester bent over backwards to save his expectations. Thus Bache (1895, p. 474–86), finding that blacks and Indians had faster reaction times than whites on his tests, concluded white subjects' "reactions were slower because they belonged to a more deliberate and reflective race." Adherence to the first assumption, that intelligence is normally distributed, is apparent today, though sometimes people mistake the assumption for an empirical fact revealed to us by IQ tests. The belief that the study of human intelligence is of necessity the study of human differences in intelligence remains unchallenged. Carl Bereiter (1970, p. 291) remarks, "One cannot expect to discover the sources of individual differences [in the development of intelligence] by investigating those developmental changes that are universal." Bereiter is assuming that the differences in IQ we see between individuals are differences in intelligence, and that there are wide variations in human intelligence. Even on his own terms, though, it is hard to see why understanding the 'universal' course of development of intelligence would tell us nothing about the sources of variations in intelligence.

I believe that the assumptions (i) and (ii), deriving as they do from an elitist and racist social theory, have seriously misled the investigation of human intelligence. What IQ testers have done is substitute the search for performance differences which they could label 'intelligence' differences for an inquiry into what intelligence is. The irony I mentioned earlier is that this substitution makes it impossible to show that the tests measuring the selected peformance differences really are measures of intelligence. There is, however, another approach for inquiring into human intelligence, one that requires abandoning, at least as initial premises, assumptions (i) and (ii).

In my own view, a more fruitful strategy for studying intelligence would begin by treating intelligence as a general capacity measure for information processing systems, whether organisms or machines. It would be a comprehensive capacity measure, probably reflecting obvious variables like memory capacity, range of retrieval strategies, quickness of processing, availability of problem solving strategies, and so on. Perhaps my suggestion is best viewed as the claim that human intelligence is a feature of the normal functioning of the human brain. That is, it is a species-typic capacity measure rather than an individual capacity measure. Specifically, I am rejecting as a starting point the assumption underlying IQ test construction that intelligence is normally distributed among individuals and has a significant variance. My view gains plausibility when it is remembered that many mental capacities and abilities, which are presumably related to intelligence when it is viewed as a capacity measure, are not normally distributed. Like the abilities to acquire spoken and written language, to exercise memory, to coordinate tactual and visual sensory inputs, they are part of the normal functioning of humans. Deficiencies in such abilities are associated with specific disorders, such as brain damage, and are not normally distributed. Similarly, differences in the normal range of human performance on tests measuring related skills cannot be assumed to reflect capacity differences rather than other learning and developmental differences, at least not without substantial argument to back up such an assumption.

Approaching intelligence in this way, tentatively assuming that humans, except in cases of serious brain malfunction, have roughly equal information processing machines in their heads, might make is possible for us to sort out the sources of the performance differences we observe in people. Performance differences, like those seen on IQ tests, might then more readily be seen as the specific effects of learning and personality inputs or histories. The incentive would be to investigate learning and personality more thoroughly, rather than chalking up performance differences to fixed differences in individual capacity, as we have seen educators all too ready to do. It is possible to suggest a general criterion to guide the development of learning theory and protect it from the nativist impulse, acute in the hereditarians we are concerned with, to abandon learning theory in favor of positing native capacity differences. Suppose that we develop complex learning and personality theories which made it possible to account for much of what happens in human learning, including the effects of a variety of personality variables. Suppose further that we can not explain a residue of differences in achievement between individuals. In the context of such a learning theory, we might then suspect that our assumption of basically equal capacity or intelligence was wrong. But, rejecting the equal capacity hypothesis before we have anything like such working learning and personality theories would be giving up too soon. It is all too easy to turn in frustration from the task of developing adequate learning theories and to postulate various built-in or innate features of the human mind. But if we resist this temptation, and also reject the elitist and racist social theory that motivated early test developers, then it seems more reasonable to begin our theory of human intelligence and learning without assumptions (i) and (ii).

As I have just pointed out, my view of how to study intelligence, with its rejection of assumptions (i) and (ii), is not just an article of faith. It has sound advantages in that it helps us to avoid the ironical dilemma of the IQ tester who cannot convince us that the performance differences he measures have any bearing on intelligence. It has the advantage of putting research effort on the more approachable components of any cognitive theory, the personality and learning components, without begging questions about what the course of performance differences may be. Nor does my suggestion make the assumption of roughly equal intelligence an unrevisable proposition. We could learn that our assumption was wrong, as I have pointed out. Moreover, my view leaves open the possibility that some specific cognitive abilities may also differ at the level of capacity and not just performance, for example, the ability to internally image things. My suggestion has another intention behind it besides the failure of the opposing strategy of the psychometrician. It is motivated by the belief that our ordinary ascriptions of degrees of intelligence to different people fudge together many quite distinct factors which it is the task of a good theory of cognitive information processing to separate and refine. In the framework of such a theory, following my strategy, we might still discover variations between people which we would want to attribute to intelligence. But until we have such a working theory, it is better to refrain from positing intelligence differences as an explanation for so many different, complex kinds of performance differences.

III

If IQ does not measure intelligence, as I believe I have shown, then the IQ Argument is false in all three of its main premises. Moreover, I have argued that the whole approach of looking at intelligence in the way IQ theorists do not only rests on racist and class biased assumptions dating to the early part of the century, but contains a serious methodological flaw of some philosophical interest. It amounts to an acute form of nativism which posits capacity differences where there is only evidence for performance differences. This nativism undercuts the search for adequate human learning and educational theories. I would now like to turn from the concepts of IQ and intelligence to the other main feature of the IQ Argument, its reliance on the concept of heritability.

I shall argue that heritability estimates, such as the claim that IQ is 80% heritable in premise (3) of the IQ Argument, lack the explanatory, and consequently the predictive, power needed to support human nature conclusions of the type expressed in 5 (a)–(d) above. First, I shall discuss briefly the general functional relationship that holds between phenotype, genotype, and environment, a relationship called the 'norm of reaction'. It will then be possible to see that heritability estimates in general contain no information about these functional relationships and so cannot be used to support claims about the fixedness of IQ or IQ differences between individuals or groups. Finally, I will argue that heritability estimates tell us virtually nothing about the mechanisms underlying phenotypic expression and that the inference, 'P is highly heritable, therefore P has a genetic basis', which Jensen and Herrnstein both make in the case of IQ, is straightforwardly invalid. This failure to tell us about causal mechanisms further undercuts the predictive power of heritability estimates. These limitations of heritability estimates are not just a feature of its application to IQ, however, and so my argument here is a general one. It warns against the limitations facing other human nature theories which attempt heritability estimates of traits such as aggressiveness or criminality, and so on.

(A) *Norms of Reaction*[8]

The 'norm of reaction' is a function which represents a phenotypic trait as a product of genotype and environment. Ascertaining the norm of reaction for a phenotype is a tricky experimental problem that requires securing a pool of genotypically identical individuals and raising them in thoroughly controlled environments that differ with regard to a well defined environmental variable. For example, suppose we had a large number of kernels of a pure strain of corn, one highly inbred to ensure (near) genotypic identity. Suppose further that we raised batches of these kernels in a range of controlled growing environments. Each environment differs from the others in containing a different per cent nitrogen in the growth medium, but is identical in every other way. We could measure the effect of environmental variation in nitrogen on a phenotypic trait, such as the height of the corn at flowering, by plotting the mean height for each batch grown in each distinct environment. Suppose we carried out such experiments for several corn strains, G_1–G_5, and plotted the results of our hypothetical research on reaction norms for corn height as shown in Figure 1. Though Figure 1 is purely hypothetical, it is not atypical of the complex gene-environment interactions scientists find in real experiments.[9]

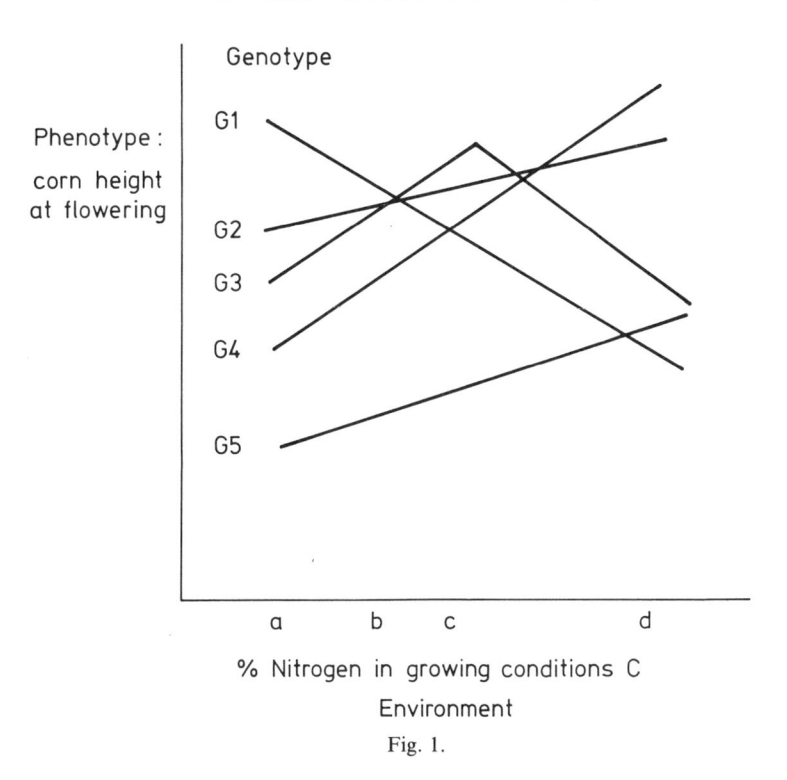

HYPOTHETICAL NORMS OF REACTION

Fig. 1.

Before explaining the relationship between the kind of functional knowledge contained in norms of reaction and the non-functional information contained in heritability estimates, it will be instructive to use our hypothetical experiments to help expose several common fallacies in thinking about the relationship between phenotype, genotype, and environment.[10] One point that should be clear from looking at Figure 1 is that the value of a phenotypic variable is always the joint product of genotype and environment acting on each other. What that means is that we cannot look at a particular value for a phenotypic traint, like height or IQ, as measured in an individual, and try to divide it into a portion due to the genes and a portion due to the environment. We cannot look at a particular corn plant and say, "Most of its height is due to its genes, though there is some environmentally produced height as well." We might call this the Fallacy of Causal Division. Unfortunately, history is full of attempts to make such a division. Lewis Terman's remark (1916) cited earlier, that the 'dullness' of a Mexican or a black is primarily a result of his family stock, seems to exemplify the fallacy of dividing the number of books printed in an hour into a number produced by the typesetter and a number produced by the press operator. Neither one, taken separately, produces any books. Neither do genes cause so many inches or so many IQ points of an individual's total height or IQ.

Knowledge of the complexity of norms of reaction also undercuts two other common fallacies in thinking about the influence of genes on phenotype. It is sometimes thought that if a phenotypic trait has a 'genetic basis', then the trait has somehow a fixed value that one can do little to alter. This Fallacy of Fixedness, as it might be called, ignores a point that is clear in Figure 1. Whenever it makes sense to speak of a 'genetic basis' for a phenotype, it also makes sense to speak of its 'environmental basis'. If one is unduly impressed by the large height differences between G_1 and G_5 at nitrogen concentration (a), one might be tempted to emphasize the 'genetic' basis of corn height, and, further, to think that height is a fixed inheritance. Different strains of corn have different inheritances. But at nitrogen concentration (d), the norms of reaction for G_1 and G_5 coincide, and this fact should lead us to see the error in the Fallacy of Fixedness.

There is always an environmental basis to phenotypic expression and so traits are not fixed, inherited values. An example from human populations also makes the error clear. Phenylketonuria (PKU) retardation is often said to have a 'genetic' basis because a particular gene form causes (in certain developmental environments) an enzyme deficiency impeding metabolism of phenylalanine. But PKU retardation also has an environmental basis, the presence of phenylalanine in the diet. Eliminate PKU from the diet and the retardation will not develop. Often we are tempted to commit the Fallacy of Fixedness because we know too little about the mechanisms underlying gene-environment interaction and do not even have a clear picture of how to define the relevant environment, let alone undertake efforts to alter it in ways that would remind us that there is always an environmental basis for traits as well as a genetic bais.

The Fallacy of Fixedness must be very tempting, because both Jensen and Herrnstein warn their readers against committing it, and yet slip into remarks that seem to embody it. Jensen, commenting on what he claims is the failure of certain compensatory education programs to boost IQ, seems to slide into the Fallacy when he remarks (1972, p. 192–193): "The techniques for raising intelligence per se, in the sense of g, probably lie more in the province of biological sciences [eugenics?] than in psychology and education." Herrnstein (1973, p. 198) commits the Fallacy in a similar context. He says that compensatory education failed because it "tried to raise IQ's, which are more a matter of inheritance than environment, and therefore not very amenable to corrective training." We have already seen that some IQ boosting programs have, in fact, been rather successful, producing significant, sustained jumps in IQ; not only is the Fallacy in general wrong, but it is explicitly misleading here in view of these results.

A third common fallacy is the Fallacy of Potential. Consider again our hypothetical plotting of norms of reaction in the low nitrogen range (a). One might be tempted to say, "The mean corn height differences for the different genotypes indicate a strong genetic influence. What this means is that genotype, even though it does not *fix* phenotypic value, at least determines the potential for phenotypic expression. Presence of the right kind of environment determines whether or not potential will be reached."

The problem with this way of retreating from the Fallacy of Fixedness to a claim about 'potential' is that it embodies essentially the same mistake. There is no determinate or fixed 'potential' associated with a given genotype in abstraction from the environment any more than there is a fixed value of the phenotype. 'Potential' phenotypic value must be viewed as a function of *both* genes and environment; the intuition behind the Fallacy of Potential is to view potential as a function of the genes alone, with environment acting only to determine whether or not potential is reached. I believe we often encounter this Fallacy in discussions of IQ. For example, it is often said that a 'disadvantaged' child is not living up to the potential set by its IQ genotype because his environment prevents full realization of the potential. The silliness of positing such a determinate 'potential' is revealed when we apply the notion to the environmentally advantaged child. Are we to assume he will surpass his potential?

Having used Figure 1 in a general way to expose some common fallacies, I would now like to look more specifically at the relation between norms of reaction and our ability to make generalizations and predictions about the relative performance of different genotypes. One point we can see from Figure 1 is that it makes no sense to ask in general, "Which genotype produces the tallest corn?" Only if we are talking about a particular environment, in this case a particular nitrogen concentration, can we try to answer the question. Moreover, as we see from the results in the middle range of nitrogen environments (c), there may be no one answer; several norms of reaction cross at that point. Similarly, from the fact that G_1 yields the tallest corn at low nitrogen concentration (a), we can say nothing about how good it is in other nitrogen environments. We can say nothing, unless, of course, we already know how its full norm of reaction compares to the other relevant norms of reaction.

These observations lead us to a very important conclusion. A given point on our graph tells us little about the other points unless we are familiar with the overall functional relations that obtain. Knowledge of how different genotypes perform relative to one another at a given point on our environmental axis tells us virtually nothing about individual or relative performances at other points. Let us call the vertical array of norm plottings associated with a narrow environmental range, like (a) or (c) on our hypothetical graph, a *slice*. Then, from what we have just seen, the information contained in a slice is quite limited since it tells us nothing about overall functional relationships between phenotype and environment. As a result, one slice cannot be used to predict plottings at another slice and so cannot be used as a basis for making claims about the relative merits of different genotypes in any environments not captured by the slice.

In the next section I shall appeal to this notion of a slice to explain the limited predictive power of the statistical concept 'heritability'. I shall argue that heritability estimates, based as they are on the linear analysis of variance, are blind to the overall functional relationships between genotype and environment in much the same way that slices are. Slices give us only a local picture of the full functional relationship revealed in Figure 1. Similarly, heritability estimates, *since they are always based on a fixed distribution of genotypes and environments*, give us only a local picture, and often a misleading one, of the influences on variation of a phenotypic trait.

(B) *Heritability and the Analysis of Variance*

The heritability of a given trait in a particular population is the proportion of total variation in the trait which can be attributed to the influence of genetic differences in the population. Variation is measured by the usual statistical notion, variance, the average of squares of deviations from the mean phenotypic value. Heritability is not, then, a property of the phenotypic trait, as one might think if one confused 'heritable' with 'inherited'. Rather, it is a statistical property of a particular population exhibiting differences in the trait. This statistical nature of the concept 'heritability' can better be understood if we look at its roots in the linear model for the analysis of variance.

We begin by selecting a determinate population P. P has a fixed distribution of genotypes and environments. Let us denote the deviation of an individual's phenotype from the population mean by $D(g, e)$, where g and e are the particular genes and environment the individual has. Let $G(g)$ be the average deviation from the phenotypic mean of all individuals with genes g, where such individuals are assumed to occupy the fixed distribution of environments found in P. Let $E(e)$ be the average deviation from the phenotypic mean of all individuals occupying environment e. Here the e-occupying individuals reflect the whole distribution of genotypes in P. Finally, let $R(g, e)$ be whatever remaining deviation from the phenotypic mean is necessary to make the following equation true for each combination of g and e in the population:

(1) $D(g, e) = G(g) + E(e) + R(g, e)$

Thus, (1) actually defines R and is tautological.

This linear model for the analysis of variance assumes, then, that we can portion the deviation from the mean experienced by an individual with gene-environment combination g, e into (i) a mean deviation G attributed to his gene value g; (ii) a mean deviation E attributed to his environment value e; and (iii) a mean deviation R attributed to various gene-environment interaction effects, that is, whatever extra 'causes' we need to postulate for (1) to be true. Equation (1) is converted into the type of equation Jensen (1972, p. 105) uses to express the analysis of variance by squaring and then averaging both sides to obtain, using the standard meanings of variance and covariance:

(2) $\text{Var}(D) = \text{Var}(G) + \text{Var}(E) + \text{Var}(R) + 2\,\text{Cov}(G, E) + \\ + 2\,\text{Cov}(G, R) + 2\,\text{Cov}(E, R).$

Heritability will be the ratio of genetic variance ($\text{Var}(G)$) to the total variance ($\text{Var}(D)$).[11]

To understand what 'genetic variance' is, it will be useful to return to the more perspicuous equation (1). For our purposes, it is crucial to see that the G and E (and R) values that are contained in (1) are expressed in phenotypic units, even though they are thought of as 'genetic' and 'environmental' deviation and give rise to 'genetic' and 'environmental' variances ($\text{Var}(G)$ and $\text{Var}(E)$). The value of G does not depend solely on g, as it might seem from the way I have written it down. Rather, it also depends crucially on the particular distribution of environments found in the population P. Calling G 'genotypic' mean deviation (and $\text{Var}(G)$ 'genetic variance') is thus somewhat misleading. What we actually are talking about is not the effect of genes g taken in abstraction from environment or in conjunction with all possible environments, but their effect *when acting in the given distribution of environments special to P*. In other words, G (and thus $\text{Var}(G)$) embodies a very important environmental specificity as well. There is a hidden environmental input determining the value of G, as G is not *just* a function of a genotypic input g. An analogous point holds for E. In thinking of it as the 'environmental' mean deviation (and of $\text{Var}(E)$ as 'environmental variance'), we must not ignore the hidden genotypic specificity that results from the particular distribution of genotypes found in P. E and $\text{Var}(E)$ have hidden genotypic inputs and are not solely functions of e. Thus the values of G and E are dependent on statistics about the population P, statistical facts about the particular distributions of genotypes and environments. And this statistical specificity carries over into 'genetic' and 'environmental' variances as well, and thus into heritability estimates.

There is an important epistemological fact about G and E which is related to the conceptual point I have just made. When we arrive at our knowledge of 'genetic' and 'environmental' mean deviations or variances, it is not as a result of any empirical knowledge of actual functions, like norms of reaction, which relate actual genotypic and environmental values to phenotypic values. Using our linear model of the analysis of variance, we are working analytically, decomposing data on phenotypic deviation or variance into 'genetic' and 'environmental' (and 'interaction') components which, as we have seen, are still necessarily expressed in phenotypic units, not real genetic or environmental units. We could not get the real genetic and environmental values without actual empirical research into the functions – but this research is exactly what the analysis of variance is intended to help us avoid having to do.

We can now see how a heritability estimate for a phenotypic trait is like the notion of a slice defined in the last section and how it shares its severe limitations. The slice-iness of heritability estimates derives from the two facts we have just discussed. It derives from the analytic point that G and E, and therefore genetic and environmental variances, depend directly on the particular genotypic and environmental distributions found in P. As a result, these variances are only a local picture of the functional relationships that hold between genotype, environment, and phenotype. The slice-iness also derives from the epistemic point that the analysis of variance nowhere draws on or gives us information about these overall functional relationships – such information cannot be recovered from data in the analysis of variance since it was not contained in it in the first place.[12] Consequently, just as a slice is no predictor of other slices, a heritability estimate for a trait is in general no guide at all to heritability estimates for the same trait if we vary either the genotypic or environmental distributions for the populations. The root of the limitation here is the same as it is for slices – the blindness of heritability estimates to the kind of functional relation expressed in norms of reaction.

Some examples taken with reference to Figure 1 might help us to see more clearly the slice-iness of heritability estimates. Suppose P consisted of an even distribution of genotypes G_1–G_5, but it had a narrow range of environments clustered around low nitrogen concentrations (a). We might suppose, just from looking at a slice from Figure 1 in the (a) area, that much of the total variation in height could be attributed to the influence of genetic differences between G_1–G_5. As a result, we would expect that genetic variance will be a large portion of total variance and we might conclude that corn height in P was highly heritable. Does this high heritability allow us to say anything about heritability in other populations, say one (P') just like P but with environments clustered around (c) not (a)? From inspection of a slice of Figure 1 around (c), we infer there is less variation attributable to differences between G_1–G_5. It is possible, then, that heritability of height would be significantly different in P' than in P. Similar slice-iness of heritability emerges when we observe how changing the distribution of genotypes and leaving environmental distribution alone also alters heritability of the trait between two populations.

I would like to turn now from this general discussion of the slice-iness of heritability to a more specific discussion of the predictive power of the concept of heritability. I want to see if the local picture property of heritability means that IQ heritability estimates can not be used in the ways our sample human nature theorists have tried to use them.

(C) *Heritability and Predictions about Human Nature*

The IQ Argument sketched in Section I uses a high (80%) estimate of IQ heritability to support a variety of claims: (a) that IQ is resistant to change; (b) that IQ inequalities are neither significantly reducible nor eliminable; (c) that race and social class mean IQ differences are probably genetic in origin; and (d) that egalitarian educational reform programs are unrealistic, since they ignore (a)–(c). For the purposes of my argument here, I will set aside Kamin's (1974) and others' conclusive arguments that the data do not support high heritability estimates. I want to argue a conceptual, not an empirical point, here: I want to show that even if we could legitimately claim that IQ was highly heritable, such a claim could still not be used to support conclusions like (a)–(d). I have already argued that the slice-iness of heritability, its local picture property, results from its lack of explanatory power at the functional level. It lacks information about functional relationships between genotype, environment, and phenotype. This lack of explanatory backing at the functional level means, as we shall now see, that high IQ heritability does not entail anything like (a)–(d). In the next section I will extend my argument to show that heritability tells us nothing at another, even more important, explanatory level. It tells us nothing about the causal mechanisms involved in phenotypic expression. This lack further undermines its predictive power.

It is pretty clear that both Jensen and Herrnstein say things rather close to (a), the conclusion that IQ is resistant to change. Jensen (1972, p. 135), for example, remarks, "The fact that scholastic achievement is considerably less heritable than intelligence... means that there is potentially much more we can do to improve school performance through environmental means than we can do to intelligence..." Herrnstein (1973, p. 198) draws a similar conclusion when he says compensatory education failed because it "tried to raise IQ's, which... are more a matter of inheritance than environment [high heritability – N.D.) and therefore are not very amenable to corrective training." Though both warn against the inference in other places, in these remarks at least they seem to support the inference, 'If trait T is highly heritable, then T is resistant to environmental change'. (Incidentally, we can view this inference as a more sophisticated variation on the Fallacy of Fixedness, discussed in III (A)).

The inference behind conclusion (a) is not valid because high IQ heritability, by itself, tells us nothing at all about how much or how little environmental change can alter IQ, either individual IQ or mean IQ. We have already seen the reason for the invalidity of the inference. Just as a slice of Figure 1 from region (a) tells us nothing about a slice at region (c), so too, heritability of height in population P with an environmental distribution in the (a) range tells us nothing about heritability of height in a population P' with environments mainly in the (c) range. Nor does heritability in P tell us whether total variance, and thus the amount of IQ inequality, will be the same in P'. In general, then, a heritability estimate cannot be used to make such projections because it has no information about the only functions, the norms of reaction, which can be used to derive such information. Conclusions of type (a) and (b) are not validly drawn from heritability estimates alone.

But not only is the *inference* from high IQ heritability to its resistance to change in general wrong, but the conclusion is in fact wrong. There is much direct evidence that IQ's can be raised significantly by changing environments. For example, when children of low mean IQ and low socio-economic status where adopted into high socio-economic status homes, Skodak and Skeels (1949) found the children's IQ's rose a mean 12 points (almost a full standard deviation) *higher* than would have been predicted by regression to the mean from their mother's low IQ's. Skeels (1969) also reports on two groups of institutionalized 'mentally retarded' children whose own IQ's as well as their natural mother's averaged 70. When one group of 13 was taken from their orphanage and placed among 'mentally retarded' female inmates of a state hospital, their mean IQ's jumped 28 points, enough so that most could be adopted. Those who were adopted showed a further mean rise of 9 points. The group left in the orphanage showed an IQ drop. Similar results indicating significantly boosted IQ's are also found in experimental educational intervention programs, such as Heber's (1972) and the Levenstein (1970, 1971) studies mentioned earlier. In the Heber studies, the child is worked with through intensive tutoring outside the home. In the Levenstein studies the home environment is changed by providing the mother with educational toys and coaching her on how to play with the child and read to it. The intention of these studies is to provide the child with an environment more like upper-middle class preschool children. The results are sustained boosts in IQ (cf. Bronfenbrenner (1974) for a review of these studies).

I am not arguing here that these successful IQ boosting studies are incompatible with IQ being highly heritable; on certain assumptions (questionable, I believe) about the intervention environments, they can be made compatible with high heritability.[13] Rather, I want to show that the inference from high heritability to the resistance of IQ to change or to the stability of IQ inequalities in general is not valid. These data on IQ boosting help make that point valid.

My argument in this general case relies on the fact that heritability itself tells us nothing about the overall norms of reaction. Yet, one must know something about these norms before projections from heritability estimates for a population with one distribution of environments can tell us anything about phenotypic means or variance in other environmental distributions. The assumption that Jensen (1972, p. 140), Herrnstein, and others usually make is that norms of reaction for IQ are roughly parallel, much as is represented by the norms of reaction for $G2$ and $G5$ in Figure 1. This assumption, sometimes called the assumption of additivity, is roughly equivalent to assuming that there is little gene-environment interaction of the type portrayed in Figure 1 by the crossing norms of reaction. It is in large part this complex gene-environment interaction which makes it impossible for us to project from one slice to another or from one heritability estimate to another. But if additivity is assumed, then we are assuming that environments that improve or diminish phenotype (height or IQ) for one genotype also improve or diminish it for other genotypes. Since we know G_2 and G_5 are parallel, then since increasing nitrogen concentration for G_2 raises mean height, we can predict it will have a similar effect on G_5. Similarly, knowing that the norms of reaction are roughly parallel, then even if environments are improved and the mean height is increased, we can rest assured that the inequality between G_2 and G_5 will be maintained. On the assumption of additivity, then we can attempt quantitative estimates about the effect on mean IQ and IQ variance that will result from stipulated changes in environmental distribution, such as improving environments for IQ.

But, is there any reason to assume additivity for IQ norms of reaction? Is the assumption non-controversial? Not at all. In fact, there is little *inductive* support for it if we look at the kinds of norms of reaction found in actual research on non-human populations. Indeed, the inductive evidence would seem to go the other way. Parallel norms of reaction seem to be rare in nature where actual research has produced any data at all (Lewontin, 1974). Moreover, the direct empirical evidence also seems to be extremely poor. The tests which have been used for gene-environment interaction have been criticized for two main defects. First, they are known to be extremely insensitive devices for detecting interaction (cf. Mortin (1974) and Rao and Morton (1974)). Second, all of the data is based on the twin studies, but these studies have been attacked effectively by Kamin (1974) and even Jensen admits (1974), at least in the case of the largest such study, Sir Cyril Burt's, that it is useless for theoretical purposes.

Since the empirical evidence for this crucial assumption is so weak, those who want to employ it often retreat to a priori arguments of various kinds, arguments which Lewontin has effectively criticized (1974, p. 408–9). One such argument is that making the assumption allows us to have a model that fits available data. Of ourse, this argument cannot be used to rule out non-additive, interaction models which might also fit the data, or do it better. Jensen has little discussion of the assumption, though he does suggest (1972, p. 126) that genetic and environmental variance add up to almost 100% of total variance so that there is little room for interaction. But this argument rests on a mistake about the basic equation.[14]

Given the central importance of the additivity or non-interaction assumption, it would seem that careful scientists would place little faith in the existing evidence or the a priori rationales for making it. Nevertheless, the assumption is made anyway, by Jensen, Herrnstein, and others. Since I am primarily interested here in the conceptual features of the IQ Argument, I will suspend my reservations about the additivity assumption, just as I have about the evidence for high IQ heritability. For the purposes of argument, then, we will assume what is contrary to empirical evidence, that IQ is highly heritable and that additivity holds. What now can be said about conclusions 5 (a)–(d) of the IQ Argument?

Consider first Jensen's and Herrnstein's remarks that high IQ heritability implies IQ is resistant to environmental change. Does assuming additivity make the inference to 5 (a) acceptable? It does not. In fact, Bereiter (1970, p. 288) has calculated that with additivity and 80% heritability, IQ could be boosted 14 points, almost a standard deviation, simply by instituting social reforms which make the worst IQ environments equivalent to the average ones existing today. Such a boost would not be insignificant, especially for Jensen and Herrnstein who believe that IQ causes success, as stated in premise (2) of the IQ Argument. Herrnstein (1973) clearly argues that IQ is a threshold indicator of ability to perform at different jobs, a given IQ level being a necessary (but not sufficient) condition for being able to perform certain tasks. A significant boost, like the one predicted by Bereiter, would mean that many types of jobs now inaccessible to below average IQ people would become accessible. Of course, my point here is *ad hominem* since, as my earlier discussion made clear, premise (2) is pretty definitely false and IQ indicates next to nothing at all about one's ability to perform on a job. It would seem that Jensen or Herrnstein would have to retreat to a position they sometimes express, that such improvement efforts would be too expensive. Their unsubstantiated claims about cost effectiveness shall not concern us here.

So even assuming additivity, conclusion 5 (a) is not implied by high IQ heritability. But what about 5 (b)? Is there support for it? It is clear that both Jensen and Herrnstein think they can infer from high IQ heritability and additivity that IQ inequalities (variance) will remain stable, regardless of improvements to environment, and, further, that the genetic variance will even become a greater portion of total variance as environments are equalized. As Herrnstein puts it (1973, p. 198), "As the environment becomes generally more favorable [sic] for the development of intelligence, heritability will increase." (Herrnstein here must mean 'equally favorable' not just 'favorable'.) Similarly, support of something like 5 (b) is what prompts Jensen's insistence that social policy makers should recognize the persistent, genetically based inequalities we will find between people. Jensen thus recommends (1972, p. 202) that we "develop the techniques by which school learning can be most effectively achieved in accordance with different patterns of ability." Jensen has in mind developing rote or associative learning techniques for low IQ people and conceptual techniques for high IQ people[15], which is his program for acting on conclusion 5 (d). Herrnstein, drawing on 5 (b) in support of 5 (d) remarks (1973, p. 59): "The false belief in human equality leads to rigid, inflexible expectations, often doomed to frustration, then to anger. Ever more shrilly, we call on our educational institutions to make everyone the same, when we should be trying to mold our institutions around the *inescapable* limitations and varieties of human ability" (my emphasis). Herrnstein appears to have in mind more rigid application of educational ability tracking, based on more extensive use of IQ and other related tests. The human nature lesson of 5 (b) and 5 (d), then, is that human inequality in ability is here to stay and we had better learn to live with it.

Unfortunately for the human nature theorist, the inference to 5 (b) is not so straightforwardly made, even granting additivity. Any heritability estimate is specific to the range and distribution of environments found in the population in question. The inference ignores the possibility of our developing new teaching and training techniques which could have a differential effect on IQ. Suppose, to raise a purely hypothetical example, that a certain learning disability is both highly heritable in our population and that it acts to depress performance on IQ tests. In fact, then, this highly heritable learning disability might contribute to the measured heritability of IQ. Suppose further that research in special education develops novel teaching techniques which correct the learning disability, with the consequence that IQ scores are also raised. Such techniques might have little positive effect on those not encumbered by the learning disability, even if they became part of standard classroom technique, because they might be superfluous for those without the disability. Indeed, these techniques might have a negative effect because they are boring or lack appeal to children unmotivated by the desire to overcome a disability. New techniques such as these might thus reduce total IQ variance and therefore IQ inequalities. Moreover, they might reduce genetic variance and thus the heritability of IQ. Thus the inference to 5 (b) depends crucially on assuming we can not find or will not look for such novel environments. Nor can the assumed high heritability of IQ or its additivity make it less likely that such environments would be found if looked for. There is no argument here, then, for abandoning research on learning theory or educational technique.

There is yet another problem with the inference to 5 (b), a difficulty which lies not with the blindness of heritability to overall norms of reaction or the failure of additivity in novel environments. Rather, the difficulty lies in the blindness of both heritability estimates and norms of reaction to actual causal mechanisms involved in phenotypic expression. This deeper level of explanatory failure further undermines the predictive power of heritability and the inference to 5 (b). In the next section I will explain this lack of explanatory power and show how it applies to the inference to 5 (b) and 5 (c).

(D) *Heritability and Causal Mechanisms*

Thus far I have argued that heritability estimates in general lack the predictive power needed by Jensen and Herrnstein to support their human nature conclusions because such estimates are blind to the functional relationships that hold between genes, environments, and phenotypes. Further, even with the empirically controversial assumption that there is little gene-environmental interaction, that is, that norms of reaction are roughly parallel, high IQ heritability still does not imply the resistance of IQ to change. Nor does it imply the ineradicability of IQ inequalities if we allow for the introduction of new environments. But the predictive power of heritability estimates is affected by another failure at the explanatory level, the failure to tell us anything meaningful about underlying causal mechanisms. I shall now argue that because of this deeper explanatory failure, the inference 'T is highly heritable in P, therefore T has a genetic basis' is best viewed as another fallacy like the three discussed in Section III (A). I shall argue that it is best viewed as the Fallacy of the Inference to Cause.

Jensen, though aware of the need for qualification (cf. 1969, p. 38), is usually quite explicit in his view that high heritability implies the operation of a genetic mechanism of a particular sort. In many of his remarks, he assumes that IQ heritability is high because people are heterogenous for alleles that act, presumably through direct developmental or physiological mechanisms, to produce changes in those brain structures which underlie human intelligence. Differences in IQ, on this view, result primarily from genes which directly cause brain and thus intelligence differences. It is this assumption that underlies Jensen's remark (1972, p. 192–3), cited earlier, that "the techniques for raising intelligence per se, in the sense of g, probably lie more in the province of biological sciences than in psychology and education." It is this same assumption that underlies his bald claim (quoted in Edson, 1969), to be discussed shortly, about a genetic basis for race differences in IQ: "There are intelligence genes, which are found in populations in different proportions, somewhat like the distribution of blood types. The number of intelligence genes seems lower, overall, in the black population than in the white." There is little doubt that Jensen at times supports some version of this Inference to Cause. Now we must see whether it should be viewed as a fallacy.

The inference from high heritability to such a specific causal mechanism is not in general valid and, in my view, not even plausible for the case of IQ. Before explaining why this is true for IQ, let me illustrate the error in the inference with an example from chickens. Suppose we had a chicken population consisting of two particular strains of chicken, A and B, raised in the normal range of temperature environments found in chicken coops. We are interested in calculating the heritability of heart disease, and we carry out appropriate chicken kinship studies. Suppose we find there is a high heritability of chicken heart disease in our population. Could we conclude that there is genetic variation in the population which acts directly through developmental and other physiological mechanisms to affect chicken hearts? For chicken strains A and B, it turns out, this Inference to Cause would lead to a false conclusion. Strain A, we find out on further study, has normal feather density. Strain B, however, has an allele which results, in that range of environments, in sparse feathers. Consequently, at low temperatures, B chickens will be colder than A chickens, will overeat to raise their body temperature, and will die as a result of metabolic consequences of gross overeating. Our genetic basis for heart disease is not what we thought. There is only an *indirect* effect on chicken hearts. Indeed, the genes that are responsible do not act on hearts at all, but on feathers. The effect on hearts is mediated through readily controlled environmental factors. But, because of the way in which heritability is calculated, any genes that act even in these quite indirect ways to produce heart disease will get counted as heart disease genes contributing to the heritability of heart disease.

A similar phenomenon may take place with regard to IQ. Suppose some honest scientist came up with a heritability estimate for IQ which showed some significant degree of heritability. Could we infer, as Jensen already has, that there are differences in 'intelligence genes', meaning differences in the genetic mechanisms underlying the direct development of human intelligence? The answer is no. In fact, there seems to be some evidence that a variety of genes may have an indirect effect on IQ differences *as a result of the way in which society acts towards people with those genes*. For example, there is some evidence that teacher expectations, which are known to affect IQ, depend on factors such as people's looks, ethnic background, height, and so on. (cf. Berscheid and Walster, 1972). People tend to assume that taller, better looking people are more intelligent, and this may have the effect of raising such people's IQ's. Many of these factors may be significantly heritable. The most extreme case would, of course, be skin color. If people with black skin get treated by society in general and by teachers in particular on the assumption they are less intelligent, then they get surrounded with an IQ depressing environment regardless of what classroom they sit in. Since skin color is highly heritable, it would contribute to IQ heritability and skin color genes would be counted as part of the IQ genotype! I am not saying that skin color underlies current heritability estimates, since they were performed on white populations (and are no good for other reasons). But the point is the same: from the point of view of heritability theory, genes get counted as 'intelligence genes' even if they act on intelligence only indirectly as a result of various social and cultural biases, such as those based on looks and skin color. In the extreme case, heritability could be as high as 1.0 and be due entirely to such indirect effects.

It should be obvious what effect undercutting this Inference to Cause has on both the predictive and explanatory force of heritability estimates. For example, suppose we intervened on behalf of all children, regardless of looks (or skin color) with IQ raising techniques, such as Levenstein's training of the mothers. Then, even if heritability was high and the additivity assumption held for existing populations, we might still find heritability would shrink to zero whenever the initial basis for the heritability estimate was indirect effects such as we described. In view of the invalidity of the Inference to Cause, we can see how terribly poor heritability estimates are as a basis for making conclusions about the intractability of human nature and the inevitability of hereditary differences in human capabilities.

So far, I have said nothing about Jensen's inference to conclusion 5 (c), the claim that there is a genetic basis for race (or class, in the case of Herrnstein) differences in IQ. Does high IQ heritability in the white population, combined with a 15 point black-white mean IQ difference, permit us to conclude anything about the reasons for, or causes of, the IQ gap? Jensen (1972, p. 163) clearly believes it does: "It is not an unreasonable hypothesis that genetic factors are strongly implicated in the average Negro-White intelligence differences." Actually, as we have already seen (Edson, 1969), Jensen's Inference to Cause in the case of race differences is quite explicit: "The number of intelligence genes seems lower, overall, in the black population than in the white."

If we construe 5 (c) as a deductive inference from high IQ heritability and the fact of an IQ gap, it is clearly invalid, as many critics (e.g., Lewontin, 1970) have effectively shown. Our discussion of the slice-iness of heritability should make it easy to see why heritability of a phenotype could be high in two populations (*assuming* both black and white IQ heritabilities are high) and there could be a significant gap in phenotypic means, yet, genotypic differences may have nothing to do with creating the difference in phenotypic means. For example, let us assume the heritability of height has been very high, say about 90%, in both 1900 and today for the U.S. population. Further, let us estimate a mean increase in height of several inches. Can we deduce that 'genetic factors' are "strongly implicated" in the mean height increase? Clearly not. Most likely, nutritional environmental changes account for the differences.

To be fair to Jensen, he does warn against our treating his inference as a deductive one. Instead, he prefers (1973, p. 144) to view it as an inference to the best explanation, a probability inference that genetic factors are implicated. But to infer that genetic factors are probably implicated would still require establishing that there are no systematic environmental differences that could be operating (cf. Layzer, 1974). But we know that in our society, with its continuing history of racism, there are systematic environmental differences between blacks and whites which very likely, in view of our earlier discussion of IQ tests, have a bearing on how well an individual performs on them. In the presence of such systematic environmental difference between subpopulations, no inference, not even to a probable implication of genetic factors, is plausible.

Finally, if we remember that the Inference to Cause is in general not valid, then we see yet another reason to block Jensen's inference to genetic factors in the case of race differences in IQ. What makes the probability inference to genetic factors plausible in the absence of environmental differences is the assumption that genetic factors are implicated *within* each of the two populations and therefore may account for the difference between them. But our discussion in this section has already shown that one can not infer even within a population that high heritability results from genetic factors acting in any direct, causal way. High heritability can be the result of quite indirect, primarily environmental, influences on genes which are really unrelated to the trait in question. But, if this is true within populations, then we cannot infer there are genetic factors available to play a role in explaining between-group mean differences.

In sum, then, I have now shown that all of the inferences 5 (a)–5 (d), which Jensen and Herrnstein both make, are not valid when based on the concept of heritability, since it lacks the explanatory and therefore the predictive power necessary to ground human nature conclusions of this type.

IV

Human nature theorists like Jensen and Herrnstein use their hereditarian thesis to political ends. They claim, as conclusion 5 (d) indicates, that there are important social implications of the contention that significant traits, like intelligence, are unequally distributed as a result of genetic factors. Such ineradicable inequality, they believe, justifies abandoning egalitarian programs in education and job training. In my own view, false and dangerous as the hereditarian or human nature position is, it is not the most dangerous part of the whole IQ controversy. A greater danger, because it is more widely believed, lies in the view that IQ measures intelligence, the view that I have attacked in Section II of this essay. I would like to conclude by explaining where that danger lies.

Many educational theorists who reject hereditarian claims about IQ in general, and about black inferiority in particular, nonetheless view the 15 point mean IQ gap between blacks and whites as important. It is important because they accept the IQ score as a measure of general intelligence, variously construed as thinking ability or cognitive capacity. Accordingly, the lower mean for black IQ scores is viewed as a measure of lower black thinking ability, although the environmentalist may explain the lower ability as an effect of cultural deprivation, cultural disadvantage, or cultural deficit, rather than of heredity. The idea here is that the child brings to school from his black or white working class home a stunted thinking capacity or level of intelligence, not just a deficit in information. If the child simply suffered a deficit in information, then we would expect the schools to be able to make it up. But, so the argument goes, the achievement gap widens as progress through the school system continues. Therefore, the initial gap must represent more than a deficiency in information; it must represent an underdeveloped intelligence. I believe this explanation is what underlies the conclusions of the famous Coleman report (1966) as well as the design of many compensatory education programs which concentrated on boosting IQ.

By ascribing to black and working class children a culturally induced, diminished thinking capacity, these environmentalist approaches end up *excusing* the schools from the task of teaching all children basic skills. They excuse the schools as adeptly as does Jensen attempts to (1973, p. 28) when he defines educability as "the ability to learn traditional scholastic subjects under conditions of ordinary classroom instruction" and suggests that educability is significantly heritable. But, even educators who are interested in finding educational techniques that reduce inequality in achievement end up with a theory of black or working class inferiority if they retain the view that IQ measures intelligence or thinking capacity. Carl Bereiter (1970, p. 279ff.), for example, does not want the prevalence of low IQ children to become an excuse for poor teaching. But his question (1970, p. 281), "Can we find ways for children [with below average IQ's] to learn that require less thinking?" clearly stigmatizes the below average IQ child as a special problem for educators. That stigma would attach with special frequency to black and working class children.

Other environmentalists are less inclined to think IQ scores reflect real differences in intelligence *between groups*, charging the test with 'bias' against blacks and working-class children. The 'bias' is found in test item selection and in the testing situation. Usually, however, these environmentalists never challenge the basic methodology and construction of the IQ test. On this view, the test, except for its bias against black or the 'disadvantaged' of any race, still has a claim to being a measure of intelligence. Some proponents of this view search for 'culture fair' tests intended to eliminate the mean performance gap between blacks and whites or other ethnic or racial groups. The standard methodology of the test remains unchallenged. This approach also has the danger of focusing too much attention on the home as the source of the cultural difference. With this focus, the problem of inequality in scholastic achievement is too easily transformed into the problem posed by a child with a different culture who has difficulty relating to mainstream culture. Such a transformation, in my view, diverts attention from a much more important source of educational inequality, namely those processes and functions of the schools which deliberately sort people into educational tracks by class and race background.

The thrust of my argument in Section I and here is that IQ tests are really political instruments, both in design and effect. They place the blame for inequalities in educational achievement squarely on the child or his home life. They posit capacity differences where, in fact, we only have evidence for performance differences, differences which are most likely the direct and indirect effects of social and educational policies. In this way, the ideology of IQ blocks investigation of the ways in which our social policies and educational institutions create educational inequality.

There has been much literature in recent years on the role of teacher expectations in determining the success and failure of different students. In a very interesting article Rist (1970) documents how class biases on the part of a kindergarten teacher provide a mechanism for sorting a group of black children into distinct achievement groups. The resulting record of unequal achievement, clearly induced by the teacher, then becomes a guide on the basis of which later teachers can make their 'objective' judgments about student capacity and potential. Children are thus set in different educational tracks, according to these induced performance differences, with profound effects on their future level of educational achievement.

Powerful as Rist's evidence is for the role of teacher expectations, it is really only part of the story. What is missing from this picture, and much of the literature in teacher expectations, is an analysis of the social forces – both within and without the school – which create and reinforce the class and race biased expectations so many teachers have. After all, the teacher is not responsible for the crowded conditions which may force him or her to attend to the students easiest to teach given the methods available. Nor is the teacher responsible for the existence of the educational tracks which are found in 85% of our secondary school systems. These tracks teach all students a basic ideological lesson – that they occupy the position they have in school and in life because of the abilities and virtues (or lack of them) that they possess. Some learn success, others learn failure and resignation. Nor is the teacher responsible for the existence of IQ test and their wide use.

The role of IQ in this process is centrally important. It gives tracking, and the meritocratic social philosophy that accompanies tracking, a veneer of scientific objectivity. Assignments to tracks seem to be determined by the scientific measurement of individual capacity. Without this veneer, the class and race biases of tracking would lose their fig leaf. And without a reliable fig leaf, tracking would not accomplish its ideological task of embedding in children's living consciousness the blinding belief that social position is determined primarily by ability or capacity. In this role as a fig leaf, then, IQ exerts a positive harm. It helps prevent us from looking carefully at the factors influencing educability by creating the illusion that we know why different children achieve more scholastically. It is time to eliminate the fig leaf, to stop pretending we know why some children are not successfully taught, and to examine more carefully those elitist and racist policies in our schools which create inequality of educational achievement.

Tufts University

NOTES

* Versions of this paper have been read at Indiana University, Ohio State University, and Tufts University, and I have benefited from discussion following these presentations. My greatest debt is to Professors R. C. Lewontin, Ned Block, and Ronald Webber for many hours of helpful discussion. Some material from Sections II and IV appears in Daniels (1975).
[1] Richard Herrnstein claims to hold a thesis only about social class differences; for an argument that he is also committed to a race thesis, see Daniels (1973, p. 34).

[2] In the Shields twin study (1962), Shields tested both twin pair members in 33 of 38 cases and found a mean within pair difference of 4.9 points, a very small difference indicating considerable genetic influence. In 5 cases, Shields tested one twin and his assistant the other in each pair, finding a 13.2 point mean difference, almost a random difference. The discrepancy was ignored. Burt (1966) reported correlations stable to the third decimal place despite sample size changes of 100%, a stability not to be found in actual experimental conditions. See Kamin (1974).

[3] My presentation of the correlation argument is here based on the comprehensive discussion in Block and Dworkin (1974, Part I, Sect. 3). Block and Dworkin also discuss a third argument, the claim that high internal intercorrelations of subtests ('g loading') permits the inference the tests measure a common underlying ability. As Block and Dworkin show, the strongest acceptable inference is that something is common to the performance required by the different subtests. But the claim that this common factor corresponds to a common *capacity*, in the sense of intelligence, would require appeal to one of the two validation arguments discussed above. See also Block in this volume, pp. 127–141.

[4] IQ correlates about .4–.6 with grades, about .68 with school years completed, about .5 with occupational status and about .35 with income. I use .5 as a composite figure.

[5] See Block and Dworkin (1974, Part I, Sect. IV) for an excellent analysis of the correlations between IQ and various measures of success.

[6] Actually, Binet was initially concerned simply to sort children into normal learners and children with serious learning problems, such as retarded children. Later test developers, like Lewis Terman, transformed the test from one with a two-value non-continuous distribution into a test that presupposed a continuous, normal distribution of intelligence, with intelligence level also correlating with social position (Cronin, Daniels, *et al.*, 1975).

[7] Of course, knowing if a student has acquired some of the measured skills may be useful for diagnostic and remedial purposes. Such uses of the IQ test may have value, though of course, they avoid the issue of measuring intelligence. It would seem advisable, however, to develop tests designed to address the specific diagnostic and remedial tasks in question rather than 'making do' with a test standardized for an entirely different purpose. The need for special diagnostic tests is particularly clear when IQ tests continue to be misused. For example, in screening for specific learning disabilities like dyslexia, an IQ score cut-off point is sometimes used above which it is assumed difficulties are due to learning disabilities and below which they are due to low intelligence. Such uses clearly discriminate against working class and minority students, as does the common use of IQ to detect retardation (cf. Mercer and Brown, 1973).

[8] I am indebted to Professors Richard Lewontin, Ned Block, and Ronald Webber for discussion of ideas contained in Section III of this paper. In addition, Lewontin (1974) makes similar points to my discussion in III-(B) and III-(C) about the blindness of heritability to functional relations (See also Hirsch, 1970, p. 93). Block and Dworkin (1974, Part II) make similar points to my remarks in III-(D) about the failure of heritability estimates to tell us about causal mechanisms. I am indebted to Ned Block for discussion of these points.

[9] For example, T. Dobzhansky and B. Spassky (1944) found such complex norms of reaction for viability of larvae in Drosophilia pseudoobscura; cited in Lewontin (1974, p. 410).

[10] Professor R. C. Lewontin remarks on the three fallacies discussed here in an unpublished lecture on heritability delivered at Tufts and various other places in 1973–1974.

[11] I am restricting my discussion here to 'broad' heritability, which includes as part of genetic variance various gene interactions, in contrast to the notion of 'narrow herit-

ability' used by geneticists concerned with selective breeding. Broad and narrow heritability need not correspond closely for the same trait.

[12] Except possibly in one case, where there is no gene-environment interaction and we have data that proves this. I return to this point in the next section.

[13] Instead of using a well-defined environmental axis, defenders of high heritability order all environments on an axis called 'IQ enriching'. They then divide this array of environments into standard deviations of environment, and can then calculate how may such standard deviations of improved environments are needed to explain a given IQ boost. Shockley (1972) assumes, for example, that the environments in the Heber study fall in the top one per cent of all environments. This procedure is ad hoc, however, since the assignment of Heber environments to the top of the environment scale is arbitrary.

[14] Jensen (1969, 1972) ignores the fact that some of the interaction terms in equation (2) above could be large and negative.

[15] Bereiter's program seems all too similar. Cf. Bereiter (1970, p. 281 ff.).

BIBLIOGRAPHY

Alexander, T.: 1972, 'Social Engineers Retreat Under Fire', *Fortune* **86**, 132–140.
Bache, R. M.: 1895, 'Reaction Time with Reference to Race', *Psychological Review* **2**, 474.
Bereiter, C.: 1970, 'Genetics and Educability: Educational Implications of the Jensen Debate', in J. Hellmuth (ed.), *The Disadvantaged Child*, Bruner/Mazel Publishing Company, New York, 1970.
Berscheid, E., and Walster, E.: 1972, 'Beauty and the Best', *Psychology Today* **5** (10), 42.
Block, N., and Dworkin, G.: 1974, 'IQ, Heritability, and Equality', *Philosophy and Public Affairs*, Part I, **3**, 331–409; Part II, **4**, 40–99.
Block, N.: 1976, 'Fictionalism, Functionalism and Factor Analysis', this volume, p. 127.
Bowles, S. and Gintis, H.: 1972–1973, 'IQ in the United States Class Structure', *Social Policy* **3**, 65–96.
Bronfenbrenner, U.: 1974, *A Report on Longitudinal Evaluation of Preschool Programs* Vol. 2: *Is Early Intervention Effective?* Washington, D.C., U.S. Department of Health, Education, and Welfare, Pub. No. OHD 74–25.
Cattell, J. M.: 1947, *James McKeen Cattell: American Man of Science*, Vol. 2, Science Press, Lancaster, Pa.
Coleman, J. S. et al.: 1966, *Equality of Educational Opportunity* U.S. Government Printing Office, Washington, D.C.
Cronin, J., Daniels, N., et al.: 1975, 'Race, Class, and Intelligence', *International Journal of Mental Health*, **3**, 46–132.
Daniels, N.: 1973, 'Smart White Man's Burden', *Harper's*, October, 24–40.
Daniels, N.: 1974, 'IQ, Intelligence and Educability', *Philosophical Forum* **6**, 56–69.
Dobzhansky, T. and Spassky, B.: 1944, 'Genetics of Natural Populations XI. Manifestation of Genetic Variants in Drosophila pseudoobscura in Different Environments', *Genetics* **29**, 270–290.
Edson, L.: 1969, 'Jensenism', *New York Times Magazine*, Aug. 31.
Heber, R. and Garber, H.: 1972, 'An Experiment in the Prevention of Cultural-Familial Retardation', in U.S. Senate Select Committee on Equal Educational Opportunity, *Environment, Intelligence and Scholastic Achievement*, Wash. D.C.
Herrnstein, R. J.: 1971, 'IQ', *Atlantic Monthly*, September, 43–64.

Herrnstein, R. J.: 1973, *IQ in the Meritocracy*, Atlantic-Little Brown, Boston.
Hirsch, J.: 1970, 'Behavior-Genetic Analysis and its Biosocial Consequences', *Seminar in Psychiatry* **2**, 89–105.
Jensen, A. R.: 1969, 'How Much Can We Boost IQ and Scholastic Achievement', *Harvard Educational Review* **39**, 1–123.
Jensen, A. R.: 1972, *Genetics and Education*, Harper & Row, New York.
Jensen, A. R.: 1973, *Educability and Group Differences*, Harper & Row, New York.
Jensen, A. R.: 1974, 'Kinship Correlations Reported by Sir Cyril Burt', *Behavioral Genetics* **4**, 1.
Kamin, L.: 1974, *Science and Politics of IQ*, Erlbaum Assoc., Potomic, MD.
Layzer, D.: 1974, 'Heritability Analyses of IQ Scores: Science or Numerology', *Science* **183**, 1259.
Levenstein, P.: 1970, 'Cognitive Growth in Preschoolers Through Verbal Interaction with Mothers', *American Journal of Orthopsychiatry* **40**, 426.
Levenstein, P.: 1971, Verbal Interaction Project: *Aiding Growth in Disadvantaged Preschoolers Through the Mother-Child Home Program*, Family Service Association of Nassau County, Inc., Mineda, N.Y., July 1, 1967–August 31, 1970, Final Report.
Lewontin, R. C.: 1974, 'Analysis of Variance and the Analysis of Causes', *American Journal Human Genetics* **26**, 400.
Lewontin, R. C.: 1970, 'Race and Intelligence', *Bulletin of Atomic Scientists* **26**, 2.
Mercer, J. R., and W. C. Brown: 1973, 'Racial Differences in IQ: Face or Artifact', in C. Senna (ed.), *The Fallacy of IQ*, Third Press, New York, pp. 56–113.
Morton, N. E.: 1974, 'Analysis of Family Resemblance I', *American Journal of Human Genetics* **26**, 318–350.
Rao, D. *et al.*: 1974, 'Analysis of Family Resemblance II', *American Journal of Human Genetics* **26**, 331–359.
Rist, R.: 1970, 'Student Social Class and Teacher Expectations. The Self-Fulfilling Prophecy of Ghetto Education', *Harvard Education Review* **40** (3), 411.
Schockley, W.: 1972, 'Dysgenics, Geneticity, and Raceology', *Phi Delta Kappan* **53**, 297.
Skeels, H. M.: 1969, 'Adult Status of Children with Contrasting Early Life Experience', Child Development Monograph **31**, 3.
Skodak, M. and H. M. Skeels: 1949, 'A Final Follow-up Study of One Hundred Adopted Children', *Journal of Genetic Psychology* **75**, 85.
Stetson, B. R.: 1897, 'Some Memory Tests on Whites and Blacks', *Popular Science Monthly* **4**, 185.
Terman, L.: 1916, *The Measurement of Intelligence*, Houghton-Miflin, Boston.
Yerkes, R.: 1921, 'Psychological Examining in the United States Army', *Memoires of the National Academy of Science* **15**.

KENNETH KAYE

THE IQ CONTROVERSY AND THE PHILOSOPHY OF EDUCATION*

Although my two colleagues on this panel are *bona fide* philosophers who clearly have a *bona fide* interest in the ongoing controversy over IQ and the educability of children (which should mean that the topic requires no further defense), I nonetheless feel constrained to say a word in defense of our topic as a viable subject for discussion in a forum devoted to the Philosophy of Science. Perhaps this is because, as the representative of the discipline which created this mess, I fear you might associate me with it. You might feel that bad science is a bad subject for study by philosophers of science, and furthermore that anyone associated with such a discipline would be the last person to have anything intelligent to say about philosophy.

While it is true that we do not need a philosophy of bad science, a kind of Gresham's Law seems sometimes to operate, in which bad science drives out the good. The reasons for which that happens, and the reasoning by which it is allowed to happen, surely are both within the purview of philosophers as well as historians of science. Furthermore, the current controversies in educational psychology have already raised a number of questions among philosophers (though not entirely philosophical questions), to which the answers will depend upon an understanding of some empirical findings, some methodological biases, and some operational definitions. Partly I wish to add to what my fellow panelists have just said; but I also wish to make a somewhat different, though not contradictory argument.

At issue, from one point of view, are a set of statistics called heritability estimates, which are supposed to tell us about the relative innateness, fixedness, determinedness of a child's potential intelligence. From another point of view, at issue are a set of social policies referred to as streaming or tracking, which have to do with the allocation of resources to children within schools, and the optimal forms of grouping for education. The reason we have a full-blown controversy is that these two major issues intersect. Or rather they are regarded as intersecting by nearly

R. S. Cohen et al. (eds.), PSA 1974, 181–188. All Rights Reserved.
Copyright © 1976 by D. Reidel Publishing Company, Dordrecht-Holland.

everyone: psychologists, educators, and the readers of magazines. It is assumed that knowing whether intelligence is innately determined will tell us how children ought to be grouped in school. Sometimes the even more outrageous converse is assumed, that knowing how we wish to organize our schools will tell us what we must believe about the nature and development of intelligence. But let us first look at the issue of heritability of IQ, and return later to the social-political context of the educational issues.

It has been pointed out, by my predecessors on this panel as well as by others, that we have no scientific reason for believing that the heritability of IQ is as high as Cyril Burt, Arthur Jensen, and Hans Eysenck have claimed. I think this is true. The twin studies are inadequately controlled and biased, etc. On the other hand, I do not think we have any basis for believing that IQ has *low* heritability. It would be astonishing, given the nature of the IQ test, if IQ were not heritable. The important point to be added to the criticisms you have just heard, without in any way denying their force, is that IQ heritability tells us nothing whatever about the educability of children. Nothing whatever.

Heritability is a measure of the amount of variance in some trait which is accounted for, in a particular environment, by genotypic variance in a population. Heritability depends upon the environment: for a given trait and a given population of genotypes, the heritability will be high in some environments and low in other environments. If the particular environment has a differential effect upon the trait in individual members of the population, heritability will be low. If the environment does not have much differential effect upon individuals with respect to this trait, the trait will be highly heritable. Heritability is just as much a measure of the environment, or the range of environments in which a tested population has developed, as it is of anything else.

Heritability also depends upon the population. A given trait may be found to have different heritabilities for different strains, breeds, races, or gene pools, even under the same range of environmental conditions. It is a measure that applies only within a gene pool, and has nothing to do with explaining between-group variance. An attempt to use heritability estimates in accounting for racial differences in any trait is, put simply, a hoax.

Finally, heritability depends upon the trait. This is a more significant point than merely that some traits are highly heritable while others are not. A trait is simply some measurable characteristic in which individuals differ. What is measured, and the way in which it is measured, determine the heritability for a given population in a given environment.

Each of these three points has been ignored by the majority of disputants on both sides of the IQ controversy. The first point alone should have ruled heritability studies out of court for any decision regarding educability. If IQ and school achievement have high heritability among present-day Americans, this may be considered an indictment of American schools. Our schools, and our society in general, are not making much of a difference in the kind of skills tested by either achievement or IQ tests. Children who go to school know more than those who do not, but the quality of particular schools, and of other supposedly educational experiences in the lives of children, do not make much difference among their test scores (Jencks, 1972; Kaye, 1973a). Both IQ and school achievement can be best predicted by knowing who a child's parents are. But this tells us nothing about either educability or heritability in another sort of environment. There is every reason to assume that in some other society the quality of schooling would make a difference, and thus heritability would be lower.

As my colleagues have pointed out, the parental variables that predict IQ are by no means all genetic variables. In all of the IQ heritability studies, chiefly by correlation among consanguineous relatives and between adopted-out twins, the common prenatal and similar postnatal environments grossly inflate heritability estimates beyond the true genotypic variance. But even if this were not the case, if we had perfect measures of true heritability, the heritability would be a function of the relative ineffectualness of a given environment.

The second point is that we can estimate (from a sample) heritability only for a particular population or gene pool. This, too, has been largely ignored in the controversy. A number of authors, including Jensen (1972), have been at great pains to argue that there is every reason to assume IQ is equally heritable among blacks as among whites. But that is off the point. Neither the heritability within one gene pool nor within the other gene pool has anything to do with the mean difference in IQ between the two populations. It is perfectly possible, indeed quite common, for a trait to be highly heritable in two or more groups, that is for the within-group variance to be accounted for genetically, while the between-group variance is entirely due to environmental differences. This is true whenever we compare two cornfields that have been irrigated differently. It would also be true if we divided a group of children in half alphabetically, and fed to one half of them the answers to IQ test items. The heritability of IQ in the A-M group would still be the same as that in the N-Z group and would quite probably be high, yet one group would be one or more standard deviations superior to the other, entirely for experiential reasons.

There is no contradiction between my first and second points. Our society could be, and apparently is, ineffectual at making much of a difference among children within gene pools, within social classes and racial groups, while at the same time it is very effective at discriminating between the major groups.

Third, heritability depends upon the measured trait. We must constantly remind each other that we are talking about IQ not intelligence. Intelligence can be defined in many ways and measured in many more, and each way of measuring it must be considered a different (though not independent) trait. The trait whose heritability has been so exhaustingly debated is IQ, the score on a certain type of test. Even if we were concerned with a single gene pool and an inexorable environment, we would still have to examine the nature of the IQ tests in order to understand and interpret the heritability of this trait.

Elsewhere I have tried to describe the systematic biases built into IQ tests, their historical development and current dangers (Kaye, 1973b). This has also been dealt with in detail by others (e.g. Cronbach, 1975). There are a few points worth reiterating briefly. Most important is that the tests discriminate between all and only those groups between which the testmakers wanted to discriminate. IQ tests do not discriminate between boys and girls because, after girls were found to be consistently superior on the 1916 Stanford-Binet, certain items were eliminated from the test until girls scored no higher than boys. The same sort of fiddling is possible, and has been successfully done, to eliminate racial differences in IQ scores. But the resulting tests were not adopted since they failed to predict school achievement (Eells *et al.*, 1951). The gross differences in *achievement* between black and white children had to be accounted for by corresponding differences on a supposedly objective test.

Another important way in which the tests were fiddled with was by eliminating items and subtests that contributed to instability of IQ's over time. Thus what was originally an hypothesis, that each individual's growth in 'mental age' was a constant function of his chronological age, became self-fulfilling. The test was constructed so that IQ's would be relatively constant. This meant eliminating items which tested skills and knowledge upon whose development the environment made a difference. Inevitably this would create a trait with relatively high heritability. If differences in experience made a difference in the acquisition of some skill, within the population on which IQ tests were standardized, then that skill was dropped from the battery of skills included in the test. Thus the genotypic variation within the population was allowed to play a relatively greater role than the environmental variation. However, this stability-increasing and heritability-increasing procedure was only applied *within* the standardization sample. It did not have the effect of eliminating items for which the experience of most white children prepared them better than did the experience of most black children. In other words, precisely the situation which I said was possible, in hypothetical examples of cornfields or alphabetically grouped children, did indeed prevail in the construction and standardization of IQ tests.

There is a basic difference between tests of the IQ or 'aptitude' type and tests of achievement. In the latter type, items are included just because the testmakers or some policy-making body regards them as important items of knowledge, whether for driving a car, programming a computer, or going on to a more advanced subject. In the IQ-type tests, on the other hand, items may have no face validity at all; their inclusion on the test is due to the fact that they contribute to total scores which meet the desired criteria of normal distribution and stability. There are similarities between the two types of test; in fact, achievement tests are often nationally standardized so that they acquire some of the worst features of 'aptitude' tests. But in principle an achievement test is potentially capable of telling us something about the strengths and weaknesses of individuals in different subjects. IQ scores can never provide such information (though the individual child's response to the IQ testing situation is often used creatively for a clinical diagnosis).

What effects has the existence of IQ tests had upon research in education and developmental psychology? First, it has perpetuated certain myths about the nature of learning and intelligence. I mentioned the fact that what was initially an hypothesis, that 'mental age' bears a constant relation to chronological age, quickly became an assumption that could no longer be tested. The tests were constructed in such a way that most children's IQ's typically were fairly constant. In fact the notion of 'mental age' lost its meaning. Since the 1940's IQ's have not been quotients at all, but simply scores read out of a table in the test manual. The table is constructed so as to yield nearly-perfect distributions of scores at each age. Unfortunately the smoothness of these curves, and of various forms of stability curves for IQ, tend to be interpreted as proving that there is something fixed, determined, and innate about human intelligence. The almost universal misrepresentation and misunderstanding of heritability compounds the problem. And as perhaps the worst consequence of all, the entrenched establishment of psychometricians and the test publishing industry (over 200 million standardized tests per year, or four per child in the United States) continually retard the prospects for investment in research into alternative conceptions of intelligence.

A second sort of effect has to do with the notion of 'upping' IQ's. At some point in history educators bought the myth that if children were educable, it ought to be possible to raise their IQ scores. This would mean, essentially, that IQ tests were nothing other than achievement tests. They do test achievement in the sense that the items require information acquired from experience, and as I have said a group of children whose experiences have been more closely oriented to the types of knowledge required on the IQ test will indeed 'achieve' higher scores than children with different kinds of experience. Nonetheless it remains true that an IQ test is the worst possible kind of test by which to measure achievement, whether one is concerned about the acquisition of important skills or the more general question whether children are educable. IQ tests are explicitly designed to be unresponsive to school experiences of varying quality. Therefore the studies which have successfully boosted the IQ's of Headstart or other children are all the more remarkable; the vast number of studies which have found negligible or only temporary effects on IQ are simply meaningless. Well-meaning educators who accept the assignment of trying to boost IQ's are taking on an Augean task. It is a Sisyphean task as well, for boosting the IQ's of some children automatically lowers the IQ's of an equal number of children as soon as the tests are restandardized.

Finally, there is tracking. IQ tests together with other tests constructed in much the same way are the principal means of dividing schoolchildren into tracks or streams. The justification for this widespread practice (beginning in the first grade in 20% of American school systems and before high school in 80%; Findley and Bryan, 1970) is that children learn more in homogeneous groups. Yet there is no adequate research indicating that any children learn more in homogeneous groups; there is some evidence that poorly-achieving children do worse; and there are clear indications that self-esteem and motivation suffer in the 'lower' streams. The popular belief that tracking defends poorly-achieving children from invidious comparisons with their betters has never received the slightest empirical support. One thing we do know is that segregating high-IQ from low-IQ children makes the prediction of school achievement by IQ a self-fulfilling prophecy. There is even some possibility that this practice aids and abets the stability of IQ scores, thus contributing to the perpetuation of the myth, but the evidence on this is not yet clear (Jencks, 1972).

A number of current educational innovations involve homogeneous grouping of pupils in one way or another. Chief among these is mastery learning, the attractive approach that allocates time and resources where they are most needed to bring the maximum number of pupils to some criterion, instead of moving the whole class along from one unit to the next while individual children fall further and further behind (Bloom, 1974). In grouping of this type, however, the tests used are necessarily achievement tests. Unlike tracking, the mastery tests and mastery groups are done independently for each school subject, so that a child can spend more time in arithmetic and less time in reading. And there is constant retesting, rather than an irreversible decision being made in the ninth, or the fifth, or even the first grade.

These mundane matters seem to have taken us far from the cleanliness of philosophy. Yet there is no clear boundary at which we could have stopped. Our purest conceptions about human intelligence and development are inseparable from the research paradigms and instruments we use. These in turn are inseparable from the applied questions which guide the research, and the policy questions are inseparable from the society's values and myths, which finally reduce to the purest conceptions about human development. Some of the critics of tracking, IQ testing, and heritability research have suggested that there is something un-

American about them, that they conflict with our egalitarian ideals. I personally see these practices and the research as very American, ideologically speaking; as the resolution of two conflicting ideals. We are said to believe, on the one hand, in equal opportunity and self-determination. Yet we clearly believe in the supremacy and inheritance of property. We need a device by which to guarantee the succession of children to their parents' status and earning power, while making it look as though they are succeeding by merit. That has been effectively accomplished by IQ testing, by tracking, and by the research purporting to investigate intelligence. This is the context in which science exists. Against this background, and yet in the abstract, please: tell us what distinguishes good science from bad.

The University of Chicago

BIBLIOGRAPHY

Bloom, B.: 1974, 'Time and Learning', *American Psychologist* **29**, 682–88.
Cronbach, L. J.: 1975, 'Five Decades of Public Controversy over Mental Testing,' *American Psychologist* **30**, 1–14.
Eells, K., Davis, A., Havighurst, R., Herrick, R., and Cronbach, L.: 1951, *Intelligence and Cultural Differences; a Study of Cultural Learning and Problem-Solving*, University of Chicago Press, Chicago.
Findley, W. G. and Bryan, M. M.: 1970, *Ability Grouping*: 1970, University of Georgia, Atlanta.
Jencks, C., Smith, M., Acland, H., Bane, M. J., Cohen, D., Gintis, H., Heyns, B., and Michelson, S.: 1972, *Inequality: A Reassessment of the Effect of Family and Schooling in America*, Basic Books, New York.
Jensen, A. R.: 1972, *Genetics and Education*, Harper and Row, New York.
Kaye, K.: 1973a, 'Some Clarity on *Inequality*', *School Review* **81**, 634–41.
Kaye, K.: 1973b, 'I.Q.: Conceptual Deterrent to Revolution in Education', *Elementary School Journal* **74**, 9–23.

CONTRIBUTED PAPERS

SESSION II

HENRYK SKOLIMOWSKI

EVOLUTIONARY RATIONALITY

I. WHY SO MANY RATIONALITIES?

It may sound strange, if not heretical to suggest, as I wish to do, that there is no such thing as the rationality of science. At best we can talk about rationalities of science. Both historically and contemporarily we have used different criteria for explaining and justifying the alleged rationality of science. By giving different criteria for the justification of rationality, we *ipso facto* constitute different scopes for rationality.

This immediately poses the question. If so, which of these criteria are more "justifiable" than others? Which rationality is preferable to other rationalities? Now the point is that in justifying a set of criteria which would elucidate for us a given concept of rationality, in other words, in outlining the scope of a given rationality, we do so by resorting to some kind of rationality. And it would be strange if the hidden rationality we resort to, in order to justify the explicit one, would be much different from the explicit one. Thus a circularity of some kind seems to be unavoidable.

We can attempt to cut through this Gordian knot of self-reference by suggesting that the "rationality" of science is, after all, established with respect to the physical world "out there," and there is nothing arbitrary, or self-referential in attempting to decipher the order of nature "out there," and then to map this order within some set of coordinates which we call rationality. Rationality here would stand for the set of abstract features, or the conceptual model within which the order of nature is being described. Such a procedure would help us with some difficulties, but would involve us in new difficulties. For then we would be equally justified in accepting the scientific rationality, as well as the religious rationality, and a post-scientific rationality. In each case there is a conceptual model – a set of coordinates – within which the order of nature is being described. Moreover, in each case, there is the world "out there," which conforms to our model of understanding of that which is out there. And what is rationality if not a model of understanding?

It would appear that a useful way to talk about rationality is not to treat it as if it were universal, essentially the same for all periods of history and all cultures, but rather recognize that rationality is paradigm-determined and paradigm-justified. I am using the term "paradigm" in a less rigid way than Kuhn, who wants to restrict it to science. If we want to limit the term "paradigm" to science, then the only rationality acceptable is scientific rationality. And then some such awkward questions appear: What about rationality before modern science had emerged? Was there no rationality before science? There obviously was. We can discern the prevailing rationality of the Greek thought vis-à-vis the prevailing rationality of Medieval Christianity, or vis-à-vis the prevailing Moslem rationality. The term "prevailing rationality" obviously refers to some kind of paradigm, which we grasp and can reconstruct and furthermore can express through some abstract features which we call the model of understanding, or rationality of this or that period.

The rationality of modern science (in my parlance rationalities of modern science) cannot claim its superiority over other forms of rationality merely because of the empirical basis of scientific rationality. All rationalities have their "empirical bases." Scientific rationality is so well confirmed by the phenomena of the physical world because it is an ingenious device, a specific set of filters, a peculiar sorting out agency, call it what you will, which admits and legitimizes those phenomena which support its validity. However, the religious rationality of medieval Christianity had equally ample confirmation in phenomena which the religious world view sought and found "out there." We cannot protest that those were "fictitious" phenomena, while ours are "real" ones, for so-called "empirical phenomena" are equally fictitious – though in a different sense. Just what are phenomena and specifically "empirical phenomena?" They are not immediate data of sense experience – we have left behind this kind of sensualism and Machism. They are not "facts" shining in their pristine ontological beauty, as Francis Bacon would imagine. Just what are "phenomena?" And what is "empirical?" The answer to these questions is simple, but embarrassing. "Phenomena," especially "empirical phenomena," as well as so-called "facts" are *consequences* of our theories. There is no way of grasping or defining them except through the mesh of our theories. Since they are constituted by our theories, no wonder that they support and confirm these theories.

The philosophy of science in the 20th century is very instructive on this point. When, with Carnap, we believed in Logical Empiricism, we also believed in the "rock bottom" of knowledge and the "ultimate empirical statements." When we ceased to believe in Logical Empiricism, we ceased to believe in the rock bottom of knowledge and in protocol statements (the ultimate empirical propositions); or conversely, when we ceased to believe in the existence of the (Platonic, because it had to reside in some Platonic heaven) rock bottom of knowledge, and in the ultimate empirical statements, we had to cease to believe in Logical Empiricism.[1]

Then, along with Popper, we invested our faith in refutability. And we believed that empirical statements and theories are those which are empirically refutable by experience. Which was, of course, a huge tautology: for as "empirical" theories we took those which were *empirically* refutable by experience, that is, this segment of reality which was capable of "empirical" treatment, or maleable to "empirical" manipulation. And the meaning of the term "empirical" in all these contexts is a bit mystifying if not mysterious.

For the point is that neither Popper, nor his followers have ever sufficiently explained what it means to be "empirically refutable." We had such definitions as: "capable of clashing with empirical reality," which does not explain much for we still do not know what is "empirical." In addition, what does the "clashing" mean in this context? And finally what is this "reality," against which we test our theories? It cannot be something firm, tangible and constant which we know beforehand, for then we would not need great pains and labors of science to describe IT. So IT must refer to something elusive and changeable. So how does this dramatic clash with something elusive ever come about? Moreover, in the laboratory we make certain measurements, and if the readings depart "too much" from the expected, there is an anomaly: there is the "clash." But the difference in reading, between the expected and the actual, which we observe in the lab, is a far cry from clashes with "reality"; from refutations of theories by "empirical reality." Furthermore, the nature of reality "out there" is infinitely flexible and complex, as ecologists and biologists tell us, and phenomena are interlinked with other phenomena, have multiple causes and effects, multiple consequences and manifestations. Are the aspects of this multiple structure which we simplify and describe in physics more important than other aspects which we perhaps will never be able to describe? Obviously not. So what we elicit and test is a tiny fragment, the fragment which is almost tautologically true.

The situation is even more difficult, in order not to say desperate, for of late Popper and his followers have renounced the principle of refutability as the corner stone of scientific validity. If theories are not refutable, or refuted in actual practice, then the "empirical" justification of science becomes even more tenuous. For what else is there to remove from science a theological stigma: like old theology science "creates" its facts and symbols, which are tied together as phenomena and theories, and then finds them "out there" because it has created them.

A closer inspection of the inner mechanism (or should we say the inner universe) of science was necessary in order to show that science is as question-begging a strategy in the generation of knowledge as are pre-scientific modes of knowledge. Each is specific to the culture by which it was originated and is governed by the cognitive paradigm characteristic of this culture. The understanding of the nature of this paradigm is tantamount to the understanding of the overriding concept of rationality specific to this paradigm.

I should mention parenthetically that rationality here discussed can be defined as a property of knowledge. It is thus *epistemological* rationality with which we are here concerned. From this rationality we have to separate sharply two kinds of rationality which are in common usage in the sociological literature. One is so-called organizational rationality (in the jargon of sociologists this is Zweckrationalität). Rationality is here defined as the pursuit of efficiency, as the choice of means for implementing desired goals, while goals are *a priori* given to us. An appropriate name for this rationality would be instrumental rationality.

The other kind is so-called "substantive rationality" (in the jargon of sociologists this is 'Wertrationalität'). Rationality is here defined as the evaluation of goals and ends of human life. Thus this rationality has to do with the choice and justification of values and ends of our lives or actions. For this reason the term "axiological rationality" would be more appropriate than "substantive rationality." Obviously, when we discuss rationality in science and in philosophy of science we mean neither of the two rationalities, but rather, as I have suggested, rationality in the epistemological sense, as a property of knowledge, a justification of knowledge, a justification of reason itself, which again is an epistemological category.

Although most philosophers would eagerly disassociate themselves from both instrumental and axiological forms of rationality, inadvertently they often subscribe to these forms. In desperation to find *any* justification of rationality of science, philosophers of science, claim that the 'success' of theories is the justification of their rationality. This is no doubt an instrumental form of rationality, and as dubious as all instrumental justifications of knowledge are. It might be argued that the Holy Inquisition that forced Galileo to recant was 'successful' in promoting its theories of the universe. Yet, we feel that this kind of "success" has little to do with the rational justification of the rationality of knowledge.

Orman Van Quine, Kazimierz Ajdukiewicz and some other 20th century philosophers attempted to give the ultimate justification of their epistemology, and thus the rationality of this epistemology, in pragmatic terms – the use and value of knowledge for the human species at large. This pragmatic justification is much broader than the instrumental justification, which blesses scientific theories on the basis of their 'success'. The pragmatic justification of knowledge may rule out that only those forms of knowledge are justified which aid the human species in the long run. On the basis of this criterion, science may fall short of the ideal.

The 20th century philosophy of science, on which I touched briefly, is an eloquent evidence of the multitudes of rationalities, sometimes competing with each other, sometimes supplementing each other, but always tied to a larger model (paradigm), or a conception of knowledge of which they are a part, which they exemplify, which they justify, and by which they are justified. If we take the current state of the debate, we may for the sake of brevity (and by no means attempting to diminish the importance of other actors, perhaps here present) limit it to a drama enacted by five principal actors: Carnap, Popper, Kuhn, Feyerabend, and Lakatos. I am singling them out as dramatis personae for they have made the 20th century philosophy of science a bit of a drama, which of course does not deny their originality.

Let me continue a bit with the narrative I already started. After the refutability principle was dethroned as the major criterion of demarcation of science from non-science, Popper did not hand over the sceptre to Kuhn. It was of course as the result of Kuhn's impact that Popper relinquished the idea of refutability. Kuhn's concept of the philosophy of science incorporates most of the essential features of Popper's philosophy of science: the dynamic and dialectical character of science, its historical nature, and the conviction that the understanding of the growth of science gives the key to the understanding of the nature of science. But Kuhn went a step further. He changed the basic units by means of which science proceeds and through which it can be best understood. Instead of theories (conjectures) to be followed by refutations, Kuhn introduced the idea of the paradigm – to be followed by normal science, puzzle solving, then to be followed by anomalies, then by a crisis, then extraordinary science, then a revolution. We obtain an altogether different model for the growth of science. Paradigm is a conceptual unit much larger than particular conjectures (theories). To accept the idea of the paradigm means:

(1) to undermine the raison d'être of the *conjecture and refutation* schema as the basic unit for the understanding of science;

(2) to undermine the Popperian idea of the rational reconstruction of the growth of science;

(3) to undermine the very idea of Popperian rationality. For if there are no conclusive refutations, we lack the ground for the justification of the rationality of science; we lack the ground, in other words, for intersubjective process of testing.

(4) to replace existing standards for the rationality of science with new standards: the concensus of the scientific community to accept or not to accept a given paradigm becomes the basis of the rationality of the paradigm.[2]

Some critics of Kuhn have suggested that this was an erosion of rationality. Others, such as Lakatos went so far as to suggest that "in Kuhn's view scientific revolution is irrational, a matter for mob psychology," which is very picturesque as an expression but really a travesty of truth. For Kuhn never advocated irrationalism as the foundation of science, let alone would dream of making it a matter for 'mob psychology'. He indeed explicitly stated that "No process essential to scientific development can be labelled irrational without vast violence to the term".[3] But there is no denying that the continuous discussion of Kuhn's model of science during the last decade has resulted in undermining all basic premises on which science is allegedly based, including the rationality of science.

Let me now summarize the situation with regard to the 20th century philosophy of science. It is my contention that:

(I) Every paradigm of the philosophy of science contains, entails or presupposes a concept of rationality through which the essential features of science within a given paradigm are expressed.

(II) Different paradigms in the philosophy of science usually (but not necessarily) outline different concepts of rationality.

(III) The continuous and successive erosion of the identity of science during the last 20 years or so signifies the erosion of the notion of the rationality of science.

II. HOW SCIENTIFIC RATIONALITY HAS BEEN DISSOLVED

Threatened by the conceptual power of the concept of 'paradigm', which explains everything, or nearly everything that Popper's philosophy of science did, but in a different way, Popper and his followers started to devise a variety of gambits which would either rescue their old position – based on theories (conjectures) as basic units – or would diminish the importance of Kuhn. First came the slogan: Revolution in permanence. Science is in the process of continuous revolution, every theory is a revolution, and aims at a revolution. (It is not quite certain who was the first to introduce the phrase: Popper or Feyerabend.) This strategy could not be employed for a long time.

So altogether new philosophies of science were worked out. I have here in mind Popper's new metaphysics called *The Third World View* and Lakatos' metaphysics called *Research Programmes*, both of which in an implicit or explicit form attempt to meet Kuhn's challenge.

In the 1950's and 1960's one heard, in the Popperian circles, a great deal about the Postscript, in which Popper's new unpublished philosophy was supposed to be contained. This postscript was never published as a separate book. What did appear in the late 1960's, however, were Popper's two important papers:

1. 'Epistemology without a Knowing Subject', *Logic, Methodology and Philosophy of Science*, III, Amsterdam 1968.
2. 'On the Theory of Objective Mind' (Vol. I, *The Proceedings of the XIVth International Congress of Philosophy*, Vienna 1968).

In these two papers, it is alleged, there is the gist of the legendary Postscript. The conceptions which Popper has developed in these papers are entirely novel and altogether surprising, for one finds in them a grand metaphysical design painted al-fresco in bold strokes, quite in contrast to his previous sober and meticulously developed philosophy. This doctrine has been further expanded by Popper in his contributions to the *Philosophy of Karl Popper*. The doctrine is based on distinguishing the three worlds: World 1, World 2 and World 3.

What are then the three worlds of Popper, and especially what is Popper's third world from which the doctrine derives its name? "I shall say there are three worlds: the first is the physical world or the world of physical states; the second is the mental world or the world of mental states; and the third is the world of intelligibles, or of *ideas in the objective sense*; it is the world of possible objects of thought." (Vienna, p. 26)

"third world structural units are intelligibles" (Vienna, p. 35)

"Theories, or propositions, or statements are the most important third world linguistic entities." (Vienna, p. 28)

"The activity of understanding consists essentially in operating with third world objects." (Vienna, p. 32)

But Popper furthermore argues: *"One can accept the reality or, as it may be called, the autonomy of the third world, and at the same time admit that the third world is a product of human activity."* (Vienna, p. 29)

"The third world has grown far beyond the grasp of any man, and even of all men. Its action upon us has become at least as important for its growth as our creative action upon it." (Vienna, p. 30)

"All work in science is work directed towards the growth of objective knowledge. We are workers who are adding to the growth of knowledge as masons work on a cathedral." (Amsterdam, p. 347)

"To sum up, I have tried to show that the idea of the third world is of interest for a theory of understanding which aims at combining an intuitive understanding of reality with the objectivity of rational criticism." (Vienna, p. 47)

Now we can clearly see that if we grant the autonomy of the third world in the way Popper requires, then we have rescued the autonomy, the objectivity and the rationality of science from the tentacles of sociologism, for then science is an aspect of this autonomous cognitive structure which "has grown beyond the grasp of any man and even all men; which is like a cathedral to which we as individual workers contribute to as masons to the construction of the cathedral." Popper's new metaphysics, without any doubt, may be interpreted as an edifice to meet the challenge of Kuhn. It is the edifice which attempts to rescue science from the jurisdiction of Kuhn's paradigm, and which gives a new justification for the objectivity and the rationality of science. And this is so *regardless* whether the main ideas were conceived before Kuhn's book appeared, or after it appeared.

By ascending to this high metaphysical level Popper has combatted the peril of the socio-historical-psychological approach to the rationality of science. But at a prize of introducing a rather complex, and often question – beginning metaphysics.

We may ask whether Popper has really combatted the peril of sociologism, if we think it to be a peril? Has he not introduced even greater difficulties in the notions of the three worlds, their peculiar existence and their interaction with each other, in order to resolve a comparatively smaller difficulty which was the relative importance of the paradigm vis-à-vis the importance of conjectures and refutations, as basic conceptual units of the philosophy of science?

Parenthetically we may observe that Popper is not quite consistent. For on the one hand he seems to justify the validity of science and the rationality of science by appealing to the autonomy of the third world, and, on the other hand, he falls back on his former criterion of rationality as inter-subjective criticism, that is, generalized refutibility.

These difficulties have not been resolved with the appearance of *The Philosophy of Karl Popper* (The Library of Living Philosophers), in which Popper, both in his intellectual autobiography and in his 'Replies to My Critics,' constantly oscillates between the two justifications of the objectivity of science (and indirectly of rationality of science, though he conspicuously avoids the term), that is between the old methodological justification: intersubjective criticisibility; and the new metaphysical justification: the autonomy of third world entities.

Quite a different strategy was chosen by Imre Lakatos. After his brilliant essays on the philosophy of mathematics, 'Proofs and Refutations' Lakatos has decided to make the whole field of the philosophy of science his domain, perhaps in more senses than one of the term 'his domain.' I am referring here specifically to Lakatos' views as expressed particularly 'Falsification and the Methodology of Scientific Research Programmes' in *Criticism and the Growth of Knowledge.* Lakatos attacks Kuhn head on. The main vehicle of this attack is Lakatos' conception of *Research Programmes.* What are the research programmes ? They are what they say – programmes for research in science (but not exclusively so). They are methodological devices which tell us how we ought to practice science in order to arrive at fruitful results. The normative nature of Lakatos' research program is quite obvious and even striking. This is not to say that Kuhn's and Popper's are not normative methodologies of science. They are. But in a more camouflaged and ambiguous way. Two important aspects of research programmes should be emphasized from the start. One is that Lakatos abandons Popper's idea of refutation. Refutations according to Lakatos belong to the stage of early, that is 'naive' falsificationism. In its more mature, more sophisticated form, falsificationism recognizes that theories are hardly ever refuted in actual scientific practice. One does not have to be very perspicuous in order to realize at once that Lakatos concedes to Kuhn a most important point: in actual scientific practice theories are hardly ever refuted.

The second important point of Lakatos' research programmes is the explicit endorsement of Feyerabend's methodology of proliferation of conceptual divices for the sake of generating new knowledge. Lakatos explicitly says "*proliferation of theories* is much more important for sophisticated than for naive falsificationism." (*Criticism and the Growth of Knowledge*, p. 121). The acquisition of knowledge, by whatever means, seem to be all important. Therefore 'anything goes.' Feyerabend has formulated the slogan, Lakatos echoes it over and over again.

The impression so far might have been created that Lakatos is rather derivative in formulating the foundations for his conception of philosophy of science. Such an impression is perhaps not far from the truth. However in developing in detail his philosophy of research programmes Lakatos at least offers new and interesting terminology. The key concepts are: the progressive problem-shift and the degenerating problem-shift. A problem-shift occurs if a series of theories brings about some excess empirical content (each greater than its predecessor), and if each predicts some novel physical facts unexpected before. Otherwise, theories represent degenerating problem-shifts. "Let us say that such a series of theories is *theoretically progressive (or 'constitutes a theoretically pro-*

gressive problem-shift') if some of this excess empirical content is also corroborated, that is, if each new theory leads us to the actual discovery of some *new fact*. Finally, let us call a problem-shift progressive if it is both theoretically and empirically progressive, and *degenerating* if it is not." (Ibid., p. 118)

Lakatos furthermore asserts: "we *'accept'* problem-shifts as 'scientific' only if they are at least theoretically progressive; if they are not, we *'reject'* them as 'pseudoscientific.'" (Ibid., p. 118).

There are endless difficulties here. For we really do not know how to interpret terms which are both italicized and put in quotes. Lakatos says: *'we accept'*. Does this mean that he *accepts*, or only pretends to ? Similar difficulties face us with the notion of *'rejecting'*, and of problem-shifts that are degenerative, thus 'pseudo-scientific'. One would have thought that if a problem-shift is degenerating, than it is at least within the realm of science, not pseudo-science. It would appear that we can trace certain problem-shifts which are fruitful, and others that are fruitless only if we accept a certain domain called science which is common to both kinds of problemshifts. Therefore, it is contradictory to call these less fruitful problem-shifts a pseudo-science, for clearly they must belong to the realm of science to be recognized as regressive, or degenerating.

The conceptual difficulties are even greater when we examine other crucial notions of Lakatos's system. Lakatos talks about the protective belt shielding theories against degenerating shifts. What does this mean – the protective belt? We can talk about the chastity belt as being really protective, but can we talk in the same sense about a protective belt of theories? Would it not imply a self-serving tendency to ignore obnoxious counter-instances?

There is also a great difficulty in the notion of progressive problemshifts leading to the discovery of "new facts". Lakatos announces with a great gusto: *"And we should certainly regard a newly interpreted fact as a new fact, ignoring the insolent priority claims of amateur fact collectors."* (Ibid., p. 157). Who is the fact collector in this context? What are 'facts'? Since when is one allowed to recognize facts independent of theories? If facts are generated by theories, as Lakatos admits, then the whole programme is tautological.

One would be inclined to agree with Feyerabend's critique of Lakatos. "Thus the standards which Lakatos wants to defend are either *vacuous*

– one does not know when to apply them – or they can be *criticized* on grounds very similar to those which led to them in the first place." (op. cit. p. 215) And Feyerabend rightly concludes that such standards are but "a *verbal ornament.*"

I do not wish to dwell on conceptual inconsistencies of Lakatos's programme of research programmes, but I rather wish to concentrate on the very foundation of his philosophy of science, and in particular his defence of the rationality of science. The astonishing fact is that after the *tour de force* of weaving various examples into his structure, when it comes to the ultimate justification of science and of its rationality, Lakatos falls back on Popper (of the metaphysical period) and completely forgets all his beautiful constructions. For Lakatos tells us that:

'the – rationally reconstructed – *growth of science takes place essentially in the world of ideas, in Plato's and Popper's 'third world'*, in the world of articulated knowledge which is independent of knowing subjects. (Ibid., pp. 179–180).

This is a complete surprise. In the final analysis Lakatos seems to have attempted two equally inconsistent strategies:

(i) He attempts to reconcile the anarchistic programme of 'anything goes' with the idea of normative philosophy of science. This would not do: if his philosophy of science is to be normative, then there *are* some norms based on such criteria as "progressive problemshifts". Therefore, it is not the case that 'anything goes'.

(ii) He attempts to reconcile the anarchistic *anything goes* with the permanence of the third world objects, as conceived by Popper. This again will not do: if there is any validity to these third world objects, then it is not the case that 'anything goes'.

The idea of proliferation is fine. But ultimately, one has to pay a price for proliferation at whatever cost. Lakatos' difficulties and inconsistencies are not accidental but rather endemic to the present state of the philosophy of science.

A very important figure in the debate over the nature of science and the rationality of science during the last two decades has been Paul Feyerabend. He has been a consistent radical sceptic, if such a thing is possible. He has attempted to remain in a coherent frame of reference while preaching his cognitive nihilism or methodology of proliferation under the banner *anything goes*. The motto *anything goes* in Lakatos is clearly an imported one. 'Anything goes' in Feyerabend is the ultimate residue of his questioning of all established laws and theories, procedures and methodologies. Since there are no sacred cannons, anything can be used as a vehicle for furthering the development of science and of human knowledge.

Among contemporary philosophers of science Feyerabend is one of the very few who have the courage to derive the ultimate conclusion from his methodological anti-absolutism: if there are no sacred cannons in science, therefore science itself is neither sacred nor absolute. Feyerabend explicitly states: "I want to argue that science both is, and should be, more irrational than Lakatos and Feyerabend, (the Popperian$_3$ author of the preceding sections of this paper and of 'Problems of Empiricism') are prepared to admit". ('Consolations for the Specialist' in *Criticism and the Growth of Knowledge*, p. 214).

He further argues that:

> The sciences, after all, are our own creation, including all the severe standards they seem to impose upon us. It is good to be constantly reminded of this fact. It is good to be constantly reminded of the fact that science as we know it today is not inescapable and that we may construct a world in which it plays no role what ever (such a world, I venture to suggest, would be more pleasant than the world we live in today). What better reminder is there than the realization that the choice between theories which are sufficiently general to provide us with a comprehensive world view and which are empirically disconnected may become a matter of taste? That the choice of our basic cosmology may become a matter of taste? (Ibid., p. 228).

Feyerabend rightly points out that science has helped man to gradually free himself from fear of unexamined systems. The question of today is: what values shall we choose to probe the sciences of today? Feyerabend's answer is: "It seems to me that the happiness and the full development of an individual human being is now as ever the highest possible value... Adopting this basic value we want a methodology and a set of institutions which enable us to lose as little as possible of what we are capable of doing and which force us as little as possible to deviate from our natural inclinations." (Ibid., p. 210).

At this point the justification of the rationality of science becomes an overall (aesthetic? ethical?) justification of the conception of good life.

Now, let us put together in Table I the various concepts of rationality held explicitly or implicitly by the various thinkers we have discussed.

TABLE I

	Given a priori or a posteriori	The nature of the justification	Rationality — Governed by	Methodological programme: descriptive/normative
Logical Empiricism	A priori	Logical	rule governed	Descriptive
Popper (methodological period)	A posteriori	Dialectical	$TT_1 \Rightarrow EE \Rightarrow TT_2$ governed	Normative
Kuhn	A priori (within the paradigm); a posteriori (in terms of the community decision to accept the paradigm)	Normal Science	Paradigm governed	Normative
Popper (metaphysical period)	A priori	The third world	The autonomy of the third world governed	Descriptive
Lakatos	A Posteriori	Research programme (the third world *and* the principle of proliferation)	Progressive problem-shifts governed	Normative
Feyerabend	A posteriori	Aesthetic/pragmatic (in the broad sense)	The principle of proliferation; (anything goes)	Normative (open ended)

III. RATIONALITY AND UNDERSTANDING

Present attempts to clarify the notion of rationality in science have resulted in deepening our confusion. We are in the midst of a deep conceptual crisis not just in a period of temporary setback which can be easily reversed with more diligence and application. Philosophy of science is at present in a worse state than Ptolomeic astronomy was at the time of Copernicus. Simply: we are in the period of the reign of (conceptual) chaos and anarchy and unreason. And yet, we pretend that all is well and produce voluminous papers, which are but epicycles built over epicycles. There is a movement and a torrent of words, but we have lost sight of ends and of the significance of these words.

During the last meeting of the Philosophy of Science Association, held at Notre Dame, there was a symposium devoted to the unity of science. For all these years one thought one knew what the thesis about the unity of science was concerned with. During this symposium however, one learned that the unity of science was about something else – an elusive examination of some very abstract problems of quantum theory, which a very few specialists are capable of grasping. What has happened to the original thesis of the unity of science? Well, it seems to have transformed itself into epicycles of quantum theory. Moreover, one thought one knew what reductionism was. One thought that it was a simple and straightforward concept. But no. During the symposium one learned that reductionism itself is nothing simple, but in fact a complex problem.

During another symposium (at the same meeting of PSA) the subject was the assessment of scientific theories. One symposiast claimed that ultimately "you must validate it [a scientific theory] in your bones." As simple and fundamental as that. Whereas the next symposiast advanced the thesis that the validation of scientific theories is a most complex process, almost beyond our reach. So on the one hand – the utter simplicity, and on the other – the utter complexity; and in the end – no criteria. Symptomatically enough, the two speakers on the panel did not seem to *see* that their respective views were entirely incompatible.

Examples could be quoted *ad infinitum*. We must look carefully at the whole situation and ask ourselves: why do we accept this conceptual chaos? For chaos undoubtedly it is! I was taken to task during my presentation for calling *it* a "juicy chaos." In my defense I might say, that it is "juicy chaos" (not a sterile one, for instance) because it still sustains us, enables us to write papers, attend conferences, and even teach and preach the subject to our poor and defenseless students!

The situation is not right. In fact, it is pathological. We generate more and more information which does not generate enlightenment, but instead breeds confusion. This is true of science, of scholarship in general, and of philosophy of science in particular. We have become intoxicated by the proliferation of theories and bits of information. We have become narrow specialists, each hoarding respective "data" and proliferating them. And the results are lamentable: grotesquely growing volume of information and proportionally growing volume of confusion. That scientists have not perceived this fact is perhaps more excusable: they are trained and conditioned to be "specialists," thus to sacrifice the whole for the sake of small parts, or parts of parts, or parts of parts of parts... For philosophers it is less excusable to become narrow specialists, for their business is that of *understanding*. Yet, we have fallen victims of the same plague: proliferation of information at the expense of understanding. The situation is especially grave with regard to such crucial concepts as rationality. Rationality is the backbone of our understanding. The crisis of rationality is *ipso facto* a crisis of our paradigm of understanding. We are willing to grant this point theoretically, but unwilling to face its consequences in practice. Since rationality is at present tottering (and science itself is perpetuating chaos), our very concept of understanding is at stake. Ultimately the debate over present conceptual chaos, and the present crisis of rationality is one over the criteria and the basis of our understanding.[4] What is rationality for, if not to aid our understanding? The business of philosophy, philosophy of science among others, has always been that of understanding. Philosophy which foregoes understanding as its basic concern loses its *raison d'être*. In this sense present philosophy of science is rapidly loosing its *raison d'être*. I suggest that most philosophers of science are somewhat aware of this fact. They are reluctant to acknowledge it because as the expression goes, "one does not talk about the dead at the funeral feast," and also because of the far-reaching consequences that follow from the fact. But acknowledge the fact we must. What follows, if we do acknowledge it? Many interesting things and a great deal of radical re-thinking which philosophers *per* philosophers should actually rejoice.

With regard to rationality, which is here our main focus, we must realize that in order to re-constitute rationality so that it serves the process of understanding, we need to go beyond present mutations of rationality – to use Toulmin's language. Evolution of knowledge and of our present understanding must be carried on beyond present mutations. No one can be certain what these new mutations may become, for there is always an element of mystery and chance in new mutations. But the essential point is that within the evolutionary framework (and who is not an evolutionist in epistemology nowadays – after Popper and Kuhn, Feyerabend and Toulmin, and, of course, Donald Campbell[5]) science in its present state must not be considered to be the ultimate one, and the present concept rationality must not be regarded as the ultimate embodiment of rationality. Put otherwise, science is just a sub-species of knowledge, not its ultimate pinnacle; scientific rationality is just a specific embodiment of reason, not the alfa and the omega of all reason. If a manifestation of reason leads to confusion and obscurantism instead of illumination, it has become a degenerate form of reason. Such is the case with the present scientific rationality, that is, the rationality which accepts science as the ultimate arbiter of our understanding of reality.

Our function as philosophers vis-à-vis science may be, on the one hand, *apologetic*: we then try to explain and justify what science does (by assuming that science is right and that it represents the "highest form" of human knowledge) – this we have done perhaps for too long; or it may be, on the other hand, *critical*: we then reflect imaginatively and critically on the role of science and scientific rationality within the scope of the whole civilization and along the vicissitudes of the entire human knowledge, and then we need not apologize for science and justify it at whatever cost, even if it is the cost of our understanding. And it is the latter function that I wish to advocate.

In short, we have to assert our critical function with regard to science, break the tethers of narrow scientism which we have inadvertently accepted, and start to develop new mutations of rationality, which will reinstate our concept of reason and of understanding, and which – no doubt – will be related to, and justified in the frame of reference of much larger than, our present science – which, anyway, is itself in shackles. At the expense of *sounding* ridiculous, one wishes to say: philosophers of science (and philosophers in general) arise; you have nothing to loose, but your conceptual epicycles.

IV. EVOLUTIONARY RATIONALITY

Although many have declared its death, positivism is still curiously alive, not so much in itself – as a coherent philosophical doctrine, but embodied and dissolved in the structure of science, and, more importantly, of the scientific world view. By positivism in this context we must not understand a set of doctrines, but rather a set of attitudes. And furthermore, a set of beliefs, of which the most important are:

(1) That all which is known can be reduced to physical laws, and that all genuinely existing phenomena have a physical foundation.

(2) That all genuine knowledge must be acquired through the scientific method which is modelled on physical science.

(3) That the pursuit of knowledge can be best conducted when we limit ourselves to simple and relatively isolated systems.

(4) That we have to give up the comprehensiveness of our inquiry for the sake of precise results; therefore of necessity, we concern ourselves with relationships which are very clearly defined although the realm of these relationships is rather limited.

(5) That there are *a priori* rules of procedure subsumed under some such principle as the objectivity postulate, which assure the validity of our cognitive pursuits.

(6) That acceptable evidence is only that which is definable and defensible in terms of physical science.

(7) That phenomena and processes which have a genuine existence are deterministic in nature or else (in evolution) are a matter of pure chance.

(8) That meaningful language must have empirical foundations and empirical consequences.

(9) That the behaviour or action of large and complex systems is the result of the behaviour of its constitutive parts so that the total behaviour equals the sum of its parts.

(10) That the explanation of the knowledge process must rule out any supernatural agencies which are beyond the reach and dominion of present science.

(11) That truths of science are the only truths, or at any rate the ultimate truths from which there is no appeal.

(12) That one of the most important justifications of the pursuit of knowledge lies in the idea of material progress.

The above set of attitudes and beliefs (which incidentally outlines a methodology and an ontology), constitutes the core of what I call scientific rationality. In contrast, we may distinguish a set of attitudes and beliefs which is the backbone of what I call evolutionary rationality and which spell out a different conception of knowledge, and even a different ontology. The difference between the two sets should not be seen in their particular points, but rather in the fact that they outline two different concepts of understanding. Here are basic principles of evolutionary rationality:

(1) That not all that is known can be reduced to physical laws; some knowledge is irreducible to physical knowledge.

(2) That the methods of physical science are insufficient for the study of the phenomena of life on the high level of complexity.

(3) That we cannot limit ourselves to simple and relatively isolated systems, for life systems are enormously complex and intricately interconnected.

(4) That the pursuit of knowledge, which gives us an understanding of complex life systems requires the inclusion of a large number of relationships which cannot be precisely defined but which enable us to comprehend the phenomena of life with more illumination than through a limited number of exact relationships.

(5) That there are no absolute, *a priori* rules that govern our quest for knowledge: any such principle as the objectivity postulate has its limitations; and that an *a posteriori* feedback is our best guide in the realm of cognition and elsewhere.

(6) That the acceptable evidence is evidence which explains phenomena under investigation regardless of whether or not it meets the requirements of the physical model.

(7) That the most significant phases or processes in evolution are simply beyond the dichotomy: chance or necessity; though not strictly determined and in a sense unpredictable, these processes exhibit patterns of cohesion and integrity which cannot be attributed to mere chance.

(8) That meaningful language must be able to express content which is significant, both empirical and extra-empirical.

(9) That the behaviour or action of many large and complex systems is often inexplicable by the behaviour of the constituents of the system; the total behaviour often equals more than the sum of its parts or differs from this sum.

(10) That those "agencies" or forces which go beyond the reach and scope of present science need not be feared or called "supernatural"; and we need not be nervous to the point of obsession or paranoid about the restitution of god and theology when we attempt to extend the reach of our present knowledge.

(11) That the truths of science are at least partially truths by convention and that there are other kinds and sources of truth over which scientific truths have no authority.

(12) That the idea of material progress is insufficient to account either for the pursuit of knowledge or for the evolutionary processes of man.

As I have mentioned, the first set of attitudes and assumptions constitutes the rationality of hard science, or, for short, (present) *scientific rationality*. The second set of attitudes and assumptions constitutes the rationality of living systems, or, for short the *evolutionary rationality*. Scientific and evolutionary rationalities may be seen as contrasting types of understanding. We must not think however that there is an unbridgeable chasm between them. Scientific rationality is a part of the evolutionary rationality, but not conversely. The characteristics of the evolutionary rationality, when simplified to the point of caricature, do become the characteristics of scientific rationality. Thus scientific rationality represents an end of the spectrum. When the variety of phenomena are reduced in number to one, when complex and open systems become simplified and closed, when language is simplified to express the physical content only, etc., etc., then we obtain a curious transformation of evolutionary rationality into scientific rationality.

We must not make a fundamental mistake and consider the end on the spectrum to be the whole of the spectrum. Because this mistake lies at the foundation of many of our conceptual troubles, we must emphasize the contrast between these two intellectual attitudes which I call: scientific rationality and evolutionary rationality. In actual scientific practice these attitudes are perhaps not held in their pure forms. But we have seen enough clashes of opinion, which are not at all concerned with matters-of-fact or empirical issues, to make us perfectly aware that these contrasting attitudes are alive and competing with each other.

The rationality of evolutionary processes is tantamount to understanding and acknowledging the properties of life. On the level of the human species, and especially on the level of *Homo symbolicus*, life is lived within the normative realm, for it embodies drives and values which are not only gratuitous elements added to the physical-chemical properties of matter, but are often the guiding forces of the whole process of evolution.

The evolution of our understanding has been frozen at the level of the static model of traditional physics. We have physical concepts, chemical concepts, electro-magnetic concepts which we recognize. In science we do not possess concepts that attempt to grasp and depict the higher levels of the complexity of matter: matter endowed with self-consciousness and with spirituality. There is thus a great discrepancy between the dynamic units of actual biological evolution and the static and petrified units of conceptual evolution. A truly evolutionary epistemology requires matching the stages of conceptual evolution with the appropriate stages of biological evolution. Ernst Mayr and others have eloquently argued that the physical interpretation of certain rudimentary biological concepts makes a caricature of these concepts.

In order to be fair to the process of evolution when we describe it on the level of *Homo sapiens*, we have to introduce into our language *open-ended* concepts, *growth* concepts and *normative* concepts. Open-ended concepts will allow us to describe without distortion living phenomena in the process of change, particularly in the process of qualitative change. Normative concepts will allow us to describe without mystification living organisms guided by specific values and directed to specific goals.

We have some ground for our hesitation to admit open-ended, growth and normative concepts to the realm of knowledge. Open-ended concepts seem to violate the principle of the invariance of meaning. Growth concepts seem to violate the law of identity $A = A$; normative concepts seem to violate the principle of objectivity and factuality. The invariance of meaning, the static identity of processes, and the factuality of phenomena are all unwritten premises on which scientific knowledge is built. The admission of open-ended, growth and normative concepts together with the admission of the evolutionist rationality would amount to a new kind of knowledge, namely normative knowledge. From Aristotle onwards, we separated the three realms once united – truth, goodness and beauty, and insisted that science is the pursuit of truth which has nothing to do with the other two realms. The so called objectivity of science came to epitomize our search for truth. The principle of objectivity, so it appears to many, seems to exclude such entities such as growth concepts and normative concepts.

Perhaps the time has come to re-examine our whole intellectual heritage from Aristotle onwards. Perhaps the separation of truth from values was premature, or only temporary. Perhaps a compassionate attitude towards living beings, and particularly to life endowed with consciousness, is as rational as an objective attitude. Indeed the development of a compassionate attitude would allow us to accommodate all the three new sets of concepts (open-ended, growth and normative) which are necessary for the understanding of evolution in its higher stages.

Our predicament can be reformulated as follows: what kind of new concepts, new modes of knowledge must we develop in order to account for the socio-cognitive-cultural stage of human evolution? If we capture the last stage of our evolution adequately, we shall be able to capture earlier stages; but not conversely.

We are animals of sorts. But is seems that in the animal kingdom, the more the animal knows, the better it can cope with the environment. Thus the enlargement of the animal's knowledge is for the animal life-promoting and life-enhancing. The knowledge animals possess is life-enhancing, thus normative; not a set of abstract categories, but normative rules for acting in order to preserve and enhance life. One wonders *why* human knowledge has ceased to be of this kind. Why does knowledge often signify mere information, why does it help the individual but little in enhancing his life, why does it so often interfere with the ends of human life? Our dilemma can be once again reformulated: What kind of knowledge will be illuminating for the understanding of life processes and at the same time serve as a guide to the good life? If we accept this reformation, then the road to a normative paradigm for knowledge is open.

In summary, rationality is embedded in the system of knowledge of which it is often seen as arbiter. Our rationality reflects the vicissitudes of our knowledge. Physical science has, during the last two centuries, been the model for all our knowledge. Present (scientific) rationality is constructed in the image of, and according to the criteria of physical science. If it is true (as I contend is the case) that physical science has become a cognitive (and in a certain sense a social and moral) straitjacket, then it is true that the rationality characteristic of physical science needs to be replaced by a more adequate form of rationality.

Alternative rationality is not a matter of imposing a set of arbitrary abstract principles on existing knowledge. Such a rationality must evolve as a function of an alternative system of knowledge. Evolutionary rationality, which I propose as an improvement on present rationality is connected with and eptomizes an alternative model of knowledge, and of understanding.

Rationality must be an aid to man's understanding of the world around him; and an aid to man's struggle for a meaningful and fulfilling existence. There is much more to our understanding of the cosmos and our quest for a meaningful life than the present (scientific) rationality allows for. The creation and implementation of an alternative mode of rationality will be but a stage in man's cognitive and spiritual development.

University of Michigan, Ann Arbor

NOTES

[1] On the post mortem of Logical Positivism see especially: John Passmore 'Logical Positivism' in which he writes: "Logical Positivism, then, is dead, or as dead as a philosophical movement ever becomes." *The Encyclopedia of Philosophy* (Paul Edwards, ed.), Vol. V., p. 56.
[2] For a further discussion of the impact of Kuhn on Popper and his school see: Henryk Skolimowski, 'Karl Popper and the Objectivity of Scientific Knowledge', in *The Philosophy of Karl Popper*, Vol. XIV of The Library of Living Philosophers, 1974.
[3] Thomas Kuhn, 'Reflections on my Critics,' in *Criticism and the Growth of Knowledge*, (Imre Lakatos, ed.), 1970, p. 235.
[4] For a further discussion of this point see Henryk Skolimowski's: 'Science and the Modern Predicament', *New Scientist*, February 11, 1972; 'Science in Crisis', *The Cambridge Review*; 'The Scientific World View and the Illusions of Progress', *Social Research*, Spring 1974.
[5] See Campbell's contribution of *The Philosophy of Karl Popper*, op. cit.

LINDA WESSELS

LAWS AND MEANING POSTULATES

in van Fraassen's View of Theories

I

In several recent papers ([7], [8]) Bas van Fraassen has suggested that the structure of a scientific theory might be more appropriately represented by his "semantic view of theories" than by the traditional "syntactic view."[1] Under the syntactic view, to characterize a theory one provides "a finite list of sentences given to count as axioms, plus a finite set of syntactic transformations, of an effective character, given to generate the set of all theorems from these axioms." ([8], p. 305) The set of all such theorems is identified with the theory, and contains the set of all claims made by the theory about physical systems. Under the semantic approach, however, one "does not view a physical theory as... a kind of Principia Mathematica cum nonlogical postulates." ([7], p. 337) Rather, "all the resources of mathematical English" are used to construct a theoretical framework, and "the theoretical reasoning of the physicist is viewed as ordinary mathematical reasoning concerning this framework." ([7], p. 337–8) In a moment I will sketch out the structure of a theoretical framework, but first let me give you a rough idea of the problem which has occasioned this paper.

As the title suggests, the problem has to do with the roles of what I shall call laws and meaning postulates in van Fraassen's semantic view of theories. Traditionally a distinction has been made in the syntactic view of theories between those postulates which are true because of the meanings of the terms contained in them, i.e., the meaning postulates, and those postulates which are true because of the way the world is, i.e., empirical postulates or laws. The distinction between these two sorts of postulates is displayed under the syntactic view by simply calling one set of postulates meaning postulates, and the other laws, but formally the two sorts function in exactly the same way – as subsets of an undifferentiated set of axioms from which the claims of the theory are deduced.

In van Fraassen's semantic view, the distinction between laws and meaning postulates is explicated by more than a mere labeling. A law of a theory is incorporated into the corresponding theoretical framework in a way quite different from the mode of incorporation for a meaning postulate, and the two function in quite distinct ways when the claims of the theory are determined. Thus the semantic view seems to imply that there is a legitimate distinction to be made between the laws and meaning postulates in an actual scientific theory. After all, one of the purposes of constructing a formal model of theories is to exhibit the way distinguishable and interestingly different parts of a theory are related. If meaning relations and laws play significantly different roles in actual theories, one should prefer a formal model of theories which reflects that difference. Conversely, if a formal model of theories exhibits a marked distinction between the roles played by two different parts of the model, one might reasonably expect to find that difference mirrored in the roles played by the corresponding parts of actual theories.

Thus we might expect that behind the semantic view lies the claim that there is an objective and philosophically significant distinction to be made between laws and meaning postulates. To be sure, this is exactly the position held by those who developed the syntactic view, but, as Quine (among others) has pointed out, the mere sorting of axioms into two categories hardly serves to explicate what this presumed distinction amounts to. The semantic view, on the other hand, displays a significant functional difference between laws and meaning postulates, grounded in the way a theoretical framework explicates logical and physical modalities. It is therefore perhaps surprising that van Fraassen takes the semantic view to demonstrate a lack of distinction between laws and meaning postulates. In his paper 'On the Extension of Beth's Semantics of Physical Theories' he asserts that a certain characteristic of theoretical frameworks supports the claim that "there is no objective distinction within physics between 'meaning postulates' and 'empirical postulates'." ([7], p. 331)

The characteristic on which this claim rests will be discussed in detail later on. What is important at this point is to note the apparent conflict between the following facts: (1) van Fraassen's theoretical frameworks do exhibit the difference between laws and meaning postulates, and (2) they also support his denial of any objective difference between the two.

My purpose in this paper is to bring this conflict more sharply into focus, and then to suggest a way in which it might be resolved.

I will begin by describing briefly the structure of a theoretical framework, and explaining the ways a law and a meaning postulate are incorporated and function in a framework. A theoretical framework is determined by specifying three things: (i) a semi-interpreted language, L, (ii) the form which a model, M, of this language must take, and (iii) the laws of the theory. A semi-interpreted language, L, is an ordered triple $\langle E, S_L, h \rangle$. The first member of this triple, E, is a set of elementary statements, i.e., the set of statements which are available in the theory for describing physical objects; S_L is the logical state space, i.e., the set of all states which a physical system could possibly be assigned; h is the satisfaction function which maps elementary statements into the subsets of the logical state space. The minimal requirement on the form of a model of L is that it be an ordered pair $\langle X, \text{loc} \rangle$, where X is a physical system and loc is the location function which associates X with a state in the logical state space.[2] Truth of an elementary statement e in a model $M = \langle X, \text{loc} \rangle$ is then defined: e is true in $\langle X, \text{loc} \rangle$ iff $\text{loc}(X) \in h(e)$. Laws of the theory, the third element of a theoretical framework, express relations (both time dependent and time independent), among the predicates of the elementary language, i.e. the language in which the elementary statements are formed.

As van Fraassen points out in his paper 'Meaning Relations among Predicates' ([5], p. 161), h and S_L determine what he calls an interpretation function for the elementary language, L_E.[3] This interpretation function, f, maps the n-ary predicates of the elementary language into the subsets of S_L^n (the set of n-tuples of members of S_L). Since the models of the semi-interpreted languages treated in this paper have as first members only single physical systems, the elementary languages of these semi-interpreted languages need contain only one-place predicates. For the purposes of this paper, then, we only need to think of this interpretation function, f, as a mapping from these one-place predicates into the subsets of S_L. Notice that this does not mean the elementary language is thereby given a semantic interpretation, for the extensions of the non-logical symbols of this language still remain to be specified. Rather, the interpretation function provides a '*semi*-interpretation' for the elementary language in the sense that the interpretation function specifies the *meaning relations*, i.e., the 'intensive relations', among the non-logical symbols of this language. The logical state space serves as a sort of mock domain of interpretation in which these meaning relations are exhibited. The inclusion and overlap relations among the subsets of the logical state space reflect the inclusion and overlap relations among the intensions of the predicates mapped onto these subsets by the interpretation function.

Now clearly, these meaning relations induce logical relations among the elementary statements. In fact, not only do the satisfaction function and the logical state space determine the interpretation function, but, as van Fraassen has pointed out, the interpretation function and the logical state space plus a truth definition for the logical terms of the elementary language determine, in turn, the satisfaction function. Thus the logical relations among the elementary statements induced by the meaning relations among predicates of the elementary language are exhibited by the inclusion and overlap relations among the sets of states which the satisfaction function assigns to these statements.

The semi-interpreted language is therefore viewed by van Fraassen as determining logical modalities which arise due to the meaning relations among the terms of the language of a theory.[4] The laws, on the other hand, determine physical modalities. The relations among properties specified by the laws are just those relations which the theory claims can hold in any physically possible system. Laws of coexistence specify the time independent relations, laws of succession the time dependent relations.[5] Thus, in van Fraassen's semantic view of theories, laws of coexistence are conditions "limiting the class of physically possible states" and, at least upon one way of viewing them, "laws of succession select the physically possible trajectories in the state space." ([7], p. 330–1)

Even this brief sketch of van Fraassen's semantic view of theories indicates the quite different ways a meaning postulate and a law are incorporated into a theoretical framework, and the resulting quite different roles they play in determining relations among claims of the theory. Meaning postulates are built into the elementary language (the language used to make claims about physical objects) by constructing an appropriate semi-interpreted language. The meaning relations thus determine what configurations of properties of physical systems are logically possible. This, in turn, determines certain logical relations among the elementary statements, as we have seen. Laws, on the other hand, appear as separate elements of the theoretical framework. They pick out certain of the logically possible configurations of properties as physically possible. Since they alter neither the structure of the logical state space nor the satisfaction function, the laws of a theoretical framework have no effect at all on the logical relations among the elementary statements. They serve merely as comments on the physical possibility of finding certain combinations of these elementary statements accruing either simultaneously or successively to a physical system.

Now how are these obvious differences between the mode of incorporation and function of laws and meaning postulates to be reconciled with the further consequence of the semantic view – that, as van Fraassen expresses it, there is no objective distinction between laws and meaning postulates? Before suggesting an answer to this question, let me clarify what this denial of objective difference amounts to. It turns out, in fact, that there is actually a little work to be done before we are justified in taking this denial in all its apparent generality as a consequence of the semantic view of theories.

Let us begin by looking at the context in which van Fraassen's denial of an objective difference appears. In that particular part of 'On the Extension of Beth's Semantics of Physical Theories' he had been illustrating the way a law of coexistence might be expressed by constructing a theoretical framework for a theory of ideal gases which contains the Boyle-Charles law as a law of coexistence. I will not go into the details of this example, but essentially what happens is that the law of coexistence is expressed by specifying as the set of all physically possible states just those states of the logical state space for which pressure, temperature and volume are related in accordance to the Boyle-Charles law. Immediately after this example, we find the following paragraph, in which appears the statement quoted earlier:

Of course, when the law of coexistence is general enough, it may be incorporated in the definition of the state space. In our example, we may take the state-space of this ideal gas to be a region in Cartesian 3-space all of whose points satisfy the Boyle-Charles Law – and not the whole 3-space. There is nothing right *or* wrong about this procedure – and this is just to say that there is no objective distinction within physics between "meaning postulate[s]" and "empirical postulates". ([7], p. 331)

In other words, we may construct a new theoretical framework in which we treat the Boyle-Charles law as if it were a meaning postulate by taking as the new *logical* state space just that subset of the old logical state space for which the Boyle-Charles law holds, i.e., that subset which we had previously designated only as the set of *physically* possible states.

This paragraph indicates that there is a certain sense in which, under van Fraassen's view, the difference between the way meaning relations and laws function in a theoretical framework is not to be regarded as a reflection of a significant difference between laws and meaning relations in actual theories. For it indicates that any general statement in an actual theory can be treated as either a meaning postulate or a law. A general statement can be incorporated into the semi-interpreted language of a theoretical framework, and thus function to determine truth and falsity *ex vi terminorum*, or it can appear among the laws of the theoretical framework, thus serving as a statement of physical possibility. Since the choice of how to incorporate such a general statement is left open, the differences which distinguish the two modes of incorporation are in some sense not significant. Thus the semantic view might be taken to indicate that there is no objective difference between laws and meaning postulates, because in this view the difference between a law and a meaning postulate is only apparent—one can always be recast as the other.

We should note, however, that this claim constitutes an extrapolation from the properties of a particular sort of theoretical framework and goes significantly beyond the contents of the quoted paragraph in an important respect. Van Fraassen illustrates and subsequently claims arbitrariness in the choice of methods for incorporating very general statements of a theory only in the case where the statements express time *independent* relations among properties, i.e., the claim is made only with respect to laws of coexistence. Now, if van Fraassen's quite general statement that there is "no objective distinction in physics between 'meaning postulates' and 'empirical postulates'" is to be taken as a consequence of his view of theories, then both modes of incorporation must also be available for statements of time *dependent* relations among properties. That is to say, it must also be true that any general statement about time dependent relations among properties which can be incorporated into a theoretical framework as a law of succession can also be incorporated as a meaning postulate, and *vice versa*. Since many important physical laws are statements about time dependent relations among properties, i.e., laws of succession, it is of more than formal interest whether one can prove that this is the case. In the second section below, this proof will be provided.

II

I will use the phrase 'succession statement' to refer to a general statement of time dependent relations among properties of physical systems (or equivalently, among the predicates of the elementary language). A succession statement viewed as a law is a law of succession; viewed as a statement expressing meaning relations, it is a meaning postulate. The task in this section is to show that a general succession statement can be incorporated into a theoretical framework either as a law of succession, or as a meaning postulate. I will leave the qualification 'general' quite vague; in what follows it is only assumed that the succession statement can be expressed by specifying an appropriate dynamical semi-group (or group). These latter notions will be clarified below.

Specifically, we want to show that any given (general enough) succession statement can either be built into a theoretical framework in such a way as to determine logical truth and falsehood, or can be incorporated into the framework as a statement of physical possibilities. I will proceed by first describing several procedures for such incorporation, and will then argue that they are of the required sort. Matters are complicated slightly by the fact that there are several ways of characterizing the evolution of a system through time. In what is commonly called the Schrödinger picture, a physical system is viewed as being in different states at different times, where each state is associated permanently with characteristic properties. In the Heisenberg picture, a physical system is associated with only one state, but the assignment of properties to states is time dependent. A combination is also possible, in which both assignment of states to physical systems and assignment of properties to states are time dependent. This combined picture, called the Dirac picture or the interaction picture, will not be explicitly taken into account in this paper, since the method for incorporating succession statements arising from it is just a "linear combination" of the methods arising from the Schrödinger and Heisenberg pictures.

The Heisenberg and Schrödinger pictures can easily be characterized in van Fraassen's view of theories. According to van Fraassen, the Schrödinger picture amounts to having "the 'location' of the system in state space [change] with time;" the state space, S_L, becomes an "instantaneous state space." For the Heisenberg picture, "the satisfaction function must be time dependent;" ([7], p. 329) assignment of a state to a physical system need be made only once.[6]

The existence of these two alternative ways of viewing the time evolution of a system suggests four different ways to incorporate a succession statement into a theoretical framework. We may (1) adopt the Schrödinger picture and incorporate the succession statement as a law, (2) adopt the Schrödinger picture and incorporate the succession statement as a meaning postulate, (3) adopt the Heisenberg picture and incorporate the succession statement as a meaning postulate, or (4) adopt the Heisenberg picture and incorporate the succession statement as a law. Below I will refer to these modes of incorporation as *LS*, *MS*, *MH* and *LH* respectively.

In explicitly characterizing these four modes, only succession statements which can be given by specifying a dynamical semi-group (or group) will be considered. A dynamical semi-group is a one-parameter semi-group of transformations, where the transformation U_0 is the identity element and the group multiplication is defined by $U_t \cdot U_{t'} = U_{t+t'}$, and where $U_t(s)$ is the state of a system at $t+t'$ if the state of the system was s at t'. (If the set of transformations needed to specify the succession statement is also such that $U_t^{-1} = U_{-t}$, the set of transformations is a group.) Since the transformations U_t are unique, and for any time period t there is but one member of the (semi-)group, U_t, which effects the transformation associated with that time period, the dynamical (semi-)group determines uniquely the 'future' of a system once it is assigned a state in S_L. If we adopt the Schrödinger picture for the moment, and therefore assume that the location function is a function of time, we can say this a bit more clearly. Suppose that a system X is assigned a state s at some time t', i.e., $\mathrm{loc}(X, t') = s$. Then the dynamical (semi-)group determines what state the system must be associated with at every time after t' (if the transformations form a group, the state of the system at every time before t' is also determined), i.e., $(t)(\mathrm{loc}(X, t+t') = U_t(s))$. Whether the 'must' is a physical necessity or a logical necessity depends upon whether the succession statement is viewed as a law or a meaning postulate.[7]

Before using this assumption about the nature of the succession statement to characterize very generally the four modes of incorporation, let us look first at a very simple example in order to see how the modes work. Suppose that we want to construct a theoretical framework for describing the motion of a classical particle with mass m under the influence of no forces. For the sake of simplicity, we shall coordinatize the space in which this particle moves by letting the x-axis coincide with the line of motion of the particle; then we need consider only one dimension in describing the position and momentum of the particle. A natural semi-interpreted language for such a theory would consist of the following elements:

E is the set of all formulae of the form '$x=r$' and '$p=r$', where x and p are intended to be real valued functions such that $x(X)$ and $p(X)$ are the position and momentum of a particle X along the x-axis.

S_L is Cartesian 2-space, i.e., the set of all ordered pairs of real numbers $\langle x_1, x_2 \rangle$.

$h(x=r)=\{\langle x_1, x_2 \rangle : x_1 = r\}$, $h(p=r)=\{\langle x_1, x_2 \rangle : x_2 = r\}$.

The succession statement for a classical particle under the influence of no forces is given by the equation:

$$\frac{dx}{dt} = \frac{1}{m} p \quad \text{and} \quad \frac{dp}{dt} = 0$$

where m is the mass of the particle. This succession statement generates a dynamical group. Using, for the moment, $\langle x, p \rangle$ to represent the state of a particle at position x and momentum p, an element of the group, U_t, is given by:

$$U_t(\langle x, p \rangle) = \langle x + \frac{1}{m} pt, p \rangle.$$

The theoretical frameworks resulting when this succession statement is incorporated by the four modes will vary mainly in the way the location and satisfaction functions are specified. For each mode of incorporation there are a number of ways to actually construct a theoretical framework; the frameworks presented below are not meant to exemplify all of these ways, but some of the interesting possible variations will be indicated. Consider the *LS* mode first. Since the Schrödinger picture is assumed, the location function is time dependent, and S_L (defined above) is an instantaneous state space. Since the succession statement is incorporated as a law it appears as a statement specifying the set of paths in state space, P_P, which it is physically possible for a location function to trace over a period of time. Such a path can be described as a function, f, from the reals, R to S_L, where $f(t)$ is the state along the path at time t. The three elements of this theoretical framework can be constructed as follows:[8]

(i) $L = \langle E, S_L, h \rangle$ as defined above.
(ii) $M = \langle X, \text{loc} \rangle$ where loc is a function with domain $\{X\} \times R$ and range S_L, i.e., loc: $\{X\} \times R \to S_L$
(iii) $P_P = \{f : f(t) \in S_L \text{ and } (t)(t')(f(t+t') =$
$$= \langle f_{x_1}(t') + \frac{1}{m} f_{x_2}(t') \cdot t, f_{x_2}(t') \rangle)\},$$

where f_{x_1} and f_{x_2} are the x_1 and x_2 components of f, i.e., if $f(t) = \langle a, b \rangle$ then $f_{x_1}(t) = a$ and $f_{x_2}(t) = b$. The truth definition for elements of E reads:

e is true at time t in $\langle X, \text{loc} \rangle$ iff $\text{loc}(X, t) \in h(e)$.

The theoretical framework resulting from incorporation of the succession statement via the *MS* mode is quite similar to that for the *LS* mode. The location function is still time dependent; a further restriction on the form of this function ensures that only the paths allowed by the succession statement can be traced over a period of time. Since the succession statement is thus incorporated as a meaning postulate, this theoretical framework has only two elements:

(i) $L = \langle E, S_L, h \rangle$ as defined above.
(ii) $M = \langle X, \text{loc} \rangle$ where loc: $\{X\} \times R \to S_L$, and
$$(t)(t')(\text{loc}(X, t+t') = \langle \text{loc}_{x_1}(X, t') + \frac{1}{m} \text{loc}_{x_2}(t') \cdot t, \text{loc}_{x_2}(t') \rangle).$$

where loc_{x_1} and loc_{x_2} are the x_1 and x_2 components of loc. The truth definition reads:

e is true at time t in $\langle X, \text{loc} \rangle$ iff $\text{loc}(X, t) \in h(e)$.

Under the Heisenberg picture, the location function is no longer time dependent; the time evolution of systems is now governed by a time dependent satisfaction function. Let us look first at how the succession statement is incorporated via *MH*. Since the succession statement is incorporated as a meaning postulate, the satisfaction function must be constructed so that the way in which the predicates of L_E are assigned to any particular state over a period of time is consistent with this succession statement. The following theoretical framework satisfies this requirement:

(i) $L = \langle E, S_L, h^+ \rangle$ where $h^+ : E \times R \to \mathscr{P}(S_L)$ (where $\mathscr{P}(S_L)$ is the set of subsets of S_L), and

$h^+(e, 0) = \{ \langle x_1, x_2 \rangle : \langle x_1, x_2 \rangle \in h(e) \}$, h as defined above,

$h^+(x = r, t \neq 0) = \{ \langle x_1, x_2 \rangle : r = x_1 + \frac{1}{m} x_2 t \}$,

$h^+(p = r, t \neq 0) = \{ \langle x_1, x_2 \rangle : r = x_2 \}$.

(ii) $M = \langle X, \text{loc} \rangle$ where $\text{loc} : \{X\} \to S_L$.

The truth definition reads:

e is true at time t in $\langle X, \text{loc} \rangle$ iff $\text{loc}(X) \in h^+(e, t)$.

Once the modes *LS*, *MS* and *MH* are understood, it is fairly clear how to construct the appropriate theoretical frameworks. The theoretical framework resulting in the *LH* mode is not quite as obvious. The major innovation in this mode is the introduction of a new logical state space. The reason for introducing this new space can best be explained by looking carefully at the role of the succession statement in *LH*. Recall, first, that in the Heisenberg picture the evolution of a physical system is given by associating the system permanently with just one state, while the satisfaction function assigns to each state a unique 'future,' i.e., it maps the elementary statements at each time t in a way that associates with each state a path in the *LS* state space. Because of the restriction placed on the sorts of succession statements being treated in this paper, a succession statement also determines a unique future for each state in S_L. This is why the *MH* mode can be implemented so easily. By building the succession statement into the satisfaction function, we both determine the required future for each state in the state space, and guarantee that the future associated with each state is both unique and the one allowed by the succession statement. But in *LH* we want to have available in the state space all of the *logically* possible unique futures, and use the succession statement to pick out the *physically* possible ones. Thus, for each state s in the logical state space for *LS*, *MS* and *MH*, we need a whole set of states for *LH* – one state in the set for each path in *LS* which could possibly originate from the state s. Since a path originating from a state in S_L (of the *LS*, *MS* and *MH* modes) is described by a function f from R to S_L such that $f(0) = s$, we construct for each $s \in S_L$ a set of states of the form $\langle s, f \rangle$, one member of this set for each possible function, f, such that $f(0) = s$, and take as the logical state space for *LH* the union of these sets. In very general terms, then, S_L^*, the logical state space for *LH*, is defined:

$$S_L^* = \{\langle s, f \rangle : s \in S_L \text{ and } f: R \to S_L \text{ and } f(0) = s\}.$$

The *LH* theoretical framework for our theory of a classical particle under the influence of no forces has the following three elements:

(i) $L = \langle E, S_L^*, h^* \rangle$ where $h^*: E \times R \to \mathscr{P}(S_L^*)$, and
$S_L^* = \{\langle\langle x_1, x_2\rangle, f\rangle : f: R \to S_L \text{ and } f(0) = \langle x_1, x_2\rangle\}$, and
$h^*(e, 0) = \{\langle\langle x_1, x_2\rangle, f\rangle : \langle x_1, x_2\rangle \in h(e)\}$,
$h^*(e, t \neq 0) = \{\langle\langle x_1, x_2\rangle, f\rangle : f(t) \in h(e)\}$.

(ii) $M = \langle X, \text{loc} \rangle$ where $\text{loc}: \{X\} \to S_L^*$.

(iii) $S_P = \{\langle\langle x_1, x_2\rangle f\rangle : (t)(t')(f(t+t') =$
$$= \langle f_{x_1}(t') + \frac{1}{m} f_{x_2}(t') \cdot t, f_{x_2}(t')\rangle)\}$$

(where S_P is the set of physically possible states).
The truth definition reads:

e is true at time t in $\langle X, \text{loc}\rangle$ iff $\text{loc}(X) \in h^*(e, t)$.

Notice that under the Heisenberg picture, a law of succession appears formally as a law of coexistence.

Clearly, the above theoretical frameworks are not general enough to describe the motion of *all* classical particles not under the influence of forces, but only those with a particular value of the parameter m. A theory of force free classical particles must therefore specify a whole set of (1) P_P's for *LS*, (2) restrictions on loc for *MS*, (3) satisfaction functions for *MH*, or (4) S_P's for *LH*. This plurality is obviously the price of generality. Thus, as van Fraassen suggests in a slightly different context, the theory of force free classical particles might be thought of as a "theory-schema" or as "the union of a family of simple theories," each specifying one theoretical framework. ([8], p. 311) The above structures are therefore more appropriately viewed as theoretical framework schemata. They indicate the form which one of these theoretical frameworks would have, but become theoretical frameworks only if a specific value is assumed for m.

The fact that a theoretical framework schema for a theory of force free classical particles can be constructed for each of the four modes of incorporation is not a proof that such schemata can be found for any succession statement of a theory. Below, then, is a general description of the theoretical framework schema resulting from the incorporation of a succession statement, S, via each of the four modes. Assume that $\langle E, S_L, h \rangle$ is a semi-interpreted language appropriate for description of the physical systems (which are the subject of the theory) before S is incorporated into the schema, where L_E includes at least all of the predicates related by S. Assume, further, that S can be expressed by specifying a dynamical (semi-)group with members U_t defined on S_L. For each mode of incorporation, the appropriate elements of the theoretical framework schema are provided, and the truth definition is given.

LS (i) $L = \langle E, S_L, h \rangle$.
   (ii) $M = \langle X, \text{loc} \rangle$ where $\text{loc}: (X) \times R \to S_L$.
   (iii) $P_P = \{f : f : R \to S_L \text{ and } (t)(t')(f(t+t') = U_t(f(t')))\}$.
     e is true at time t in $\langle X, \text{loc} \rangle$ iff $\text{loc}(X, t) \in h(e)$.

MS (i) $L = \langle E, S_L, h \rangle$
   (ii) $M = \langle X, \text{loc} \rangle$ where $\text{loc}: \{X\} \times R \to S_L$, and
     $(t)(t')(\text{loc}(X, t+t') = U_t(\text{loc}(X, t')))$.
     e is true at time t in $\langle X, \text{loc} \rangle$ iff $\text{loc}(X, t) \in h(e)$.

MH (i) $L = \langle E, S_L, h^+ \rangle$ where $h^+ : E \times R \to \mathscr{P}(S_L)$, and
     $h^+(e, 0) = \{s \in S_L : s \in h(e)\}$,
     $h^+(e, t \neq 0) = \{s \in S_L : U_t(s) \in h(e)\}$.
   (ii) $M = \langle X, \text{loc} \rangle$ where $\text{loc}: \{X\} \to S_L$.
     e is true at time t in $\langle X, \text{loc} \rangle$ iff $\text{loc}(X) \in h^+(e, t)$.

LH (i) $L = \langle E, S_L^*, h^* \rangle$ where $h^* : E \times R \to \mathscr{P}(S_L^*)$, and
     $S_L^* = \{\langle s, f \rangle : s \in S_L \text{ and } f : R \to S_L \text{ and } f(0) = s\}$.
   (ii) $M = \langle X, \text{loc} \rangle$ where $\text{loc}: \{X\} \to S_L^*$.
   (iii) $S_P = \{\langle s, f \rangle : (t)(t')(f(t+t') = U_t(f(t')))\}$.
     e is true at time t in $\langle X, \text{loc} \rangle$ iff $\text{loc}(x) \in h^*(e, t)$.

A different approach to incorporation of succession statements under the Schrödinger picture should also be mentioned, since it is probably closer to the way laws are actually handled under the Schrödinger picture in quantum theory. On this alternate approach, an auxiliary state space, S_t, is employed to carve out paths in the instantaneous state space S_L. The location function is no longer time dependent, but maps a physical system into S_t; the state of S_t thus associated with the system is a function from R to S_L, and thereby determines for the system a path in S_L. If the succession statement S is incorporated under this approach as a law of succession, the theoretical framework schema would be:

LS' (i) $L = \langle E, S_L, S_t, h \rangle$ where E, S_L and h are those assumed for the above schema, and
$S_t = \{f : f : R \to S_L\}$.
(ii) $M = \langle X, \text{loc} \rangle$ where $\text{loc} : \{X\} \to S_t$.
(iii) $S_P = \{f : f \in S_t \text{ and } (t)(t')(f(t+t') = U_t(f(t')))\}$.
e is true at time t in $\langle X, \text{loc} \rangle$ iff $(\text{loc}(X))(t) \in h(e)$.

As in *MH*, the law of succession appears formally as a law of coexistence (cf. van Fraassen's remark in [7], footnote 10, p. 331). The schema resulting from incorporating the succession statement as a meaning postulate under this alternate Schrödinger approach is the same as *LS'*, except there is a new auxiliary state space, S_t^*, where $S_t^* = S_P$ of *LS'*, and no third element appears in the schema.

Now it remains to be argued that the schemata above are of the appropriate sort. Recall that the purpose of this section is to demonstrate that any succession statement, S, can be incorporated as either a law or a meaning postulate. This purpose is accomplished if the following claim is true: when incorporated via *LS* or *LH*, S is incorporated as a law, and when incorporated via *MS* or *MH*, it is incorporated as a meaning postulate.

Let us examine the second clause of this claim first. A meaning postulate determines truth and falsity *ex vi terminorum*. Thus when S is construed as a meaning postulate, it must be incorporated into a theoretical framework in a way which makes it impossible to use this framework to ascribe to a physical system an evolution incompatible with S. Under the Schrödinger picture this amounts to the requirement that S be incorporated in a way which ensures that there is no model $\langle X, \text{loc} \rangle$ of the semi-interpreted language $\langle E, S_L, h \rangle$ in which loc traces a path in S_L not among the paths allowed by S. The MS mode meets this requirement by restricting the class of models of $\langle E, S_L, h \rangle$ to those with location functions which only trace such allowed paths. (The MS' mode meets this requirement by restricting the range of any location function of a model to a set of functions which carve out only allowed paths in S_L.) In the Heisenberg picture this amounts to the requirement that S be incorporated in a way which ensures that there can be no model $\langle X, \text{loc} \rangle$ of the semi-interpreted language $\langle E, S_L, h \rangle$ in which the location function assigns to X a state not allowed by S, i.e., in which loc assigns to X a state associated with a future corresponding to a disallowed path in LS. Since futures are associated with a state in the Heisenberg picture by the time dependent satisfaction function, h, MH meets this requirement by building S into h itself. In this way h assigns to each state of S_L only that future determined for it by S, and hence, a location function can do nothing but associate X with a state (and therefore a future) allowed by S. Thus the modes MS (and MS') and MH incorporate S as a meaning postulate. The second clause of the above claim proves true.

The first part of this claim was that "when incorporated via LS or LH, S is incorporated as a law." Recall that a law of succession designates as physically possible a certain subset of all the logically possible futures available to a physical system. The set of logically possible futures has been identified with the set of all paths in a state space S_L, where a path is determined by a function from R to S_L, and S_L is the state space of a semi-interpreted language appropriate for description of physical systems (which are the subject of the theory) as long as S is not incorporated into the theoretical framework. This is exactly the set of paths available to a location function of a model in LS. The succession statement, S, is then used in LS to define a subset of this set of logically possible paths, where the subset contains all and only the paths allowed by S. (In LS', S is used to define a set of states in the auxiliary state space, where this set contains all and only those functions which trace out allowed paths in S_L.) In the Heisenberg picture we no longer characterize futures by paths in the state space, but by states, each of which has a future associated with it by the satisfaction function h. The logical state space for LH is constructed to contain enough states so that each state represents one of the logically possible futures described above. The succession statement, *S*, is used to define a subset of these states, where this set contains all and only states associated with futures allowed by *S*. The modes *LS* (and *LS'*) and *LH* therefore incorporate *S* as a law, as is asserted by the first clause of the claim above.

Thus we may conclude that any given succession statement can be incorporated into a theoretical framework either as a law or as a meaning postulate, and can therefore assert quite generally that in van Fraassen's view of theories both modes of incorporation are available to *any* general statement expressing relations among properties.

III

It is the existence of this option in a theoretical framework which, according to van Fraassen, reflects the lack of objective distinction between laws and meaning postulates. But, as we saw earlier, a theoretical framework also indicates that there *is* a distinction between laws and meaning postulates, and this distinction rests not on just labeling them as such, but on the different sorts of modalities they determine. Thus, when a general statement is incorporated as a law, there exists in the resulting theoretical framework a certain number of logically possible states (if we consider a law of coexistence) or paths traceable in the state space through time (if we consider a law of succession) which are judged by the law as not physically possible. When that general statement is incorporated as a meaning postulate, these same states or paths in state space do not appear as logically possible – they do not even appear in the resulting theoretical framework at all. This characteristic of the semantic view therefore determines a criterion which does allow one to distinguish between those general statements being used as laws, and those being used as meaning postulates. If a general statement is being used as a law, there are some states of affairs which are considered by the users of the theory to be describable in a hypothetical way, but are denied by the law to be physically possible; if that same statement were being used as a meaning postulate, it would be deemed logically impossible to even talk about these states of affairs. And notice that this criterion allows one to make an *objective* distinction; any number of investigators using this criterion would categorize general statements as laws and meaning postulates in the same way – their distinction between laws and meaning postulates would not depend on their own subjective opinions about the matter. The fact that any general statement found to be functioning as a law *could* be incorporated into a theoretical framework as a meaning postulate or *vice versa* does not nullify the fact that that statement *does* function as a law rather than as a meaning postulate, for surely the proper way to model a theory under investigation would be to incorporate as laws those general statements found to function as laws, and to incorporate as meaning postulates those found to function as meaning postulates.

Clearly then, the fact that *formally* any general statement can be incorporated as either a law or a meaning postulate cannot be taken to mean that there is not an objective way to distinguish one from the other. I would suggest that this option of alternative modes of incorporation reflects not the absence of an *objective* distinction between laws and meaning postulates, but rather the lack of any *non-pragmatic* distinction between the two. That is to say, there are no characteristics of the structure of a theory as it stands abstracted from its use by human beings that determine which of the general statements of the theory must be taken as laws and which as meaning postulates. While the distinction between laws and meaning postulates, and the distinction between logical and physical modalities underlying it, can be explicated in the abstract structure of a theoretical framework, the actual partitioning of the general statements in a particular theory into laws and meaning postulates is relative to the way a particular community uses that theory at a particular time. The semantic view is therefore capable of reflecting the fact that during the historical life of a theory, a general statement may be for a period regarded as a law and thus used to determine physical modalities, and then later taken as a meaning postulate and used as an incontrovertible determiner of logical possibilities. As van Fraassen remarks in a footnote to the paragraph quoted above, the distinction between laws and meaning postulates "may be an *historical* distinction; before a certain stage in the development of a theory, the law may not yet be an inherent principle of the language game." ([7], p. 331)

One might feel uneasy about this invocation of pragmatics in order to understand the significance of some of the characteristics of the semantic view of theories because it seems to blur the distinction between semantics and pragmatics. For van Fraassen's view of theories is billed as an extension of Evert Beth's "use of formal semantic methods in the analysis of physical theories" ([7], p. 325), and now I am suggesting that it be understood to mirror pragmatic distinctions among parts of a theory as well. There is no need for such uneasiness, however, since this semantics-pragmatics distinction is already a fuzzy one, at best. Attacks in the last quarter century on such notions as meaning, synonomy and analyticity have reminded us that it is really language-in-use, that product of uncountable historical vagaries, which underlies the traditional semantic distinctions. Furthermore, I am not grafting something foreign onto the original intent behind the semantic view, for van Fraassen argued in an early paper that "the demand for and interpretation of [semantic] concepts... can only be construed as a request that we show the concepts in question to have a significant role," and went on to suggest that "there is in fact an obvious field which may provide such a grounding for semantic concepts, namely pragmatics." ([5], p. 167)

This move to pragmatics is perhaps not the only way one can understand the implications of the semantic view of theories for the structure of actual scientific theories, but it does provide a consistent way, and it seems to me the most plausible way, of understanding the two characteristics of a theoretical framework which I have been concerned with here. The fact that in a theoretical framework laws and meaning postulates are incorporated in different ways and serve to define two different sorts of modalities is taken to reflect the fact that in actual theories, certain states of affairs are deemed impossible because of the very meanings of the relevant terms, and thus certain general statements are regarded as true *ex vi terminorum*, while other states of affairs are deemed impossible because of the way the world is, and certain corresponding general statements are regarded as true but vulnerable to any contrary evidence which might surface in the future. The fact that, as I have proved, any law of a theoretical framework can be reincorporated as a meaning postulate, and *vice versa*, is taken to reflect the fact that there is no non-pragmatic reason why certain general statements appear as laws and others as *ex vi terminorum* truths, and that, furthermore, the status of any particular general statement may change, i.e., the scientific community may come to change its mind about what is physically and logically possible. A resolution to the apparent conflict with which we started is thus provided, and some insight into the implications of the semantic view for the structure of actual scientific theories is gained.

University of Wisconsin-Milwaukee

NOTES

[1] The 'standard view' was first proposed by N. R. Campbell in [4] and by R. Carnap in [1]. It has since been developed by Carnap, C. Hempel, H. Reichenbach, E. Nagel and others. The account given here reflects that version of the standard view which van Fraassen often places in opposition to his own semantic view of theories. It ignores, however, a later development of the view by Carnap. In his final version of the standard view, Carnap abandoned the idea that a certain subset of the postulates of a theory can be identified as meaning postulates. See [2], pp. 965–966, and [3], Chapter 28, for example. He defines a single (theoretical) meaning postulate for a theory as the conditional sentence $^RTC \supset TC$, where ^RTC is the Ramsey sentence of TC, and TC is the conjunction of all the theoretical postulates and correspondence rules of the theory. The set of all deductive consequences of this meaning postulate is taken as the set of all non-observational analytic statements of the theory. The fact that no theoretical postulates or correspondence rules appear in this set (see [9]) indicates that in this later version, meaning relations are assigned a much more subtle role. As in van Fraassen's view, the meaning relations lie beneath the surface of a theory; they are reflected in the logical relations among the statements of the theory but are not explicitly expressed as a subset of these statements.

[2] In general, a model of a semi-interpreted language can be defined as an ordered pair, $\langle D, \text{loc} \rangle$, where D is a *set* of physical entities and loc assigns to each of these entities in D a state in S_L. In this case it makes sense to include n-ary predicates in the elementary language. In this paper I will consider only the simple case, where D contains only a single element. Thus we need consider only elementary languages with one-place predicates, and define a model as an ordered pair, where the first member of the pair is a single physical system. No generality is lost here, since any set D can itself be looked upon as a single physical system by appropriately recasting the relevant semi-interpreted language.

[3] If the language contains the existence predicate included by van Fraassen in [5] and [6], then one also needs a model M to determine h from S_L and f. See [6], pp. 160–161.

[4] This is not to say that *all* logical relations among elementary statements arise because of the meaning relations. In [6] van Fraassen distinguished two levels of logical necessity. The meaning relations determine truth and falsity *ex vi terminorum*, while a broader logical necessity is determined by the possible transformations of the structure of a logical state space of a semi-interpreted language. The latter modalities, van Fraassen's "strict modalities," will not be discussed here.

[5] A third category of laws, laws of interaction arise when the first member of a model is a *set* of physical systems (see Note 1 above). Such laws are not considered here since they can always be recast as laws of succession by viewing the set of physical systems itself as a single physical system.

[6] In quantum theory this is not strictly true; in many cases a physical system must be reassigned a state after a measurement is made on that system. The necessity of such reassignment is not easily explained, however, as the mounting literature on the 'problem of measurement' indicates.

[7] Limitation of our discussion to succession statements of this sort may appear overly restrictive, since many interesting succession statements are probabilistic rather than deterministic, i.e., they determine 'transition probabilities' rather than unique futures for a system in a particular state. The discussion is actually quite general, however, since all probabilistic succession statements can be recast as succession statements of the sort treated in this paper by constructing an appropriate logical state space. ([7], pp. 333–334)

[8] See p. 224 below for another approach to incorporation of a succession statement under the Schrödinger picture.

BIBLIOGRAPHY

[1] Carnap, R.: 1923, 'Über die Aufgabe der Physik und die Anwendung des Grundsatzes der Einfachstheit', *Kant-Studien* **28**, 90–107.

[2] Carnap, R.: 1963, 'Replies', in P. A. Schilpp (ed.), *The Philosophy of Rudolf Carnap*, Open Court, pp. 359–999.

[3] Carnap, R.: 1966, in M. Gardner (ed.), *Philosophical Foundations of Physics*, Basic Books.

[4] Campbell, N. R.: 1920, *Physics: The Elements*, Cambridge University Press.

[5] van Fraassen, B.: 1967, 'Meaning Relations among Predicates', *Nous* **1**, 161–179.

[6] van Fraassen, B.: 1969, 'Meaning Relations and Modalities', *Nous* **3**, 155–167.

[7] van Fraassen, B.: 1970, 'On the Extension of Beth's Semantics of Physical Theories', *Philosophy of Science* **37**, 325–338.

[8] van Fraassen, B.: 1972, 'A Formal Approach to the Philosophy of Science', in R. C. Colodny (ed.), *Paradigms and Paradoxes*, University of Pittsburgh Press, pp. 303–367.

[9] Winnie, J.: 1973, 'Theoretical Analyticity', in *Boston Studies in the Philosophy of Science*, Vol. VIII, D. Reidel, Dordrecht, pp. 289–305.

PAUL FITZGERALD

MEANING IN SCIENCE AND MATHEMATICS

I

Philosophers working in the logical empiricist tradition, such as Rudolf Carnap and Carl Hempel, have held that theoretical terms get their meanings from their connections with observational terms in the postulates of scientific theories. They have tried to explain exactly how this happens, in the sense of giving a logical reconstruction of it. For various reasons, to be discussed below, the belief arose that it is impossible in general to give explicit definitions of theoretical terms. Attention thus shifted to showing how correspondence rules relating theoretical and observational terms give 'partial interpretations' to the theoretical terms.

Three objections can be raised against the way in which this enterprise was conceived and carried out. These objections involve what will here be called (1) 'The Problem of the Theoretical-Observational Dichotomy', (2) 'The Meaning-Magic Problem', and (3) 'The Problem of Unwanted Interpretations'. The purpose of this article is to set forth the second and third of these objections and develop a solution to them.

The first objection is the by now familiar one that the theoretical-observational dichotomy as understood by the logical empiricists is misconceived.[1] The assumption that the two kinds of terms are mutually exclusive has come under fire, along with the assumptions that the distinction is a sharp one, and that the terms which apply to the observable entities of ordinary life are always observational, never theoretical; whereas those which apply to the unobservable micro-entities of physics are always theoretical, never observational.

The second objection concerns the particular account given by logical empiricists of how theoretical terms acquire their meanings, and involves the 'Meaning-Magic Problem'. According to their logical reconstruction of scientific practice, as distinguished from its psychological-historical description, we start as scientists with a handful of strictly meaningless predicate letters which are destined for the high office of theoretical terms. We sprinkle them into various slots throughout a set of quantificational schemata in which there also appear observational predicates and in which all of the individual variables are bound by quantifiers. Presto! The meaningless predicate letters somehow acquire meaning by contact with the observational terms; the schemata become statements and the set of them turns into a scientific theory, if (dis)confirmable by observation statements.[2]

The problem is to see how the magic is done. How can we start with a meaningless predicate letter 'N', let it appear in a closed quantificational schema, such as '$(x)\ Nx$', form a truth-functional compound of it and an observational statement such as '$(Ey)\ Oy$', and expect the result '$(Ey)\ Oy \in (x)\ Nx$' to have enough meaning to be true or false? If the original schema '$(x)\ Nx$' is neither true nor false, then to assign a truth-value to formulas in which it appears, we need either a three-valued logic or one with truth-value gaps. There are two standard ways of giving a three-valued truth-table for an operator analogous to the 'ε' of two-valued logic.[3] On one way, the conjunction of a true statement and a 'neuter' statement is neuter, whereas the conjunction of a false and a neuter is false. On the other way, a conjunction of a neuter statement with anything is neuter. The point is that a 'scientific theory' with dummy predicate letters doing duty for theoretical terms in search of meaning should be assigned a truth-value according to the rules of some three-valued logic. And the schematic letters in question do not have semantic meaning or 'partial interpretation' rubbed off on them through being jostled by their observational neighbors in the schemata of the theory. First, last and always, a schema like '$(x)\ Nx$' is neuter, even if deduced from a schema containing observation terms from which it is supposed to have received its meaning or partial interpretation. The last state of that letter 'N' is no better than the first.[4]

The third objection to the traditional approach is that its glossing of theoretical terms does not distinguish them from trivial expressions which have no place in scientific theories. Scientists know that the empirical significance, perhaps even the semantic meaning, of theoretical terms and properties has to do with their connection with observable phenomena, expressed in hypothetical 'if... then' statements.[5] Being hard-headed, we read statements of the form 'if p then q' as our familiar and well-loved '$p \supset q$'. This simply means '$-p \vee q$'. If we regard it as a correspondence rule which gives a partial interpretation to any theoretical terms appearing in it, why not also admit that any truth-functional or quantificational compound containing a theoretical and an observational term does the same? Thus such trivia as '$(x)(Nx \vee Ox)$', where 'O' is observational, somehow suffice to put the gibberish expression 'N' on the same semantic footing ('partially interpreted') with a theoretical term of science.[6] Such are the penalties of overgenerous construal of what counts as a correspondence rule, together with over-attachment to extensional logic. Even where one resists this *reductio* by demanding that the correspondence rule which gives meaning to the theoretical terms be a whole complicated scientific theory, we are not much better off. First, scientists often want to introduce new theoretical terms before they have much of a theory. Complicated theories are often built up sequentially, piece by piece. Insistence on complicated correspondence rules to begin with might thwart theory formation.

Second, consider the case where we do have a considerable mass of theory to use in defining a theoretical term. In the logical empiricist approach there are two ways to express the theory. One is to replace the theoretical terms by dummy predicates wherever they appear in the conjunction of the purely theoretical part of the theory with its correspondence rules. Call the resulting quantificational schema 'TC', meant to suggest the conjunction of the theory's correspondence rules, C, with its purely theoretical part, T. The other way to express the theory, for purposes of logical reconstruction, is by the Ramsey sentence corresponding to TC. This is formed by quantifying existentially over all of the dummy letters of the schema TC. Call the Ramsey sentence 'R'. Pretend for the moment that the purely extensional schema TC has the same truth-value as the genuine true-or-false scientific theory from which it was derived in the course of our logical construction. It is highly probable that for each theoretical expression E_t in our original theory there is a contrived molecular predicate, consisting of irrelevant observational and/or mathematical expressions which, when substituted for E_t in TC, preserves the truth-value that TC was supposed to have. For all of the theoretical predicates, or rather their dummy letters in TC, substitute appropriately contrived, trivial counterparts. You then have some statement S which is a substitution instance of the Ramsey sentence R and which has the same truth-value as the original scientific theory.

Of course, the theoretical expressions in the real-life scientific theory do not have the same meaning as their contrived counterparts in *S*. And yet the 'partial interpretation' provided by *TC* does not suffice to distinguish the two. So contrary to the logical empiricist claim, *TC* does not adequately explicate the meanings, partial or complete, which the theoretical expressions in the scientific theory possess. This problem with the *TC* approach will be called the 'Problem of Unwanted Interpretations'. Now let us see what can be done to solve these problems.

II

In what follows I will not be concerned with the first objection, which challenges the theoretical-observational dichotomy. But even if we admit the cogency of the objection, as I am inclined to do, we still have a problem, and the work of the logical empiricists is relevant to its solution. How do we give meaning to new scientific terms when these are not just abbreviations of old expressions but represent new concepts? How are new theoretical concepts and properties defined, specified or made intelligible? I want to suggest a natural way for a logical empiricist to give explicit definitions of theoretical terms in a way that avoids my second and third objections. Moreover, I think that the way suggested is not very far from the truth about how theoretical terms get their meanings.

Scientists define a new quantity, such as entropy, enthalpy or gravitational mass, by positing laws governing its relations with other, previously known quantities. Roughly speaking, something has the new quantitative property just in case it behaves in the way that things having that property should behave, according to the suggested laws. If nothing, or not enough things, do behave in the required way, then the property is regarded as uninstantiated, like *being-dephlogisticated* or *being-rich-in-caloric*.

This introduction of new properties looks puzzling at first. Theoretical properties are expressed by theoretical predicates. Those predicates are presumably not ostensively definable, nor mere abbreviations for observational predicates. Somehow the theory or the putative laws in which they appear give them their meanings, but in order for the theory or the laws to be meaningful the theoretical predicates which appear in them must have semantic meaning. Do they lift themselves by their own bootstraps into meaningfulness? Some logical empiricists thought that the marriage of meaningless predicate letters and a semantically under-interpreted schema *TC* gives birth to a theory with enough meaning to be true or false. But what is that theory... *TC* again? This is the Meaning-Magic Problem. Theories do give meaning to theoretical predicates. But how, exactly?

The answer which I want to explore is that any theoretical term can be explicitly defined using the theory in which it appears. The intuition to be exploited is that an entity has a scientific theoretical property just in case it behaves in the way that entities having that property should behave, according to the theory. Corresponding to that theoretical property is the theoretical expression '*M*', which appears as a dummy predicate in the disinterpreted schema *TC*. To form the theory's Ramsey sentence we would quantify existentially over all of these dummy predicates. Let us quantify over all of them except '*M*', thus forming a 'near-Ramsey sentence', symbolized by '$R(...M...x...y...)$'. This is supposed to indicate that '*M*' is the only free dummy type-predicate in the near-Ramsey sentence, and '*x*' and '*y*' are the only individual type-variables of which it is predicated. Several tokens of each type-variable may of course appear. Now the idea is to use the near-Ramsey sentence '$R(...M...x...y...)$' to give an explicit definition of the theoretical terms corresponding to *M*, which we can call 'T_m'. The theoretical property corresponding to 'T_m' is by definition possessed by anything which has some property *M* which is such that anything having it behaves as the theory says that it should. Explicitly, the definition of T_m runs this way:

(i) $\quad (z)\,[T_m z \equiv (\exists M)\,(Mz \in R(...M...x...y...))]$

If desired, a definition-schema can be introduced instead. Simply drop the initial universal quantifier in formula (i). The theory's Ramsey sentence will tell us that any entity has *M* only if such-and-such is true of it. We then define the property T_m as the property which an entity possesses just in case there is a property *M* which it has and anything with *M* behaves as the theory says. My illustration was in terms of a monadic property T_m, and obviously we would have to generalize to *n*-adic properties, where '*n*' stands for any positive integer. This should pose no problems of principle.

I use the theory's Ramsey sentence simply as a way of relating this approach to the traditional literature. Suppose that we are not concerned to distinguish observational from theoretical terms and to show how all theoretical terms get their meaning from observational terms. We only want to find out how one or more *new* theoretical concepts are introduced by means of previously understood concepts, be they theoretical, observational, or whatever. Then instead of using the theory's Ramsey sentence we can use the counterpart of it in which all terms already understood appear *in propria persona*, and the only ones which are replaced by bound predicate variables are the one or more new predicates which we seek to define. Future references to 'the theory's Ramsey sentence' are intended to cover this case too.

We have shown how to use a theory's Ramsey sentence to give explicit definitions of the terms it introduces. In virtue of these definitions the theoretical terms have their full share of meaning, and so the Meaning-Magic Problem has been solved. But it is not clear that the terms which one defines explicitly in this way really are the theoretical scientific terms whose meaning was to be explicated. In particular, what about the common logical empiricist teaching that theoretical terms are only 'partially interpreted', unlike the explicitly defined terms being offered here? And is it the case that explicit definitions of the kind here sketched really avoid the Problem of Unwanted Interpretations?

The answer is that they do not. And, as will be seen later, the reason why they do not has an intimate connection with the logical empiricist doctrine that theoretical terms are only partially interpreted. For that doctrine seems forced on us only because we insist on confining ourselves to extensional logic and, in particular, we insist on translating the non-truth-functional 'if... then...' of ordinary English into material conditionals. This procedure leads either to the Problem of Unwanted Interpretations or to the doctrine of partial interpretations. To see why, consider a nice modest term such as 'brittle', and imagine that we are teaching it to someone for the first time. We tell him that something is brittle if and only if it breaks into fairly sharp pieces if it is struck (brittle things do not *crumble* when struck). This is roughly correct; we might want to soften it by saying that if struck it *probably* will break into fairly sharp-boundaried pieces. But if this is translated into extensional logic it says that something is brittle if and only if it is either not struck, or else it breaks into sharp-boundaried pieces. Formally, this reads:

(ii) $(x)(x \text{ is brittle} \equiv x \text{ is not struck} \lor x \text{ breaks into sharp-boundaried pieces})$

But this has the absurd consequence that all unstruck entities are brittle, which is certainly an unwanted interpretation of *brittle*! To avoid it, one introduces the doctrine of partial interpretations, according to which we surrender any attempt to give an explicit definition of brittleness and rest content to specify its meaning partially, by the following sort of formula:

(iii) $(x)(x \text{ is struck} \supset (x \text{ is brittle} \equiv x \text{ breaks into sharp-boundaried pieces}))$

This tells us what it is for struck things to be brittle, but not what it is for unstruck things to be brittle, so it is regarded as providing only a partial interpretation of 'brittle'. But, alas, this approach via partial interpretations runs aground on the Meaning-Magic Problem, so we had best try to go back to giving full interpretations in terms of Ramsey sentences.

It might be thought that the Problem of Unwanted Interpretations is solved by insisting on the use of complicated Ramsey sentences in defining theoretical terms. Start with a simple Ramsey sentence admitting many unwanted interpretations for a theoretical term being defined. Complicate it by adding further clauses containing observational terms and imposing ever tighter empirical constraints on the world. This causes scores of unwanted interpretations to fall by the wayside at each step. But we have no guarantee that other hosts of them will not always remain tagging along at our heels.

The more important reason why a problem remains is that we have no guarantee, in general, that we can always find a complicated Ramsey sentence for use in defining our new predicate. In the early stages of an investigation a complicated theory has not yet been constructed; we have simply some lawlike correlations and perhaps a sketch of a theory. Take as an example Clausius's definition of entropy, mentioned in an earlier note. This property was defined decades before the statistical mechanical theory of entropy had been developed. The network of theory on which the original definition of entropy was constructed had very few strands indeed. Its Ramsey sentence would not have been strong enough to rule out unwanted instantiation conditions for the property being defined. Nor is this an exceptional situation. *All* developed science proceeded from an originally undeveloped state in which theoretical concepts were being introduced on the basis of rather modest resources. No view of how theoretical expressions initially acquire meaning can afford to overlook this fact. To take account of it we must find a way to rule out unwanted instantiation conditions (the analogue of unwanted interpretations) *without* relying on the presence of an obligingly complicated Ramsey sentence. For that complicated sentence is usually available to perform this service only *after* some theoretical terms have been defined without it.

Notice that we are still much better off on my proposal than we would be on the partial interpretation approach. The latter gives us no distinction in meaning between the partially interpreted term and other terms which are coextensive with it in the domain for which it is interpreted. My suggested definitions specify the differences in meaning between theoretical terms and other terms which happen to be coextensive with them. For the defined theoretical terms, unlike the other terms, are instantiated only if the part of the theory used to define them is true. That constitutes a difference built into the very meanings of the theoretical terms. So what problem remains?

David Lewis has independently worked out an approach to defining theoretical terms which shares some similarity with the present one in that both involve use of the theory's Ramsey sentence. But his approach suffers from a problem akin to that of unwanted interpretations, and much more severe than any which afflicts the present proposal. For as Lewis defines theoretical terms, no other property could (logically) be coextensive with a theoretical property. Since it is virtually certain that each theoretical property appealed to in actual science is coextensive with at least one other contrived property, perhaps a complicated set-theoretic one, Lewis's proposal does not adequately capture the meanings of theoretical terms as actually used. Even if each theoretical property happened not to be coextensive with any other property, the objection would remain. For surely it is not part of what we mean by the theoretical term in question that the property which it expresses not be coextensive with any other property. But that demand is built into the meanings of theoretical terms, as Lewis defines them.

Some would say that the present proposal faces no further problem. Others would disagree. Their rationale is this. We think that a given theoretical property is really instantiated in the world only if it plays a role in

MEANING IN SCIENCE AND MATHEMATICS

some genuine *laws*. Nothing has phlogiston because the supposed laws governing phlogisticated bodies turned out not to be laws after all. A statement can fail to express a law either through failing of truth or through failing to be lawlike.

A scientific theory's pretensions to being true and lawlike, i.e., being naturally necessary, are found in the occasional use of modal expressions which render its content in a natural language, as well as by non-truth-functional 'if... then' conditionals. When, as philosophers, we omit the modal locutions and form the theory's Ramsey sentence in extensional predicate logic, we get a statement with wider truth-conditions than the original theory. For all sorts of bizarre properties may be available to make the Ramsey sentence true, even if the theory is false. And the Ramsey sentence may turn out true because certain material conditionals in it are vacuously satisfied, whereas the corresponding non-truth-functional 'if... then' conditionals of the theory are false. Theoretical properties presuppose that there are laws; laws differ from accidental truths; non-modal reconstructions of the theories do not capture that difference; and therefore they cannot adequately capture the meanings of theoretical terms.

It would be too tedious here to discuss whether modal operators, non-truth-functional connectives, or some mix of the two, best captures the meaning of a theory and its distinctive predicates. It will be assumed here that we will not fall too far short of what is required if we add nothing more than a modal necessity operator, symbolized 'Nec'. This is a statement-forming operator on statements. It should be read 'It is naturally necessary that...' or 'It is a natural law that ...' True conjunctions of laws and true scientific theories are to count as naturally necessary, as well as single laws. If we need non-truth-functional sentence connectives also, then take our operator 'Nec' as a shorthand indicator of their presence in the resulting sentence.

Our earlier definition (i) of a sample theoretical property T_m incorporated the demand that the theory's Ramsey sentence be a truth if the property T_m is to be instantiated. We now add the demand that the Ramsey sentence be a natural law as well. This is done by taking the theory's full Ramsey sentence, symbolized 'RS', and prefixing the modal operator to it to get 'Nec·RS'. We then add that to the definiens in (i) to get the new definition, which reads,

(iv) $\quad (z)\,[T_m z \equiv (\text{Nec} \cdot RS \in (\exists M)\,(Mz \in R(\ldots M \ldots x \ldots y)))]$

The effect of the revision is that the theoretical property T_m is so defined that it is instantiated only if the Ramsey sentence which is its natural habitat (strictly, the natural habitat of its dummy replacement) is a natural law.

III

Having sketched the strategy for defining theoretical terms, we must deal with certain objections to it. The first is that this strategy gives us complete definitions for theoretical terms, whereas it is widely believed that such terms are only partly defined or partly interpreted. So is not the strategy defective?

I think that the strategy is not defective on this count. Rather, we have an objection here to the idea that theoretical terms are only partly interpreted. How did this belief arise?

One reason why it arose is that Carnap and others discovered that theoretical terms, dispositional terms, and the like could not be explicitly defined using only extensional logic and observation terms. The reason is that one then lacks the non-truth-functional 'if... then' conditionals which are required. It was concluded that they could not be explicitly defined using any respectable means. So they must be only partially interpreted, through reduction sentences. This problem surfaced at least as early as Carnap's 'Testability and Meaning' (1936).

But if modal devices or a non-truth-functional 'if... then' is used then this reason for thinking that theoretical terms are only partly interpreted disappears. Practicing scientists are the ones who give meaning to theoretical terms, and they do so in non-truth-functional language. While I have no particular brief for modal apparatus, I do not think that dislike of it should cause us to distort the conceptual situation as it exists in real-life science. If someone wants to improve that situation by eschewing modal notions and trying to get truth-functionally characterizable versions of scientific theories, more power to him. But if a strictly truth-functional version forces us to say, that theoretical terms are only partially interpreted, then it neither accurately captures the meanings of terms in real-life scientific theories, nor can it serve as a satisfactory substitute for those theories.

The second reason for thinking that theoretical terms are only partially interpreted looks plausible, for it rests on a basic truth about science. But that basic truth is compatible with the strategy here outlined for giving explicit definitions. The basic truth is that as science advances our scientific concepts change as well. We forge new ones which are both more precise and more general in scope. We also discover new lawlike facts about old theoretical properties. All of this *can* be construed as involving a progressive expansion of the meanings of scientific terms. So it seems that in the beginning, and at most other stages as well, they are only partially interpreted, and that the interpretation gets filled in progressively.

But scientific meaning-change does not really drive us to the view that at any given time the meanings of our theoretical terms are somehow only partial, and that those expressions are living a semantic shadow life, faint and gibbering, like the ghosts in the afterworld of the ancients. All that we need conclude is that certain expressions which at one time had one meaning are later given another related meaning to make them more useful.

Consider as examples some familiar thermodynamical quantities, such as internal energy, entropy, enthalpy, or temperature. These are originally defined only for thermodynamic systems in equilibrium, in accord with the limitations of phenomenological thermodynamics. Later on, other, broader concepts are defined, to which we give the old names 'energy', 'entropy' and so forth. The entities which have internal energy or entropy in the old sense constitute a proper subclass of those which have the new, more general energy or entropy. (At least this sometimes holds, provided the old theory is a special case of the new one, and not incompatible with it.) For systems not in thermodynamic equilibrium can instantiate the new properties but not the old ones. The fact that the same expression may be used for both kinds of properties should not lead us to think that the same concept or property is involved early and late, and that its 'partial interpretation' is progressively filled in.

Our definitional strategy makes clear one way in which an older theoretical property, such as phenomenological entropy, and its newer, high-powered counterpart, may be related. The older entropy was defined in such a way that only thermodynamic systems in equilibrium possessed it. But that definition involved quantifying over a certain predicate in the old theory's Ramsey sentence. The Ramsey sentence says that there is some property or other, at *least* one, which plays such-and-such a role. The new high-powered entropy may well be such a property, as long as the role does not by definition include the proviso that the system possessing the new entropy be in equilibrium. And since it is more interesting, once discovered, than the old property, and really explains why the old property is instantiated at all, it takes on the name of the old property.

This raises another point, however. Suppose that we reach some sort of rock-bottom level of science. We have isolated basic theoretical properties which have no further *fundamentum in rebus*, so to speak, than themselves. By this I mean that whether or not we know it, there is no further theoretical property which provides an acceptable substitution-instance for the relevant quantified variable in the definition's modalized Ramsey sentence. We then have a definition which looks like our earlier (iv) on page 244. And the only property which constitutes an acceptable substitution-instance for 'M' in the modalized Ramsey sentence 'Nec·RS' is T_m itself, the property being defined. Here we have a demand for an impredicative definition. If there is an ultimate level in nature whose properties are irreducible in this way, then such definitions are required. Perhaps today's concept of rest mass, or of electric charge, are just such irreducible properties. I do not see that the impredicativeness of the definitions is an objection to them.

IV

The method described so far for defining theoretical scientific terms has another interesting application. Some have found it a cause of sorrow that contemporary reconstructions of mathematics seem to commit us to the existence of sets, conceived as Platonic entities, i.e., as non-spatiotemporal and non-individuals. We seem driven to this because mathematics is reconstrued via set theory, and the sets in question are not naturally regarded as denizens of space-time. The null set, for example, seems to have no natural location or duration assignable to it.

How did we get into this mess? Presumably, we started out with an ordinary notion of *set* as some sort of assemblage or ensemble or group or bunch. Sets in that sense are as spatio-temporal as their members or parts. There is an ensemble of cups in the cupboard, and yesterday the whole group of Rotarians went on an outing up the Hudson. Then mathematicians found it helpful for their own dark purposes to lay down certain restrictions. For example, for any two sets there is a set which is their intersection. This introduces a null set as the intersection of a pair of disjoint sets. The term 'member', which usually applies to people connected with organizations, was pre-empted to designate any relatum of the relation symbolized by 'ε', and axioms were laid down which stipulated its behavior. 'Group', 'class' and 'set' were given different, specialized meanings. And the extensional identity conditions for sets determined that there would be only one null set (unless a theory of types is introduced, in which case we get one for each type... and so on for other complications). Now ordinarily we would say that since there are no fairies there is no such thing as the set of all fairies. Or, if we are arm-twisted into admitting that there is such a set (with purely possible, non-actual members), it is certainly not the null set, for unlike the null set it has pure possibilia as members. We do not say *both* that there are no fairies and that there is a set of them, with no members, which moreover is also the set of all chimaeras. However, the usual philosophical view would have us surrender these intuitions when speaking of the sets dealt with in set theory.

But when sets are conceived in this artificial and Platonic fashion we cease to have very strong reason to believe that they exist. We do have strong reason to believe that the sets and groups and aggregates of which we do ordinarily speak do exist, for they are sum-individuals of the things which make them up. We know that the group of Rotarians went up the Hudson because we know that several Rotarians did so together. Not so for the mathematical set of those persons; it, being non-spatio-temporal, is precluded from travelling anywhere. Why believe that there is such an entity?

The usual answer given is that we want to believe that mathematics is true and the truth of mathematics requires the existence of such sets. That last claim is the one which I will now show to be questionable.

Mathematics requires the existence of entities satisfying the axioms of whatever versions of set theory we deem to be true. This is trivial, since set theory is part of mathematics. But the entities required by set theory need not be *Platonic* sets, any more than they need be sets in the ordinary pre-set-theoretic sense of the term. For we can use the strategy previously outlined to define what it is to be a 'set', for the purposes of a given version of set theory, in a way that *leaves open* the question of whether sets so defined are Platonic. Set theory is usually construed philosophically in such a way that none of the familiar spatio-temporal objects of our world could be sets, for they do not have *members*, only *parts*. But we can construe set theory more abstractly. Its subject matter consists of entities, any entities, which bear to one another certain relations which are captured in the quantificational structure of our axioms. These relationships are definable in terms of a basic 'ε-relation', modelled originally after ordinary talk of what it is to be a member of something (the NAACP, say), but with some additional restrictions imposed, which of course vary from one version of set theory to another. *Any* relation R which behaves, with respect to a given universe of discourse, in the way that the 'ε' relation behaves in Zermelo-Fraenkel set theory, for example, can be regarded as *a* Zermelo-Fraenkel ε-relation. Corresponding to it there is *just one* entity in that universe to which nothing bears R. This is 'the Zermelo-Fraenkel null object' for that universe and that relation R. It can be as abstract as a mathematical theory or as concrete as your left hand. Talk about *the* Zermelo-Fraenkel null set is like talk about the whale in 'The whale is a mammal.' *The* Zermelo-Fraenkel null object and *the* Zermelo-Fraenkel epsilon relation are type entities. What we say about them can be paraphrased into talk about *any* Zermelo-Fraenkel null object or any Zermelo-Fraenkel epsilon relation, just as talk about The Whale is really talk about any whale.

We must define what it is for something to be a Zermelo-Fraenkel epsilon relation and a Zermelo-Fraenkel null object. The intuitive idea is that a relation R is a *ZF*-epsilon relation just in case its role in some universe of discourse is mirrored by the epsilon symbol in the *ZF* axioms. Alternatively, it plays the role of the epsilon relation in some standard model for the *ZF* axioms. We can turn this intuitive description into a definition as follows. Take some version of the *ZF* axioms, say the one in which there are no 'non-sets', and the axiom schema of Replacement appears. Form the conjunction of the axioms and axiom-schema. Quantify universally over the dummy predicate appearing in the Axiom of Replacement. Replace the predicate constant 'ε' with the free dyadic relational variable 'R'. Call the result '$ZF(...R...)$' Then define what it means to be a Zermelo-Fraenkel epsilon relation as follows:

$(R)(R$ is a ZF-ε-relation iff $ZF(\ldots R \ldots))$

Alternatively, we can leave the dummy predicate in the Axiom-schema of Replacement unbound, thus getting a definition schema. Or we can pick a language L, relative to which the Axiom of Replacement is stated. Then there will be a natural number n which is such that no 'specifying relation' in the Axiom is more than n-ary, though one at least is n-ary. Then we define:

$(R)(R$ is an L-ZF-ε-relation iff $ZF(\ldots R \ldots))$

where the Axiom of Replacement is expressed in n conjuncts, $A_1 \ldots A_i \ldots A_n$, with the free variable in each A_i universally quantified over, and substitution instances limited to predicates in L (or, extralinguistically, to properties in L's universe of discourse).

Corresponding to each ZF-ε-relation R there is a unique ZF null-object. This object is the only entity (or the only entity in the universe of discourse in question), to which nothing bears R. Once again, we can if we like relativize this notion to a language L and set of predicates $A_1 \ldots A_n$.

We can then define what it is to be a ZF-object generally ('set' in the old terminology with the misleading overtones). To be a ZF-object is to play the role of one in a model for set theory à la Zermelo-Fraenkel. We can define the notion this way:

(x) $\quad$ (R) $[x$ is an R-ZF-object iff $(R$ is a ZF-ε-relation$)\in$
$\quad\quad\quad \in (x$ is the R-ZF-null-set $\vee$ $(Ey)(yRx))]$.

In short, to be a ZF-object is either to be a ZF null object or to be a relatum of a ZF-ε-relation.

The above definitions do not force on us the usually intended interpretation of Zermelo-Fraenkel sets. This is in one way an advantage, for ZF-objects so defined need not be Platonic. But it might be thought a defect in another way. For, as we know from the Löwenheim-Skolem theorem, the ZF-axioms so abstractly construed do not even force the Zermelo-Fraenkel objects to be nondenumerably numerous. One who demands this can perhaps add it as an axiom expressed *in additional primitive extralogical vocabulary* to the old axioms. This circles in more closely on the intended interpretation of the Zermelo-Fraenkel axioms while yet not committing us to believing that their subject matter must be the 'neither fish nor fowl' Platonic sets of the received philosophical tradition.

It is natural to object that avoiding Platonic sets at the expense of quantifying over Platonic properties is jumping from the frying pan into the fire. This objection has some force. But I think that despite first appearances it should be rejected.

The usual argument for what might be called 'grudging Platonic-set-theoretic realism' is this. We would like to avoid commitment to Platonic entities altogether. But physics requires mathematics. And mathematics requires Platonic sets, both as the subject matter of set theory and in reconstructions of the theory of real numbers, etc. Given sets, we need no *other* Platonic entities, such as numbers construed as non-sets, properties, propositions, or facts. So sets are the only Platonic entities we should admit. But vast indenumerably infinite legions of them should be admitted; enough to meet the requirements of the usual axioms of set theory on their usual interpretation.

The first reply to this is that if all that is wanted is enough to satisfy set-theoretic axioms, construed broadly as having a customarily large, indenumerable model, then physical universes can do that – our own four-dimensional space-time, and additional hosts of space-times, not spatiotemporally related to ours. You object that we have no evidence for their existence? I agree. We have little reason to posit them, except that set theory demands for its usual models appropriately numerous entities. But surely it is preferable to posit more of the same old familiar physical stuff if positing without reason is our game, than to posit lots of unfamiliar Platonic sets! Better still, why not simply admit that we lack sufficient reason to posit *either* more physical worlds *or* Platonic sets? *If* set theory must be regarded as true and as talking of indenumerably numerous entities, posit them, but do not posit that they are Platonic without independent reason. To satisfy set theory's axioms, the 'sets' need not be Platonic.

But, it can be objected, the properties quantified over in my definitions of 'sets' are Platonic! So why not just stick with Platonic sets?

Why not? First, because we need not construe our definitions as committing us to Platonic properties. We would impose a substitutional interpretation on the quantifiers. If this is done for the quantification over individual variables in the Zermelo-Fraenkel axioms, we cannot follow the usual custom of favoring an indenumerably infinite model for them. If the substitutional interpretation is confined to the quantification over non-individual variables, such as 'ε', then this objection falls. But we still cannot guarantee that our language contains a predicate with the structural properties demanded by the *ZF* axioms. Quite so. And to that extent we should be unsure of their truth as so construed.

But surely we know that there exist numbers, and that the reals, for example, are indenumerably numerous. We know also that they need not be taken as primitive but can be regarded as Platonic sets. Here is a good reason for admitting sets, and rejecting other Platonic entities as unnecessary.

That brings me to my second reply which is that (i) we need other Platonic entities, such as properties, propositions and/or facts anyhow, and (ii) given them we can dispense with Platonic sets! The argument for (i) consists partly in pointing out the utility of properties in ontology generally, e.g., to individuate individuals, and is impossible to establish here. The other part of it consists in pointing out that, for example, there are more truths, or facts, about the real numbers than could be expressed in any natural language. These truths, or facts, cannot possibly be construed as Platonic sets. They have intensional identity conditions, like propositions. Since we need these anyhow, we cannot limit ourselves to Platonic sets. The next step, (ii), is to argue that we can construe the mathematical talk about numbers as talk about numerical *properties*. This gives us a more natural reading for the mathematical truths of elementary number theory, for instance, than is provided by any set-theoretic reconstruction. And it enables us to do without Platonic sets.[7] Moreover, the numerical properties, such as those of *being a double* or *being a $\frac{1}{4}$-tuple*, which on this reconstruction are expressed by '2' and '$\frac{1}{4}$', can themselves serve as the entities in the nondenumerable universes of discourse of our set theories! And whatever may be the case for properties generally, these numerical properties are no more unclear than our mathematical talk about numbers.

V

Actually, a double construal can be given to statements about numbers.

On the one hand, they can be viewed as talking about whatever quantitative aggregates there may be in reality, be they physical, mental, platonically set-theoretic, or whatever. On this construal, an equation, such as '$2+2=4$', says roughly that any double, of fingers for example, joined to any other double, is a quadruple. On the second construal, '$2+2=4$' says that the property of being a join of a double to another double is identical to the property of being a quadruple.

The main motivation here is to find a more natural and straightforward reading of statements about numbers than that provided by the various artificial and unintuitive set-theoretic approaches. I would have liked to have given a purely nominalistic reconstruction but have not succeeded in doing so. Though some of my ostensible quantifications over properties are compatible with nominalism, others, as far as I can see, are not. In particular, general existence claims, such as 'There are more real numbers than natural numbers', seem to demand platonic numerical properties. This is perhaps disappointing, but not such as to rob the program of interest. For the alleged advantage of sets over properties in having purely extensional identity conditions is suspect, in my view, and certainly does not justify the greater arbitrariness and unnaturalness of the competing set-theoretic reconstructions. This is not to deny that those reconstructions can be formally correct, of course; nor that they reveal interesting structural similarities between numbers and certain set-theoretic entities. It is only to deny that they give us as much *philosophical* insight into the nature of numbers and the meanings of numerical statements as does the program to be sketched here. Also, I believe that this program points the way to a satisfactory epistemology of mathematics, if any is to be found. It comes closer than its competitors to vindicating the claim that all mathematical truths are analytic, and focusses more clearly what remains to be done to establish that claim, over and above making an honest woman of analyticity generally. If she simply cannot be made respectable, then at least we are on the way to showing why truths of number theory come close to being as epistemologically sound as the usual candidates for analyticity, like 'All vixens are foxes'. Not that we are yet home free. We still have to worry about those statements positing the existence of numerical properties, for example. But there is hope that this can be shown to be no more objectionable than talking about possible instances.

My strategy is to construe the statements of pure number theory as simply the most generalized form of the statements of applied number theory. So '2 plus 2 equals 4' is the generalized form of '2° Kelvin plus 2° Kelvin equals 4° Kelvin', '2 lbs. added to 2 lbs. make 4 lbs.', and '2 fingers and two fingers equals 4 fingers'. Some statements of applied mathematics deal with intensive magnitudes, like the first above; others with extensive magnitudes or discrete aggregates, like the second and third respectively. The meanings of the mathematical operations, such as addition, subtraction and multiplication, vary from one kind of application-statement to another. Here only statements about aggregates of individuals and about extensive magnitudes will be treated, and the operations will be interpreted in terms of mereology, the general logic of part-whole relations developed by Lesniewski and others. In the statements of pure mathematics the operations have a meaning which covers this interpretation but is more general than it.

I will try to construe mathematical truths and falsehoods in such a way that whatever existential commitments are needed to give them the proper truth-values will automatically be satisfied by the commitments of the rival set-theoretic approaches, though not demanding them. For example, if we read '1000 equals 998 plus 2' in such a way that it demands the existence of 1000 entities to turn out true, then at least the sets posited by the Russell-Whitehead or the Zermelo-Fraenkel reconstructions will fill the bill. But the bill may be filled even if those sets do not exist, as long as *some* 1000 tuple of something-or-others does.

Let us proceed to the addition and subtraction of signed integers. It has been explained that '$2+2=4$' means roughly that any quadruple is a $(2+2)$-tuple, and vice versa. The general operation of addition covers several special cases, such as augmentation (pouring 2 gallons of water into a barrel already containing 2 gallons), and mereological summing. A mereological sum of two or more entities is the possibly scattered individual which consists of them as its parts or components, albeit sometimes non-adjacent, non-connected parts. Thus, we can talk about the scattered individual consisting exactly of my eyeglasses and Cleopatra's last sneeze. It is assumed, correctly I believe, that given any pair of entities, or N-tuple generally, there is an entity consisting of their mereological sum.

An equation of signed integers, like '$2+3=5$', cannot quite be construed as saying that the mereological sum of any double with any triple is a quintuple. For the double chosen might share a part or member with the triple, thus giving us only a quadruple. If there are 3 pens and 1 pencil on the table, and the pencil and one pen are colored indigo, then we could sum the 3 pens and the 2 indigo objects to get a foursome of objects, not the desired quintuple at all. So our mathematical equation must be viewed not as talking about mereological sums generally, but about what I'll call 'joins', which are sums of entities which are discrete, that is, share no common parts in a sense shortly to be explained.

Also, we do not want to be forced to say, for example, that '$2+2=128$', on the grounds that a twosome of (complete) dentures comprises 64 teeth, and so the join of it to a discrete twosome of dentures constitutes a 128-some of teeth. One and the same entity, it seems, may be both a triplet of Xs and a sextet of Ys (e.g., if each X is a pair of twins, the join of them is a triplet of pairs and a sextet of twins.) So our mathematical statements must speak of tuples of entities *qua* being of some fixed kind. '$2+2=4$' says that any join of a double of something or others, say BLOOPS, with another double of bloops, is a quadruple of bloops, and vice versa. The two doubles must share no bloops or bloop parts, though they may include other entities and even share them in common. Our apparent quantification over kinds can be handled without ontic commitment to kinds or universals; they rear their heads only later. At this stage we can treat them as follows. We allow our language to form such predicates as 'is a finger-tuple', 'is a pencil-tuple', and as shorthand, 'is a bloop-tuple'. Here 'bloop–' is a placeholder for a common noun. We treat it as a kind of variable which can be bound by quantifiers, thus: '$(B)(x)(x$ is a B-tuple...)'. Then adopt a substitutional approach to this quantification. If the quantifier is universal, the statement in question is true iff all statements formed from it by dropping the quantifier and substituting a common noun wherever the quantified variable appears are true.

Ordinary subtraction forces us to backtrack and refine somewhat our ideas about what arithmetical statements are saying. Take '$9-7=2$'. We might try to construe this as a statement about what is left when you form some kind of a *dis*join of a ninesome and a septuple, where the septuple is completely included in the ninesome. But we would have difficulty in guaranteeing that *every* double is in fact the disjoin of a ninesome and a sevensome. What if the universe, past, present, and future, contains only two bleeps? There is no ninesome of bleeps from which we can take away a sevensome to get that twosome. Moreover, we have to face the disheartening fact that some subtractions land us in the realm of negative numbers, such as '9 − 12 = − 3'. We want these cases to be treated without our having to burble about anti-beings or deficits of existence.

Men often have occasion to assign either of two mutually exclusive statuses to entities of the same kind. We have cows owned versus cows owed, birds flying north versus birds flying south; objects above sea level versus objects below sea level. So a given tuple of bloops can sometimes be regarded in addition as a tuple of *positive-status* bloops or one of *negative-status* bloops. Why not regard talk of negative numbers as roundabout talk concerning tuples of negative-status bloops? Of course, whether a given bloop has positive or negative status is relative to the interests of a given person on a given occasion, and to the conventions which he adopts to express those interests. Attributions of negative-status to a bloop are as indexical as the claim that one thing is to the left of another. Statements of pure mathematics codify in general terms the particular truths expressed by these indexicals, where the relevant variables are assumed to be fixed, that is, we are agreed on which bloops are positive and which are negative.

We have two further problems to solve, which I call the 'Nobel Prize-winner Problem' and the 'Zero Problem'.

The Zero Problem arises naturally when we try to explain what it is to form a *dis*join of positive- and negative-status bloops; the mereological interpretation of subtraction. To form the disjoin of a positive tuple A of bloops with a negative tuple $-B$ of them, do this. Take the tuple with the smaller number of bloops in it, if there is one. Otherwise, take either tuple. Match each bloop from the chosen tuple, say $-B$, with just one bloop in the other tuple, so that each bloop in B is matched with a different mate in A. In this way you get a bunch of bloop-pairs, monogamous marriages, each consisting of one spouse from A and one from B.

Now form the join of the two tuples A and B, *but omitting all married tuples*. The result is a disjoin of our positive tuple A and the negative tuple $-B$. A disjoin of a foursome of positive bloops with a duo of negative ones is a double of positive ones, as we are told by the time-honored truth '$4-2=2$'. And a disjoin of a threesome with a minus-foursome is bound to be a single negative-status bloop.

The Zero Problem arises when we ask what a disjoin of N positive bloops with N negative bloops looks like. There would seem to be no such entity. For if you omit all married couples from the join of the two tuples, there seems to be nothing left. WE WANT SOMETHING TO BE LEFT, namely, a *nultuple* of bloops. A nultuple of bloops is any entity which has no bloop or bloop-part as a part. Thus, my hand is a nultuple of handkerchiefs. We want statements like '$2-2=0$' to say that a disjoin of a positive and a negative double of bloops is a nultuple of bloops, but not just nothing at all. To ensure this result, let's reconstrue what we are going to mean by a tuple of bloops. No longer is a 5-tuple of bloops, for instance, going to be any old sum of 5 discrete bloops. It must include in addition something which is neither a bloop nor a bloop-part, a foreign element. The disjoin of a positive 4-tuple with a negative 4-tuple of bloops consists of the sum of the two foreign elements, one from each 4-tuple, and is thus a perfectly respectable nultuple of bloops.

The Nobel Prize-winner Problem arises from the fact that we are inclined to read a statement like '2 equals $10000-99998$' as saying that any double of bloops is a disjoin of a 100000-tuple of positive bloops to a 99998-tuple of negative bloops. But no double of Nobel Prize-winners, for example, is such a join, since there aren't enough Nobel Prize-winners to constitute a 100000-tuple of them. But perhaps we can make sense of the idea that any twosome of Nobel Prize-winners, though not such a disjoin, *is* nonetheless a ($100000-99998$)-tuple of Nobel Prize-winners.

Say that a tuple of Nobel Prize-winners is a ($100000-99998$)-tuple of them if and only if there are as many Nobel Prize-winners in it as there are bloops in any disjoin of 100000 positive bloops and 99998 negative bloops. This can be spelled out in mereological terms.

Now we can develop our two slightly different ways of reading equations and inequality statements, and two correspondingly different ways of reading the numerical expressions that flank either side of an equals-sign or an inequality sign. The first way to read '$2+2=4$' is to read it as saying:

For any B, any $(2+2)$-B-tuple is B-equal to any 4-B-tuple. And an inequality, such as '$2+4>3$', is really an abbreviated way of talking about one bloop-tuple's being greater than another with respect to the number of bloops in the two. One entity may have more toothbrushes composing it than does another, and yet have fewer bristles than that other. So the first is toothbrush-greater-than the second, but bristle-wise-less-than it. The equals-sign and the various inequality-signs, such as '>' and '<', are general shorthand for such contrived expressions as 'is toothbrush-greater than', 'is bristle-wise-less than', 'is pencil-equal to', and so forth. So '$2+3>4$' is shorthand for 'For any B, any $(2+3)$-B-tuple is B-greater than any 4-B-tuple.'

The second way to read equations and inequalities is more like the way they are usually read in traditional philosophy of mathematics. An equation, such as '$2+2=4$', says that $2+2$ is *literally identical to*, the same thing as, 4. Those entities are usually identified with Platonic sets of some sort. I prefer to regard them as numerical properties, predicable of tuples. The equation is read as saying that the property of being a $(2+2)$-tuple of bloops is identical to the property of being a 4-tuple of bloops. The inequality signs must be reconstrued. To say '$2+2>3$' is to say that the property of being a $(2+2)$-tuple of bloops is of a higher bloop-order than the property of being a 3-tuple of bloops. This just means that any $(2+2)$-tuple of bloops is bloop-greater than any 3-tuple of bloops. No mystery.

Before wading into the deep waters of multiplication and division involving negative integers and fractions, let's look at those operations on positive integers alone. To say that $2 \cdot 3 = 6$ is to say that any double of triples of bloops is bloop-equal to any sextuple of bloops. A double of triples is an entity consisting of two other entities, each of which is in turn a threesome of bloops. It is obvious that commutativity of multiplication holds; a double of triples is indeed a triple of doubles, as the diagram shows.

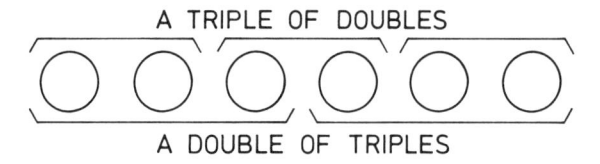

A TRIPLE OF DOUBLES

A DOUBLE OF TRIPLES

And multiplication can be iterated indefinitely. We can have a double of triples of doubles, for example. This is the same as a quadruple of triples, or a double of sextuples.

The fun begins with negative numbers. A first, naive way of reading '$-2 \cdot 6 = -12$' would be to read it as saying that any negative-status double of positive-status sextuples of bloops is bloop-equal to any negative-status twelve-tuple of positive bloops. But why should this be true? For that matter, why should a negative 12-tuple of positive bloops be bloop-equal to a positive 12-tuple of negative bloops? As long as we allow the status of tuples to be independent of the status of their component bloops we can't answer this question. So let's rule that a negative-status N-tuple of bloops is simply an N-tuple of negative-status bloops. Status attaches primarily to the bloops; only derivatively to the tuples.

To read arithmetical statements, first change all negative numbers to positive numbers multiplied by minus-ones. Then gather all of the minus ones at the end of the term in which they appear. If there is an odd number of them there, replace them all with a single '-1'. If an even number is there, erase them all. To illustrate: '$(-1)(-3)(-5)$' goes to '$(-1)(-1)(3)(-1)(5)$', which becomes '$3 \cdot 5 \cdot (-1)(-1)(-1)$', which goes to '$3 \cdot 5 \cdot (-1)$', which is then read, 'a (any) triple of quintuples of negative bloops'. The binary subtraction-operator is simply read as a notational variant of the binary addition operator followed by the unary '(-1)' operator, which then hooks onto the following term. So '$5 - 3 = 2$' says '$5 + (-1) \cdot 3 = 2$', which says '$5 + 3(-1) = 2$'. This implies that the join of any quintuple of positive bloops to a triple of negative bloops is a double of positive bloops, any such join of oppositely signed bloops being, of course, a *dis*join. The artificiality of treating negatively signed numbers in the way just suggested is mitigated when we see how the same mode of treatment appears when we deal with the complex numbers, quaternions, and other extensions of the real number system.

Zero requires special treatment, not surprisingly. Any term of the form '$M \cdot \ldots \cdot 0 \cdot \ldots$', in short, any term with one or more zeros as factors, must be processed before its meaning can be read off. First, move all the zeros to the end of the term, then replace them with a single zero. Then read ahead: 'Any M-tuple of N-tuples of... of zero-tuples of bloops...'

Before turning to division, let us introduce fractions. A one-half tuple of bloops is anything which is such that the join of it to anything bloop-equal to it is bloop-equal to any single bloop. We are now going to count any bloop-part as itself a fractional bloop-tuple, provided that there is some rule for assigning a definite fractional magnitude to it *qua*-bloop-part. That is, it may not make sense to speak of one-half of a cloud, a revolution, or an ocean, for lack of precise boundaries and/or plausible conventions for assigning fractional numbers to the mereological parts. But we can speak of one-half of a pie, or of a chunk of mass, or of a distance of 50 miles. So the arithmetic of fractions applies here. We are faced with the unsettling fact that not every join of a one-half-some of pies with another halfsome of pies is a pie. For the two halfsomes may have been cut off from pies of different sizes, and moreover be separated by miles. Standard set-theoretic approaches to foundations require us to reconstrue the nature of the entities and operations in arithmetic as soon as rational numbers are introduced. '2 + 2 = 4' says something different if taken to be about rational numbers from what it says if taken to be about signed integers, or natural numbers. Analogously, we must reconstrue either our notion of a join or our notion of a tuple, or both, as we move from the arithmetic of signed integers to that of rational numbers. To handle the above-mentioned Pie Problem we can either demand that joins of rational tuples to one another be permitted only where the tuples are suitably spatially situated; or generalize the notion of *tuple* so that the above join of two scattered half-pies is ruled a one-tuple of pie, though not a pie; or only allow joins of *bloop*-tuples where the 'bloop' stands for a quantitative stuff-name, like 'mass-in-grams', which so behaves that joins of scattered tuples sum in accord with the tenets of traditional arithmetic. We can probably get well-behaved realizations of arithmetic by using *any* of the above strategies, all of which are mereological. So the statements of pure mathematics are simply the generalizations summing up the features common to all of these mereological systems of applied mathematics, and the non-mereological ones as well.*

To be a $1/N$ tuple of bloops, for any integer N, is to be such that any join of N such entities, all bloop-equal to one another, is bloop-equal to any one-tuple of bloops.† (This holds no matter which of the strategies sketched above is adopted.) To be an M/N tuple of bloops, for integers M and N, is to be bloop-equal to any M-tuple of $1/N$ tuples of bloops.

---

* Such as the system which interprets addition and subtraction in terms of augmentation and diminution of intensive quantities.
† See Appendix.

Something is a rational-tuple of bloops iff it is either a nultuple of bloops or, for some integral M and integral $N > 0$, it is an M/N tuple of bloops. This does not yet require quantification over numbers or numerical properties. For we assume that we have in our language expressions for each integer, which function as predicate-forming operators on predicates. Quantification over these operators is then given the substitutional interpretation.

We want to be able to read standard mathematical statements in which rational numbers are raised to rational powers, as in

(i) $\frac{3}{4}$ is greater than $(\frac{1}{2})^{17}$
(ii) The square root of 2 is irrational
(iii) $(\frac{1}{2})^{1/2}$ equals $(.5)^{.5}$

To this end we must develop some machinery.

Suppose that we have two individuals in the mereological sense, 'Roger' and 'Damian', each of which is a rational-tuple of bloops. Roger is a bloop-double of Damian just in case Roger has twice as many bloops as Damian. Similarly, we can see what it is for Roger to be a bloop-R-tuple of Cosmas, for rational R. It means that for some rational M, Cosmas is a bloop-M-tuple, and Roger is a bloop-$R \cdot M$-tuple.

Now to exponents. We'll find a way to make 'exponential comparisons' of individuals, e.g., the number of bloops in Roger is the square of the number of bloops in Rufus. Using that, we will then get a construal of equations and inequalities in which no specific individuals are mentioned, and also a way to deal with irrational-tuples of bloops.

To say that Roger is a bloop-(square-tuple) of Damian is to say that for some rational M, Damian is an M-tuple of bloops, and Roger consists of M discrete M-tuples of bloops. In other words, Roger can be dissected into M different M-tuples of bloops, no one of which shares a bloop or bloop-part with any of the others. To say that Roger is a bloop-(cubetuple) of John, a '3-power' of John, for short, is to say that for some M, John consists of M bloops, and Roger consists of M discrete bloop-(square-tuples) of any M-tuple of bloops.

In general, for any positive integer $N > 1$, Roger is a bloop-$(N+1)$-tuple of Damian iff for some rational M, Damian is an M-tuple of bloops and Roger consists of $|M|$ discrete bloop-(N-powers) of any $|M|$-tuple of bloops. The bloops in Roger are positive if M is positive.

Otherwise their status S is given by the rule $S = (-1)^{N+1}$. Roger is a bloop-1-power of Damian iff Roger is bloop-equal to Damian. And Roger is a 0-power of Damian iff Roger consists of just one bloop.

For any rational number M and positive integer R, something is a bloop-M^R-tuple iff it is a bloop-R-power of any bloop-M-tuple. Thus, to be a 23^{14} tuple of mass-in-grams is to be a 14-power of any tuple of 23 grams.

When rational numbers are raised to rational powers irrational numbers sometimes appear, as the Greeks discovered about 2½. We now have the means to express claims about algebraic numbers, which include all of the rationals plus the irrationals which are expressible as rationals raised to rational powers. To start with, take fractional exponents with a numerator of one. To say that John is a ½ power of a rational-tuple John is to say the following:

(i) Joan is B-(greater-than-or-equal-to) any B-rational tuple which is such that any B-square-tuple of it is B-(less-than-or-equal-to) John; and

(ii) Joan is B-($\leq$) any B-rational-tuple which is such that any B-square-tuple of it is B-($\geq$) to John.

The general formula, for $1/N$, is easily derived. And in terms of it we can see what it is to be an $R^{1/N}$ B-tuple, where R is a positive rational number and N is a positive integer. It is simply to be a $1/N$ B-tuple of any R-tuple of Bs.

More generally, for positive integers M and N, to say that Joan is an M/N power of a rational-tuple John is to say

(i) Joan is B-($\geq$) any rational-B-tuple any N-power of which is B-($\leq$) any M-power of John; and

(ii) Joan is B-($\leq$) any rational-B-tuple any N-power of which is B-($\geq$) any M-power of John.

Negative exponents can be treated as follows. For any rational number R and positive integers M and N, to say that something is an $R^{-M/N}$ tuple of bloops is to say that it is a $1/R^{M/N}$ tuple of bloops, which is simply to be a $1/N$ power of any $1/R^M$ tuple of bloops.

We now understand what it is to be an irrational-tuple of bloops, in those cases where the irrational in question is algebraic, but not generally. And we know what it is for one B-tuple to be an R-power of another, for any rational R, provided that the tuples are themselves rational-B-tuples. We do not yet know what it is to be a real-tuple generally, or to be an R-power of an irrational tuple.

One of the standard ways to define the real numbers is by regarding each as as an equivalence class of equal Cauchy sequences of rationals. That is not for us, of course. But we can take some cues from the idea. Corresponding to a Cauchy sequence of rationals we can talk about a Cauchy sequence of rational-tuples of bloops. The sequence is not a set but a mereological join. And the properties of real numbers, or better, of real-tuples of bloops, will be explained in terms of the relations between those real-tuples and the corresponding Cauchy sequences of rational-tuples.

For us, a Cauchy-sequence of rational-tuples of bloops will be a join of bloops which are simply ordered by any irreflexive, asymmetric and transitive relation R. The bloops must of course bear to one another the Cauchy-type relations. For any rational number I there must be a tuple N in the sequence which is such that any pair of tuples beyond ('R-beyond') N bloop-differ from one another by less than R. This can be spelled out in mereological terms. As an example, we might take a tuple of volumes of space, each of which is a certain number of cubic centimeters, or is a 'cm^3-tuple'. The first tuple in the sequence is *here*, and is 1 cm^3. The tuples all stretch out endlessly in a certain direction, say galactic North, or 'G' for short. Let the Nth tuple have a volume of $1 + 1/N^2$ cm^3. A 'limit-tuple' of the sequence is any tuple with a volume of 2 cm^3.

To be a real-tuple of bloops is to be a limit-tuple of a Cauchy-sequence of rational-tuples of bloops. A real number can be thought of as the property of being such a limit-tuple. This can be defined as follows. We are assuming that bloop-tuples can be compared with respect to bloop-magnitude; i.e., that we know how to apply the relational predicate '$B \geqslant$'. And we can say that the bloop-difference between a pair of tuples is greater than (or less than), some rational number, such as $\frac{1}{10}$. Mereologically, this means that any disjoin of the two is greater than (less than) any $\frac{1}{10}$ tuple of bloops. A Cauchy-sequence of B-tuples ordered by the relation G will be called 'a G-B-sequence'. A B-tuple, Ralph, is a limit-tuple of a G-B-sequence iff for any rational number R there is a positive integer N such that the B-difference between Ralph and any B-tuple beyond the Nth in the sequence is less than R.

If space, or time, is infinite and infinitely divisible then lengths, areas and volumes ordered sequentially by a directional relation will provide us with all of the Cauchy sequences which we need to circle in on all of the real numbers. It will also provide us with a limit-tuple to instantiate each real number. Thus, there is a pi-tuple of cubic centimeters, an e-tuple, a $2\frac{1}{2}$ tuple, and so forth.

The standard mathematical operations, addition, subtraction, multiplication, division, and exponentiation of reals, should be definable in terms of the corresponding operations on Cauchy sequences which circle in on those reals. For example, suppose we want to define what it means for one real-B-tuple to be a square of another. We take any B-sequence of which the first is a limit-tuple, and any B-sequence of which the second is a limit-tuple. We say that the first sequence is a square of the second iff the Nth B-tuple in the first is a square of the Nth in the second, for all N. That is what it means for a limit-tuple of the first to be a B-square of a limit-tuple of the second.

It is well-known that there are more real numbers than are expressible in English, or in any natural language which has a finite number of primitives and in which all expressions are of finite length. So our substitutional interpretation of quantification over numbers fails when we get to the real numbers. We seem to be committed to numbers, construed as numerical properties, as the domain of discourse of our quantifiers. As suggested before, I find these no more unacceptable than Platonic sets; indeed, more acceptable, if it can be agreed that the construal of mathematics in terms of them is more natural than the standard set-theoretic construals. But perhaps there are ways of avoiding commitment to Platonic number properties which are compatible with our treatment of mathematics.

One such way might be simply to assume that for any number R there exists a rational-tuple of a non-Platonic kind which is an instance of it. We might then try to translate talk of numbers and their relations into talk of tuples and the relations among them. If the translating can be done, then the requisite tuples will exist if the standard set-theoretic apparatus exists, and may exist without that apparatus. We can form mereological sums of sets. We can take the real numbers as construed set-theoretically in one of the standard reconstructions, and make sums of them serve as rational-tuples and real-tuples. We do this by assigning a measure to them. For example, corresponding to the set of real numbers between zero and two there is a mereological sum of such numbers. To this sum we assign the property of being a measure-theoretic 2-tuple. These tuples then provide the subject matter of real number arithmetic, construed in our mereological fashion. My point is to try to strengthen the conviction that if set-theoretic reconstructions work, then so does my reconstruction, but not vice versa. For the tuples which I require may exist even if sets do not.

Let us now treat the complex numbers. Each complex number is expressible in the form $R_1 + iR_2$, where R_1 and R_2 are reals and i is $\sqrt{-1}$. We treat i in a way similar to the way we treated -1, as indicating a status, which we can call 'imaginary status' or 'i-status'. Negative status of bloops comes to the fore when we add bloop-tuples. Imaginary status of bloops lies dormant until multiplication is done. Then it comes to life, and results in the reassigning and juggling of positive and negative statuses to the bloops which we started with. To illustrate: we have an $A + iB$ tuple of bloops, which is simply an $(A + B)$-tuple in which B bloops have i-status. To take a $C + iD$ tuple of $A + iB$ tuples, take $(C + D)$-tuple of $(A + B)$-tuples, but assign negative and i-status so as to get a tuple of this kind: $(AC - BD) + i(AD + BC)$. We assign a negative-status to a BD-tuple and form a disjoin of it and the AC-tuple. That disjoin is joined to an $(AD + BC)$-tuple, to each bloop in which is assigned an i-status. Do not add or subtract i-status bloops from bloops which lack that status... though one can form mereological joins of the two kinds of bloops, as long as the i-status is of each i-bloop is retained, like a flag on its back.

Further extensions of the number system, the hypercomplex numbers, are feasible, and can be treated analogously to complex numbers, in terms of introducing new statuses. In any given system of hypercomplex numbers, each number is representable in the form '$aR_1 + bR_2 + \cdots + nR_n$', where the Rs are real numbers and the lower case letters represent 'statuses' or 'flags'. Addition and subtraction follow the usual rules, with the proviso that only bloops of the same status are subtracted from one another, and bloops retain their flag when joined or disjoined. Multiplication is defined by specifying what happens when flags are multiplied.

E.g., we must specify that *a* times *b* will equal *d*, and then translate this into rules for assigning flags to the bloops in the product-tuple. Division can be defined in terms of multiplication.

Transfinite arithmetic can be treated in terms of transfinite tuples and their mereological behavior when joined and disjoined. An aleph-null tuple of bloops is no less conceivable than a 72-tuple, and similarly for tuples of higher transfinite cardinality. Two *B*-discrete *B*-tuples, Ted and Ned, stand in 1 — 1 correspondence iff the mereological sum of the two is dissectable into

(a) the sum of the foreign or non-*B* parts of both, and

(b) a tuple of non-overlapping entities, each consisting of a *B* from Ted and a *B* from Ned.

Lehman College
City University of New York

NOTES

[1] For such charges see, for example, Hilary Putnam's 'What Theories Are Not', in E. Nagel, P. Suppes, and A. Tarski (eds.), in *Logic, Methodology and Philosophy of Science*, Stanford 1962; and Peter Achinstein's *Concepts of Science*, Johns Hopkins Press, 1968.

[2] For a fairly recent account along these lines, see the late Rudolf Carnap's eminently readable *Philosophical Foundations of Physics*, Basic Books, Inc., New York and London, 1966, p. 267.

[3] In constructing a three-valued logic we should keep in mind that in two-valued logic conjunction and alternation are intimately connected with one another, in that they are mutually distributive and are De Morgan transforms of one another. By speaking of an operator analogous to the 'ε' of two-valued logic, I mean a binary connective which is associative and commutative; whose truth-table coincides with the two-valued truth-table for 'ε' in those cases where only the two standard truth-values are assigned to its conjuncts; and which is such that it and the three-valued analogue for '$\vee$' are mutually distributive and are De Morgan transforms of one another. There are two ways of giving joint truth-table definitions to 'ε' and '$\vee$' in three-valued logic so that as defined they turn out to be associative, commutative, mutually distributive and De Morgan transforms of one another.

[4] The difficulty is found though not acknowledged in Carnap's above-mentioned work. Carnap supposes that the theoretical terms ('*T*-terms') are uninterpreted until they get their empirical meanings via correspondence rules ("*C*-postulates"). He says: 'The *C*-postulates cannot, of course, be taken alone. To obtain the fullest possible interpretation (though still only partial) for the *T*-terms, it is necessary to take the entire theory, with its combined *C*- and *T*-postulates" (ibid., p. 267). He refers to that combination as 'TC'. The Ramsey sentence formed by existentially quantifying over each of its theoretical predicate letters is called '$^R TC$'. He then seeks an analytic meaning postulate ('*A*-postulate') from which all of the analytic meaning relations involving the theoretical terms can be defined. His suggestion is that

> ... the simplest way to formulate an A-postulate A_T for a theory TC is:
>
> (A_T) $\qquad {}^R TC \supset TC$
>
> It can easily be shown that this sentence is factually empty.... All the factual content is in the Ramsey sentence F_T, which is the Ramsey sentence ${}^R TC$. The sentence A_T simply asserts that *if* the Ramsey sentence is true, we must then understand the theoretical terms in such a way that the entire theory is true. It is a purely analytic sentence, because its semantic truth is based on the meanings intended for the theoretical terms (ibid., p. 270).

We have a problem here. The sentence A_T looks like a sentence in the object language, not the meta-language. And it looks as though it may be neither true nor false, being infected with meaningless predicate letters. That depends partly on the three-valued truth-table which we give to the conditional sign '$\supset$', partly on the truth-value of the Ramsey sentence ${}^R TC$, and partly on the truth-value which our three-valued logic assigns to 'TC'. There is in any case no question of A_T's *automatically* turning out analytic, independently of which of the possibilities just mentioned are realized. Suppose that the Ramsey sentence ${}^R TC$ is true, but the consequent TC is neuter. If we take the conditional form '$p \supset q$' to be equivalent to '$-p \vee q$', then regardless of which strategy we use to define '$\vee$', the conditional A_T must be assigned a truth-value *neuter*, and is thus not analytic (*true* simply in virtue of the meanings of its terms).

But Carnap also describes A_T as though it were a metalinguistic formula which tells us how we must understand TC if ${}^R TC$ is true. Is this metalinguistic formula analytic? Or is it an exhortation; neither true nor false? Or a resolution about how we are going to interpret the schematic letters in TC? How then is the resolution to be carried out? Certainly A_T by itself does not carry out the resolution; that is, it does not impart the intended meanings to the dummy predicate letters appearing in the TC part of it. Some of these difficulties were independently pointed out in James K. Derden's doctoral dissertation 'Analyticity and Scientific Theories with Special Reference to the Work of Rudolf Carnap' (University of Toronto, 1971, unpublished). I hope to resolve them here.

[5] Example: "If a thermodynamic system is brought reversibly from one thermodynamic equilibrium state to another, then, regardless of the path, the change of entropy is given by $\int dq/T$." Here the term 'change of entropy' has its significance explained in terms of the way in which entropy change is related to observable facts, such as the temperature of the system and increments or decrements of heat energy.

[6] Even dispositional terms such as 'brittle' had to be construed as only partially interpreted, for well known reasons reviewed on pages 240 and 241 of the text.

[7] In 'How To Define Theoretical Terms', *Journal of Philosophy* **68**, 13, July, 9, 1970.

APPENDIX

Suppose that we want to develop our reconstruction of mathematics without using any non-truth-functional connectives. Then the definition given of what it is to be a $1/N$-tuple of bloops will give us trouble. At least some quantities in the world are such that only a finite amount of the corresponding stuff exists. Suppose that there is less than one ton of platinum in the entire universe. So there is no one-ton platinum-tuple, or P-tuple, for short. In this case, our definition of what it is for something to be $\frac{1}{5}$ of a ton of platinum will apply to everything!

The definition runs:

(x)   (x is a $\frac{1}{5}$ P-tuple iff (y) (y is a one-P-tuple ⊃ (Ez) (z is a join of 5 entities each P-equal to x∈z is P-equal to y))

The antecedent of the definitions is not satisfied by any entity. So the definiens conditional is true of each entity, having a false antecedent for each. Disastrous, for then sneezes and suspicions and ice cream cones are all ruled by the definition to be $\frac{1}{5}$ P-tuples.

The problem here is the same as that encountered by Carnap in trying to define dispositional terms. I think that it shows that some sort of non-truth-functional 'if...then...' is built into the meanings of the theoretical quantitative predicates of science, such as 'has mass' and 'has electric charge'. The content of science is no more clear than the non-truth-functional connectives which must be used in specifying what its terms mean. But those meanings are not obscure, comparatively speaking. If we use a non-truth-functional connective in place of the conditional in the definition above, the result is certainly not obscure, even though we may not be able to give an account of its meaning in terms of our favorite truth-functional, extensional, etc. vocabulary.

We can do a bit better than this if we allow ourselves to accept certain locutions as already understood. Suppose that we allow ourselves the assumption that, for any two kinds P and Q, we know what it means to say that x has as much P-stuff as y does Q-stuff. For example, we know what it means to say that this chunk has as much platinum-in-tons as that ruler has length-in-meters. Then we can say that anything is a P-$\frac{1}{5}$ tuple iff there is as much P-stuff in it as there is Q-stuff in anything which is such that any join of 5 discrete entities each Q-equal to it is a Q-one-tuple. As an abbreviation, say that if there is as much P-stuff in z as Q-stuff in y, then x is P-Q-equal to y. We can then get a definition using only truth-functional connectives which will rule all and only the $\frac{1}{5}$ P-tuples to be such, provided only that there is a one-tuple of *anything*, not necessarily P-stuff.

First, we define that x B-one-tuples y iff x is B-equal to y. In terms of this, we define recursively what it means for x to B-(N+1)-tuple z, for any positive integer N. The definition:

(x)   (z) (B) (N + 1) (x B − (N + 1)-tuples z iff (Ew) (Ev) (w B-N-tuples z & v is B-discrete from w & the join of v to w is B-equal to x))

We now use these definitions to define what it is for Rufus to be a $\frac{1}{5}$ P-tuple.

Rufus is a $\frac{1}{5}$ P-tuple iff (B) (y) (z) [y is a B-one-tuple & y B-5-tuples z ⊃ Rufus is P-B-equal to z]

As long as, for some B, there is a B-one-tuple, the antecedent of the definiens will have one true substitution instance, at least in the case of extensive divisible quantities. For in those cases, if a one-tuple exists then so do all of its fractional parts. (For intensive magnitudes that is not necessarily true.) Given a single true substitution instance for the antecedent, the definiens will not simply be true of everything because of the universal falsity of the antecedent. It will be true of all and only the $\frac{1}{5}$ P-tuples, as desired. The main text explains how the usual set-theoretic apparatus can provide us with the necessary one-tuple, though it is not needed for this purpose, since any extensive and divisible one-tuple will do.

There is another way to maximize the chances of mathematical truths being preserved while yet construing all connectives in them as truth-functional. The idea would be to construe them as being about only those quantities which are instantiated in the ways required. To give a rough example, we would define what it means for something to be a $\frac{1}{5}$ B-tuple in such a way that an entity can be such only if there actually exists a B-one-tuple. This is the definition:

(x)   (B) (x is a B-$\frac{1}{5}$-tuple iff (Ey) (y is a B-one-tuple & (z) (y B-5-tuples z ⊃ x is B-equal to z))).

This strategy is tantamount to saying that mathematics is not directly about all of the tuples that we have been discussing so far. It is only directly about tuples of an ideal kind, the tuples of 'requisitely instantiated bloops'. In the unlikely event that there are *no* bloops of this kind, certain statements ordinarily regarded as mathematical falsehoods would turn out true, and vice versa. For example, if there fails to exist a one-tuple of anything then nothing is even a $\frac{1}{20}$ tuple of stuff. Then the statement that $\frac{1}{20}$ plus $\frac{18}{20}$ equals one would turn out true! For it would say that anything is a one-tuple of bloops iff it is bloop-equal to any join of a $\frac{1}{20}$ tuple of bloops to a discrete $\frac{18}{20}$ tuple of bloops. Every instance of this universal generalization is true, for each side of the biconditional is false. So the whole universal generalization would be true. Yes.

But if the above situation were to obtain then the mathematical truths as reconstrued in the usual set-theoretic fashion would also in at least some cases turn out false. For if no one-tuple of bloops of any kind exists, then no measure-theoretic one-tuple of the kind described on p.   exists. This can only be because the requisite real numbers, construed set-theoretically, do not exist. So once again, our situation is no worse than that confronting the standard set-theoretical philosopher of mathematics.

PHILIP A. OSTIEN

OBSERVATIONALITY AND THE COMPARABILITY OF THEORIES*

I. INTRODUCTION

Feyerabend and others have been defending for some time the thesis that even observation sentences depend for their meanings on the theories in which they play a role. According to the most radical version of this thesis, if two theories contain incompatible theoretical statements, then even if the observation terms and sentences associated with the two theories are the same, these terms and sentences *mean* something different, according as they are associated with one theory or the other. This claim has made it hard to see how, on this view, any two theories could be said to compete, or be incompatible, or be comparable on the basis of observation.

The whole dispute over this claim seems highly suspect to begin with. For look at the philosophical terms in which it is couched. On the one hand there is the notion of the "meaning" of a term or sentence, whose status as a useful tool of philosophical analysis has been in serious question for the past two or three decades. The problems with the notion, which have been highlighted rather than resolved by the dispute over the comparability of theories, remain what they always were: an inability on the part of philosophers who use the notion to say what meaning is supposed to be; an inability even to specify clear and plausible conditions for likeness of meaning; and, as a consequence of the second problem, an inability to distinguish between changes in the meaning of a term and mere changes of belief about the objects the term denotes.[1]

On the other hand there is the notion of observationality, which used to be counted on to deliver up those sentences which were to be the impartial and objective source of evidence for science. There is of course Feyerabend's problem with this notion, the problem about the apparent dependence of observation sentences for their "meanings" upon the theories with which they are associated, which has been seen, by both Feyerabend and others, as calling into question the impartiality of observation

R. S. Cohen et al. (eds.), PSA 1974, 271–289. All Rights Reserved.
Copyright © 1976 by D. Reidel Publishing Company, Dordrecht-Holland.

sentences in their role as adjucators of disputes between rival theories (see [6], [8]). Potentially even worse, however, are the problems raised against the notion of observationality by others, problems having to do with our very ability to recognize which the observation sentences and terms of a language or theory are. One such problem is simply that no one (apparently) has been able to characterize the notions of 'observation sentence' and 'observation term' in a general, clear way; all we have had are *lists* of such sentences and terms, and vague characterizations subject to easy counterexample.[2] Worse, it has begun to appear that no general characterization of these notions can be given at all which answers to the usual conception of the role of observation sentences in the epistemology of science. The reasons for this, I think, are the following. One clearly wants the observation terms to be those which apply to what can be seen, felt, etc.; observation terms are supposed to be terms which apply to what can be observed. Thus most attempts to characterize the notion of 'observation term' have been couched, at least partly, in terms of the observability of things to which terms apply. But several considerations seem to militate against the possibility of any general, epistemologically interesting characterization of observation terms and sentences along these lines. First, some have argued that there is just no clear, nonarbitrary line to be drawn between those things which are observable and those which are not (cf. [10]). Second, it has been pointed out that what is observed in a given situation differs, depending on who is doing the observing, what his purposes are, and what his background knowledge happens to be; thus what terms and sentences count as observational, as opposed to theoretical, must vary in the same way.[3] As Quine puts it, in summing up this problem:

The veteran physicist looks at some apparatus and sees an X-ray tube. The neophyte, looking at the same place, observes rather 'a glass and metal instrument replete with wires, reflectors, screws, lamps, and pushbuttons.' One man's observation is another man's closed book or flight of fancy. The notion of observation as the impartial and objective source of evidence for science is bankrupt. ([11], p. 88)

A third problem about observation sentences – one that has been little noticed in the literature – is that it is hardly possible to state one of them in the English language without using a *sortal* term, or term of divided reference: *objects* are observable too, not just qualities. Yet if you look at the more or less standard lists of observation terms you will notice that sortals are conspicuous by their absence, and for good reason. Sortals, unlike quality-terms, are pretty clearly theory-laden in an important sense: their correct use carries with it certain principles of individuation: the objects stand out and can be counted, they remain the same from time to time, they take on a life of their own. In short, our full-fledged talk of objects, even objects of garden variety and even talk in observation sentences, can get well out beyond its evidential basis. And yet it seems clear that such sentences as 'This table is brown' are observational. Some account must be given of such sentences in any theory of observationality.

Fourth, and relatedly, there is a serious question about what would count as an epistemologically interesting characterization of observationality. In particular, can any notion characterized in the full light of scientific day – naturalistically, that is – be epistemologically interesting? This question faces, but to my knowledge has not been faced by, those who would invoke observable objects in defining 'observation term'.

It seems to me that no serious study of the epistemology of science can get along without the notion of observationality; observationality is what ties language down, wherever you locate it, and observation sentences must be theory-neutral in some sense; the problem is to locate that sense. If, on the other hand, we can get along without the notion of meaning, all to the good. Now Feyerabend himself has suggested that we may be able to rescue the notion of observationality, at least to the extent of saying which the observation sentences and terms are, by giving a pragmatic theory of this notion.[4] He has also suggested that when it comes to a question of comparing theories on the basis of observation, questions about the meanings of terms "play a negligible part, and that attention to the 'variety of kinds and degrees of dependence' [of meanings on theories] while certainly populating the semantical zoo, does not solve a single philosophical problem." ([7], p. 267) I agree, on both counts. In the remainder of this paper, then, I will suggest, following Quine, what a pragmatic theory of observationality might be like. The theory suggested avoids reference to what is observed in its characterization of the class of observation sentences, and thus avoids many of the problems mentioned above which seem to face any such approach. Next I will discuss some features of the sentences characterized as observational by Quine's theory; here I will be concerned both to justify the claim that these sentences are indeed the "observation sentences" we need to supply the epistemological foundations of science, at least if epistemology is construed naturalistically, and to show that Feyerabend's thesis about the dependence of the meanings of observation sentences on the theories with which they are associated is in fact true. Finally, I will argue that despite this latter fact, theories are still comparable on the basis of the observational evidence.

II. SKETCH OF QUINE'S THEORY OF OBSERVATIONALITY

In *Word and Object*, Chapter 2, Quine is concerned with the epistemology of radical translation. The problem is to see what objective sense can be made of the notion of translating a hitherto unheard of native language into one's own. As classically conceived, such translation would amount to matching up, word for word or sentence for sentence, native utterances with one's own, on the basis of *sameness of meaning*. Unfortunately, the problem so conceived is as vague as the notion of meaning itself, and its solution is as far off as a solution to the problem of how to assess likeness of meaning even within one's own language, let alone translinguistically.

Quine thus reassesses the problem of radical translation. He proposes that we conceive language to be a "complex of present dispositions to verbal behavior"; translation is then a problem of matching up native utterances with one's own on the basis of the range of stimulations likely to activate these dispositions. Clearly, the first native utterances to be translated on this basis will be the ones keyed most directly to current stimulations of the sensory organs. As Quine puts it,

The utterances first and most surely translated in such a case are ones keyed to present events that are conspicuous to the linguist and his informant. A rabbit scurries by, the native says 'Gavagai', and the linguist notes down the sentence 'Rabbit' (or 'Lo, a rabbit') as tentative translation, subject to testing in further cases. ([14], p. 29)

The testing in further cases amounts to collecting evidence for the general hypothesis that the native will assent to 'Gavagai?' under just those non-verbal stimulations which would prompt our assent to 'Rabbit?' if this sentence were queried.[5]

Quine now defines a notion of *meaning* which is purely experimental. He points out that for such utterances as 'Gavagai' we have the makings of a crude conception of *empirical meaning*: "For meaning, supposedly, is what a sentence shares with its translation; and translation at the present stage turns solely on correlations with non-verbal stimulation." ([14], p. 32) Thus, we define "affirmative stimulus meaning":

The affirmative stimulus meaning of a sentence S for a speaker A is the class of all stimulations that would prompt A's assent to S.

It is useful to spell this out in more detail:

A stimulation σ belongs to the affirmative stimulus meaning of a sentence S for a given speaker if and only if there is a stimulation σ' such that if the speaker were given σ', then were asked S, then were given σ', and then were asked S again, he would dissent the first time and assent the second. ([14], p. 32)

Negative stimulus meaning is defined similarly, with 'assent' and 'dissent' interchanged; stimulus meaning is then taken to be the ordered pair of positive and negative stimulus meaning.

So the linguist's hypothesis mentioned above about 'Gavagai' comes to this: 'Gavagai' and 'Rabbit', construed as one-word sentences, have the same stimulus meaning.

The affirmative stimulus meaning and negative stimulus meaning of a sentence are mutually exclusive sets of stimulations; but they are not necessarily jointly exhaustive of the set of all stimulations; for we must allow some stimulations to *prompt* neither assent to nor dissent from some sentences. The more nearly exhaustive the two are for given sentence S, however, the more nearly is S an *occasion sentence*, as opposed to a *standing sentence*. A standing sentence is one like 'The newspaper has come', to which a speaker may be prompted to assent by current stimulation, but to which he may repeat his old assent unprompted by current stimulation when we ask him again later. An occasion sentence, on the other hand, is one like 'Red' or 'Rabbit', which commands assent only if queried after an appropriate prompting stimulation. ([14], §9)

Which, now, among the sentences of a language, are the observation sentences? We have already seen that we want the observation sentences to be those in closest touch with what we actually observe; as Quine puts it, we want them to be the ones "in closest causal proximity to the sensory receptors." ([11], p. 85) Clearly these will be among the occasion sentences as opposed to the standing sentences of a language, for the former are those keyed most closely to current stimulation. Are we then to regard all occasion sentences as observational? Clearly not. For consider the one-word sentence 'Bachelor'. This is surely as un-observational as you please; but it is, nonetheless, an occasion sentence; it will command assent from a speaker of English only if queried after an appropriate prompting stimulation. The trouble with it is that who assents to it, and after what stimulations, depends heavily on the background knowledge of the speaker; on his knowledge of the marital status of various people, as opposed to his knowledge "merely" of the English language. Here the problem about background knowledge comes in again; the problem is reminiscent of the one mentioned above over whether to count 'X-ray-tube' as observational or not.

We would like our observation sentences to be those to which assent and dissent depend only on present stimulation, to the exclusion of stored information. But as Quine points out, the very fact of knowing how to speak the language evinces much storing of information. The problem we now face is that of distinguishing experimentally between information that goes into *understanding* a sentence, and information that goes beyond this. Quine's suggestion is that we can make our definition of observationality sensitive to this distinction at least in those cases where the extra information, the background information, is not community-wide: we do this by characterizing as observational all and only those occasion sentences which have the same stimulus meaning for all members of the linguistic community. This cuts out 'Bachelor'; and it cuts out 'X-ray-tube' if we count the community as all fluent speakers of English. The observation sentences are those, like 'Red' and 'Rabbit', which command the same response from all speakers of the language after the same stimulations. They are the ones which vary none in stimulus meaning within the linguistic community under the influence of collateral information.

This characterization of observationality is not sensitive to differences between the information that goes into knowing the language and extra information, where this extra information is community-wide. Is it therefore defective on this count? Quine thinks not; he thinks that no objective sense is to be made of this distinction; that it is, in fact, illusory:

It is simply a question whether to call the transitivity shortcuts (§3) changes of meaning or condensations of proof; and in fact an unreal question. What we objectively have is just an evolving adjustment to nature, reflected in an evolving set of dispositions to be prompted by stimulations to assent to or dissent from sentences. These dispositions may be conceded to be impure in the sense of including worldly knowledge, but they contain it in a solution which there is no precipitating. ([14], pp. 38–39)

There remain two points worth mention. We would like to know, first, how to gauge the width of the "linguistic community". Quine suggests that general fluency of dialogue will serve here as a criterion, and points out that of course this criterion admits of degrees; this allows us to take the community more narrowly for some studies than for others. Thus, "what count as observation sentences for a community of specialists would not always so count for a larger community." ([11], p. 87). We might, then, be willing to count 'X-ray-tube' as observational for the community of physicists; we would not so count it, as already noted, for the whole English-speaking community.[6]

Just as the width of the linguistic community admits of degrees, so does the notion of observationality. For even such sentences as 'Red' can be subject in point of stimulus meaning to background information which is not community-wide. The less subject a sentence is in this way to differences in background information, the higher it is in observationality.

Viewing the graded notion of observationality as the primary one, we may still speak of sentences simply as observation sentences when they are high in observationality.... It is for observation sentences in some such sense that the notion of stimulus meaning constitutes a reasonable notion of meaning. ([14], p. 44)

We shall return to this last point in a moment.

III. QUINE'S OBSERVATION SENTENCES AND THE USUAL CONCEPTION OF OBSERVATIONALITY

Let us generalize a bit on the foregoing. What Quine has supplied us with is an (admittedly rough) method of recognizing the observation sentences of a given language as spoken by a given group of people. Even better: he has supplied us with a method of recognizing the observation sentences of a given language as used by a given group of language-users, whether people (i.e., human beings) or not. The method is this: figure out what kind of sense organs the creatures have (i.e., what kinds and ranges of stimulations their sense organs are sensitive to); identify the sorts of behavior which for them count as the utterance of a sentence and as the querying of a sentence; identify the sorts of behavior which for them count as assenting to and dissenting from a sentence; and then by feeding them doses of the proper sorts of stimulation and querying their sentences for assent or dissent, identify those of their occasion sentences which have the same stimulus meaning for all members of the group. These sentences are the observation sentences of the language as used by that group.

Putting it this way makes clear how much in the way of theory is already presupposed when one comes to an alien tongue with the aim of translating it. In the case of alien human languages most of these presuppositions go without saying, and may pass unnoticed. But it is worth remarking how heavily theory-laden any self-conscious attempt to apply Quine's criterion of observationality would be. The criterion is fully naturalistic: it embodies a theory about which sentences constitute the evidential basis of science which is already fully scientific. It behooves us, then, to examine closely the question whether the sentences Quine's criterion picks out as being observational can play the role at the foundations of science always envisioned for observation sentences.

As advertised above, the class of sentences Quine calls observational has not been distinguished on the basis of the notion of *observable object*; observation sentences have not been *characterized* as those which apply to, or are "about" what is observable. But given the way they have been characterized, it is clear that these sentences will be, in the case of the English language at least, about the sorts of things we speak of observing – publicly accessible physical objects and their observable qualities. As Quine sums up the matter, "Since the distinguishing trait of an observation sentence is intersubjective agreement under agreeing stimulation, a corporeal subject matter is likelier than not." ([11], p. 87)

Another way in which our observation sentences accord with the usual conception is in point of the unproblematicality of their meanings. Observation sentences, classically conceived, are supposed to provide the epistemological foundations for science, both "doctrinally" and "conceptually"; that is, they are supposed to supply both the means by which we understand the meanings of the terms which comprise the vocabulary of science, and the evidence for the truth of the statements couched in that vocabulary. But the first of these roles requires that the meanings of the observation sentences themselves somehow be clear as sunshine, completely unproblematical. It has always been difficult to see how this unproblematicality was to be guaranteed; even difficult to see what it meant to say that observation sentences are unproblematical in point of meaning. This problem has become especially acute in view of recent discussions of the dependence of observation sentences for *their* "meanings" on the theories with which they are associated. One begins to wonder whether we have any semantical access to language at all.

Viewing the matter in the light of the characterization of the observation sentences given above, we can see that the old intuition is vindicated, at least to the following extent: observation sentences are as unproblematical in point of their meanings as any sentence can be – provided by 'meaning' we mean *stimulus meaning*. For learning the meaning (=stimulus meaning) of an observation sentence is on the present account merely a matter of the direct conditioning of the sentence to the appropriate range of non-verbal stimulation; observation sentences are just those keyed directly to such stimulations, the same stimulations for the same sentences across the whole linguistic community. They are just the sentences about which, where disagreement arises over their applicability under uniform stimulation, we naturally say that someone does not know the meaning of the sentence, or is using it in a different meaning; thus it is that "for observation sentences... the notion of stimulus meaning constitutes a reasonable notion of meaning." For other sentences of the language, of course, stimulus meaning does not come close to approximating anything we would wish to call meaning; learning to use non-observation sentences – thus, learning their meanings, in the old sense of the word – involves (by Quine's account) the conditioning of these sentences to both non-verbal stimulation and to other sentences. Still, observation sentences are surely basic to this more sophisticated language-learning process; observation sentences provide us with our point of entry into a language; for they, unlike other sentences, wear their meanings on their sleeves.

Their stimulus meanings, that is; for observation sentences as presently conceived do not wear anything like their "meanings", in the old, weighty sense of this term, or even the referents of the terms occurring in them, on their sleeves. In seeing how this is, we shall see how it is that given Quine's view of observationality, Feyerabend's thesis about the dependence of observation sentences for their "meanings" on the theories with which they are associated turns out to be true.

Suppose, then, that we determined that some native occasion sentence – say 'Gavagai' – was highly observational, and had the same stimulus meaning as our one-word sentence 'Rabbit'. Given just this much, can we conclude that the native *term* 'gavagai' and ours, 'rabbit', either "mean the same" or, even, have the same referents? Not at all. Here I quote Quine:

> For, consider 'gavagai'. Who knows but what the objects to which the term applies are not rabbits after all, but mere stages, or brief temporal segments, of rabbits? In either event the stimulus situations that prompt assent to 'Gavagai' would be the same as for 'Rabbit'....
>
> A further alternative... is to take 'gavagai' as a singular term naming the fusion, in Goodman's sense, of all rabbits: that single though discontinuous portion of the spatio-temporal world that consists of rabbits. Thus even the distinction between general and singular terms is independent of stimulus meaning. ([14], pp. 51–52)

The general problem here is that we can translate, in the sense of pairing off sentences having the same stimulus meanings, all the observation *sentences* of a language, without thereby having translated any of the *words* of the native language, particularly those which incorporate that part of the apparatus of a language so crucial to reference, the individuative apparatus. Whereas the notion of observationality has been characterized *transcendentally*, that is, in terms definable for and in common to all languages, and is thus a linguistically transcendental notion, the notion of reference is linguistically immanent, local to a given conceptual scheme, or language, or, as Feyerabend might put it, theory.[7] This is because the notion of reference can be characterized for a given language only in terms of the individuative apparatus of that language; but this apparatus is local to the language, not possessed in common by all languages. Thus it is that the "meanings", and even referents of the terms of observation sentences are dependent upon encompassing "theory". Or, to relate this to Feyerabend: Insofar as we construe the observation sentences associated with a theory to be distinguished in Quine's way, and assume that Feyerabend means to interpret the term 'theory' in a way broad enough to include the individuative apparatus of the language in which the theory is couched, then we can see that Feyerabend is right in his claim that the "meanings" of observation sentences and terms depend upon encompassing theory, at least in the following sense: you can know what all the observation sentences associated with a theory are; and you can even know the stimulus meanings of these sentences; but you still might not know even the referents of the terms which occur in them. To find out what the "observation terms" refer to, you will have to master more of the theory. Observation terms depend for their referents, and thus surely for their "meanings", on the theories with which they are associated.

I think of the matter in this way. Observation sentences have a sort of dual loyalty, or dual role to play in any well-developed language. On the one hand they play their role as observation sentences, pure and simple. In this role their loyalty is to the range of stimulations to which they have been (or could have been) conditioned, and viewed in this role they are merely conditioned responses to their several ranges of stimulations; there is no connection between them or between any of them and any other sentences of the language. Observationality, in this sense, is *flat*. Talk in a purely observational language would be very flat; it would be that "fancifully fanyless medium of unvarnished news" that Quine speaks of – if, indeed, one can even speak of "news" in such a context. In such a language each sentence is conditioned to its own range of sensory stimulation, and that's it. In such a language there would seem to be no *describing* the world, no *reference* to objects; no distinction could be made, relative to such a language, between "verbal" and "factual" error; there would be no *room* for it.

As Quine has pointed out, talk of a purely observational language sounds very much like talk of a pure sense-datum language. This is no accident: Both sorts of language are candidates for the role of supplying the foundations of science, and perhaps especially for the role of "tying language down to the world". Proponents of a pure sense-datum language have thought that this tying down was to be done in terms of the notion of reference; they have thought that their sense-datum language could at once be "pure", free of any *theoretical* burden, and could also secure reference to objects or qualities of some sort. The problem was always to see how such a language could be learned and also to see how it could maintain its theoretical innocence, especially if it was to be construed as talk of objects.

The advantages of locating observationality where Quine locates it are that we at once are able to specify plausible learning mechanisms for observation sentences, *and* give a plausible account of how language is tied down to the world, *and* see how it is that observation sentences, as such, carry no theoretical burden, and are thus theory-neutral. Observation sentences are (or could be) learned by straightforward induction, with a little help from your friends.[8] What ties language down to the world is not reference, which is always a matter of articulate theory, but the conditioning of some sentences as verbal responses to certain ranges of sensory stimulation. And it is because we see observation sentences, as such, as innocent of any burden of reference or description – as innocent of any connection with other sentences – that we are able to see them as free of any burden of theory which "goes beyond the evidence", and so as verbal behavior which is neutral as between the adoption of one theory or another.

Language as we know it rises out of observationality, via the interanimation of sentences; and in particular, the whole apparatus of objective reference rises out of observationality, as do the other components of the meanings, ordinarily so called, of sentences. It is by their inclusion in an articulate language, or articulate theory, by the various sorts of connections they acquire with other sentences in such language or theory, that observation sentences acquire their other loyalty or other role and at the same time take on the burdens of reference and description, ordinarily so called.[9] In their role within theory observation sentences comprise the repository of hard data on which the theory thrives; but it is always because of their loyalty to the ranges of stimulations to which they have been conditioned – a loyalty which a higher need can sometimes override – that they can perform this function.

A higher need within articulate theory can sometimes, but not often, pry observation sentences loose from the ranges of non-verbal stimulation to which they have been conditioned, so that observation sentences, as they play their role within theory, are corrigible. But theoretical considerations can also override the other loyalties of observation sentences, those they have come to acquire to other sentences of the theory, while leaving their loyalties to their several ranges of non-verbal stimulation intact. This, I take it, is what happens when, in times of scientific revolution, one changes one's theory about one's instruments and what they measure, and perhaps more generally about what the objects are to which observation terms refer. As I picture it this is a matter of leaving the observation sentences intact in their role as "mere" observation sentences – leaving their stimulus meanings intact – while making systematic changes in the way the observation sentences are associated with other sentences within the theory. Such theory change can work wholesale changes in the meanings and referents of observation sentences and terms, insofar as meaning (and surely reference) is a matter of the connections between sentences in articulate theory; but the point to notice is that in such cases central components of the "meanings" of observation sentences, their stimulus meanings, abide unchanged. This is what supplies the bedrock of continuity and stability upon which the shifting sands of our theories rest. The observation sentences, as we have noted, are corrigible but not highly so; to deny an observation sentence in circumstances in which it is obviously true is to risk being accused either of not knowing what you are talking about or of changing the subject. But our theories are highly corrigible. Our instruments abide, but our theories about what they measure and even about what they are continually change.

Observationality, in its pure flatness, is a matter of the conditioning of sentences to ranges of non-verbal stimulation. The stimulations have to be of our external sense organs, because observationality requires intersubjective agreement under like stimulations. But there are sentences, which we should perhaps call 'quasi-observational', which are like observation sentences in that they can be conditioned in the flat observational way to certain ranges of "stimulations", but which are unlike observation sentences in that the "stimulations" to which they are conditioned are not stimulations of our external sense organs but rather internal states, most likely states of our central nervous systems. These are the so-called first-person reports that we learn to make on our subjective states. Somehow, with the aid of judicious coaching in roughly describable public circumstances, we learn that the similarity basis for 'It hurts' lies not in the public circumstances themselves but rather in the internal states those circumstances usually occasion; and 'It hurts' gets conditioned directly to the internal states in just the way ordinary observation sentences get conditioned directly to their ranges of stimulations. The point I want to make here about these quasi-observation sentences is that they, like the observation sentences, come to have dual loyalties within articulate language; they come to be associated in various ways with other sentences as well as with the ranges of internal states to which they are conditioned. Insofar as they are "merely" quasi-observational, they are not properly called reports; insofar, they cannot be said to refer to or describe anything; they are merely conditioned responses of a certain sort, of a kind with the salivation of Pavlov's dogs. On the other hand, insofar as the quasi-observation sentences are seen as reports, as vehicles for reference and description, they must be seen in the context of an articulate theory with whose other sentences they have become associated. This theory, as before, can change, leaving the "stimulus"-meanings [10] of the quasi-observation sentences unchanged.

Quasi-observation sentences are even less corrigible than ordinary observation sentences. This, I take it, is because the theory to which they bear loyalty – the net of other sentences with which they are associated – is far less articulate and well-established than the theory to which ordinary observation sentences bear loyalty, so that it hardly ever has the strength to pry one of the quasi-observation sentences away from the range of internal states to which it has been conditioned. On the other hand, since this is so, this theory, or part of the language which we might call mentalese, seems ripe for overthrow and replacement by something better. It is important to realize that such overthrow and replacement would be very much in the spirit of scientific revolution. The quasi-observation sentences would remain, conditioned as always directly to their several ranges of internal states, retaining the clearest and most important component of their meanings, their "stimulus"-meanings; what would change would merely be the theory we build upon this quasi-observational bedrock, the theory about what the quasi-observation sentences are reports on, what they describe, what they refer to. One can hope that any such new theory, if and when it is achieved, will link up in clearer and more direct ways with the physical theory that rises out of full-fledged observationality than does our current mentalistic theory.

I have been trying to suggest to the reader how fruitful Quine's approach to observationality is. We have just been seeing how this approach might be extended so as to give an account of what I have called the quasi-observationality of the first person reports of subjective states. Before bringing this section of the paper to a conclusion I wish to suggest three further advantages of Quine's approach.

I suggested early in this paper that any adequate theory of observationality should take account of the undeniable role that sortal terms play in observation sentences. I believe there is ample room within Quine's theory of observationality for such an account. For consider the central characteristic of observation sentences according to the theory: they are sentences which could be learned by direct conditioning to some range of non-verbal stimulation. Such conditioning depends on some prior inclination that people have to count one stimulation as more similar to another than to a third – some innate "quality spacings" that come with the rest of our natural equipment. Now the point about sortals is that we are evidently as easily conditioned to (what we later come to recognize as) objects of various shapes and sizes and kinds as we are to qualities which vary after their kinds. Indeed, if my experience as a parent is any guide, we are more easily conditioned to kinds of objects than to qualities – man is a body-minded animal. My children learned to respond with 'Telephone' to telephones long before they learned unfailingly to respond with 'Red' to red things. Of course these terms come to play different roles in articulate language, and within such language come to be seen as susceptible to misapplication in different ways; but the fact remains that each can be learned as an observation sentence in just the way Quine has sketched.

Quine's theory also opens up the possibility of fruitful speculation on somewhat the sort of thing we found Feyerabend deriding early in the paper, speculation on the "variety of kinds and degrees of dependence" of the "meanings" of terms on encompassing theory. What I have in mind here is the fact that our "quality spaces", which manifest themselves in our similarity judgments or susceptibility to conditioning, can change as we learn, so that a little overlay of theory can render sentences observational for us that were previously not so. "It's Mozart" is keyed as an observation sentence to certain ranges of auditory stimulations for some of my friends but not for me. "It's publishable!" is keyed as an observation sentence to certain ranges of visual stimulations for some of my colleagues but not for me. I believe that further study of the ways in which learning modifies our dispositions to be conditioned to ranges of stimulation will shed much light on how it is that genuine learning distinguishes itself from idle speculation by constantly opening up new dimensions of observationality, and how it is that some sentences can come to depend upon encompassing theory even for their observationality.[11]

The fact that the sentences of a language that count as observational by Quine's criterion are just those sentences that could be learned as directly conditioned responses to certain ranges of sensory stimulation has been central in the last few pages. Now let me suggest one important advantage that the criterion itself seems to me to have over criteria that attempt to characterize the observation sentences or terms of a language in terms of the objects referred to by those sentences or terms (the *observable objects*). (cf. [3], pp. 64–71, and [5], III.) The problem with any such approach to observationality, I believe, is that it can only supply us with a linguistically immanent characterization of a concept which we would like very much to be linguistically transcendent. The observation sentences of an unknown language provide our only access to the language; so we would like to be able to recognize the observation sentences (on the basis of our theoretical characterization of them) independently of knowledge of any of the specific characteristics of the language itself. Quine's characterization of observationality provides such a basis for the recognition of observation sentences. But any characterization of observationality which requires that we know what the terms of a language *refer to* before we can pick out its observation terms and sentences fails to provide such a basis; for to pick out the observation terms and sentences on the basis of such a theoretical characterization of them would require that we already be able to tell what the terms of the language refer to, and thus that we already have mastered the individuative apparatus of the language, before we even know which its observation sentences are. Such a characterization would seem to put the epistemological cart before the horse.

IV. THE COMPARABILITY OF THEORIES ON THE BASIS OF OBSERVATION

But consider, finally, the problem of the comparability of theories on the basis of observation. This was the problem with which we started, the problem which is supposed to be so severe once Feyerabend's thesis about meaning-dependence and meaning variance is granted. How shall we now view the problem?

I take it that the problem here can be no worse than that which faces the linguist attempting the radical translation of a native language into his own. Indeed, the problem is usually much less severe than that; for the case of two rival scientific theories is usually a case in which both theories are at least couched in the same background language – say scientific English – in which the same individuative apparatus is used. So what could the problem be? Suppose we can fix the observation sentences of both the theories we want to compare, and match them up on the basis of stimulus synonymy insofar as this is possible. Let us suppose that the sentences so matched are in fact grammatically identical, but that as used in the one theory or the other, a given sentence turns out to "mean" something different, even to refer to something different. E.g., 'Rabbit' might refer from the point of view of T_1 to rabbits, and from the point of view of T_2 to undetached rabbit parts. Given all this, I submit, there is still no problem about comparing the two theories on the basis of observation. One simply proceeds as usual here: get the two theories to make contradictory predictions about the truth of some observation sentence stimulus synonymous for the two on the basis of other observation sentences stimulus synonymous for the two. Contradictory predictions? But the observation sentences do not *mean* the same from the point of view of both theories. It does not matter: for by hypothesis, partisans of both theories assent to and dissent from each of the observation sentences on the basis of the same stimulations, so that no matter which way things turn out when the test is made, someone is bound to admit trouble in his theory.

It has been widely held, in the literature on these topics, that two theories could not be compared except on the basis of observation sentences fully synonymous as used in the two theories. We now see that this is by no means necessary. Stimulus-synonymy is enough. Let meaning, in any other sense than stimulus meaning, and even reference, fall where they may; we can still compare two theories on the basis of observation; provided, of course, we can settle which are the observation sentences of the two theories. The sentences characterized above as observational thus turn out to have a further important characteristic that has classically been ascribed to observation sentences: they are the "datum sentences" of science, the "court of appeal of scientific theories". For

the observation sentences as we have identified them are just the occasion sentences on which there is pretty sure to be firm agreement on the part of well-placed observers. Thus they are just the sentences on which a scientist will tend to fall back when pressed by doubting colleagues. ([14], p. 44)

We have seen that a class of sentences of a language can be characterized clearly whose members share many of the characteristics of observation sentences as usually conceived. Thus one of the important problems with the notion of observationality mentioned above would seem to have been solved. We have seen also that these sentences are dependent for their "meanings", indeed, even for the referents of the terms occurring in them, on the theories with which they are associated; thus is vindicated a characteristic Feyerabendian thesis about observationality. We have seen finally, that despite this dependence of "meaning" upon encompassing theory, there is still no serious epistemological problem about comparing theories on the basis of observation; at least there is no problem here more serious than that facing any attempt at radical translation and comparison of native beliefs with our own.

The University of Iowa

NOTES

* My indebtedness to the work of W. V. Quine will be observable throughout this paper, at least to those familiar with his work. I shall not signal every borrowed phrase or idea with an explicit reference, but let this note stand as a general acknowledgement of my indebtedness. Any mistakes, either of exposition or of application, are of course my own.

[1] Achinstein, one of the more prominent participants in the dispute over meaning variance and the comparability of theories, has made some effort toward the resolution of these problems, invoking to this end his notion of "semantical relevance". (See [1], *passim*.) Unfortunately, this notion is as suspect as the notion of "meaning" itself; for at its core it involves the notions of "logical necessity and sufficiency", which turn out to be, on inspection, explicable only in terms of analyticity, another member of the tight little circle of terms with which "meaning" is bound up and out of which it seems impossible to break in order to gain a clear understanding of any of its members.

[2] This sort of problem is urged in [1], ch. 5, and in [15].

[3] For discussions of this problem, see [1] and [15]. Sometimes this objection to the observation/theoretical distinction comes dangerously close to the *irrelevant* objection that the notion of observationality is used *ambiguously* in ordinary scientific discourse. Cornman, [2], has dealt with both sorts of problems I have mentioned here.

[4] Feyerabend has even sketched such a theory, in his 'An Attempt at a Realistic Interpretation of Experience', *Aristotelian Society Proceedings* 58 (1957–58). I find Feyerabend's theory far less clear than Quine's.

[5] It is a touchy problem to characterize clearly "non-verbal stimulation which would prompt assent to a given sentence" and "same stimulation". Quine discusses this in [14], § 8, and again in [12]. Some philosophers have argued that these notions are as badly off as the notion of observationality ever was, and will thus not get us to a clear characterization of observationality; see [9] and [16]. This is a point worth a good deal more study than I can give it here, and for present purposes I shall just assume the salvagability of these notions.

[6] See below, p. 285, for a bit of speculation on the sort of further study of observationality it might be fruitful to undertake if Quine's theory of observationality is on the right track.
[7] For a discussion of transcendence and immanence of this variety, see [13], pp. 19, 20. There is, as I pointed out in Note 5, some question whether Quine has succeeded in characterizing observationality clearly, given that he uses the notion of 'same stimulation'. Part of this question is whether the notion of 'same stimulation' can itself be characterized transcendentally.
[8] See *Word and Object*, ch. 1, and *The Roots of Reference*, II.
[9] Here of course I am only reporting Quine's conclusions, without benefit of the supporting evidence and articulation of the view so admirably set forth in *Word and Object* and *The Roots of Reference*.
[10] As the internal states, or "stimulations", to which the quasi-observation sentences get conditioned, are to stimulations of the organs of sense, so "stimulus"-meanings are to stimulus-meanings.
[11] Observationality, or something very like it, goes very deep in the language. This fact is surprising to those of us who have been accustomed to thinking of observationality in terms of such sentences as 'Blue here now', but it is readily observable once your eyes are open to it.

BIBLIOGRAPHY

[1] Achinstein, P.: 1968, *Concepts of Science*, Johns Hopkins Press.
[2] Cornman, J. W.: 1972, 'Craig's Theorem, Ramsey-Sentences, and Scientific Instrumentalism', *Synthese* **25**.
[3] Cornman, J. W.: 1971, *Materialism and Sensations*, Yale University Press.
[4] Dretske, F.: 1964, 'Observational Terms', *Philosophical Review* **63**.
[5] Dretske, F.: 1969, *Seeing and Knowing*, University of Chicago Press.
[6] Feyerabend, P. K.: 1962, 'Explanation, Reduction, and Empiricism', in *Minnesota Studies in the Philosophy of Science*, Vol. III, University of Minnesota Press.
[7] Feyerabend, P. K.: 1965, 'On the "Meaning" of Scientific Terms', *Journal of Philosophy* **63**.
[8] Feyerabend, P. K.: 1965, 'Problems of Empiricism', in R. Colodny (ed.), *Beyond the Edge of Certainty*, Prentice-Hall.
[9] Martin, Jr., E.: 1973, 'The Intentionality of Observation', *Canadian Journal of Philosophy* **3**.
[10] Maxwell, G.: 1962, 'The Ontological Status of Theoretical Entities', in *Minnesota Studies in the Philosophy of Science*, Vol. III, University of Minnesota Press.
[11] Quine, W. V.: 1969, 'Epistemology Naturalized', in *Ontological Relativity and Other Essays*, Columbia University Press.
[12] Quine: 1969, 'Propositional Objects', in *Ontological Relativity, op. cit.*
[13] Quine, W. V.: 1970, *Philosophy of Logic*, Prentice-Hall.
[14] Quine, W. V.: 1960, *Word and Object*, M.I.T. Press.
[15] Spector, M.: 1966, 'Theory and Observation (I)', *British Journal for the Philosophy of Science* **17**.
[16] Wallace, J.: 1971, 'A Query on Radical Translation', *Journal of Philosophy* **68**.

SYMPOSIUM

SCIENCE EDUCATION AND
THE PHILOSOPHY OF SCIENCE

MICHAEL MARTIN

THE RELEVANCE OF PHILOSOPHY OF SCIENCE FOR SCIENCE EDUCATION

What relevance – if any – does philosophy of science have for science education? Unfortunately, this question has been largely unexplored. To be sure a great deal has been written on the philosophy of science; perhaps even more has been written in science education. However, surprisingly little has been written on the relation between the two areas.[1] In this paper I will suggest some ways in which philosophy of science can have relevance for science education.

I. THREE TYPES OF PHILOSOPHY OF SCIENCE

Before one considers how philosophy of science can have relevance to science education it is important to consider what is meant by 'philosophy of science'. Some contemporary philosophers of science have distinguished two different approaches to philosophy of science: analytic and speculative.[2] However in recent years a view of the philosophy of science has become popular that does not seem easily to fit into either the analytic or speculative categories. Although this view may be new in contemporary philosophy of science its proponents argue that their view of the role of philosophy of science is closely analogous to the view held by Bacon, Mill, Whewell and other historically important philosophers of science. According to this view philosophy of science proposes, defends, and criticizes theories of scientific growth and development, theories of how scientific knowledge progresses. I take Popper's method of conjecture and refutation, Feyerabend's epistemological anarchy and Lakatos' methodology of research programs as contemporary examples of this type of view. This third type of philosophy of science, for want of a better title, I will call 'growth theory philosophy of science.'

I cannot consider here the relevance of speculative philosophy of science for science education. However, I have argued elsewhere that it may have relevance.[3] I will consider the relevance of both analytic philosophy of science and growth theory philosophy of science.

II. THE VALUE OF ANALYTIC PHILOSOPHY OF SCIENCE FOR SCIENCE EDUCATION

There is one obvious contribution that analytic philosophy of science can make to science education. Analytic philosophy of science can provide a clarification and analysis of some of the major concepts and methods of science education namely the concepts and methods of science itself. However, it is important to notice that such an analysis can enter into science education in at least two ways:

(1) Students of science can learn the analyses and criticisms of scientific concepts and methods.

(2) Science educators – teachers, curriculum planners, text-book writers, researchers – can learn the analyses and criticisms of scientific concepts and methods.

Notice that these two ways in which analytic philosophy of science can enter into science education are independent. Science educators may be well advised to study cognitive psychology, but it surely does not follow from this that science students would be well advised to study this topic. Similarly it may not be pedagogically advisable for science students to study analytic philosophy of science but it may still be a good idea for science educators to study it. The failure to keep the distinction in mind may be responsible for the fact that some science educators too quickly reject the use of philosophy in science education. They seem to assume that since the study of philosophy of science is pedagogically unfruitful for science students, philosophy of science has no relevance in science education. Now whether they are correct about the pedagogical inadvisability of science students studying philosophy of science is at least debatable.[4] However, it is surely a mistake to suppose that having students study the philosophy of science exhausts the uses of philosophy of science in science education.

I have already mentioned the relevance of analytic philosophy of science for clarifying and criticizing the concepts of methods of science for science educators. Closely connected with this clarification and criticism is something else that has immense value for science educators. Clarification and criticism may suggest new pedagogical insights and perspectives, new ways of looking at and examining traditional approaches, new research problems and goals. This heuristic value for science educators is perhaps the most important contribution analytic philosophy of science can make to science education.

III. THE SCIENCE TEACHER

A typical high school biology teacher will present different hypotheses about why the dinosaurs became extinct without evaluating or having the class try to evaluate these different hypotheses. One basic reason for this, I believe, is that the average biology teacher simply does not know the criteria used in evaluating scientific explanatory hypotheses, something commonly discussed by philosophers of science.[5] This is unfortunate, since it generates a very uncritical and unilluminating approach to science teaching. A science teacher who knew such criteria could not only help his or her students evaluate hypotheses about the extinction of dinosaurs but would be able to suggest illuminating parallels and contrasts which the class could profitably follow up. For example, is the explanation presented in the text for the extinction of the dodo bird more adequate (according to certain specified criteria and purposes) than the explanations presented for the extinction of the dinosaurs? How does the adequacy of the dodo-bird explanation compare with the adequacy of the explanation given in yesterday's newspaper for the probable extinction of a certain species of whales?

IV. THE SCIENCE CURRICULUM PLANNER

Recent science curriculum theorists, e.g., Bruner[6] and Schwab[7], have stressed that the structure of science should be taught. What exactly 'structure' means is not exactly clear but people who emphasize structure in science education seem to have in mind the basic concepts and methods of particular sciences and perhaps of all or most sciences.

Bruner, for example, has suggested that a basic concept in biology is function. He suggests that biological research, as well as biological education, progresses by asking more detailed questions about function.[8] Whether Bruner is correct about biological research and education is questionable. Nevertheless, it may still be true that a functional approach to biology has intuitive appeal and pedagogical value for beginning biology students. Consequently the concept of function may serve well as a basic concept in a beginning biological curriculum. Philosophical analysis of the concept of function would surely be helpful for curriculum planners in devising this curriculum.

V. THE SCIENCE TEXTBOOK WRITER

The science textbook writer often deals with certain general concepts in his work that have been analyzed and discussed at length by analytic philosophers of science, but is usually ignorant of these analyses or discussions. The textbook which is the result of his work suffers accordingly.

Consider the notion of the confirmation of a hypothesis. This notion has been widely discussed and analysed in philosophical literature. It is also introduced in science textbooks. Typically however, the presentation is not in keeping with the level of scientific sophistication presented in the same book. Put in a different way, rather complex and difficult scientific theories are presented with some sophistication along with simple-minded views about confirmation of these theories.

Consider, for example, the discussion of Redi's experiment of spontaneous generation in BSCS *Biological Science: An Inquiry into Life*. It is argued:

What he [Redi] had shown was that under the conditions of his experiment magots did not arise spontaneously in decaying meat.[9]

However, as every philosophical student of confirmation knows, this is not so. Hypotheses are not tested in isolation but against a background of auxiliary hypotheses.[10] Redi's evidence was disconfirmatory of the spontaneous generation hypothesis only relative to certain auxiliary hypotheses about how spontaneous generation works. Indeed, there is no explicit discussion at all in *Biological Science* of the role of auxiliary hypotheses in Redi's experiment. Yet without such a discussion, the tentativeness of the result of Redi's experiment – stated in the text – remains completely obscure. The whole discussion of Redi's experiment would have been enormously clarified by the explicit use of different types of logical arguments and a little formalism, the very sort that is utilized in philosophical analysis of theory testing.[11]

Thus knowledge of confirmation theory would have helped in clarifying the basic ideas in *Biological Science*. But it takes little imagination to see how such knowledge might suggest different approaches to the material. For example, knowledge of the function of auxiliary hypotheses might have suggested to the authors how people like the 'flat-earthers' can still maintain – despite the evidence – that the earth is flat so long as certain complex auxiliary hypotheses are maintained. A discussion of the flat-earthers and their strange beliefs and modes of reasoning might be an illuminating contrast to scientific beliefs and reasonings.[12]

VI. THE RESEARCHER IN SCIENCE TEACHING

Researchers in science teaching are empirical scientists who use concepts such as experiment, test, evidence, explanation, and theory in their thinking and writing. Since analytic philosophy of science clarifies such concepts, it would not be surprising that researchers in science teaching who studied analytic philosophy of science would have some of their own concepts clarified.

Moreover, one can imagine that analytic philosophy of science could suggest new hypotheses for researchers in science education to investigate. For example, recent philosophical analysis has suggested that there are at least two ways that a scientist may accept a hypothesis.[13] A scientist may accept a hypothesis in the sense of believing that the hypothesis is true; alternatively a scientist may accept a hypothesis as a useful research tool, without necessarily believing that the hypothesis is true.

Now some philosophers of science have suggested that the acceptance of a hypothesis by a scientist may influence the scientist to overlook negative evidence for the hypothesis and thus that acceptance creates bias in theory testing. An important research question suggested by the distinction between the two types of acceptance mentioned above is whether both types of acceptance bring about equal bias. One can easily imagine researchers in science education setting up controlled experimental conditions in which one group of science students is taught to accept a hypothesis in the sense of believing it to be true, another group is taught to accept the hypothesis as simply a useful research tool; and these two groups are tested for their differential biases with respect to negative evidence. The results of such an experiment would be significant for both science education and science.

VII. THE VALUE OF GROWTH THEORY PHILOSOPHY OF SCIENCE FOR SCIENCE EDUCATION

Let us now consider the relevance of growth theory philosophy of science for science education. Time limits my discussion to some brief considerations of science teachers and researchers in the teaching of science.

One reason for science teachers studying theories of scientific growth is obvious. Since science teachers influence future scientists, it is important for them to know how science proceeds for maximum growth. For unless they have such knowledge they will be unable to instill in their students research habits that will ensure maximum scientific growth.

Of course, not all science students are future scientists. Does the study of theories of scientific growth have relevance to the teacher of these students? It might in that such a study might suggest approaches to teaching that help to maximize the *science student's* (rather than science's) growth.

Consider, for example, a growth theory called theoretical pluralism, which I have discussed elsewhere.[14] Very roughly this view says that maximum scientific growth is achieved by working with a number of different and conflicting scientific theories or research programs. Whatever the obscurities and difficulties with this theory as a theory of scientific growth, it does suggest an analogy in science education, namely that maximum intellectual growth for science students will be achieved by working with many different and conflicting theories or research programs. One might call such a view Pedagogical Theoretical Pluralism.[15]

Now whether such an approach is justified is another question. This is where, I believe, researchers in the teaching of science come in. I suggested earlier that researchers in science teaching might test the truth of some hypotheses suggested by analytic philosophy of science. Here we have a case of empirical researchers testing a hypothesis suggested by growth theory philosophy of science. Again we can imagine testing groups of students – one group working with many alternative theories and another group working with one theory – with respect to their intellectual growth measured in terms of their increased understanding of science and increased ability. (More refinements of this experiment, of course, are possible.) The results of such an experiment would prove very useful to science teachers in deciding how and what to teach.

VIII. CONCLUSION

Although my time is up I believe I have said enough to indicate that philosophy of science – interpreted as either analytic philosophy of science or what I have called growth theory philosophy of science – is very relevant to science education. I would like to urge that philosophers of science (in this audience and elsewhere) apply some of their insights to science education. Here at least, I believe, philosophers of science can have an important practical impact.

NOTES

[1] There have been a few works. See Michael Martin, *Concepts of Science Education: A Philosophical Analysis* (Scott-Foresman, Glenview, Ill., 1972); James T. Robinson, *The Nature of Science and Science Teaching* (Wadsworth Publishing Co., Belmont, Cal., 1968).
[2] For a similar contrast see I. Scheffler, *Anatomy of Inquiry* (Alfred A. Knopf, New York, 1963), Chapter 1; May Brodbeck, 'The Nature and Function of the Philosophy of Science', *Readings in the Philosophy of Science* (ed. by H. Feigl and M. Brodbeck) (Appleton-Century-Crofts, New York, 1953).
[3] Michael Martin, 'Philosophy of Science and Science Education', *Studies in Philosophy and Education*, 1972, pp. 210–225.
[4] See for example, Barney M. Berlin and Alan M. Gaines, 'Use Philosophy to Explain the Scientific Method', *The Science Teacher* (1966) p. 52; see also Merritt E. Kimball, 'Understanding the Nature of Science: A Comparison of Scientist and Science Teacher', *Journal of Research in Science Teaching*, (1967–68), pp. 110–120. Kimball's study showed that philosophy majors scored higher on *The Nature of Science Scale* than science majors.
[5] See for example, Peter Achinstein *Law and Explanation* (Clarendon Press, Oxford, 1971), pp. 78–84.
[6] Jerome S. Bruner, *The Process of Education* (Harvard University Press, Cambridge, 1961). For an insightful critique of Bruner's views on structure see James Hullet 'Which Structure?', *Educational Theory* **24** (1974), 68–72.
[7] Joseph J. Schwab, 'Structure of the Disciplines: Meaning and Significance', in G. W. Ford and Lawrence Pugno (eds.), *The Structure of Knowledge and the Curriculum*, (Rand McNally, Chicago, 1964).
[8] Bruner, *op. cit.*, p. 28.
[9] Biological Sciences Curriculum Study, *Biological Science: An Inquiry into Life* (Harcourt, Brace and World, New York, 1968), p. 25.
[10] See, for example, Hempel, *Philosophy of Natural Science*, pp. 22–28. See also Michael Martin, *Concepts of Science Education*, Chapter 2.
[11] See for example, Hempel's discussion of Semmelweis, *ibid.*, Chapters 2, 3. Also Martin, *Concepts of Science Education*, Chapter 2.
[12] For an interesting discussion of flat earthers see Martin Gardner, *Fads and Fallacies in the Name of Science* (Dover, New York, 1967), Chapter 2. For a discussion of the use of pseudo-science in science education see Martin, *Concepts of Science Education*, Chapter 2.
[13] Israel Scheffler, *Science and Subjectivity* (Bobbs-Merrill, Indianapolis, 1967), p. 86.
[14] Michael Martin, 'Theoretical Pluralism', *Philosophia* **2** (1972), 341–349.

[15] Such a position was suggested, although not called by this name, by Noretta Koertge in 'Theoretical Pluralism, Criticism and Education' in a paper read at the 1970 OISE Conference, 'New Directions in Philosophy of Education'. See also her paper 'Towards an Integration of Content and Method in the Science Curriculum' *Curriculum Theory Network* **4** (1969–70), 26–43.

HUGH G. PETRIE

METAPHORICAL MODELS OF MASTERY: OR, HOW TO LEARN TO DO THE PROBLEMS AT THE END OF THE CHAPTER OF THE PHYSICS TEXTBOOK

Without question, one of the most important cluster of issues in recent philosophy of science has centered around the attack on the rigid positivist distinction between theory and observation or between a theoretical language and an observational language. Kuhn, Feyerabend, Hanson, Toulmin, and Polanyi are all names closely associated with one version or another of this attack.[1] All have argued that observational categories are essentially theory-determined and there is no determinate observational base, or neutral observational language. Thus at least the positivist account of the objectivity of scientific knowledge would seem to be seriously threatened by the thesis of the theory-ladenness of observation. For without an independently accessible observational base against which to test scientific theories, wherein would objectivity consist?

A number of philosophers of science have rallied to the defense of objectivity against the threat posed by the thesis of the theory-ladenness of observation. One of the earliest defenses and still one of the most reasonable and persuasive was offered by Israel Scheffler in his book, *Science and Subjectivity*.[2] Scheffler by no means defends a phenomenalist or positivist account of observation or objectivity, but rather grants a good deal of the thesis of the theory-ladenness of observation. His strategy is to attempt to save objectivity while granting that observation is essentially theory or cognitive-laden.

Whether Scheffler is successful or not in this attempt will not be my concern here; indeed that controversy still rages. Rather I shall attempt to show that the philosophical thesis of the theory-ladenness of observation even in the attenuated form accepted by those such as Scheffler provides an extremely difficult problem for science education. Indeed, if one adds to the theory-ladenness thesis the very plausible assumption that common-sense categories of observation are not identical with the categories of observation associated with scientific theories,[3] then one can pose the Kantian-like question, 'How is science education possible?' It is only with such cases of non-identical sets of observational categories that I shall be concerned in the remainder of the paper.

What I wish to do, then, is this: first, I shall examine Scheffler's two main moves with regard to the theory-ladenness of observation in order to show that whatever his success in saving objectivity in the abstract, he seems to leave the science student in an extremely precarious situation. Sketching the nature of this situation vis-à-vis the yet to be learned subject and vis-à-vis the teacher will, hopefully, illustrate the force behind the question, 'How is science education possible?' Finally, I shall suggest an answer to this question which utilizes metaphor as a key feature of science education.

I

Scheffler's defense of objectivity has two main prongs. First, he urges that in considering the theory-ladenness or conceptual nature of observation, we proceed by

> ... distinguishing categories from hypotheses, and contrasting the general ordering imposed by the former with the particular categorical assignments predicted by the latter; observation "determined by," "dependent on," or "filtered through" categories is thus quite conceivable as independent of any special hypothesis under test, expressible through reference to such categories. The general view thus advocated seems to me to preserve a tenable notion of the objectivity of observation, and to do so, moreover, without presupposing that the given is ineffable, uninfluenced by categorization, or reported by statements that are necessarily certain.[4]

Thus while different people may have different categorial schemes, these schemes *may* overlap, and, in any event, the *hypotheses* which describe the relations into which categorized experience will fall can be tested by seeing whether or not our experience does so order itself. Furthermore, note that Scheffler holds this view without holding that the given is ineffable, pure, or certain.

Now while the observational categories of science student and scientist or science teacher *may* overlap to some degree, it is highly unlikely that they will overlap as regards those categories peculiar to the scientific theory. Thus, even if observation in terms of a category schema provides an objective check on hypotheses for those who already share the category schema, the poor science student who has not yet acquired that schema remains in a quandary. How is he to learn the observational category schema when that schema depends on the theory which he does not know?

It might be suggested here that the way out is to teach the student the theory and then the category schema will follow. This suggestion leads me to the second prong of Scheffler's defense of objectivity, namely the defense of the objectivity of meaning. Scheffler's problem is this: the meaning of categorial terms is language dependent. Thus for the student to learn the theory, he must, in effect, already have learned the language in terms of which the theory is expressed. But this seems to be impossible. Furthermore, Scheffler has already abjured the traditional way out of this difficulty, namely, that the student and teacher share the same basic observations and neutral observation language, and, hence, could build up the meanings of the new terms out of already shared meanings along with ostensive definition. In short, Scheffler must account for the meaning independence of some categories and experimental laws from their role in specific theories without utilizing neutral observation.

Scheffler's solution to this problem lies in an appeal to the distinction between sense and reference. Differing senses of categories of observation and of experimental laws are indeed possible in different theories, although some synonymies may persist from theory to theory. Of even more importance, however, even in cases of varying senses of terms, reference can be the same from theory to theory.[5] Furthermore, different theorists can agree on the constancy of referential interpretation by application to specific cases. Surely both Aristotle and Galileo could point downward and agree that was the earth.

Yet this appeal to agreement of reference in specific cases is not without its problems, for Scheffler also acknowledges the possibility of a multiplicity of schemes of reference, thus allowing only a relative independence of observation from theory. In effect, the appeal to agreement of reference in special cases is open to change, reinterpretation, and charges of misinterpretation. In essence Scheffler accepts the Wittgensteinian attack on the ultimacy of ostensive definition. Thus while the logical possibility of scientific objectivity may have been salvaged by Scheffler, the epistemological problem, and even more importantly, the pedagogical problem, seem to remain. How can we know when we have commonality of reference? How can we teach the student the language of reference peculiar to the science in question?

II

Scheffler seems aware of the general thrust of these problems as is revealed in his description of a new theory with its own unique scheme of reference emerging within a different, given, referential tradition. His description is worth quoting at length.

> A new theory arising within a given referential tradition cannot command initial consensus on presumably confirming cases of its own, but must prove itself against the background of prior judgments of particulars. It must acknowledge the indirect control of accumulated laws and theories encompassing already crystallized judgments of cases. Even if some such judgments are to be challenged from the start, the challenge needs to be expressed in a form that is intelligible for the received descriptive mode, and special motivation for the challenge must, of course, be adduced. The inventor of a new theory cannot, at the outset, motivate his new forms of discourse by simply saying "Look and see!" Now, it may turn out that this new theory, having won an initial place largely through indirect forms of argument against the background of acknowledged facts, eventually forces a revision of older judgments of cases and, what is more significant, perhaps, opens up new ranges of evidential description, thereafter developing consensus on relevant instances of its own. It remains true that, at the outset, its advantage needs to be shown in the context of judgments of cases already available, and in relation to the scheme by which such judgments are formulated.[6]

Now it seems to me that this situation is precisely that which obtains most of the time between science teacher and beginning science student. The student is the bearer of the 'received descriptive mode' and the teacher is attempting to make plausible a 'new theory' with new schemes of reference. Scheffler admits the teacher cannot simply say, 'Look and see!' Rather he must couch his instruction in relation to the student's existing referential scheme. Of course, one would not wish to deny this rather novel way of expressing the hoary dictum that one must start with where the student is at, but the question remains, just how does that aid the student's understanding? Recall in this connection that building up new categories out of neutral, shared, independent, categories has been ruled out. What kind of 'indirect argument' for the new theory can be made to the student? How is science education possible? For that matter, how is any education which utilizes radically different schemes of reference possible?

III

There are a number of science educators who believe strongly in one version or another of the thesis of the theory-dependency of observation, and, furthermore, utilize this cluster of ideas in their recommendations for what science teachers should be taught to do. One of these is my colleague, Charles Weller, at Illinois.[7] Now although Weller's total view lends itself in places to the kind of subjectivist interpretation attacked by Scheffler, nevertheless, some of his suggestions as to how standard science education is possible seem to me to be fairly illuminating. He lists four recommendations.

(1) The teacher must assess the student's frame of reference as quickly as possible.

(2) If the student has no prior experience with the relevant phenomena, then direct experience with the phenomena should be provided.

(3) The student should be involved with the phenomena attempting to organize them in terms of the accepted model.

(4) The student should be communicating actively with others, at least some of whom already have a good working knowledge of the model in order to test out their tentative articulations.[8]

Now while these suggestions seem most plausible and several of them compatible with Scheffler's argument, they seem to me to lack a coherent unifying perspective. What I will do in the remainder of the paper, therefore, is to try to sketch such a perspective by using the notion of a metaphor as the key pedagogical device.[9] In brief, I want to say that science education, indeed any education involving differing schemes of reference, is possible because of the existence of metaphor.

In making this claim, I shall be using a fairly standard account of metaphor due to I. A. Richards and modified by Max Black.[10] I shall speak of the linguistic term being used metaphorically, as the 'vehicle.' That about which the metaphorical assertion is being made I shall call the 'topic,' whereas the 'ground' will be that which the topic and the ordinary literal referent of the vehicle have in common. The dissimilarity between the ordinary reference of the vehicle and the topic is called the 'tension.' On Max Black's view there are two types of metaphorical assertion which are relevant for my purposes.[11] *Comparative* metaphors are to be understood as basically analogies between topic and ordinary referent. *Interactive* metaphors, on the other hand, involve an interaction between the system of categories and beliefs clustering around the ordinary referent of the vehicle and the system of categories and beliefs clustering around the topic. Black's reason for also recognizing the interactive metaphor is that in some cases one simply cannot understand how a set of analogies as used in comparative metaphors can give us the insight which a metaphor often provides.

It would be more illuminating in some of these cases to say that the metaphor creates the similarity than to say that it formulates some similarity antecedently existing.[12]

Thus, at least with interactive metaphors, there is a unique cognitive role played by the metaphor. Significantly, my colleague, Andrew Ortony, titles a recent paper of his, 'Why Metaphors Are Necessary and Not Just Nice.'

Note well that metaphorical assertions necessarily *show* us something about the topic rather than describing the topic literally. Metaphorical assertion is thus peculiarly suited to cases where we do not wish to or cannot literally describe the topic.[13] This seems to me to provide the key to bridging the gap between the student with his frame of reference and the to-be-learned frame of reference of the scientific theory. Aspects of the topic can be *shown* to the student rather than described, and this is crucial since by hypothesis the student is unfamiliar with the appropriate mode of description in the science. Note too, that the 'showing' in the use of metaphor is *not* simply ostensive definition. If it were, we would not need the metaphor to direct our attention to crucial aspects of the topic.

How does this work in detail? Well, the to-be-learned scheme of reference may overlap partially or not at all with the student's existing scheme of reference. If it overlaps partially, then the student can perform some of the tension-elimination for himself, i.e., he can see, to some extent what the dissimilarities between topic and literal referent are. To the extent that this is possible, the metaphor may be a comparative one. I shall concentrate, however, on the extreme case in which the two schemes of reference at least as regards the scientific categories are disjoint.

In this case the teacher introduces the key metaphor to the student. The vehicle must be part of the student's linguistic and cognitive scheme, hence the necessity for Weller's point that the student's frame of reference be assessed by the teacher. But, furthermore, since the student has, by hypothesis, no familiarity with the topic, he must, as Weller says, be put into direct contact with the phenomena, even though he may be familiar with the phenomena under a different categorization. In Weller's terms the student must be involved with trying to organize the phenomena by means of the model (the metaphor) he has been given. Initially this will involve his organizing the topic in accordance with the literal meaning and implications of the vehicle.

But since on Weller's view the student is in the presence of someone already familiar with the topic, the tension involved in these naive inferences can gradually be shown to the student and thus eliminated. Books could, in principle, also serve this weeding out or tension-elimination function.

Furthermore, in some cases, the environment itself may be sufficient to show the need for the appropriate tension-elimination as the student acts on his initial understanding of the topic wherein the vehicle is used literally in speaking of the topic. In other words, the ground is taken by the student to be complete prior to the weeding out by the environment of false predications and inferences. Logically, the research scientist at the frontiers of his field is in the same position as the student without a teacher but with an environment to help him understand the topic. The claim that nature is science's teacher may not be so metaphorical after all!

Let me now consider one possible objection. Black asserts that *both* comparative and interactive metaphorical assertion require simultaneous awareness of both topic and ordinary referent.[14] Yet I have claimed that the student's frame of reference need not overlap at all with the to-be-learned frame of reference. This would appear to be equivalent to denying student awareness of the topic. In an important sense, this logically possible case of total disjointedness of systems of reference probably almost never occurs in fact. The physics student at least sees the cathode-ray tube as a glass and metal object. Even the superstitious native sees his own photograph as magic of the gods. So in most cases there are at least some referential categories common to the student's system of reference and the new system of reference, and in this sense there is an awareness of both systems. So too will there be some tension resulting from the metaphor's invitation to apply the familiar categories to the new situation, along with a fairly vaguely apprehended ground. The teaching-learning situation then proceeds by making explicit more of the tension and progressively eliminating it thus making the ground more and more precise until, at some unspecifiable point, we would probably cease to count the metaphor as a metaphor. The term has a new meaning and science education has been seen to be possible. At the same time, however, it does seem possible that occasionally the systems of reference are totally disjoint, and it is here also that I am suggesting that metaphor plays an essential role.

As I have described the situation, however, there is something which serves a role similar to that of the interaction of two ideational systems in awareness. It is at this point that the influence of the world can make itself felt as an independent cause of objectivity. The teacher, or environment in the ultimate case, edits out those categorizations and inferences which are carried over whole with the vehicle and applied to the topic. One literally does not know that about which one enquires, but one can, at least, eliminate false guesses.[15] The interaction can be between idea and world in a strictly causal sense. The awareness that occurs may be no more than the realization that there is something wrong with organizing phenomena in the accustomed way. I think this kind of ideational-environment interaction satisfies the spirit if not the letter of Black's requirement. Objectivity needs no more of a foothold.

All this talk of the metaphorical extension of schemes of reference may have left the impression that I have been dealing with something extremely rare, revolutionary, and mysterious. This impression was almost unavoidable given that I have been dealing with the relatively self-contained theories and models of contemporary science. Learning such material is largely confined to fairly standard courses and, indeed, appears quite mysterious to many students – those who succeed as well as those who fail. Then too I have, by stipulation, been discussing the situation in which the scheme of reference of the student is radically disjoint from the scheme of reference of the science to be learned. However, it may be worthwhile to point out that the extension and change of schemes of reference by means of metaphor is simply an extreme case of a process of conceptual change and modification which goes on all the time in learning.[16]

Consider the child who has come to know about soft, fuzzy animals called dogs. One of them, called Socrates, is her pet, and there are several others in the neighborhood. Now this child, call her Ann, is taken to visit her Aunt Louise where she is fascinated by a tiny porcelain figure she can hold in her hands. 'That, too, is a dog,' says her father. Is the phrase 'china dog' a metaphor? Probably not in the ordinary sense. But then there's an awful lot of difference between Socrates and Aunt Louise's dog, too. Have Ann's conceptual scheme and referential categories been changed? Almost surely, but there is nothing rare, nor revolutionary, nor mysterious about it. My point is that learning a new categorial scheme in a science is very much like, although more extensive than, learning about china dogs.

I think that Kuhn has also recently made this same point regarding the possibility of learning a completely new categorial schema without any necessarily linguistic translation of the new into the old. In 'Second Thoughts on Paradigms,' Kuhn says,

... I suggest that an acquired ability to see resemblances between disparate problems plays in the sciences a significant part of the role usually attributed to correspondence rules.[17]

This suggestion occurs in the context of Kuhn's discussion of exemplars as an extremely important sense of 'paradigm.' Briefly, exemplars are shared exemplary 'problem solutions' – actually basic observational categories which are learned *without* necessarily learning criteria of application.

As Kuhn says

... I continue to insist that shared examples have essential cognitive functions prior to a specification of criteria with respect to which they are exemplary.[18]

Kuhn distinguishes the 'exemplar' sense of paradigm primarily for the historical reason that he could not find sufficient evidence of rule and criteria assimilation in the training of scientists to explain the similarity of judgments made by scientists in actual uses. Thus, Kuhn posits the exemplar as a non-rule-governed manner of accounting for similarity judgments.

My preceding discussion has made explicit another, logical, reason for requiring something like a learned exemplar. Barring a neutral observational base or the possibility of translating between referential schemes, only something like an exemplar *could* do the trick. But what may appear mysterious when spoken of in the language of exemplars becomes, I suggest, quite plausible in the language of metaphor – especially interactive metaphor. For metaphors have been used for a very long time to organize new fields into similarity sets prior to our ability to state formally the criteria with respect to which the similarity grouping is made.

The new element, seemingly forced on us by the theory dependency of observation, is the essential role of metaphor or exemplars. Metaphors have generally been viewed as merely heuristic devices to aid learning, and, indeed, they may have no more of a role than heuristic in other areas of science education. However, if my preceding characterization of the predicament of the science student is at all correct, the logical role of metaphor *must* be played if the student is to learn a radically new scheme of reference.

Let me conclude then by noting that Paul Feyerabend's fascinating account of Galileo's Tower Argument seems to illustrate beautifully all of my points.[19] The Tower Argument was advanced by proponents of a stationary earth. It consists essentially of arguing that a rotating earth would mean a stone dropped from a tower would have to land some distance from the base of the tower. Clearly this does not happen and so the thesis of a rotating earth must be false.

Feyerabend notes here that Galileo's problem is to replace one natural interpretation of motion, i.e., one theory-laden categorial system, with another. People must be brought to see the operative nature of only relative motion. And how does Galileo qua science teacher attack this problem? By introducing the metaphor of an artist drawing a picture in a boat during a long trip and placing the observer outside the boat and noting the relative motion of boat (as earth), paper (as tower) and pen (as stone). Galileo also uses other boat and carriage metaphors to teach the new system of natural interpretations or observational categories. Feyerabend admiringly refers to this process as 'psychological trickery' but I suspect he would not object too strenuously to my stronger claim that such 'psychological trickery' is essential for learning radically new systems of reference.

In summary, I have urged that the pedagogical question, 'How is science education possible?' remains as a legacy of the theory-dependency of observation even after objectivity has received its due. I have further urged that an interactive view of metaphor provides the key to answering this pedagogical question in that it allows one to bridge the gap between alternative systems of reference even when these are disjoint. In short, metaphor plays an essential cognitive role in learning. I have also expanded the notion of the interaction of ideational systems in metaphor to the interaction of ideational systems and environment and urged that this interaction is very similar to Kuhn's idea of learning non-rule governed perceptual similarity relationships by means of exemplars. Finally, I have hinted that this interaction may prove to be the material locus of the objectivity formally saved by Scheffler's analysis. Science education is possible with the aid of the humanistic tool of metaphor, and consideration of current philosophy of science seems to force such a conclusion on one.

University of Illinois

NOTES

[1] See for example, Thomas Kuhn, *The Structure of Scientific Revolutions*, 2nd ed. (Chicago University, Chicago, 1970); Paul Feyerabend, 'Against Method: Outline of an Anarchistic Theory of Knowledge', in M. Radner and S. Winokur (eds.), *Analyses of Theories and Methods of Physics and Psychology*, Minnesota Studies in the Philosophy of Science, Vol. IV (University of Minnesota, Minneapolis, 1970); N. R. Hanson, *Patterns of Discovery* (Cambridge University, Cambridge, 1958); Stephen Toulmin, *Human Understanding* (Princeton University, Princeton, 1972); and Michael Polanyi, *Personal Knowledge*, revised edition (University of Chicago, Chicago, 1962).
[2] Israel Scheffler, *Science and Subjectivity* (Bobbs-Merrill, New York, 1967).
[3] Both Scheffler and one of my co-symposiasts do seem to grant this very assumption. See Scheffler, *Ibid.*, e.g., pp. 40–41 and 64–65. See also Michael Martin, *Concepts of Science Education* (Scott-Foresman, Glenview, Ill., 1972), p. 127.
[4] Scheffler, *op. cit.*, p. 43.
[5] *Ibid.*, p. 62.
[6] *Ibid.*, pp. 65–66.
[7] Charles M. Weller, 'A Psycho-Epistemological Model for Teaching Science and Its Articulation with Classroom Activities'. A position paper delivered at the Association for the Education of Teachers in Science Meeting, Chicago, March, 1974.
[8] *Ibid.*, pp. 26–27.
[9] In the following, I am extremely indebted to my student, Felicity Haynes, and my colleague, Andrew Ortony, for having opened my eyes to the crucial importance of metaphor in learning situations where a new scheme of reference is being learned. See Andrew Ortony, 'Why Metaphors are Necessary and Not Just Nice', *Educational Theory* **25**, Winter 1975, pp. 45–53.

[10] See I. A. Richards, *The Philosophy of Rhetoric* (Oxford University, London, 1936) and Max Black, *Models and Metaphors* (Cornell University, Ithaca, New York, 1962).

[11] *Ibid.*, pp. 25–47.

[12] *Ibid.*, p. 37.

[13] I am indebted for this point to Felicity Haynes. Ortony, *op. cit.* makes a similar point in speaking of the inexpressibility thesis of metaphors. Ortony urges that there literally are things which cannot in fact be expressed in a given language. For my purposes, I only need admit that some things are not now expressible literally in the language presently at the command of one speaker, namely, the student.

[14] Black, *op. cit.*, p. 36, 46.

[15] This is, in roughest outline, my own version of Popper's method of conjectures and refutations. See Karl Popper, *Conjectures and Refutations*, 2nd ed. (Basic Books, New York, 1965).

[16] I am indebted for this point to my colleague, F. L. Will.

[17] Thomas Kuhn, 'Second Thoughts on Paradigms', in F. Suppe (ed.), *The Structure of Scientific Theories*, (University of Illinois, Urbana, Illinois, 1974), p. 471.

[18] *Ibid.*, p. 477.

[19] Paul K. Feyerabend, 'Against Method: Outline of an Anarchistic Theory of Knowledge', in Michael Radner and Stephen Winokur (eds.), *Analyses of Theories and Methods of Physics and Psychology*, Minnesota Studies in the Philosophy of Science, Vol. IV (University of Minnesota, Minneapolis, 1970).

ROBERT PALTER

PHILOSOPHY OF SCIENCE, HISTORY OF SCIENCE, AND SCIENCE EDUCATION

Since I have had the privilege of reading the remarks of my fellow-symposiasts before writing out my own, I should like to begin by making the briefest of comments on each of their contributions. Professor Petrie raises the Kantian-like question of how science education is possible – or, more precisely, how, if all observation is theory-laden, science education is possible. His intriguing answer is that "Science education is possible with the aid of the humanistic tool of metaphor." I find it difficult to asses this formulation without further examples of its concrete application to actual pedagogical problems in the natural sciences. In Kantian terms, one might say that Professor Petrie, having made use of the *synthetic* method in deducing the possibility of science education, should now turn around and use the *analytic* method (the method of Kant's *Prolegomena*), that is to say, he should take a broad range of representative examples of successful science education and show how their success depends critically on certain key metaphors which bridge the gap between the students' frame of reference and the frames of reference of the respective scientific theories being learned/taught.

About such enterprises as this latter, Professor Martin would, I feel sure, say that in the interests of improving science education we must test empirically the value of different key metaphors in the learning/teaching process for a given scientific theory. Here, as in the case of Professor Martin's other suggested research projects dealing with science education, I must confess myself rather at a loss in imagining possible experimental designs which might enable one to discriminate significantly among alternative educational techniques. My difficulty stems from the fact that no one to my knowledge has even established convincingly the main parameters upon which 'scientific understanding' and 'scientific ability' depend, so that any attempt to measure these traits would seem to be premature at this time.[1] But I would go further: I am not even sure it makes sense to set out deliberately to produce better scientists by improving science education – any more than it makes sense to set out deliberately to produce better artists by improving art education. (Not that 'better' has the same meaning in the scientific and in the artistic context – I am sure it does not – but it does seem equally futile in either context to attempt to institute contingency planning for future needs, whether for theoretical or experimental scientists on the one hand or for naturalist, abstractionist, or expressionist artists on the other.) Producing better understanding of science in students by improving the science curriculum is a somewhat different matter, and here I shall myself later propose some modest measures which – even in the absence of a deeper understanding of scientific understanding – might be expected to contribute toward the desired goal.

Let me turn now to some more constructive remarks. Like my two fellow-symposiasts I want to refer to some of the views of two of the most influential contemporary writers on science, Karl Popper and Thomas Kuhn. Now, I hope that I am properly appreciative of the many insights into the nature of science to be obtained from Popper's and Kuhn's writings; and I also hope I am properly aware of the desirability if not the necessity of a systematic epistemology of one's own if one is to engage in any useful criticism of the philosophical presuppositions of their work. Such a systematic epistemology and such a critique I am not now prepared to formulate and defend, beyond the claim that there is at least this much truth in inductivist theories of scientific inquiry (as expounded by such philosophers of science as Bacon, Newton, and Mill): it is possible to generalize from a set of observed data without presupposing the resulting generalization, or some logically stronger principle, in the collection of the data. On the basis, then, of a set of objectively ascertainable historical facts which Popper and Kuhn either overlook or deliberately ignore, I wish to call attention to the *incomplete consensus* which generally characterizes the full set of opinions on any given scientific topic at any given time during the history of active inquiries into that topic. (Even excluding the opinions of demonstrably incompetent or cranky individuals, there will generally remain a plurality of divergent opinions.) 'Active inquiries' here I mean to be construed very generously so as to include, for example, the composition of new textbooks on the given topic. Put otherwise, my point is simply that active inquiry has always in the past generated diversity if not conflict of views on a given scientific topic, so that today's consensus on many, if not all, important scientific topics is neither perfect nor static: such consensus as has been achieved is usually enlivened (not, I would insist, marred) by dissent and is in a state of continuing evolution. Diversity or pluralism – at least in my extended sense – I therefore take to be a well-substantiated characteristic of the history of science.

Take, for example, the case of textbooks on the topic, or better the cluster of related topics, often referred to as 'classical mechanics.' Each such textbook represents a more or less orthodox variant on this topic; more or less significant variations will be played by each author on the way foundational questions are treated (i.e., the choice of primitive terms, definitions, and axioms), on the manner in which the theory is given an empirical interpretation, on the use of specific problems to illustrate problem-solving methods, and on the inclusion of exemplary historical episodes. Those committed to Kuhn's conception of normal science supposedly characterized by its dogmatic adherence to fixed paradigms will, no doubt, want to reply that a new textbook merely serves to reinforce the prevailing massive orthodoxy; and Popperians will, no doubt, want to discount the vast majority of such textbooks as unadventurous, unconjecturing, even subscientific, specimens of a literary genre whose existence is an unfortunate though perhaps inevitable consequence of the growth of genuine scientific knowledge. In the present context, my rejoinder consists simply in asserting as hypotheses that scientific ability, possibly, and scientific understanding, almost certainly, are functions of the textbooks (and other ingredients) of a student's scientific education. For a science educator the proper attitude toward textbooks is, after all, neither celebratory nor contemptuous; rather, the necessity for such pedagogical devices must be seen as simply a part of the facts of life in the complex set of social institutions which constitute scientific inquiry today.

I do not wish to be understood, however, as advocating a merely passive or uncritical attitude toward science textbooks; far from it. Although in some respects these books reflect an achieved consensus, in other respects they reflect the author's special and idiosyncratic variant of that consensus. And obviously not all idiosyncrasies are equally valuable. One criterion that I consider of some importance in evaluating a science textbook is how accurately it reflects the current state of research in relevant scientific topics. In point of fact, science textbooks often do not reflect this state at all; the impression fostered by too many of them –

especially the elementary ones – is, rather, that no research at all, or at least none which need be taken seriously, is being done on many of the topics covered in the textbook. Classical mechanics again provides us with a ready illustration. Over the past two decades or so a large number of new and important results have been discovered in the (classical but non-linear) mechanics of diffusion, elasticity, viscosity, and hydrodynamics.[2] Furthermore, deeper insights into the character of the fundamental laws of classical mechanics have been attained by expressing these laws in terms of that generalized geometry of space-time first introduced in formulations of relativistic mechanics.[3] It seems to me that the existence of such developments as these ought not to be hidden from our students, both because of their intrinsic interest and because of their larger significance as vital clues to the way in which scientific knowledge grows.

This question of just how scientific knowledge grows has lately been much discussed, with two of the leading accounts being those of Popper (conjectures and refutations) and Kuhn (normal scientific problem-solving punctuated occasionally by a scientific revolution). One thing I find lacking in both accounts is *verisimilitude*, that is, agreement with the actual details of the history of actual scientific investigations. Both accounts, whatever their authors' intentions, tend to call up in one's mind the image of scientific growth in any one field as a single progression or stream in which one hypothesis is continually being replaced by another (Popper), or in which a large number of paradigmatic problem-solutions and a much smaller number of paradigm-violating anomalies accumulate until a new paradigm accommodating both the earlier problem-solutions and the anomalies (not to mention new problem-solutions) emerges to replace the earlier paradigm (Kuhn). One striking thing about these formulations is the way they, surely inadvertently, resemble the Whiggish account of scientific progress allegedly offered by so-called positivist or inductivist historiography and so scorned by both Kuhn and the Popperians; in each case, a *unilinear sequence* – whether of facts and laws, or of paradigms, or of conjectural hypotheses – provides the central metaphor. An alternative metaphor for scientific growth might be derived from the course of biological evolution; and, indeed Toulmin has worked out a version of this line of thought in some detail (1972). Now, it is notorious that biological metaphors can be highly misleading when applied to historical matters; but, if we are properly sensitive to the dis-analogies, there may, nevertheless, be some heuristic value in illustrating my incomplete consensus thesis concerning the history of science by the evolutionary development of organisms. What I have in mind is simply the multi-linear – and indeed the often highly-branched – structure of biological evolution.[4] Consider, for example, the most interesting – though far from the best-documented – case of man himself. Remembering that most of the details of human evolution are still highly speculative, let us recall the main outlines of the process as it is understood today. Between 10 and 15 million years ago the earliest known primate with man-like traits (Ramapithecus) emerged along one branch of a hominoid line whose other branches represent ancestors of the living large apes (gorillas, chimpanzees, and orangutans). The hominoid line represented by Ramapithecus then branched to give rise to several new species (including Australopithecus, Homo erectus, and Homo habilis) some of which became extinct and one of which evolved further to become eventually Homo sapiens. The latter species then branched further some 50 000 years ago to give rise to the various races of modern man. One thing to note about this evolutionary sequence is that the living apes represent a line of evolution that has by now diverged very considerably from the hominoid line that gave rise to modern man (so that it is risky to draw inferences about modern man from the study of living apes). More generally, we see that when a single line of evolution branches it is not necessary for all but one of the resultant divergent lines to become extinct: though closely related, the various new species may not be in direct competition for survival (i.e., they may occupy different ecological niches). So with scientific theories: the advent of relativistic mechanics and quantum mechanics did not necessarily mean the extinction of classical mechanics as a viable – that is, living and growing – discipline; not did the advent of molecular biology mean the extinction of classical Mendelian genetics; nor – to take just one more example – did the advent of statistical thermodynamics mean the extinction of classical, so-called phenomenological, thermodynamics. Of course, the older discipline in some of these cases has been more or less transformed by the presence of the newer discipline but not – at least not yet – transformed out of existence.

My earlier point concerning the prevalence of incomplete consensus in the history of science is easily taken account of in the multilinear historical model: a given theory must be represented not by a single line but by a sheaf of lines originating in some common point, itself perhaps the confluence of several earlier lines. This last property clearly differentiates the structure of biological evolution from that of scientific growth: in the former, branching usually diverges with time, whereas in the latter convergence also occurs (as when several independent theories are unified or synthesized by a later theory).[5] Thus, to take the case of classical mechanics, each current version of this discipline (some principle of identification of distinct versions being presupposed) would be represented by a distinct line and all these lines would originate (to oversimplify drastically) in Euler's paper of 1750, 'Discovery of a New Principle of Mechanics,' in which the so-called Newtonian equations of motion,

$$F_x = Ma_x, \; F_y = Ma_y, \; F_z = Ma_z,$$

are formulated for the first time as applying to mechanical systems of all kinds (particles, extended bodies, fluids, etc.). Euler's paper was itself, of course, a confluence of early work, in particular the *Principia* of Newton and a paper by James Bernoulli of 1703, "second only to the *Principia* itself in influence on the later growth of the discipline" (Truesdell, 1960, p. 15). Naturally, there were dead ends in the history of classical mechanics and these will be represented in our model by terminating lines; but there are numerous lines today which have not yet terminated and which appear to be capable of further (perhaps indefinite?) extension into the future.

Much of what I have been saying about the history of science is perhaps of little interest from the point of view of high school or introductory college science courses. The incomplete consensus of scientific disciplines *today* is, I believe, another matter. Selected cases of contemporary conflict and disagreement among equally competent scientists in some specialized field might be extremely illuminating even to students in an elementary science course. An example of what I have in mind would be the current dispute as to the explanation of the red shift in quasars (Field *et al.*, 1973). An even 'farther out' dispute would be that concerning the existence of tachyons (or particles moving faster than light) (Kreisler, 1973). Finally, I share Professor Martin's conviction that one might profitably raise with students the question of how to formulate criteria for distinguishing science from pseudo-sciences like astrology. I also agree with Professor Martin's view that a pseudo-science cannot be characterized simply in terms of its theories (or other types of statements): untested, refuted, even untestable statements may very well – probably often do – belong to a science (Martin, 1972, pp. 40–43). It is rather the *manner* in which such statements are tested or left untested and the *way* in which they are believed or disbelieved which serve to distinguish science from pseudo-science. (Thus, for example, belief in the cosmic scope and the infallibility of his principles is characteristic of the pseudo-scientist.) And, I might add, these distinguishing criteria of test and belief may well be historically conditioned: perhaps astrology was not a pseudo-science at all in Ptolemy's time.

I have just been proposing a role for history of science in science education. Is there not also a role for philosophy of science? Of course; but it too would profit from some historical perspective – even the perspective of just the last few decades. To put it bluntly, science educators must be careful not to be carried away by the latest novelty in philosophy of science, whether it be the operationalism of the '30s, the testability of the '40s, the covering law model of explanation of the '50s, the paradigms of the '60s, or the theory-ladenness of the '70s. Let our students by all means know of the recent controversy between Popper and Kuhn; but let them also know of the important work of Grünbaum and Reichenbach, of Hempel and Carnap, not to mention such older figures as Einstein, Poincaré, Duhem, Bernard, Mill, Whewell, and Newton. In the absence of a fuller understanding of scientific understanding, we can perhaps do no better – but it may well be good enough – than to exhibit for our students the unvarnished details of how science has actually developed and what the best critical minds have said about this development.

University of Texas at Austin

NOTES

[1] For a recent attempt to characterize scientific understanding, see Friedman (1974). For a general discussion of understanding in the context of teaching and learning, see Martin (1970, Part III); she stresses the "open-endedness" of the concept of understanding and distinguishes several varieties. Again, the title essay of Ziff's volume *Understanding Understanding* concludes with the words: "... one can no longer avoid the dismal conclusion that

to understand understanding is a task to be attempted and not to be achieved today, or even tomorrow" (1972, p. 20).

The situation with respect to our understanding of understanding recalls that with respect to our understanding of intelligence: much research effort has been expended on measuring IQ and correlating it with other variables but not enough thought has been devoted to clarifying the concept of intelligence itself. (See Block and Dworkin, 1974, pp. 354–9.) We all, of course, share to some extent a common concept of understanding but surely prudence demands that we clarify that concept considerably prior to devising tests to measure it.

[2] For introductory – but not up-to-date – accounts, see Truesdell (1952, 1968). The fullest systematic account to date is in the treatise by Truesdell and Noll (1965); new developments are published in the *Journal of Rational Mechanics and Analysis*.

It is difficult to explain in a few words the general character of these results in non-linear (continuum) mechanics; eschewing mathematics makes it even more difficult. For what it's worth, however, let me try. To formulate an adequate theory of any mechanical system (e.g., a set of interacting particles, a perfect fluid, etc.) requires a precise specification of the forces – more generally, the forces and stresses – which characterize that system. Until fairly recently the vast majority of such mechanical theories were linear in the sense that two forces or stresses applied simultaneously were assumed to have the same effect as when they were applied successively. This assumption of linearity simplifies the mathematics but is, unfortunately, highly unrealistic for many mechanical systems of great interest, e.g., materials exhibiting such phenomena as plastic deformation, turbulent flow, purely elastic behavior, or purely viscous behavior. Hence, non-linear methods have had to be developed and these have led to mechanical theories which are more realistic, more general, and in some cases more rigorous than any previous ones. (Occasionally, a non-linear theory has striking practical applications, as in the explanation – in terms of the theory of non-linear fluids – of why rotary stirrers are ineffective in stirring paint.)

[3] See Trautman (1965). In the light of work such as Trautman's I find rather pointless remarks like the following: "... in some fundamental ways Einstein's general relativity resembles Aristotle's physics more than Newton's" (Kuhn, 1970, p. 265). Aristotle didn't use second order differential equations – a rather fundamental difference from both Newton and Einstein!

[4] Rensch (1966, p. 97) has coined the term "kladogenesis" for "phylogenetic branching and splitting in general". He also introduces the term "anagenesis" for "the development toward higher phylogenetical levels" (*ibid.*) Anagenesis corresponds to that aspect of the growth of scientific theories which we might characterize as 'progressive'; and it is this aspect which has been of most interest to philosophers, such as Toulmin (1972), who make use of biological metaphors for scientific growth. But see Note 5.

[5] This difference is remarked on by Popper (1972, p. 262). Of course, biological evolution sometimes exhibits hybridization, a process in which two species give rise to a new species (of particular importance in the evolution of plants).

BIBLIOGRAPHY

Block, N. and Dworkin, G.: 1974, 'IQ: Heritability and Inequality, Part 1', *Philosophy and Public Affairs* **3**, 331–409.

Field, G., Arp, H., and Bahcall, J.: 1973, *The Redshift Controversy*, W. A. Benjamin, Inc., Reading, Mass.

Friedman, M.: 1974, 'Explanation and Scientific Understanding', *J. of Philosophy* **71**, 5–19.

Kreisler, M.: 1973, 'Are There Faster-than-Light Particles?', *American Scientist* **61**, 201–8.
Kuhn, T.: 1970, 'Reflections on My Critics', in I. Lakatos and A. Musgrave (eds.), *Criticism and the Growth of Knowledge*, Cambridge U.P., Cambridge, pp. 231–78.
Martin, J.: 1970, *Explaining, Understanding, and Teaching*, McGraw-Hill, New York.
Martin, M.: 1972, *Concepts of Science Education*, Scott, Foresman and Co., Glenview, Illinois.
Popper, K.: 1972, *Objective Knowledge*, Oxford U.P., Oxford.
Rensch, B.: 1966, *Evolution Above the Species Level*, John Wiley and Sons, New York.
Toulmin, S.: 1972, *Human Understanding*, Vol. I: *The Collective Use and Evolution of Concepts*, Princeton U.P., Princeton, N.J.
Trautman, A.: 1965, 'Comparison of Newtonian and Relativistic Theories of Space-Time', in B. Hoffmann (ed.), *Perspectives in Geometry and Relativity*, Indiana U.P., Bloomington, Indiana, pp. 413–25.
Truesdell, C.: 1952, 'A Program of Physical Research in Classical Mechanics', *Zeit. f. Angewandte Math. u. Physik* **11**, 79–95; reprinted in C. Truesdell, *The Mechanical Foundations of Elasticity and Fluid Dynamics*, Gordon and Breach Science Publishers, New York (1966), pp. 187–203.
Truesdell, C.: 1968, 'Recent Advances in Rational Mechanics', in C. Truesdell, *Essays in the History of Mechanics*, Springer-Verlag, New York, pp. 334–366.
Truesdell, C.: 1960, 'A Program Toward Rediscovering the Rational Mechanics of the Age of Reason', *Archive for History of Exact Sciences*, **1**, 3–36.
Truesdell, C. and Noll, W.: 1965, *The Non-Linear Field Theories of Mechanics, Encyclopedia of Physics* (ed. by S. Flügge), Vol. III/3, Springer-Verlag, Berlin.
Ziff, P.: 1972, *Understanding Understanding*, Cornell U.P., Ithaca, N.Y.

CONTRIBUTED PAPERS

SESSION III

RAIMO TUOMELA

CAUSES AND DEDUCTIVE EXPLANATION*

1. EVENTS AND SINGULAR CAUSAL CLAIMS

According to the backing law account of causation a singular causal claim is to be analyzed (or "justified") by reference to a suitable nomic theory which, together with the given singular statement describing a cause, deductively supports or explains the statement describing the effect. This backing law (or deductive-nomological) account of singular causation has recently become the target of several kinds of criticism. First, the possibility of giving a detailed and elaborate account of the required nomic or explanatory backing has been doubted. Secondly, it has been claimed that the deductive-nomological account is bound to lead to unacceptable ontological difficulties (see Kim [1969] and [1973a]). Thirdly, it has been argued that the backing law account fails to give a satisfactory analysis of certain conditional or counterfactual aspects of causation (see Lewis [1973b]). Fourthly, it has been claimed that this approach does not work at least within the social and historical sciences as there are no laws and theories of the required sort to be found there.

In this paper I shall try to constructively defend the backing theory account of scientific causation against some of the mentioned criticisms. More specifically, I shall be concerned with the second and third type of criticisms (see, e.g. Tuomela [1972] and [1974b] for an attempt to deal with the first and the second).

There has been much disagreement concerning which kind of ontological entities causes and effects are. Thus, are they extralinguistic entities such as events, states, conditions, etc., or are they rather linguistic (or semilinguistic) entities such as propositions or facts, etc.? Without further justification, we shall below assume that it is primarily *events* (broadly understood so as to include short-term states and conditions) which in the first place qualify as causes and effects in scientific contexts.

Mackie (1974) makes a distinction between *productive* and *explanatory* causes (also see Davidson [1967]). This distinction seems to correspond at least in part to the above distinction between extralinguistic and linguistic entities. More specifically, for Mackie (1974) events qualify as productive causes whereas facts (which involve occurrence of an event) qualify as explanatory causes. We shall later see that there is in fact a closer correspondence between productive causes and explanatory causes than has been thought, provided that events are construed as structured, and not bare, particulars.

In the view to be accepted in Sections III–VI below an event consists of a locusobject, a point of time, and a generic property exemplified by the locus object (cf. Kim [1973a]). The property in question can be called the *aspect* property (or the *constitutive* property) of the event. To take a simple monadic example, the event of Socrates' drinking hemlock at time t is represented by $\langle\langle \text{Socrates}, t\rangle, \text{drinks hemlock}\rangle$, or, more formally, by $\langle\langle x, t\rangle, P\rangle$. The property P is the aspect or constitutive property of this event. On the other hand, this event is said to *merely* exemplify the property of consuming poison, say Q, which is not constitutive of the event of Socrates' drinking hemlock at t. The identity condition for monadic events now may be construed as follows: $\langle\langle x, t\rangle, P\rangle = \langle\langle y, t'\rangle, Q\rangle$ if and only if $x=y$, $t=t'$, and $P=Q$. As in our above example $P \neq Q$, the events in question are deemed different.[1]

The second hard problem in discussing the ontology of causality is this: What kind of statements qualify as *event-descriptions*? If we want to make sense of locutions like a 'redescription of an event' this problem has to be solved. Davidson has argued (partly on the basis of a paradoxical result discussed in the next section) that the attempt to characterize event-describing statements is futile and that (singular) events should rather be referred to by nominalizing verb phrases (singular terms) (see e.g. Davidson [1969]). Many ordinary language statements which may seem to describe particular events are on a closer look existential. Thus, for example, the correct logical form of the action sentence 'Shem kicked Shaun' is not, according to Davidson, given by anything like '*Kicked* (Shem, Shaun)', which sentence is based on a two-place predicate, but rather by '(*Ex*) *Kicked* (Shem, Shaun, *x*)', which is based on a three place predicate incorporating the variable *x* for singular events. According to Davidson's suggestion, if a kicking of Shaun by Shem occurred at midnight this is taken care of simply by '(*Ex*) (*Kicked* [Shem, Shaun, *x*] & *At midnight* [*x*])'. But what if this kicking was intense? If we add the conjunct 'Intense (x)' we are saying that the event was intense whereas what we wanted to do was to modify the verb 'kicked'. However, it seems that problems like this can be handled, and attempts to develop a theory of verb-modification have been made. Therefore the Davidsonian framework seems to us acceptable provided the structured nature of events can properly be accounted for.

Singular events are referred to and picked out by singular terms of the scientific language employed. Therefore we may employ singular causal claims such as: e caused e'. Here e names the singular event e and e' names e'. But singular causal claims like this do not display the structure of events and hence not the generic properties in virtue of which the singular events can be shown to be causes or effects. The justification of causal claims is to be made by reference to a backing theory, which in our view is best formulated linguistically. It follows that we need *event-describing statements*. Of their form and nature we can here say at least the following.

Given an event $e = \langle\langle x, t\rangle, P\rangle$ in most cases of interest the property P can be designated by a predicate P in our scientific language. Then the occurrence of e can be represented linguistically by '$P(e) \& (Ex)(x=e)$'. (It is not necessary here to consider linguistically the locus and the time of the occurrence of e.) The true statement '$P(e)$' is a *constitutive* statement for e, and it may be considered (in a sense) analytically true of e (see Tuomela [1974b]). Notice, however, that the backing theory need not contain P at all as the subsumption can be effected by reference to other predicates true of e.

If what I just said can be accepted it seems that we may move from a singular causal statement like 'e caused e'' or 'e is a cause of e'' to a corresponding linguistic formulation of the type 'That e has P caused that e' has Q' without affecting at least the truth of our singular causal claims. Thus we seem to be able to go back and forth between ontological or material talk and linguistic fact-talk.

But there are troubles here. First, we may not know under which predicates e and e' are causally related even if we may have reasons for claiming that e caused e'. Secondly, the linguistic formulation must be taken to entail the truth of both the antecedent and the consequent, I think. But then it is immediately seen to be non-truth-functional (switching antecedent and consequent does not preserve the truth of the causal claim). If 'caused' then represents a non-truth-functional conditional what could its logical form be? We shall discuss this problem below in Section VI after a digression into some ontological and other issues involved in the analysis of singular causal claims.

II. ONTOLOGICAL DIFFICULTIES CONNECTED WITH CAUSAL STATEMENTS

Let us now consider a certain analysis of singular causal claims. It is the most sophisticated version of those discussed and criticized in Kim (1969). For the purposes of this section we consider a purely extensional set-theoretic construal of events (thus aspect properties become sets and relations). This causal analysis can be formulated as follows:

(C) Event e is a cause of event e' if and only if there are singular statements D and D' such that D and D' describe e and e' respectively, and D' is a logical consequence of D together with some causal law S but not of D alone.

It may seem initially plausible to assume that event-describing statements are referentially transparent and that the class of event-describing statements is closed with respect to logical deduction. This amounts to accepting the following two innocent looking principles (cf. Kim [1969], pp. 206–207):

(I) If a statement D' is obtained from a statement D by replacing some referring expression in D by a co-referential expression, then D and D' describe the same event.

(L) If a statement D entails another statement D' the event described by D is said to include the event described by D'.

As an immediate corollary of (L) we then have

(L') Logically equivalent statements describe the same event.

But the acceptance of these principles leads to the result that singular causal statements are truth-functional, after all. Hence we get an "ontological" paradox. This well known paradox has many faces, but we shall be mainly interested in the following version of it: Any cause can be described by means of any arbitrary true event-describing statement, and any effect can be described by means of any arbitrary true event-describing statement. (See, e.g. Mackie [1974] and Tuomela [1974a] for a detailed discussion of this result.)

Quite analogously one can show that any law that (C)-subsumes at least one pair of events (C)-subsumes every pair.

Our above paradox can in fact be "ontologically" strengthened if we, in accordance with our earlier considerations, require that singular causal statements entail the existence of the cause-event and the effect-event. For then we get the following result: If an event e causes an event e' then e causes any arbitrary event e'' and e' is caused by any arbitrary event e'''.

Obviously something is deeply wrong with our referential apparatus, but what? Recall that we were discussing some singular statements which were supposed to be event-describing. But it is by no means clear that such singular statements, not even those we have been considering, are event-describing. Thus, even if we may plausibly assume that we start in (C) with some event-describing statements there is no guarantee that the application of conditions (I) and (L) always yields event-describing statements. One may now suspect that the class of event-describing statements is not closed under logical deduction, and that hence (L) is not acceptable for this reason. Similarly one may have doubts concerning (I) in the case of *predicative expressions* (e.g. 'the tallest man in this room') as compared with *genuine referring* expressions (e.g. 'John') in this condition (cf. Kim [1969] and Mackie [1974]). I mainly agree with such doubts and critical remarks concerning (I) and (L). But below I shall concentrate on different aspects of the situation.

However, some philosophers see no fault in these conditions. Thus, Davidson (1967), for one, considers (I) and (L) (or at least (L'), if that matters) and hence the resulting paradoxical theorem, acceptable. But as he on the other hand admits that causal statements are non-truthfunctional, he goes on to conclude that causal connections between events cannot be described by means of singular statements, but that rather it has to be done by means of a two-place causal predicate. For the reason mentioned earlier, and for some additional reasons related to partial causation (see e.g. Davidson [1967], pp. 698–699), I have accepted this conclusion even if I do not accept Davidson's main reason (i.e. that singular causal claims treated by means of *statements* are "shown" to be both truth-functional and non-truth-functional, which "result" then is considered absurd).

My approach to solving the ontological paradox expressed basically consists of the rejection of condition (C). But, taken in conjunction with some plausible semantical assumptions, it also follows that at least condition (L) will have to be rejected. (See Tuomela (1974a) for an argument concerning this and for a broader discussion concerning "ontological" paradoxes.)

We may notice here that our rejection of (I) and (L) means that at least the linguistic notion of eventhood is to be construed as an *intensional* (i.e. non-transparent and ultra-intensional) notion. This of course fits well together with the analysis of events sketched in Section I.

III. CAUSALITY AND DEDUCTIVE EXPLANATION

We shall below analyze causality in terms of nomological causal explanation, and this requires some modifications in (C). Now I consequently suggest that nomic backing is to be considered equivalent to explanatory backing. Speaking in general philosophical terms, I here understand explanation simply as an argument providing reasons for the truth of the explanandum (see e.g. Tuomela [1973], Chapter VII for my underlying intuitive ideas). Let us now write $C(e, e')$ for 'e was a cause of e' or 'e caused e''. Instead of (C), I then propose the following principle as our new starting point:

(C*) $C(e, e')$ is true only if there are singular statements D and D' such that D and D' describe e and e' respectively, and there is a causal law (or theory) S which jointly with D deductively explains D' (but D alone does not explain D').

Condition (C*) gives a necessary condition for the truth of the causal predicate $C(e, e')$. To give a sufficient condition as well we have to add some conditions. We shall soon discuss these additional conditions as well as what is required of the causal theory S as to its form and content (to e.g. avoid the apparent circularity in (C*)).

To clarify my new proposal (C*) I shall start by sketching the account of the logical (i.e. deductive-inferential) aspects of deductive explanation I have in mind for (C*).[2]

One central requirement for deductive explanations is that they should not be more circular than is "necessary." (Of course there must be some common content between the explanans and the explanandum; otherwise there could not be a deductive relationship between them.) As recent discussion on explanation has shown, the following kind of general condition has to be accepted: In an explanation the components of the explanans and the explanandum should be noncomparable. We say that two components or statements P and Q are *noncomparable* exactly when not $\vdash P \to Q$ and not $\vdash Q \to P$. (See e.g. Ackermann [1965] and Tuomela [1972] for a discussion leading to the acceptance of this general condition.) Actually our analysis of noncomparability needs some refinement, mainly because of the vagueness of the notion of a component. To accomplish this we use the technical notions of a sequence of truth-functional components of an explanans and of a set of ultimate sentential conjuncts of an explanans (cf. Ackermann and Stenner [1966]).

A sequence of statemental well formed formulas $\langle W_1, W_2, ..., W_n \rangle$ of a scientific language $\mathscr{L}$ is a *sequence of truth-functional components* of an explanans (theory) T if and only if T may be built up from the sequence by the formation rules of $\mathscr{L}$, such that each member of the sequence is used exactly once in the application of the rules in question. The W_i's are thus to be construed as tokens. The formation rules of $\mathscr{L}$ naturally have to be specified in order to see the exact meaning of the notion of a sequence of truth-functional components of a theory finitely axiomatized by a sentence T. A *set of ultimate sentential conjuncts Tc* of a sentence T is any set whose members are the well formed formulas of the longest sequence $\langle W_1, W_2, ..., W_n \rangle$ of truth functional components of T such that T and $W_1 \,\&\, W_2 \,\&\, ... \,\&\, W_n$ are logically equivalent. If T is a set of sentences then the set Tc of ultimate conjuncts of T is the union of the sets of ultimate sentential conjuncts of each member of T. We may notice here that although by definition the Tc-sets of two logically equivalent theories are logically equivalent they need not be the same. (Also notice that there are in principle no restrictions which would exclude e.g. the use of a causal or nomic implication in $\mathscr{L}$.)

Now we are ready to state a better version of the noncomparability requirement for a Tc of a theory T constituting an explanans (cf. Tuomela [1976]): For any Tc_i in the largest set of truth-functional components of T, Tc_i is noncomparable with the explanandum.

In addition to this condition we require that the explanans and the explanandum of an explanation are consistent, that the explanans logically implies the explanandum, and that the explanans contains some universal laws.

Finally, there is a nontrivial logical condition for our explanation relation, call it E, which condition guarantees that an explanans provides a proper amount of relevant information. This is condition (5) below. (The reader is referred to Tuomela [1972] and [1976] for a discussion of its acceptability.) Now we can state the basic logical conditions of our model of explanation (termed the *weak* DEL-*model* in Tuomela [1972]). Let T be a statement, Tc a set of ultimate sentential components of T (or actually a conjunction of components in the context $E(L, Tc)$), and L a statement to be explained. Then we say that the relation $E(L, Tc)$ satisfies the logico-inferential conditions of adequacy for the deductive explanation of (singular or general) scientific statements if and only if

(1) $\{L, Tc\}$ is consistent;

(2) $Tc \vdash L$;

(3) Tc contains at least some universal statements;

(4) for any Tc_i in the largest set of truth-functional components of T, Tc_i is noncomparable with L;

(5) it is not possible, without contradicting any of the previous conditions for explanation, to find sentences $S_i, ..., S_r$ ($r \geq 1$) at least some of which are essentially universal such that for some $Tc_j, ..., Tc_n$ ($n \geq 1$):
$Tc_j \& ... \& Tc_n \vdash_p S_i \& ... \& S_r$
not $S_i \& ... \& S_r \vdash Tc_j \& ... \& Tc_n$
$Tc_s \vdash L$,
where Tc_s is the result of the replacement of $Tc_j, ..., Tc_n$ by $S_i, ..., S_r$ in Tc, and '$\vdash_p$' means 'deducible by means of predicate logic but not by means of universal or existential instantiation only'.

Condition (5) is not quite unambiguously formulated as it stands. The reader is referred to Tuomela (1976) for its clarification (note especially conditions (V) and (Q) discussed in that paper) and also for an alternative interpretation of '$\vdash_p$'. See Tuomela (1972) for a detailed discussion of the formal properties of the notion of explanation that this model generates. Here it must suffice to make the following general remarks only.

In the above model of explanation an explanandum may have several explanantia differing in their quantificational strength (depth). On each quantificational level, however, only the weakest explanans-candidate qualifies. Our model thus generates an explanation-tree for each explanandum such that the explanantia in different branches may be incompatible whereas the explanantia within the same branches are compatible and increasingly stronger.[3]

More exactly the weak DEL-model has the following central logical properties (see Tuomela [1972]):

(a) $E(L, Tc)$ is not reflexive.

(b) $E(L, Tc)$ is not symmetric.

(c) $E(L, Tc)$ is not transitive.

(d) If $E(L, Tc)$ and if, for some Tc', $\vdash Tc' \to Tc$ (assuming not $\vdash_p Tc' \to Tc$), then $E(L, Tc')$, provided every $Tc' \in Tc$ is noncomparable with L.

(e) $E(L, Tc)$ is not invariant with respect to the substitution of either materially or logically equivalent explanantia nor explananda.

(f) $E(L, Tc)$ and if, for some T' such that $\vdash T \equiv T'$, T and T' possess identical sets of ultimate sentential components (i.e. $Tc = Tc'$), then $E(L, Tc')$.

(g) If $E(L, Tc)$ and for some L', $\vdash L \equiv L'$, then $E(L', Tc)$, provided that for all Tc_i in Tc, Tc_i and L' are noncomparable.

(h) If $E(L, Tc)$ and $E(L, Tc')$, then it is possible that Tc and Tc' (and hence the corresponding theories T and T') are mutually (logically) incompatible.

What interests us especially much in this context is the property (e). The lack of linguistic invariance exhibited by it shows or expresses the fact that explanation is a *pragmatic* notion: How you *state* your deductive argument may make a great difference concerning the explanatory value of the argument.

Whether an argument is an explanation should *prima facie* be decided on its own right, so to speak, and not indirectly on the basis of its having been somehow derived from another acceptable explanatory argument.

This is the case in our model of explanation. If an argument satisfies our logical arguments it becomes a potential explanation. To qualify as a materially valid explanation still some other requirements have to be imposed, as is well known. For instance, the generalization (of theory) used in the explanans should be lawlike, which we take to entail that it contains only predicates expressing natural kinds. Furthermore, the explanāns and the explanandum should be acceptable as true, etc. Some of these further demands will concern the "pragmatic" features of the explainer and his scientific community.

On the basis of some features, especially property (e) above, of our model it can be argued that the principle (L) is not acceptable. In addition, some further assumptions concerning the semantical relationships between event-descriptions and events described are needed. That argument is given in Tuomela (1974a) to which the reader is referred. In that paper also the case of composite events and the possibility for paradox free event descriptions in the light of our model for deductive explanation are considered. However, our main concern in this paper is to discuss some other aspects of causality to which we now turn.

IV. THE BASIC COMPONENTS OF CAUSALITY

Our scheme (C*) may be taken to express the main idea of the backing law theory. Still it leaves many aspects of this view unclarified. In addition, (C*) totally ignores some central features of causality and fails to give a sufficient condition for the truth of the singular causal predicate $C(e, e')$. Let us now try to improve our analysis in some respects towards a fuller backing law account of causality.

As we know, causal phrases like 'cause', 'effect', 'consequence', 'result', 'bringing about', 'producing' 'nomic necessitation', etc. are used in a great variety of ways both in ordinary and in scientific discourse. Therefore one may doubt whether one can say many things which are universally applicable even to all notions of *scientific* causation (which is our topic). For instance, accepting the backing law idea, it seems that, depending on science, these backing laws may vary greatly as to their form and not only as to their substantive content. We shall, however, try to find some features which are common to causation, as discussed within science.

Let us thus ask the following question: What does it mean to say that a singular event e was a cause of another singular event e'? We can broadly answer this question by saying that e *produced* e' (in those circumstances). However, producing is also a causal notion, and thus it may not seem to give much further elucidation of the notion of causation. But one may consider this producing-relationship also from a, so to speak, generic point of view. Thus the producing-relationship exhibited by $C(e, e')$ really involves a nomic correlation between two *generic* events which e and e' exemplify. Let us assume that e exemplifies the property P and e' the property Q. (P and Q do not necessarily have to be those aspect properties of e and e' with which we originally characterized these events.) So what is really involved here is that events of the kind P produce events of the kind Q. As we shall mostly be concerned with *deterministic* causation here this means that under normal conditions every singular event of the kind P produces an event of the kind Q.

Causal producing is thus primarily a relationship concerned with *kinds* of events. However, I think that at least when P and Q are aspect properties of e and e' the nomic necessity involved in this generic production and necessitation is exhibited or "seen" also on the singular level between e and e'. Thus $C(e, e')$ also may have or acquire some reportive (evidential) uses. Scientists so to speak internalize the rules of usage of the predicate $C(e, e')$, that is, the right hand side of (C*) or analogous account, and start applying it in their experiments as describing what is going on in singular event sequences. Thus the originally (i.e. from a conceptual point of view) purely metalinguistic predicate becomes a kind of theoretical predicate with possibly some direct evidential uses. Able and reliable scientists working within a certain scientific community learn to use the predicate $C(e, e')$ closely analogously to how they learn to use such theoretical predicates as 'electron', 'radio star', or 'subconscious wish'. Thus they also learn to make statements like "Giving subject *S* a certain kind of electric shock caused the trembling of his left hand."

It can be argued on the basis of examples like this that even singular causal statements can be taken to express a kind of necessity (see Mackie [1974], ch. 8). We shall later return into the "necessity" in singular causal processes. Let me here only notice two consequences of accepting this view. First, even if the predicate $C(e, e')$ can sometimes be used for direct evidential reporting, it is not an observational predicate (in any strict sense of 'observational'; see Tuomela [1973], ch. I). Instead it is highly theory-laden in the obvious sense of depending on a backing law or theory. Secondly, the acceptance of some kind of objective necessity to be exemplified even in *singular* causal processes clearly means the rejection of a Humean (regularity) account of causality as Hume did not accept any objective necessity either on the level of singular or generic events.

Any scientific theorizing seems to involve a "modelling" of reality in a way which involves idealization, schematization and some distortion of reality. The investigation of causal connections between objective events and processes serves as a good example of this. When a scientist studies a system's behavior he must somehow distinguish between the relevant and irrelevant aspects and properties of the objective situation. To use a commonly accepted phrase, he must construct a *causal field*. Such a causal field selects some features as possible causes and effects, whereas other aspects or properties are only "standing conditions" (preconditions) or are left out of consideration almost completely by including them into *ceteris paribus* conditions. I thus take the construction of a causal field to involve a pragmatic element which involves a departure from the idea of a purely objective notion of productive causation (in something like the sense of Mackie [1974]).

Another pragmatic element in the backing theory analysis comes into the picture in the formulation of theories. For the backing theories often employ predicates (concepts) which do not quite correctly represent the objective properties in the causal field; and these theories may be wrong in how they express relationships between the properties represented by these predicates.

As the backing theory account justifies the truth of a singular causal claim by requiring an appropriate backing theory or "theory in work" to exist several difficulties are seen to be involved. Our condition (C*) quantifies existentially over theories, i.e. over the set of all possible theories containing predicates for the properties of the underlying causal field. But this set of theories need not be a well defined set at all, it seems. It cannot *a priori* be limited in any way for otherwise we may lose the "law of working" which represents the underlying objective causal producing relation. It should also be noticed that even if our model of (actual) explanation, and hence our (C*), requires the backing theory to be *true*, there may still be several theories characterizing the underlying causal process equally well and considered true with equally strong reasons. How can one choose the "correct" law or theory of working among these? The difficulties are still the more difficult as we cannot require in our (C*) that the backing theory somehow be algorithmically constructible. (Notice that, *ceteris paribus*, to epistemically justify a singular causal claim it suffices to have enough *evidence* for the claim that there is a backing theory S.) We cannot either in principle require that the backing theory be finitely axiomatizable, although we shall, for simplicity, below proceed as if it were.

I shall not here try to solve the raised difficulties point by point. Let it suffice to make the following two general remarks. First, as our discussion has shown, even if some of the above problems could be solved it is very likely that some distortion will always remain in the knowledge our backing theory gives of the underlying causal process. (Perhaps in a mythical Peircean limiting science all these distortions disappear and producing causes and explanatory causes become identical. But to note this is about as helpful as to say that in the long run we are all dead.) Secondly, the above difficulties affect more or less any current account of explanation and confirmation (even if such an approach would not *explicitly* quantify over theories), so that we are all in that same big boat.

Our schema (C*) for singular causation is rather incomplete as it states only some central necessary conditions for the truth of a singular claim. To make our analysis look more complete we shall now isolate some other aspects of causality.

First, we are in this paper concerned with *direct* (as opposed to indirect or mediate) causation. I take this directness to be analyzable primarily in terms of spatiotemporal contiguity. Here, however, no attempt to analyze the content of this factor can be made (see e.g. Kim [1973a] for a good recent discussion of this Humean aspect of causality).

Secondly, causes and effects are clearly *asymmetric*. But what is involved in this asymmetry? Time does not seem to be a definite condition for causal priority, as recent discussions have shown (see v. Wright [1971] and Mackie [1974]). Instead it would seem that causes somehow fix effects in a way effects do not fix the occurrences of cause-events. Speaking in anthropomorphic terms, at least when causes are either sufficient or necessary for events, by means of a cause-event one can either produce or, in the latter case, prevent an effect-event. It seems to us that a more objectivistic analysis in terms of fixity conditions may give a satisfactory account of causal priority. In this paper, in any case, we must omit a deeper analysis of this aspect of causality as well (except for a later comment on *explanatory* asymmetry in Section VI).

Let us next consider a simple example. There was a fire in a house and an inspector looks for its cause. He notices that the fire had something to do with a defect in the electrical system: a short-circuit somewhere was at least a necessary condition for the occurrence of the fire. Other necessary conditions were that there was current in the wire and that there was inflammable material. Which of these, if any, is the cause of the fire?

It may be that it is true to say of the short-circuit, but not of the other two factors, that had the fire not occurred then the short-circuit had not occurred. If so, the short-circuit was in the above sense a counterfactually sufficient condition of the fire whereas the other two factors are merely necessary conditions. This might be taken to justify to call the short-circuit *the* cause, *ceteris paribus*. The short-circuit represents a change in the causal field, whereas the other two factors are *standing conditions* (preconditions). This indicates that in a search for *the* cause it is of importance first to distinguish between potential causal factors and standing conditions.

However, the distinction between *the* cause (or *the total* cause) and *a* cause (or standing condition, either) cannot always be made in terms of the above counterfactual sufficient conditionship criterion. Clearly, at least non-deterministic causes (e.g. radium atom's decay causing the occurrence of a radon atom) do not satisfy the above criterion of sufficiency. In addition, there seem to be more homely examples within deterministic causation in which causes do not satisfy this kind of strong sufficiency condition (see Section V).

An attempt to distinguish *the* cause from other causal determinants seems to essentially make reference to scientists' interests and "values" (involved in selecting "comparison situations" for "effect situations" and the like). Their consideration here would take us too far aside from our main concern, which is the analysis of direct productive causal determinants of events.

V. A BACKING LAW ACCOUNT OF SINGULAR CAUSATION

After this discussion concerning the isolation of the central features involved in singular causation we can complement our previous scheme (C*) by mentioning some new necessary conditions so that we obtain a set of conditions which are individually necessary and jointly sufficient for the truth of our singular causal predicate C(e, e'). The rest of our discussion will then be devoted to the elaboration of some special features of our scheme.

As said earlier, I take 'e caused e'' to mean the same as 'e produced e''. An analysis of producing again brings into the picture the backing law. Thus a scheme like our earlier (C*) and our (C**) below can be regarded as a kind of meaning-explication, in terms of truth conditions, of C(e, e'). However, as was pointed out, the question is primarily of justifying the possible uses of the predicate C(e, e'), and such a justification also involves epistemological and metaphysical considerations (e.g. considerations of truth lawlikeness, and necessity) which transcend the boundaries of a strict meaning analysis.

As I will be concerned with causality as a kind of determination, I will not be concerned with anything like probabilistic singular causation. But I do not assume the validity of any universal principle of determination. Nor do I have to assume even that the backing law is deterministic, as the determination may be effected in part through some auxiliary assumptions and contextual information, and not merely through the backing law.

We can now summarize our above discussion in terms of the following amended scheme where we let C(e, e') read 'e is a deterministic direct cause of e'':

(C**) The statement C(e, e') is true if and only if there are singular statements D and D' such that D and D' describe the occurrence of e and e', respectively, and there is an applied causal theory S such that
(1) S jointly with D gives an actual explanation of D' in the sense of the weak DEL-model, viz. there is a Tc for $T = = S \& D$ such that $E(D', Tc)$ in the sense of the weak DEL-model of explanation;

(2) e is causally prior to e';
(3) e is spatiotemporally contiguous with e'.

There are several features in our scheme (C**) which need clarification and elaboration. To start, we assume that D and D' are true statements describing the occurrence of e and e', respectively. This is to be understood in the strong sense that D and D' contain rigid designators for these events. (Notice the difficulty this entails for creating a non-circular causal theory of reference.) Our constitutive statements discussed in Section I serve as examples for this kind of singular statements. The existence of such true statements D and D' guarantees the occurrence of the cause-event e and the effect-event e'. Thus the truth of a singular causal claim entails the occurrence of both the cause-event and the effect-event, which of course should be the case.

Before discussing the nomic nature of the backing law S we still have to say something about the requirement of explanation (condition (1)). First, the explanans-theory S is said to be an *applied* theory. This means roughly that, in addition to a nomic core theory, S may contain general auxiliary assumptions as well as singular statements, which take into account contextual features of the explanatory situation (cf. Tuomela [1973]). The applied theory S *cum* D then explains D' in the sense of the weak DEL-model. (Notice that in this model of explanation the statement D cannot alone explain D', whence such a requirement would be superfluous.) It should be emphasized that the conditions (1)–(5) were taken to explicate those aspects of scientific explanation which have to do with explanation as a special kind of logical inference (argument). What we need here is a "full" account of actual explanation. Therefore, I propose that the relation E of explanation really should be taken to satisfy all the conditions of adequacy of Tuomela (1972) relevant to singular explanation. One such central additional condition is that the explanans and the explanandum should be accepted as *true*.

At this point at the latest the reader may voice the objection that the analysis given by (C**) is circular as the word 'causal' occurs even twice in the analysans. But this objection can be avoided (to a great extent at least). First we recall that causal priority (condition (2)) was claimed to be analyzable in terms of objective fixity conditions, and hence without reference to causal notions. But the word 'causal' also occurs as an attribute of the backing theory S. Is this not fatal for our account? It should be noticed to begin with that the backing theory does not contain the predicate $C(e, e')$ nor its "counterpart" in the language of the backing theory, so that no strong circularity in this sense obtains. Furthermore, we do not attempt to give a reductionist analysis of causality at all. That would be a philosophical mistake. Notice also that a causal theory is something more than merely a *nomic* theory, for there are non-causal nomic theories (e.g. "The formation of the respiratory system in a human embryo always succeeds the formation of the circulatory system"). Thus causal laws cannot be analyzed entirely in terms of nomicity.

Nagel gives as his paradigm example of a causal law: 'Whenever a spark passes through a mixture of hydrogen and oxygen gas, the gases disappear and water is formed' (Nagel [1961], p. 74). The defining conditions for such a causal law are in Nagel's analysis the following four. (1) The causal law expresses an invariable and uniform connection between cause-events and effect-events. The cause-events and effect-events are (2) spatially and (3) temporally contiguous. (4) The generic causal relation is asymmetric. But it seems to me unnecessary to require that a backing causal law expresses spatial and temporal contiguity. Even Nagel's own example of a causal law does not (at least explicitly) take into account these features. In our (C**) these features are included in those conditions of causality which directly characterize singular cause-effect pairs. In other words, spatiotemporal contiguity has to be required only on the singular but not necessarily on the generic level.

I shall not try here to present a uniform analysis for causal theory applicable to all sciences, for I believe there are hardly any such common articulated features of causal laws except for their nomicity. What a causal theory looks like depends on the particular science in question, and it may also depend on the paradigmatic values and beliefs held by the scientists of a given community.

What does necessary connection between generic events amount to? As any hard and difficult philosophical problem this matter can and has to be approached from various points of view. As we know the problem indeed has interesting logical, semantical, ontological, metaphysical, epistemological and pragmatic aspects. I can here only briefly touch on some of the involved features, and I do so only in order to put our analysis of causation in a proper philosophical perspective.

One can claim that the necessity exemplified by pure *causal* "laws of working" is the *persistence* and (qualitative and structural) *continuity* to be found in *objective* causal processes (cf. Mackie [1974]). For instance, the uniform movement of a single particle, free from interference, would be an example of this. The above features clearly go beyond mere regular connection of events (in some Humean sense) and they may also be regarded as transcending "mere" nomicity. Furthermore, such persistence and continuity can be claimed to be objectively present in any *singular* cause-effect connection as well (contrary to Hume).

I would like to distinguish between the *descriptive* and the *prescriptive* sides of lawlikeness or lawhood. The descriptive aspects of a law or theory have to do with a statement or a piece of linguistic discourse describing the world (or a certain idealized model of the world). The prescriptive aspect is concerned with the theory-users linguistic (and other) behavior (especially argumentation) and thus with the so-called pragmatic aspects of science.

A scientific law may be considered to nomically connect two or more concepts (on the conceptual level) and thus to express a nomic connection holding between real properties or generic events. Thus we may come to express nomic laws by means of modal locutions like, e.g. 'Necessarily, if something a has property P then a has property Q'. What do such locutions mean? Modal logicians have recently been busy investigating causal necessity and have done much in clarifying the semantics of such necessity statements, and we shall discuss it in more detail later.

One widely accepted criterion for nomicity is the support of "corresponding" counterfactual statements (e.g. 'If a had P, when in fact it does not, it would have Q'). This is certainly an important feature belonging to nomic theories. However, some remarks are due here. First, as soon as we have decided to call a statement nomically necessary it will automatically support counterfactual statements (at least) if e.g. Lewis' (1973a) or a related analysis is accepted. So on that analysis at least there is a conceptual connection between these two features. The problem then becomes in finding other criteria for calling a statement nomic. Thus I think the meaning-problem is not so central in the case of lawlike statements, but rather the important and interesting problems lie in finding epistemological and metaphysical criteria for lawhood.

One may now be interested in either the descriptive or the prescriptive dimensions of lawlikeness in looking for such criteria. As far as one is interested in laws merely as general descriptions of the world one should in principle be able to get along without modal necessities (be they *de dicto* or *de re*) and hence without tools for expressing them. I have elsewhere defended a view which can be summarized as follows: A statement is a law if and only if it is genuinely corroborable and corroborated both *empirically* and *theoretically* (see Niiniluoto and Tuomela (1973), ch. 11). This view comes relatively close to Popper's view with two important modifications. For Popper a statement is a law if and only if it is deducible from a statement function which is satisfied in all worlds which differ from our world at most with respect to initial conditions. If we (a) use the concept of a pure law of working (cf. Mackie [1974]) instead of Popper's too general notion of a statement function and (b) substitute 'deducible or inducible' for 'deducible' we seem to get an acceptable idea of a nomic generalization. (Inducibility is here to be understood in the sense of Niiniluoto and Tuomela [1973].) Notice also that our notion of lawhood is further qualified so that it does not satisfy any strong logical principles of transformation (cf. Section III). As a consequence none of the "usual" arguments requiring the use of modal implications in laws apply. We shall still below make some specific remarks concerning the logical form of a causal law to get an idea of the logical aspects of lawhood.

As to the prescriptive side of causal laws, we may say that a law entitles the user of the law to certain expectations, if not directly to some actions. Thus if we have a law which says that property P necessitates property Q, a user of the law is entitled to expectation statements like 'If a were P then one is entitled to expect that a would be Q'. One can also put this in explicitly deontic terms. 'If a were P one ought to expect that a would be Q'.

This kind of modal force of laws essentially comes from the role of laws in *backing singular explanatory inferences*. We shall technically implement this in terms of the conditional $\triangleright\!\!\!\rightarrow$ in Section VI below. Our $\triangleright\!\!\!\rightarrow$ can be interpreted as a conditional obligation operator analogously with the strictly variable conditional $\square\!\!\!\rightarrow$ of Lewis (1973a).

Next we make a few remarks concerning the logical form of causal laws from the extensionalist or descriptivist point of view. I think that rather little can be said about the problem in general, i.e. without relativization to a specific science or type of theorizing. However, some generality can be claimed for the following two types of deterministic causal laws formalized within standard first-order logic (cf. Kim [1973a]):

(i) $(x)(y)(P(x) \& R(x,y) \rightarrow Q(y))$
(ii) $(x)(P(x) \rightarrow (Ey)(Q(y) \& R(x,y)))$.

Here the predicates 'P' and 'Q' are complex interpreted predicates of singular events (with a "pragmatic commentary"). They include predicates for initial conditions of a sufficiently abstract level. The predicate 'R' is a (universal) pairing or coordinating predicate. It is supposed to correctly pair the instances of the generic properties (or events) P and Q. The following example shows the need for such pairing. If two adjacent guns are fired simultaneously and if these two events cause two deaths, how can the causes and effects be paired so that a cause becomes correlated with *its* effect and an effect with *its* cause. It often seems plausible to require of the pairing relation that it makes causes and effects unique (*vis-à-vis* the other element in the pairing relation).

The main difference between the laws (i) and (ii) is that (ii) guarantees the existence of an effect for every cause, whereas (i) does not. Both of these laws represent sufficient causes only. The symbol '$\rightarrow$' can here represent either a material implication or a more complex connection, like in the Simon-Kron theory to be discussed below.

We now turn to two examples of causal theories, especially relevant for the social sciences, both of which take into account some necessity and sufficiency aspect of causation, though in different senses. We now turn to these examples.

Mackie has analyzed singular causes in terms of generic necessary and sufficient conditions (Mackie [1965], cf. also Mackie [1974]). While the full generic cause of a property or generic event P may be its necessary and sufficient condition, what we normally call "a cause" is not. Mackie's analysis is based on the view that a cause is an *insufficient* but *necessary* part of a condition which is itself unnecessary but sufficient for the result (see Mackie [1965], p. 245). Such a cause is called an INUS-condition.

Mackie's approach is thus concerned with the form of a causal theory as his conditions belong to generic events (broadly understood). His idea seems useful for explaining causation in some ordinary life situations where knowledge (and beliefs) are formulated in rough terms by means of necessary and sufficient conditions. E.g. a cause for the occurrence of a car accident or for a fire or perhaps even for the declaration of war by a nation might conceivably be given in these terms.

Mackie explicates his idea of an INUS-condition in terms of the disjunction of all the minimal sufficient conditions of a given (generic) event constituting a necessary and sufficient condition of it: "A is an INUS condition of a result P if and only if for some X and for some Y (AX or Y) is a necessary and sufficient condition of P, but A is not a sufficient condition of P and X is not a sufficient condition of P" (Mackie [1965], p. 246). Here the term X stands for the conjunction of terms which together with A constitute a minimal sufficient condition of P. The term Y represents the disjunction of other minimal sufficient conditions. For example, P might represent the generic event of a fire taking place, A the occurrence of a short-circuit at the place in question (necessary condition for P), and X other conditions which jointly with A make the conjunction AX sufficient for P. (Of course Y contains other additional INUS-conditions for P.)

There are a couple of central remarks that can be added here. First, the INUS-theory represented above quantifies existentially over the predicates X and Y. In other words, it is a theory in the form of a Ramsey sentence. If some predicate constants are found to substitute for X and Y we get an "ordinary" first-order theory (see, e.g. Tuomela [1973], ch. III on Ramsey sentences and theoretical concepts). In any case, a backing law or theory can very well be only a Ramsey sentence.

Secondly, if the backing law in Mackie's analysis really is nomological it will sustain the following counterfactuals:

(i)   For any x, if x had (AX or Y), when in fact it does not, there would have been a y which is P.
(ii) For any x, if x did not have P, even if it in fact does have P, there would not have been any y which has (AX or Y).

However, the notion of singular cause we are analyzing does not prescribe that the backing theory must give either sufficient or necessary conditions for the generic effect-event in the above manner. But our analysis comes closer to requiring that certain singular conditionals should be true. Using our earlier symbols D and D' (which here correspond to 'e has A' and 'e' has P') we may consider the following three singular conditionals:

(a) If D, then D', given the circumstances.
(b) If $\sim D$, then $\sim D'$, given the circumstances.
(c) If $\sim D'$, then $\sim D$, given the circumstances.

The circumstances are here taken to include statements concerning the occurrence or non-occurrence of all the factors in the causal field presupposed. Obviously it includes all the factors the theory speaks about (e.g. X and Y) but also contextual factors. We shall later discuss the acceptability of these and related weaker conditionals. What interests us in this connection is that the conditionals (i), (ii), and (a)–(c), if true, provide a roundabout and indirect way of characterizing the backing theory. For instance, (i) and (ii) may be true also for more complicated theories than considered above, and they may in some cases provide our only "information" about the backing theory. Similarly, (a)–(c) give still a much more indirect way of characterizing the backing theory. For instance, in many cases these conditionals may be true just in the case of Mackie's INUS-theory considered above.

The second example I wish to take up briefly originally comes from economics. Simon (1953) has presented an analysis of causality in terms of *identifiability* of certain *variables* by means of certain other variables. These variables represent conditions, and hence generic events in our terminology. We might be simultaneously interested in the causal relationship between variables such as the growth and price of wheat, and the effects of rain and fertilizers on them, etc. One may try to analyze all this within first-order logic and develop a full blown logical theory of causality. This is what Kron (1975) has done in an interesting way on the basis of Simon's ideas. It is not possible to present Kron's important results here for limitations in scope. Let me just say a few words about it in order to have an idea of how our explanatory causal theory S might look like in some of the social sciences.

The basic idea is to start with a theory S which contains a "*linear*" subtheory, say K. This subtheory consists (in the finite case) of formulas $K_1, K_2, \ldots$ which are in a certain sense linearly ordered and, in that sense, also causally ordered. The principle of ordering is the identifiability of certain variables in terms of certain other variables, which ordering carries over to the formulas $K_1, K_2, \ldots$ in which these variables occur and also to the models in which these formulas are satisfied. For instance let $K = \{A(v_1, v_2), B(v_2, v_3), C(v_1, v_3, v_4)\}$. Here $v_1, \ldots, v_4$ are just the variables in question, and they are essentially ordinary individual variables of standard first-order logic. Now, if K is a causally ordered set, we might have the ordering on the basis of the exogenous variable v_2. Then we would have

$$S \vdash (v_2) \, (E!v_1) \, (E!v_3) \, (E!v_4) \, A(v_1, v_2) \, \& \, B(v_2, v_3) \, \& \\ \& \, C(v_1, v_3, v_4).$$

Thus for any value of the variable v_2 the values of v_1, v_3, v_4 are uniquely determined, i.e. identified.

Generally speaking, the variables v_i are the entities between which causality holds, and this is justified by reference to the theory S which contains the causally ordered sequence K. Thus we have causality between generic events. The principle of identifiability between variables represents a kind of pairing relation between causes and effects (cf. our discussion earlier in this paper). Notice also that Simon's idea of causality represents controlability of outputs by means of inputs in systems-theoretic terminology. Hence it can easily be given an *interventionistic* interpretation.

Singular causes are obtained by considering the satisfaction of a K_n in some model M_n and its successor formula K_{n+1} in a M_{n+1} (K_n and K_{n+1} may contain appropriate names for the objects in the models). Such a singular "effect" K_{n+1} is derivable from S and the singular "cause" K_n.

As this Kron-Simon theory of causality is open to "ontological" paradoxes such as those discussed in Section I. The obvious suggestion then is to complement it with the restrictions expressed by our scheme (C**).[4]

VI. CONDITIONAL ANALYSES OF SINGULAR CAUSATION

In last section we noticed that the backing law account of singular scientific causation is closely connected with counterfactuals in the sense that causal laws seem to support conditionals (including counterfactuals) such as (i), (ii), (a), (b), and that also singular causal claims support certain kinds of singular conditionals, such as (a) and (b). Indeed the analysis of causal statements by means of conditionals has been claimed to be the central feature of causality. Thus, for instance, Lyon (1967) and Lewis (1973) base their analysis of causality entirely on conditionals (see also Mackie [1974], ch. 2). (These authors seem to mean by 'analysis' only the analysis of the meaning of the concept of a cause.) As I find Lewis' account the best available in this line of thinking, I will mostly concentrate on discussing his results and claims.

Lewis combines his analysis of singular causation with a strong criticism of (a version of) the "regularity" account which he takes to cover also the backing law approach, it seems. I shall below try to constructively defend the backing law account against Lewis' critical remarks. In addition some criticisms against Lewis' approach will be made.

In Lewis' approach the term 'counterfactual' is taken to cover any conditionals irrespective of the truth values of their antecedents and consequents. Let us consider any two propositions A and C and the counterfactual "If it were the case that A, then it would be the case that C." In his book, *Counterfactuals* (1973a), Lewis has given a possible world analysis of such counterfactuals. He assumes that possible worlds can be compared in terms of a *comparative* overall *similarity relation* (which he takes as an unanalyzed primitive concept). In terms of this concept we can now give a semantical analysis of such counterfactuals as that above. If we symbolize it by $A \square\!\!\rightarrow C$ ($\square\!\!\rightarrow$ is a "variably strict conditional") we can give truth conditions for this counterfactual operator as follows: $A \square\!\!\rightarrow C$ is true (at world w) if and only if either (1) there are no possible worlds in which A is true (in which case $A \square\!\!\rightarrow C$ is vacuous), or (2) some world where A is true and in which C holds is closer (to w) than is any world in which A is true but where C does not hold (see Lewis [1973b], p. 560). Thus we can say that a counterfactual is nonvacuously true just in case it takes less of a departure from the actual world to make the consequent true along with the antecedent than it does to make the antecedent true without the consequent. (For other characterizations of the truth conditions of $\square\!\!\rightarrow$ see Lewis [1973a].) I shall not here quarrel about whether counterfactuals can be regarded as true or false in the same sense as indicatives are true or false. Instead I shall in this paper speak of the truth of counterfactuals in the inclusive Sellarsian sense as 'assertability according to the semantic and material rules of the language used'.

The conditional $\square\!\!\rightarrow$ has the following central properties. First, $A \square\!\!\rightarrow C$

implies the material conditional $A \to C$. Secondly, A and C jointly imply $A \square\!\!\!\to C$. (Note the symmetry with respect to the actual world!) Thirdly, it is worth emphasizing that $A \square\!\!\!\to C$ does *not* entail the contraposition $\sim C \square\!\!\!\to \sim A$, although one is entitled to infer $\sim A$ from the two premises $A \square\!\!\!\to C$ and $\sim C$.

Now we can give Lewis' definitions for some basic causal notions. As we shall be interested only in causation between two distinct singular events, say c and e, we can explain Lewis' notions as follows. Let now $O(c)$ and $O(e)$ be propositions expressing the occurrence of c and e, respectively. Analogously $\sim O(c)$ and $\sim O(e)$ express that c and e, respectively, did not occur. Then we say that e *depends causally* on c if and only if the counterfactuals $O(c) \square\!\!\!\to O(e)$ and $\sim O(c) \square\!\!\!\to \sim O(e)$ are true, that is, if and only if the family $O(e)$, $\sim O(e)$ *depends counterfactually* (in Lewis' terminology) on the family $O(c)$, $\sim O(c)$. Lewis thinks that causation is always transitive. Hence the above notion of causal dependence will not quite do as an explicate of 'a cause', for the former notion clearly is not transitive. Let thus c, d, e, ... be a finite sequence of actual singular events such that d depends causally on c, e on d, and so on throughout. Then this sequence is called a *causal chain*. Now Lewis defines a singular event c to be a *cause* of another singular event e if and only if there exists a causal chain leading from c to e. (See Lewis [1973b], pp. 561–563 for a fuller discussion.)

Before commenting on Lewis' (1973b) analysis of causation let us summarize his basic criticism against the backing law account. First we notice that Lewis' notion of a regularity (or backing law) account is essentially that formulated by our analysis (C) of Section II (provided that S is regarded as an *applied* theory in our earlier sense and that no singular part of S entails D').

Lewis now goes on to define a notion of nomic dependence: A family $C_1, C_2, \ldots$ of pairwise exclusive propositions *depends nomically* on a family $A_1, A_2, \ldots$ of pairwise exclusive propositions if and only if there is an applied theory S such that S implies (but that no singular part of S alone implies) all the material conditionals $A_1 \to C_1, A_2 \to C_2, \ldots$ between the corresponding propositions in the two families. Here we also say that nomic dependence holds *in virtue of S*.

Next Lewis defines that a proposition B is *counterfactually independent* of the family $A_1, A_2, \ldots$ of alternatives if and only if B would hold no matter which of the A's were true – that is, if and only if the counterfactuals $A_1 \square\!\!\rightarrow B$, $A_2 \square\!\!\rightarrow B, \ldots$ all hold.

Lewis then makes two observations directed against the backing law account (see Lewis [1973b], pp. 564). The first is (essentially) this:

(1)   If the C's depend nomically on the A's in virtue of S and if all the conjuncts (parts) of S are counterfactually independent of the A's, then it follows that the C's depend counterfactually on the A's.

Lewis concedes that often, perhaps always, counterfactual dependencies may be nomically explained in the sense of (1). However he emphasizes that the requirement of counterfactual independence is still indispensable here.

Secondly, Lewis shows that nomic dependence in his sense is in a sense reversible, whereas counterfactual dependence is not:

(2)   If the family $C_1, C_2, \ldots$ depends nomically on the family $A_1, A_2, \ldots$ in virtue of S, then also $A_1, A_2 \ldots$ depends nomically on the family $A \,\&\, C_1, A \,\&\, C_2, \ldots$, in virtue of S, where A is the disjunction $A_1 \vee A_2 \vee \ldots$.

Let us now proceed to an evaluation of Lewis' counterfactual account of causation. I shall start by some relatively general critical remarks directed against the philosophical foundations of this account and then go on to defend the backing law account.

While I shall not in this context go into a discussion of the clarity and explanatory value of the notion of a set of possible worlds any more than of the semantical adequacy of $\square\!\!\rightarrow$ as elucidating the meaning of counterfactual statements in general, I would like to make a remark on Lewis' concept of *similarity*. (Recall that his account of the truth conditions and thus the meaning of $\square\!\!\rightarrow$ depends on that notion.) My criticism is that I find it hard to understand similarity (at least in causal contexts) without at least partially elucidating it in terms of nomic or causal necessity. Obviously, if my claim really is acceptable then Lewis' attack on the backing law account cannot even get off the ground.

To illustrate what I have in mind consider our earlier example in which the occurrence of a short-circuit in a house caused a fire. Let s and f represent these singular events. As s and f occurred $O(s)$ and $O(f)$ (to use Lewis' terminology) are true and hence is $O(s) \square\!\!\rightarrow O(f)$. But how about the truth of $\sim O(s) \square\!\!\rightarrow \sim O(f)$? This statement is nonvacuously true just in case it takes less of a departure from the actual situation to make $\sim O(f)$ true, given the truth of $\sim O(s)$, than it does to make $\sim O(f)$ false, given the truth of $\sim O(s)$. Now how do we account for the fact that $\sim O(s)$ goes together with $\sim O(f)$ rather than with $O(f)$ and hence that the world representing the former case is closer to our actual world than the world represented by the latter construction? The obvious answer to me is that this relies on a complex nomological connection between short-circuits (with some other relevant factors) and fires. In the situation in question the other possible causal candidates have been blocked somehow, we think, such that the short-circuit is the sole causal candidate left. If an account like this is acceptable similarity becomes dependent on nomicity (rather than *vice versa*). Indeed, it seems to me almost impossible to get an elucidation of objective similarity in cases of causation except along the lines sketched above. (To declare similarity an unanalyzable and unelucidable similarity would of course be question-begging.)

Let us now consider Lewis' definition of causal dependence and cause. First, I do not accept that (indirect) causes are transitive as Lewis requires. However, I shall not here argue for this claim (although it is an obvious consequence of my account) as I will be interested only in the kind of direct causation Lewis' notion of causal dependence explicates. Lewis identifies causal dependence with counterfactual dependence, as we saw. But his analysis is both too wide and too narrow. It is too wide, for instance, because there are counterfactual dependencies which are not causal. If we accept that only events which have occurred can qualify as causes and effects, then the truth of $\sim O(c) \square\!\!\rightarrow \sim O(e)$ entails under Lewis' analysis not only that e depends causally on c (and that c thus is a direct cause of e) but also that c also is a cause (in Lewis' full sense) of e. But consider the following three statements the first of which is true on semantical and the other two on factual grounds:

(a)   If John had not gotten married before or in 1974 he would not have divorced in 1974.

(b) If the sun would not have risen yesterday morning it would not have set last night.

(c) If the earth had not cooled a few billion years ago then I would not have written this paper.

Thus (a) shows that John's getting married is a cause of his divorce, (b) that sunrise is a cause of sunset, and finally (c) that the earth's cooling caused my writing this paper. But to make the notion of causation this broad is to deprive it of any scientific interest. (In Kim [1973b] some other additional counter-examples to Lewis' analysis can be found.) We shall later show that Lewis' counterfactual analysis is too wide for other reasons as well.

On the other hand, one may claim that Lewis' account is too strict. Thus, in the case of causal prediction, given a singular event c (a cause) this event is supposed to cause another specific singular event e (rather than, for instance, an unspecified singular event exemplifying a generic event). But this is a very strong requirement. Why should my particular punching you in the nose be required *ex ante* to cause your nose bleeding (etc.) in a certain specific way rather than in another, very closely similar way? Thus, if the aspect property of e is E, should we not satisfy ourselves by requiring only that there is some further unspecified singular event y in E such that c causes y (without requiring y to be e or any other specified member of E)? An analogous remark can be made for cause-events, given a certain effect-event.

In terms of Lewis' conditions we may also say this. If c and e have occurred why should we here require the truth of $\sim O(c) \square\!\!\rightarrow \sim O(e)$ rather than something more general such as e.g. the truth of 'If the generic event C had not been exemplified the generic event E would not have been exemplified' where c and e, respectively, exemplified C and E?

Analogously in the case of a non-occurrent cause, to examine the counterfactual aspect $\square\!\!\rightarrow$ is supposed to capture, we may consider our house fire example. Suppose that we in the last moment succeeded in preventing a fire by removing some inflammable material from the place where the defect in the electric wire was. Thus the generic short-circuit event C was not exemplified by any singular event c nor was the generic fire event E exemplified by any singular event e. Thus, for all $c \in C$ and $e \in E$, $\sim O(c) \square\!\!\rightarrow \sim O(e)$ is true. However, the requirement that $O(c) \square\!\!\rightarrow O(e)$ now seems extra-ordinarily strict and hard to satisfy, no matter which specific singular events c and e we pick, be they members of C and E, respectively, or not.

Let us now consider the following three singular conditionals:

(a) $O(c) \square\!\!\rightarrow O(e)$ is true at the world w
(b) $\sim O(c) \square\!\!\rightarrow \sim O(e)$ is true at the world w
(c) $\sim O(e) \square\!\!\rightarrow \sim O(c)$ is true at the world w.

Here the world w is understood to represent "the circumstances" or a certain state of the adopted causal field. Our above considerations seem to make an unqualified universal acceptance of any of these conditionals rather questionable. Considered alone none of them can be regarded as analytically true about any respectable scientific notion of a cause. In conjunction with $O(c)$ and $O(e)$, however, (a) sounds acceptable, and in conjunction with $\sim O(c)$ and $\sim O(e)$ (b) and (c) may seem better acceptable.

It must then be said in defense of the counterfactual analysis that if we restrict ourselves to cases where c and e have actually occurred the acceptance of either (b) nor (c) may seem plausible if c and e are thought of only as representatives of the corresponding generic constitutive events C and E. That seems possible. But in the case of non-occurrence of c and e this alternative does not seem viable in view of our remarks above.

There are several other problems for a counterfactual analyst concerning the acceptance of (a), (b), and (c). For instance, (b) seems to conflict with cases where there is a plurality of causes (either quantitative or alternative overdetermination, with or without linked cause-events). (See e.g. Lewis [1973a], p. 567, and Mackie [1974], pp. 44-46 for this.) As in my analysis (C**) neither (b) nor (c) is an analytic truth about singular causal statements and as I think this to be a problem for scientists, rather, I shall not here discuss it much further.

There are, however, some remarks to be made concerning these conditions. First, condition (c) is a kind of strong sufficiency condition. Mackie (1974) requires (b) and (c) to be true of singular causes.[5] But as (a) is sufficiency condition why should one accept (c) rather than (a)? As contraposition operation is not valid for conditionals such as $\square\!\!\rightarrow$ (a) and (c) do not imply each other either way. Consider the statement that, *ceteris paribus*, rain causes wheat to grow. It would now seem that (a) and (b), but not (c), are acceptable here. (E.g. some insects might have destroyed the crop; hence (c) is not true.)

A second remark is that Lewis requires his counterfactual to be true only in the actual world (or the world where c and e occur). However, within a deductive-nomological or backing law account one might want to require more. It is a feature of scientific theories that they have some stability properties, *viz.* their truth is invariant with respect to a great variety of "irrelevant" changes in the universe. So what one might suggest to a counterfactual analyst to be worked out is the idea that singular causal statements hold not only in the actual world (given that c and e did occur) but also in worlds in which c and e occurred and which are *nomologically close* to the actual world.

Finally we shall try to defend the backing law account against Lewis' specific criticisms, i.e. his theorems (1) and (2) mentioned earlier. What is wrong with them? The matter seems to me straightforward. What Lewis calls a regularity analysis of causation (see the earlier given definition) simply represents an overly simplified and untenable position. What I mean by this is that that particular version of the regularity theory is (at least formally) identical with an account of deductive explanation rejected long ago (see e.g. the discussion in Tuomela [1972]). Accordingly, one may argue that Lewis' definition of nomic dependence, central for the derivation of theorems (1) and (2), is also unacceptable. To be a little more informative and constructive about this I shall now point out some particular reasons why Lewis' claims are unacceptable, and I shall also suggest a better definition for nomic dependence.

First we notice that Lewis' notion of nomic dependence is unacceptable for the reason that it is, at least, too wide: it does not capture relevant nomic dependence. There are in fact lots of examples to demonstrate this, but let it suffice to consider only one.

Let $F(a)$ be an arbitrary singular statement. Consider some law of the form $(x) G(x)$ such that 'G' belongs to a completely different field of science. Now $F(a)$ (and any disjunction in which $F(a)$ occurs as a disjunct) can be seen to depend nomically on the singular statement $\sim G(a) \vee \vee F(a)$ in the sense of Lewis' definition. But such "nomic dependence" is of course hardly relevant to questions of causality or explanation or any other philosophically interesting matter.

I suggest now that we instead explicate nomic dependence in terms of explanation. Certainly within deductive-nomological explanation the explaining law is nomically relevant to the explanandum. I also think we can for our present purposes deal with a strong idea of nomic dependency which entails (deductive) explainability. We then define:

A singular statement D' *depends nomically* on a singular statement D (in symbols $D \triangleright\!\!\rightarrow D'$) if and only if there is an applied theory S such that S *cum* D deductively explains D' in the full sense of the weak DEL-model of explanation, that is, in the sense of condition (1) of our characterization (C**).

The generalization of this definition for families of statements is of course obvious. It is immediately seen that my notion of nomic dependency entails Lewis' notion (but of course the converse is not true).

Now we can consider Lewis' first theorem (1). (We restrict ourselves to finite families below.) Lewis claims that the assumption of counterfactual independence is indispensable. (This assumption is that the components of the applied theory S should be counterfactually independent of the initial conditions A_i.) However, within *actual* explanation all the initial conditions must be true and so must the theory S be. Thus in any world in which the relevant nomic dependence holds $A_i \square\!\!\rightarrow S$ is true. Thus we get the result that not only is Lewis' theorem (1) false for my notion of nomic dependency but also the following theorem holds:

(1*)   If the C's depend nomically (in the sense explicated by $\triangleright\!\!\rightarrow$) on the A's in virtue of S, then the C's depend counterfactually on the A's.

What we have here is thus an "explanation" of counterfactual dependence by means of nomic dependence *without* any additional counterfactual independence assumption.

What happens to Lewis' theorem (2)? It is immediately seen that it does not hold true as it cannot be proved without violating the non-comparability requirement of the weak DEL-model (for we would have A in the explanans for each A_i). It follows instead that nomic dependence is irreversible:

(2*)   If the family $C_1, C_2, \ldots$ depends nomically (in the sense of $\triangleright\!\!\rightarrow$) on the family $A_1, A_2, \ldots$ in virtue of S then the family $A_1, A_2, \ldots$ does *not* depend nomically on the family $AC_1, AC_2, \ldots$ in virtue of S.

This result gives some additional content to condition (2) of (C**) which says that a cause is "prior" to its effect and that causation hence does not represent a symmetric relationship.

In Lewis (1973a) an interesting "correspondence" theorem is established to exhibit the close logical connection between metalinguistic accounts of counterfactuals (used in contexts of singular explanation and causation). To explain this connection we need the notion of *cotenability* of premises. Lewis gives it the following semantical definition: Proposition A is *cotenable* with proposition B at a world w if and only if (1) A holds at all worlds accessible from w, or (2) some world in which B is true is closer to w than any world in which A is true. (See Lewis [1973a], p. 57 for more details.)

Now the following result can be obtained: $A \square \!\!\rightarrow B$ is true at world w if and only if A together with finitely many premises $P_1, P_2, ..., P_n$, each cotenable with A and w, logically imply B. Thus a counterfactual is true (or assertable, if you like) on an occasion if and only if there exists an argument backing it.

What interests us here is of course the case where the singular statement $A \square \!\!\rightarrow B$ is interpreted as 'B because of A' or 'B depends causally on A'. In those cases the argument backing $\square \!\!\rightarrow$ must be deductive-explanatory. To discuss this, we first notice that cotenability can be characterized in terms of comparative possibility as follows: the denial of a cotenable premise is less possible than the counterfactual antecedent itself, unless the antecedent is already minimally possible. In view of this we agree with Lewis in his claim that laws tend to be cotenable with singular antecedents which are not strictly logically inconsistent with the law in question (cf. Lewis [1973a], p. 72). But given this cotenability assumption it immediately and trivially follows that if $A \triangleright \!\!\rightarrow B$ is true at a world w then $A \square \!\!\rightarrow B$ is true at w.[6] However, the converse does not hold.

This result of course is based on the fact that our $\triangleright \!\!\rightarrow$ is backed by an *explanatory* deductive argument which involves logical deducibility but which is stronger than mere deducibility. What this result again shows is that Lewis' $\square \!\!\rightarrow$ is far too weak to capture notions like explanation and causation. Rather I suggest that my $\triangleright \!\!\rightarrow$ can be taken to explicate singular explanation. That is, $A \triangleright \!\!\rightarrow B$ can be interpreted as 'B because of A', and hence the former is acceptable exactly when the latter is (provided we are concerned with *scientific* explanation). If the backing theory S is required to be a causal theory the 'because of' relation is explicitly a causal relation. Thus the truth of a singular causal claim $C(e, e')$ entails the existence of an assertable statement $D \triangleright\!\!\rightarrow D'$ (the converse holds if the additional clauses of (C**) are satisfied).

What we have here is a correlation between "ontological talk" (in terms of objective events) and linguistic "fact talk" (in terms of statements involving the occurrence of events). Here we also see what kind of connections there can be between explanatory causes (expressed by $D \triangleright\!\!\rightarrow D'$) and productive causes (*in part* expressed by $C(e, e')$). $C(e, e')$ can be taken to fully represent a productive cause only in the Peircean limit science, where the backing theory fully captures the underlying objective causal process.

It should be noticed here that as $\triangleright\!\!\rightarrow$, assuming cotenability, entails Lewis' $\square\!\!\rightarrow$, it can be given the same applications as $\square\!\!\rightarrow$ has. Thus $\triangleright\!\!\rightarrow$ can be interpreted as a conditional obligation operator, although it is much stronger than $\square\!\!\rightarrow$. Notice also that $\triangleright\!\!\rightarrow$ takes into account the nomic force of laws in the sense discussed in Section V.

We have not yet connected our $\triangleright\!\!\rightarrow$ to Lewis' defining conditions of causation (causal dependence). First, we notice that while the truth of D and D' automatically makes $D \square\!\!\rightarrow D'$ true (assertable), still $D \triangleright\!\!\rightarrow D'$ is not yet assertable, which fact is just to the good. However, the latter of course has to become assertable in order to make $C(e, e')$ true. This means that I accept that also $D \triangleright\!\!\rightarrow D'$ is assertable. This statement concerns singular events, but, given that e and e' have indeed occurred, it can still be viewed as primarily concerned with the generic events the backing theory speaks about.

Secondly, whereas Lewis requires that $\sim D \square\!\!\rightarrow \sim D'$ be true, it might seem too strong to require analogously that $\sim D \triangleright\!\!\rightarrow \sim D'$ be assertable. We have to notice that "it is not the case that e occurred" cannot meaningfully be said to explain or cause that "it is not the case that e' occurred." But if we interpret $\sim D \triangleright\!\!\rightarrow \sim D'$ as 'without D, D' is not explainable' it may seem acceptable. If we then use 'assertable' in a broad Sellarsian way covering this usage, we can interpret $\sim D \triangleright\!\!\rightarrow \sim D'$ as 'If D were not assertable, then D' would not be assertable'. Under this interpretation $\sim D \triangleright\!\!\rightarrow \sim D'$ will probably be often assertable in cases when $C(e, e')$ holds, although my account of causality does not require it. For instance, if there are genuine cases of overdetermination, $\sim D \triangleright\!\!\rightarrow \sim D'$ is not in this broader sense assertable. Thus consider an example where two switches a and b are flipped simultaneously such that both were sufficient for the light to go on. Then neither flipping a nor flipping b was individually necessary (at least under that description). Both were "sufficient parts" of the total cause. – If we, however, are interested in necessary and sufficient causes only, we can do it by requiring that both $D \mathbin{\vrule height1ex width.1em depth0pt \vrule height.1em width.6em depth0pt}\!\!\to D'$ and $\sim\! D \mathbin{\vrule height1ex width.1em depth0pt \vrule height.1em width.6em depth0pt}\!\!\to \sim\! D'$ be true.

There is a related counterfactual concerning generic events which seems to be always assertable in view of the nomic character of the backing theory. If the generic events or properties it connects are C and E we seem to be in a position to assert 'If C were exemplified, E would be exemplified'. However, this clearly is not yet sufficient for the assertability of $\sim\! D \mathbin{\vrule height1ex width.1em depth0pt \vrule height.1em width.6em depth0pt}\!\!\to \sim\! D'$ even under the above final liberal interpretation of it, given that the cause and effect occurred.

There are many problems of causality which I have not touched at all above.[7] Still, it seems to me that the backing law account of scientific causation I have tried to defend can cope with these problems at least as well as any other account. I hope to have shown that the prospects for a full backing law account look all but dark.

University of Helsinki

NOTES

* My research for this paper was supported by the Academy of Finland.

[1] The hard question in discussing event identity is how to characterize the identity of aspect properties. I am willing to assume the possibility of contingent identities holding between them. I cannot, however, in this paper discuss at any length the difficult problems involved here.

[2] Note here that unless deductive explanation is given a stringent formulation all kinds of paradoxes ensue already on the linguistic level of statements quite independently of any ontological and semantical assumptions (such as (I) or (L)). For instance, even within the original Hempel-Oppenheim model it is possible to explain practically any singular statement by means of almost any explanans. This same remark applies to all attempts to clarify philosophical notions by means of some kind of subsumption. For instance, of the definitions of *determinism* at least the one by Berofsky, but also that by Goldman, are affected and trivialized by exactly this paradox (see Goldman [1970], p. 172 and Berofsky [1971], p. 168)). Similarly this linguistic deducibility paradox seems to affect Goldman's definition of level-generation of action in Goldman (1970), p. 43.

[3] Thus, for instance, both of the following two arguments are valid *potential* explanations in our (weak) DEL-model:

(i) $(x)(y)(F(x, y) \to G(x, y))$
        $\dfrac{F(a, b)}{G(a, b)}$

(ii) $(y)(F(a, y) \to G(a, y))$
        $\dfrac{F(a, b)}{G(a, b)}$

[4] Let us here mention a criticism, which, if true would be unpleasant for a backing law account. Namely, it has been claimed that causes cannot be regarded as conditions or properties which are both necessary and sufficient for the effect condition (see Brand and Swain (1970)). However, as has been shown, Brand's and Swain's analysis is based on certain unacceptable assumptions on the logical properties of necessary-and-sufficient conditions. Especially their principle (P3) requiring *transitivity* as well as their (P4) requiring *uniqueness* of a necessary and sufficient causal condition for the effect seem unacceptable (cf. Hilpinen [1974] and the papers referred to in it). Also in the analysis by Kron [1975] a cause can be a necessary and sufficient condition without contradiction.

It should be mentioned here that Lewis (1973b) analyzes a (kind of direct) cause just in terms of necessary and sufficient conditions in terms of the counterfactual $\Box\!\!\to$ (see Section VI of this paper). That analysis fails to satisfy just Brand's and Swain's conditions (P3) and (P4), as is easily seen. It should also be pointed out that the analogous notion of causality definable in terms of our analysis (C**) (and more exactly in terms of our $\triangleright\!\!\to$ to be discussed in Section VI) also fails to satisfy the problematic conditions (P3) and (P4).

[5] Mackie seems to take (a) and (c) to be equivalent, however (see Mackie [1974], p. 49). There are good reasons for thinking that that cannot be the case for any reasonable interpretation of a non-material conditional (cf. e.g. Lewis' [1973a] arguments).

[6] In more like Goodman's idea we may define a weaker notion of cotenability by saying that S is cotenable with A at w if and only if S is true at w and A $\Box\!\!\to$ S is true at w. The correspondence theorem for the metalinguistic and counterfactual accounts discussed on p. 356 above also holds for this second notion of cotenability (cf. Lewis [1973a], 70). But the premises of any actual (weak) DEL-explanation are easily proved to be cotenable in this sense (as A and S are true), whence $\triangleright\!\!\to$ is seen to entail $\Box\!\!\to$ directly, too.

[7] For instance, I have not here dealt with problems such as a) the problem of effects, b) the problem of epiphenomena, and c) the problem of pre-emption (see e.g. Lewis [1973b], pp. 565–567). However, solutions of a) and b) are easily and directly on the basis of what has been said earlier about the causal relation C(e, e'). In the case of c) one cause-event supersedes another cause-event, but no formal contradiction for my account is involved. To make the problem more interesting the nature of the interaction between these events would have to be specified in more detail, and I think that would in fact bring backing laws into play.

BIBLIOGRAPHY

Ackermann, R.: 1965, 'Deductive Scientific Explanation', *Philosophy of Science* **32**, 155–167.

Ackermann, R. and Stenner, A.: 1966, 'A Corrected Model of Explanation', *Philosophy of Science* **33**, 168–171.

Berofsky, B.: 1971, *Determinism*, Princeton University Press, Princeton.

Brand, M. and Swain, M.: 1970, 'On the Analysis of Causation', *Synthese* **21**, 222–227.

Davidson, D.: 1967, 'Causal Relations', *The Journal of Philosophy* **64**, 691–703.

Davidson, D.: 1969, 'The Individuation of Events', in N. Rescher *et al.* (eds.), *Essays in Honor of Carl G. Hempel*, Reidel, Dordrecht, pp. 216–234.
Goldman, A.: 1970, *A Theory of Human Action*, Prentice-Hall, Englewood Cliffs.
Hilpinen, R.: 1974, 'A Note on Necessary-and-Sufficient Causes', forthcoming in *Philosophical Studies*.
Kim, J.: 1969, 'Events and Their Descriptions: Some Considerations', in N. Rescher *et al.*, *op. cit.*, pp. 198–215.
Kim, J.: 1973a, 'Causation, Nomic, Subsumption, and the Concept of Event', *The Journal of Philosophy* **70**, 217–236.
Kim, J.: 1973b, 'Causes and Counterfactuals', *The Journal of Philosophy* **70**, 570–572.
Kron, A.: 1975, 'An Analysis of Causality', in Manninen, J. and Tuomela, R. (eds.), *Essays on Explanation and Understanding*, Reidel, Dordrecht, pp. 159–182.
Lewis, D.: 1973a, *Counterfactuals*, Blackwell, Oxford.
Lewis, D.: 1973b, 'Causation', *The Journal of Philosophy* **70**, 556–567.
Lyon, A.: 1967, 'Causality', *British Journal for the Philosophy of Science* **18**, 1–20.
Mackie, J.: 1965, 'Causes and Conditions', *American Philosophical Quarterly* **2**, 245–264.
Mackie, J.: 1974, *The Cement of the Universe*, Clarendon Press, Oxford.
Nagel, E.: 1961, *The Structure of Science*, Harcourt, Brace and World, Inc., New York.
Niiniluoto, I. and Tuomela, R.: 1973, *Theoretical Concepts and Hypothetico-Inductive Inference*, Synthese Library, Reidel, Dordrecht and Boston.
Simon, H.: 1953, 'Causal Ordering and Identifiability', in W. Hood and T. Koopmans (eds.), *Studies in Econometric Method*, Wiley, New York, pp. 49–74.
Tuomela, R.: 1972, 'Deductive Explanation of Scientific Laws', *Journal of Philosophical Logic* **1**, 369–392.
Tuomela, R.: 1973, *Theoretical Concepts*, Library of Exact Philosophy, Springer-Verlag, Vienna and New York.
Tuomela, R.: 1974a, 'Causality, Ontology, and Subsumptive Explanation', forthcoming in the proceedings of the Conference for Formal Methods in the Methodology of Empirical Sciences, held in Warsaw, Poland in 1974.
Tuomela, R.: 1974b, *Human Action and Its Explanation*, Reports from the Institute of Philosophy, University of Helsinki, No. 2.
Tuomela, R.: 1976, 'Morgan on Deductive Explanation: A Rejoinder', forthcoming in *Journal of Philosophical Logic*.
von Wright, G. H.: 1971, *Explanation and Understanding*, Cornell University Press, Ithaca.

LAIRD ADDIS

ON DEFENDING THE COVERING-LAW "MODEL"

Hempel in *Aspects of Scientific Explanation and Other Essays* tells us that "all scientific explanation involves, explicitly or by implication, a subsumption of its subject matter under general regularities; that it seeks to provide a systematic understanding of empirical phenomena by showing that they fit into a nomic nexus."[1] This I take to be an informal statement of his covering-law "model" of scientific explanation. In defense of his "model" against apparent examples of scientific explanations which do not fit it, Hempel's almost instinctive reaction has been to patch them up in some way so that they do conform. This, I believe, is the wrong way to defend the covering-law "model". In the three parts of this brief essay I shall do the following: (1) show that Hempel's attempt to patch up a certain version of so-called "rational" explanation does not succeed of its purpose, (2) generalize that result with respect to all dispositional explanations, and (3) reflect momentarily on philosophic method and another way to defend the covering-law "model".

I

William Dray in *Laws and Explanation in History* argues that "the explanation of individual human behaviour as it is usually given in history has features which make the covering law model peculiarly inept".[2] In particular he is concerned with what he calls *rational* explanation; explanation, that is, in terms of motivating reasons. Hempel, after rehearsing some of Dray's considerations for this view, presents the following schema as being Dray's account of the "real" form of a typical rational explanation:

> A was in a situation of type C.
> In a situation of type C the appropriate thing to do is x.
> Therefore, A did x.[3]

Hempel, taking this account to be at odds with his theory of the covering-law "model", patches it up by defending an alternate schema as being the "real" form of a typical rational explanation:

> *A* was in a situation of type *C*.
> *A* was a rational agent.
> In a situation of type *C*, any rational agent will do *x*.
> Therefore, *A* did *x*.[4]

And about his amendment Hempel says the following:

> This schema of rational explanation differs in two respects from what I take to be Dray's construal: first, the assumption that A was a rational agent is explicitly added; and second, the evaluative or appraising principle of action, which specifies the thing to do in situation C, is replaced by an *empirical generalization* stating how rational agents will act in situations of that kind. Thus, Dray's construal fails just at the point where it purports to exhibit a logical difference between explanations by reference to underlying reasons and explanations by subsumption under general laws, for in order to ensure the explanatory efficacy of a rational explanation, we found it necessary to replace Dray's normative principle of action by *a statement that has the character of a general law*. But this restores the covering-law form to the explanatory account.[5]

In short, Dray's account is on Hempel's view inadequate in the first instance because the resulting schema is not that of a deductively valid argument. When it is patched up in a plausible way to correct that defect, we have the further happy consequence that the schema is of the covering-law type and all is once more right with the world.

Hempel fails to achieve what he thinks he must. Whether his schema for rational explanation is the correct one or not, it is not in any case of the covering-law type – that which he is most anxious to establish. It is not of the covering-law type because, contrary to Hempel, the third premise is *not* (the form of) an "empirical generalization" or a "statement that has the character of a general law". It is in fact neither empirical nor a generalization any more than 'If anything is brittle, it will break when struck with force' is empirical or a generalization although both have the grammatical form of genuine empirical generalizations. Like 'All brittle objects break when struck with force' the proposition (-form) 'All rational agents will do *x* in a situation of type *C*' is, rather, a definitional truth.

Let us label as *P* the proposition 'If anything is brittle, it breaks when struck with force'. Further, let us label as *Q* an instance of the proposition (-form) in question. Let that instance be 'If anyone is thirsty, healthy, knows that fresh, tasteful, healthful water is immediately available, is physically capable of drinking the water, and is rational, he will drink of that water'. My argument will be that Q is of the same logical character as P, and that it is easily shown that P is a definitional truth and not an empirical generalization. Let us see.

That brittle objects do break when struck with force is not an empirical discovery. It is not an induction from experience, and no experience could falsify it. If it were an empirical generalization (and if we are good Humeans as, at any rate, Hempel and I are) then the world might have been such that nothing which is brittle breaks when struck with force. The absurdity is patent. But do not exactly the same considerations apply to Q? That a rational person who is thirsty, healthy, knows that fresh, tasteful, healthful water is immediately available, and is physically capable of drinking the water would drink of the water is not an empirical discovery. It is not an induction from experience and no experience could falsify it. If it were an empirical generalization, the world might have been such that no rational person would drink water under the described circumstances and the world be otherwise the same.

I conclude that Hempel is mistaken in believing that 'In a situation of type C, any rational agent will do x' is (the form of) an empirical generalization and therefore also mistaken in believing that his schema of rational explanations is of the covering-law type. Before I proceed, however, it is only fair to point out that Hempel has himself dealt at some length with precisely this objection, that propositions of the sort in question are not empirical generalizations but definitional truths.[6] Rather than address myself to his arguments directly, I have simply given independent reasons for supposing that he is mistaken.

II

Rational explanations, then, would appear to be a species of dispositional explanations. Let us forget rational agents now and turn to brittle windows. Consider the following dispositional explanation:

>This window was struck with force.
>This window is brittle.
>Brittle objects break when struck with force.
>Therefore, this window broke.

This explanation, while both deductively valid and informative, is not, I maintain, of the covering-law type. My reasons are by now obvious. What may first appear to be an empirical generalization, the third premise, is really a definitional truth. And since the first two premises are statements of particular fact, the premises do not contain an empirical law. And indeed, to generalize, no genuinely dispositional explanation is of the covering-law type. (This is not to say that dispositions cannot be causes or cannot enter into genuine empirical laws. It is only to say that dispositions as dispositions are not the causes of what they are dispositions *to*. But they may be causes by way of entering into genuine lawful connections with other, definitionally unrelated properties.)

But even though no dispositional explanation is of the covering-law type, each is uniquely related to a potential explanation of that type. In order to explain this idea, however briefly, I must introduce a general analysis of dispositional terms. I do this, however, by considering the particular case of 'x is brittle'. Letting 'B' stand for the dispositional property of being brittle, 'R' for the property of breaking, 'S' for the property of being struck with force, and 'f' as a predicate variable which takes only (the names of) non-dispositional properties for its values, I propose the following analysis:

'x is B' is analyzed as '$(\exists f)\,[(fx \cdot (y)\,(fy \cdot Sy) \supset Dy]$'

In general, the use of a dispositional term as applied to a particular thing is to be understood as signifying that that thing possesses some not further specified non-dispositional property *and* that anything possessing that property and some further specified property will also possess yet another specified property. No doubt many objections can be raised to this proposal. I am presently convinced that it is at least as defensible as any of its extant competitors. But one objection I will consider briefly. The objection is that in our use of dispositional terms we don't *mean* any law of nature (or form of such a law). Furthermore, it might be insisted that it is a mere matter of fact whether dispositions have any connection with laws at all. So laws (or law forms) cannot enter into the correct analysis of dispositional terms. My reply is two-fold. First, analysis *of this sort* is not grounded in what people have in mind when they use the notions to be analyzed. Rather, *such* an analysis is a precise specification of the truth conditions for any statement of the relevant sort. Second, to refuse to admit the lawful side of the analysis, at least for the reason of there being no "necessary" connection between dispositions and laws, is to affirm that two objects with precisely the same non-dispositional properties might have different dispositions. This is, to be sure, not a logical contradiction. But if it were the case, then the ideal of scientific explanation would have broken down altogether.

Permit me to proceed, then, on the assumption that my analysis is adequate. If it is, then the (correct) attribution of a disposition to a given object presupposes the truth of a certain, though perhaps as yet unknown, law of nature. For the predicate *variable* 'f' occurs in the analysis of any particular case, and what the non-dispositional property is which when (its name is) substituted for the variable results in a true law of nature may be completely unknown. Yet we now can see that for every dispositional explanation there does correspond a constructible potential explanation of the covering-law type, one which becomes a full-blown explanation of that type upon substitution for the predicate variable. So, for example, for the dispositional explanation

> This window was struck with force.
> This window is brittle.
> <u>Brittle</u> objects break when struck with force.
> Therefore, this window broke.

we can construct the following potential explanation of the covering-law type:

> Anything which has a certain (non-dispositional) property and is struck with force breaks.
> <u>This</u> window has that property and was struck with force.
> Therefore, this window broke.

Thus while no dispositional explanation is of the covering-law type, any phenomenon which is so explained can, in principle, be explained by one which is of the covering-law type. Yet dispositional explanations are of a type to be met with in science. It would appear then that Hempel is also mistaken in believing that "all scientific explanation involves, explicitly or by implication, a subsumption of its subject matter under general regularities."

III

If Hempel is mistaken in this respect, must we also conclude that the covering-law "model" is false or inadequate or somehow defective? But then, what is Hempel doing in first place in putting forth his "model" of scientific explanation? Let us hear from him again:

> This construal [the covering-law model]... does not claim simply to be descriptive of the explanations actually offered in empirical science; for – to mention but one reason – there is no sufficiently clear generally accepted understanding as to what counts as a scientific explanation. The construal here set forth is, rather, in the nature of an *explication*, which is intended to replace a familiar but vague and ambiguous notion by a more precisely characterized and systematically fruitful and illuminating one.[7]

This conception of his undertaking, however, given the appearance of being simply the standardization and making precise of the meaning of a certain notion. In that case, no substantive claim is being made and no argument is relevant. Definitions are not true or false. Yet Hempel tells us explicitly what kinds of arguments he does consider to be relevant to his "model":

> Like any other explication, the construal here put forward has to be justified by appropriate arguments. In our case, these have to show that the proposed construal does justice to such accounts as are generally agreed to be instances of scientific explanation, and that it affords a basis for a systematically fruitful logical and methodological analysis of the explanatory procedures used in empirical science.[8]

This makes the "theory" sound descriptive once again. If Hempel is not making the substantive, descriptive claim that every explanation generally regarded as scientific either is or can be patched up to be of the covering-law type, then what substantive claim is he making? Surely the mere fact that there is no simple, general, and widely accepted "definition" of 'scientific explanation' does not entail that his claim, if there is a claim, is not a descriptive one. I have already argued that this descriptive claim is false, for dispositional explanations are respectable scientific ones and they are not of the covering-law type. I am unable to discover, much less to decide for Hempel, exactly what he is claiming with respect to the covering-law "model".

But I do actually believe that given the philosophical method employed by Hempel the vagueness is irreparable. For myself, I should make a clear, analytic distinction between explication of various notions on the one hand which I take to be essentially stipulative or non-controversially descriptive and, on the other hand, the evaluation of substantive claims which employ the notions as explicated. The first of these tasks is merely preparatory to the second which is genuinely philosophical. Hempel, however, like the majority of contemporary philosophers, either conflates these two analytically distinct activities or simply takes philosophy to begin and to end with the first. He calls it "explication", others "conceptual analysis". Yet, of course, such philosophers are not content to describe *themselves* as merely stipulating how a notion might be employed or describing how in fact some notion is employed. So they imagine they are *claiming* something which both can be and is worth arguing about. This, to my mind, is simply confusion.

Be all that as it may, I now shall employ the very distinction I have just insisted upon in order to indicate how I believe the covering-law "model" can and should be defended. The explication, in the case at hand, is merely a precise stipulation as to the meaning of 'explanation of the covering-law type'. The substantive philosophical claim the defense of which I believe constitutes the appropriate defense of the covering-law "model" is the following proposition: 'There is no good reason to believe that any event occurs which cannot be given an explanation of the covering-law type'. It is worth noticing, before further remark of this proposition, that Hempel explicitly rejects a similar conception of what he is about. Concerning his own claim that "all adequate scientific explanations and their everyday counterparts claim or presuppose at least implicitly the deductive or inductive subsumability of whatever is to be explained under general laws or theoretical principles,"[9] Hempel remarks in a long footnote that

This idea needs to be sharply distinguished from another one, which I am not proposing, namely, that any empirical phenomenon can be explained by deductive or inductive subsumption under covering laws. The idea here suggested is that the logic of all scientific explanations is basically of the covering-law variety, but not that all empirical phenomena are scientifically explainable, and even less, of course, that they are all governed by a system of deterministic laws. The question whether all empirical phenomena can be scientifically explained is not nearly as intelligible as it might seem at the first glance, and it calls for a great deal of analytic clarification. I am inclined to think that it cannot be given any clear meaning at all; but at any rate, and quite broadly speaking, an opinion as to what laws hold in nature and what phenomena can be explained surely cannot be formed on analytic grounds alone but must be based on the results of empirical research.[10]

Hempel gives no reason for suspecting that what appears to be a perfectly intelligible proposition may not really be. Perhaps my proposition, similar though it be to his, is sufficiently different not to suffer the same problems Hempel sees in his. In any case, Hempel rightly affirms that it is an empirical and not a philosophical question "what laws hold in nature and what phenomena can be explained." Indeed the defense of that proposition itself is part of the defense of mine. For many philosophers have believed and have argued that there are *a priori* or so-called "conceptual" grounds for maintaining that some phenomena do not even possibly admit of covering-law explanations. This has been argued for mental phenomena in general and moral choices in particular, for actions in general and rational actions in particular, for such events as the emergence of life or the evolution of mind, and so on. In some cases it is argued directly, as it were, that certain phenomena are not amenable to lawful explanation; in others it is argued that since the phenomena are explainable in some other manner – say by reasons or goals or dispositions – they cannot also be explained by laws. Showing that none of these arguments succeeds of its purpose is to show that there is no good reason to believe that any event occurs which cannot be given an explanation of the covering-law type. In short, the proper defense of the covering-law model concerns its applicability and not its employment.

University of Iowa

NOTES

[1] Carl G. Hempel, *Aspects of Scientific Explanation and Other Essays*, The Free Press, New York, 1965, p. 488.
[2] William Dray, *Laws and Explanation in History*, Oxford University Press, London, 1957, p. 118.
[3] Hempel, *Ibid.*, p. 470. This is not a precise quote insofar as Hempel does not write the conclusion immediately following the premises.
[4] *Ibid.*, p. 471.
[5] *Ibid.*, p. 471, my emphases.
[6] *Ibid.*, pp. 457–463. The essential premise of Hempel's argument is that "multi-track" dispositions such as *being magnetic* cannot be analyzed wholly into conjunctions of "single-track" dispositions. If that is so then it is an empirical and not a definitional matter what the possessor of a given multi-track disposition would do in any given situation. Rationality, of course, is a multi-track disposition. I believe the essential premise to be false.
[7] *Ibid.*, pp. 488–489, Hempel's emphasis.
[8] *Ibid.*, p. 489.
[9] *Ibid.*, pp. 424–425.
[10] *Ibid.*, p. 425.

CARL G. HEMPEL

DISPOSITIONAL EXPLANATION AND THE COVERING-LAW MODEL: RESPONSE TO LAIRD ADDIS

The following considerations are offered in response to the critical observations and constructive proposals set forth by Laird Addis, in his paper 'On Defending the Covering-Law "Model"', concerning my explications of "rational" and dispositional explanation and concerning the claims associated with the covering-law model of explanation.

1. My main objection to Dray's construal of rational explanation, as characterized by the first schema in Addis's paper, was not that it is at odds with the covering-law model, but that the second explanans sentence, 'In a situation of type C, the appropriate thing to do is x', expresses a norm and therefore cannot possibly explain why A did in fact do x: to do that, we need, not a normative sentence, but a descriptive one, roughly to the effect that A was disposed to act in accordance with the normative principle. Accordingly, I proposed this modified construal for rational explanations of the kind considered by Dray:

(Schema R)
(P_1) A was in a situation of type C
(P_2) A was a rational agent
(P_3) In a situation of type C, any rational agent will do x
(E) (Therefore,) A did x.

Here, the explanans sentences P_2 and P_3 jointly imply the required descriptive sentence just mentioned[1].

2. Addis argues that, thus construed, a rational explanation is not a covering-law explanation, on the ground that the only sentence of general form occurring in the explanans, namely P_3, is not an empirical law, but a definitional truth. For reasons to be suggested in Section 3, I do not think that a clear and defensible construal can be given to this claim. But suppose, for the sake of the argument, that the term 'rational agent' has been introduced explicitly as an abbreviation of 'agent who does x

R. S. Cohen et al. (eds.), PSA 1974, 369–376. All Rights Reserved.
Copyright © 1976 by D. Reidel Publishing Company, Dordrecht-Holland.

whenever he is in a situation of kind C'. Then, in virtue of this convention, the sentence P_2 attributes to A a general behavioral disposition, namely, to do x whenever faced with a situation of kind C; and the explanation represented by schema R accounts for A's action by showing that it was but a particular manifestation of A's general disposition to do x when in C. In this "dispositional explanation", the role of the covering law is played by the sentence P_2. This sentence has nomic force; it is what Ryle calls a law-like sentence, ascribing a nomic behavior pattern to a particular individual[2].

3. But, as mentioned before, I must take exception to Addis's view that the sentence P_3, and analogous sentences in other dispositional explanations, are definitional truths and therefore not empirical.

Suppose that the term 'temperature' is originally introduced by the convention to add to the set of accepted physical sentences a new one saying that a physical body has a temperature of r degrees centigrade just in case a regular mercury thermometer that is in contact with the body will show a reading of r on its scale. Though adopted by convention, this dispositional "criterion sentence" for the term 'temperature in centigrades' is not treated in physics as devoid of empirical content and immune from challenge by empirical findings. For suppose that physical research, making use, among other things, of the thermometer criterion, leads to the establishment of the laws of heat exchange. These laws imply that when a mercury thermometer is inserted into a body of water, then – unless thermometer and water happen to have the same temperature before insertion – heat will flow from the warmer to the colder of the two bodies, and the resulting reading of the thermometer will not show the temperature that was to be measured. Thus, if those laws are accepted as (presumably) true, the criterion sentence for temperature must be rejected as (presumably) false. In fact, considerations of just this kind have led to modifications of the criterion sentence which provide for specific ways of compensating for the disturbing effect. Thus, the original criterion sentence, though introduced by terminological convention, has been abandoned in response to empirical findings, namely, those that led to the acceptance of the laws of heat exchange.

In similar fashion, dispositional or operational criterion sentences for many other scientific terms have come to be abolished or revised even though they were originally accepted by linguistic convention. The preceding considerations apply equally, of course, to any dispositional criterion sentence, such as P_3 above, that might be used to characterize an agent as rational.

The reply might suggest itself that the abolition of the criterion sentence for temperature does not amount to the retraction, as presumably false, of an empirical belief or assertion which has come into conflict with empirical evidence: that it is rather a matter of abandoning, as inexpedient, a terminological convention which could have been retained, but at the price of making it practically impossible to establish general and simple laws of heat exchange and of other thermodynamic processes. But there is no fundamental difference between the two kinds of theoretical change thus distinguished. For, as has been pointed out by Duhem, and repeatedly emphasized by Quine, a scientific hypothesis will normally conflict with empirical findings, not when taken in isolation, but only when combined with a more or less extensive body of further hypotheses; and in this case, the conflict does not unambiguously establish the falsehood of the given hypothesis and can be resolved not only by abandoning the latter, but alternatively by a variety of other changes within the theoretical system. And which of the possible adjustments is chosen will depend not just on the conflicting evidence, but also on the effect the contemplated modification would have upon such general features of the theory as scope, simplicity, and closeness of empirical fit. In sum, both the so-called retraction of an empirical hypothesis as presumably false and what is described as the withdrawal, as inexpedient, of a "truth-by-linguistic-convention" are based on considerations of empirical evidence and of certain overall features of the theoretical system within which the given hypothesis functions. Consequently, criterion sentences such as P_3 above cannot properly be qualified as non-empirical, purely definitional, truths.[3]

4. For brevity of exposition, the preceding argument treats dispositional terms as standing for specific, "single-track"[4], dispositions. But this is a considerable oversimplification: such expressions as 'has a temperature of 7 degrees centigrade', 'electrically charged', 'greedy', 'rational agent', must be regarded as dispositional in a broad sense, i.e., as each standing for a large bundle of dispositions, corresponding to the diverse ways in

which electric charge, greed, a temperature of 7 degrees centigrade, etc., can manifest themselves under different conditions. In particular, any of the various "operational" criteria of application for a term may be regarded as specifying one of the dispositions the term stands for.

But this conception requires further refinement. What determines the set of dispositions that a given term stands for? Surely, it is not specified by some sort of explicit enumeration; rather, it is determined by a system of hypotheses or a theory in the context of which the term is used. To put the point schematically: a theory T in which a term, say 'D', occurs may imply sentences of the following kind:

$(H) \qquad (x)\,[Dx \supset (t)\,(Sxt \supset Rxt)]$.

This is then one of the dispositional criteria provided by T for the presence of D; it states that if an object has the property D, then it will respond in manner R whenever it is in a stimulus-, or test-, situation S. In many cases, especially when D is a quantitative feature, T will imply infinitely many such such dispositional criteria for D. And when the theory changes – e.g., by the addition of new laws linking the temperature of a body to other quantitative characteristics of the body or its environment – then the set of dispositions that T links to D changes accordingly.

But the term 'D' cannot be said to stand just for the conjunction, as it were, of the various dispositions the theory associates with it; for T will normally include further, nondispositional, hypotheses about D. For example, T will typically contain theoretical principles that link D to other attributes each of which has a set of dispositional aspects: thus, thermodynamic theory links the temperature of a gas to its pressure and volume; and a theory of rational action may link an agent's decisions to his objectives, values, and beliefs, all of which have dispositional aspects.[5] In short, what I called a broadly dispositional term is more adequately conceived in analogy to the "theoretical terms" of physics, as standing for a theoretically characterized property, which is linked to other such properties by the basic principles of a theory T, and which can manifest itself in a vast variety of ways, as indicated by the corresponding dispositional criteria that T implies for it. And if such a property, say D, is invoked for explanatory purposes, it is the theoretical principles concerning D that do the explaining. Dispositional explanation is basically of a kind with theoretical explanation.

There is a certain affinity, I think, between this conception and Addis's claim that every dispositional explanation is uniquely related to a potential covering-law explanation. The law in question is made evident, according to Addis, if the dispositional property, say D, of reacting in manner R under circumstances of kind S is analyzed as follows:

(Schema A) 'x has the property D' is tantamount to 'there is a non-dispositional property f such that x has f and anything that has f reacts in manner R when in situation S'.

Proposals to analyse or define dispositional properties by expressions of just this form have, in fact, been made by some earlier writers, especially Eino Kaila and Arthur Pap.[6] They were prompted by Carnap's observation[7] that the apparently obvious definition of 'Dx' by 'whenever Sx then Rx' would attribute the disposition D to, among other things, all those objects which are never exposed to situation S: for them the antecedent of the defining conditional is false, and the conditional thus trivially true. Thus, any rock that is never immersed in water would trivially satisfy the definiens for 'x is soluble'.

Now, definitions conforming to schema A yield the same consequence unless the range of the permissible properties f is suitably restricted: a rock never placed in water would still qualify as soluble because it has the property of dissolving-whenever-placed-in-water; and, as required in schema A, anything that has this property does dissolve when placed in water.

Addis's restriction of f to non-dispositional properties is intended to rule out such undesired consequences. In the case of dispositions like solubility, elasticity, or electric conductivity, the relevant non-dispositional properties might be identified with certain microstructural characteristics which account for a substance having the dispositions in question. But the general requirement that f be "non-dispositional" might be questioned on the ground that any property may be amenable to dispositional characterization.[8]

Pap, in his definition, requires f to be a "kind-property", i.e., a property "with respect to [the possession of] which a thing cannot meaningfully be said to change"[9]; but the notion of meaningful assertability faces obvious difficulties of its own.

However, as I have tried to show, an analysis of rational and other dispositional explanations need presuppose no "definitions" of dispositional terms. Theoretical principles of the character of H above are sufficient for the purpose. D is a property dealt with by a theory, and the explanatory principle H is part of what the theory asserts about D; and, as a general theoretical principle, it expresses a nomic claim.[10]

5. In sum, then, I think that rational and dispositional explanations do conform to the covering-law conception of explanation. As for the general claim to be made for that conception, I would persist in limiting it to the thesis that any adequate scientific explanation accounts for its explanandum phenomenon by subsuming it under general laws or theoretical principles. I do not wish to assert that all empirical phenomena admit of scientific explanation; for what empirical phenomena are explainable depends upon the nomic structure of our world: and the tightness and reach of the nomic net cannot be determined by philosophical analysis. Another reason for my position will be indicated in Section 6.

Addis suggests another claim, stronger than the one I have put forward, but weaker than the one just spurned; it asserts: "There is no good reason to believe that any event occurs which cannot be given an explanation of the covering-law type". I certainly agree with Addis in the view, suggested at the end of his paper, that arguments that have been offered to prove the absolute unexplainability of certain aspects of the world are flawed; indeed, I think that no conclusive proof of this kind is possible.

But Addis's thesis seems to me open to question. Consider the probabilistic laws that specify the half-lives of radioactive elements. They may be said, in the sense of the inductive-statistical form of the covering-law model, to explain such facts as that, of a certain initial amount of the element Ra^{223}, only very close to half is left after a period of 11.2 days – its half-life; that close to one-quarter remains after 22.4 days, etc. But, one may argue, these laws cannot account for certain other kinds of event: they cannot explain, for example, why atom a_1 decayed exactly two hours after the start of the test, or why atom a_2 survived for exactly 200 hours and 41 minutes; for to events thus characterized, the decay law ascribes zero probability.[11] An explanation of individual events of this kind would presumably require laws of strictly universal form. But suppose there are grounds for doubt that there are universal laws governing such events: would this not constitute a good reason to believe that there are events which cannot be explained? Now, just such grounds for doubting the explainability of the individual decay events, and of various other kinds of elementary event, would seem to be provided by the probabilistic aspects of quantum theory and by the various considerations that throw into question the possibility of supplementing the theoretical system by a hidden-variable theory which would restore a form of determinism to the level of elementary events.

6. On closer examination, the idea of the explainability of an event proves to be frustratingly obscure. Events which form the subject of any explanation are always events-under-a-sentential-description; the explanandum of an explanatory account is a sentence describing the occurrence to be explained, not an expression naming it. Perhaps, then, the thesis that every event can be explained should be construed as claiming that for every true sentence E describing an event in some language L, there exists a set of true sentences expressible in L which jointly explain the event described by E. But thus construed, the thesis is surely false: the expressive means of L may be too limited for the formulation of suitable explanatory laws or theories. So one would have to claim, perhaps, that L can always be extended or modified in ways that permit an explanation of the given event, or that there is some other language in which the event can be described and explained. But this proposal needs supplementation by an account of the conditions under which a sentence E' in a new language L' can be said to describe the same event as E in L: and this project faces familiar difficulties.

Analogous problems stand in the way of a sharp formulation of the principle of causality and of the idea of universal determinism. The thesis that all events are explainable is very closely related to these two ideas; and like them, it had perhaps best be viewed, not as an elusively comprehensive claim about the world, but as a heuristic maxim for scientific inquiry, calling for unflagging efforts to find explanatory principles where none are known so far.

Princeton University

NOTES

[1] For fuller details, see C. G. Hempel, *Aspects of Scientific Explanation and Other Essays in the Philosophy of Science* (Free Press, New York, 1965), pp. 469–472.

[2] See G. Ryle, *The Concept of Mind* (Hutchinson's University Library, London, 1949), pp. 88–90; note Ryle's remark "The imputation of a motive for a particular action is... the subsumption of an episode proposition under a law-like proposition." (p. 90) The character of dispositional explanations and of law-like sentences is discussed in some detail in Hempel, *op. cit.*, pp. 457–463.

[3] These considerations are closely akin to ideas developed by Quine – for example, in "Carnap and Logical Truth", in P. A. Schilpp (ed.), *The Philosophy of Rudolf Carnap* (Open Court, La Salle, Illinois, 1963), pp. 385–406.

[4] This characterization is used by Ryle, who then goes on to stress that there are "many dispositions the actualisations of which can take a wide and perhaps unlimited variety of shapes" (*op. cit.*, pp. 43–44).

[5] The issue is examined more fully in Hempel, *op. cit.*, pp. 472–477.

[6] See A. Pap, *Analytische Erkenntnistheorie* (Wien, Springer, 1955), pp. 140–142 (reference to Kaila on p. 141), and the amplified discussion in A. Pap, *An Introduction to the Philosophy of Science* (Free Press, New York, 1962), pp. 278–284.

[7] R. Carnap, "Testability and Meaning", *Philosophy of Science* **3** (1936), pp. 419–471 and **4** (1937), pp. 1–40; see pp. 439–441.

[8] Cf. N. Goodman, *Fact, Fiction, and Forecast* (2nd edition, The Bobbs-Merrill Company, Indianapolis, 1965), p. 41.

[9] Pap, *An Introduction to the Philosophy of Science*, pp. 281–282. Pap's analysis also differs from Kaila's and Addis's by invoking causal implication where the other two authors rely on the material conditional.

[10] A series of illuminating and suggestive observations on the issues touched upon in this section will be found on pp. 4–15 of W. V. Quine, *The Roots of Reference* (Open Court, La Salle, Illinois, 1974).

[11] The point is not affected by the fact that limitations of measurement do not permit the assignment of an event to an instant, but only to a finite time interval, and that the decay law yields some finite probability for the latter case. The point at issue hinges on the logical form of the explanatory laws: limitations of measurement notwithstanding, strictly universal laws do permit the deduction of consequences concerning what happens at some particular instant.

Moreover, even the disintegration of a given atom during a specified time interval may have an extremely small probability according to the decay law – depending on the length and location of the interval; accordingly, the law tells us that the decay was almost certain not to occur during that interval: and this hardly qualifies as an explanation of why it did occur all the same. My earlier analyses of statistical explanation do not qualify the statistical arguments associated with such cases as explanatory; but important alternatives to my way of viewing this issue have since been proposed by R. C. Jeffrey in 'Statistical Explanation *vs.* Statistical Inference' and by W. C. Salmon in 'Statistical Explanation'; both of these studies are included in W. C. Salmon (ed.) *Statistical Explanation and Statistical Relevance* (University of Pittsburgh Press, Pittsburgh, 1971). I doubt, however, that Addis would take comfort from these views in regard to his thesis about explainability.

JAMES H. FETZER

THE LIKENESS OF LAWLIKENESS*

ABSTRACT. The thesis of this paper is that extensional language alone provides an essentially inadequate foundation for the logical formalization of any lawlike statement. The arguments presented are intended to demonstrate that lawlike sentences are logically general dispositional statements requiring an essentially intensional reduction sentence formulation. By introducing a non-extensional logical operator, the 'fork', the difference between universal and statistical laws emerges in a distinction between dispositional predicates of universal strength as opposed to those of merely statistical strength. While the logical form of universal and statistical laws appears to be fundamentally dissimilar on the standard account, from this point of view their syntactical structure is basically the same.

A fundamental problem of the theory of science is to develop an adequate explication of the concept of a physical law. Theoreticians such as Goodman and Hempel especially suggest that a sentence S is a physical law if and only if (a) S is lawlike and (b) S is true, where S is supposed to be lawlike only if (i) S is logically general and (ii) S is capable of providing support for counterfactual and subjunctive conditionals.[1] Attempts to formalize lawlike sentences by employing the resources of such wholly extensional languages as the first-order predicate calculus, however, provide insufficient syntactical and semantical criteria for distinguishing between genuinely nomological and merely accidental true generalizations, leading other theoreticians – including Braithwaite and Ayer – to the conclusion that this difference has to be viewed as pragmatical, e.g., as a question of context or as a matter of attitude.[2]

The thesis of this paper, by contrast, is that extensional language alone affords an inadequate foundation *in principle* for the logical formalization of any lawlike statement. The following arguments are intended to establish that lawlike sentences are logically general dispositional statements requiring an essentially intensional reduction sentence formulation. By introducing a non-extensional logical operator, the 'fork', the difference between universal and statistical laws emerges in a distinction between dispositional predicates of *universal* strength and dispositional predicates of *statistical* strength. While the logical form of universal and statistical laws appears to be fundamentally dissimilar on the standard account, from this point of view their syntactical structure is basically the same. My purpose here, therefore, is to demonstrate the benefits of an intensional explication.

I. REFUTATIONS

In order to exhibit the deficiencies of extensional language for this role, the distinguishing features of extensional operators, such as the horseshoe (or material conditional), require explicit recognition, namely:

(a) that any molecular statement constructed through the application of extensional operators is truth-functional (in the sense that its truth-value is exhaustively determined by the truth-values of its component sentences); and,

(b) that the truth-values of atomic statements within such an extensional language are determined in each case by the history of the physical world (when they are not expressed in self-contradictory predicates).

The translation of 'if _____ then ...' sentences as material conditionals, therefore, fulfills this conception provided the truth-values of these statements are determinate for *every* combination of truth-values for their atomic constituents, whose own truth-values do not vary with the occasions of their use.[3] Quine has observed, "There is reason to believe that none but truth-functional modes of statement composition are needed in any discourse, mathematical or otherwise; but this is a controversial question".[4] Indeed, as we shall discover, the problems encountered in formalizing lawlike sentences present at least three reasons for rejecting that belief altogether.

On an extensional conception, lawlike sentences – universal or statistical – characterize some relation that may obtain between the members of two classes, i.e., a *reference* class and an *attribute* class. These classes must be logically general in the sense that they must not be limited to a finite number of members on the basis of syntactical or semantical considerations alone; the definitional properties of these classes, however, are otherwise logically arbitrary, i.e., they are pragmatically contrived.[5] A universal generalization thus merely asserts that every member of a reference class **K** is also a member of an attribute class χ; and a statistical generalization similarly asserts that a certain proportion p of members of **K** are also members of χ. The appropriate syntax for a universal lawlike statement (of the simplest kind) employing the predicate calculus is

(1) $(x)(\mathbf{K}x \supset \chi x)$

which, of course, will be true just in case every nameable thing participating in the history of the physical world either does not possess the property $\mathbf{K}$ or does possess the property χ.[6] Correspondingly, the appropriate syntax for a statistical lawlike statement by employing probability notation is

(2) $P(\mathbf{K}, \chi) = p$

which receives a finite frequency interpretation, if the world's history is finite, and a limiting frequency interpretation, if it is not.[7]

The criterion for distinguishing between merely accidental and genuinely nomological true generalizations of either kind, of course, is supposed to be that the latter do, while the former do not, provide support for counterfactual and subjunctive conditionals. The difficulty, however, is that there are no apparent syntactical or semantical criteria for separating the sentences that provide such support from those that do not. The universal generalization,

> Students who attend universities in tropical climates are less than 7′ tall

for example, which might be formalized as

(3) $(x)[(Sx \cdot Ux) \supset Lx]$

is supposed to be distinguishable from the universal generalization,

> Water at sea level atmospheric pressure boils at a temperature of 212 °F.

which might be formalized analogously as

(4) $(x)[(Wx \cdot Ax) \supset Bx]$

on the grounds that (4) supports the counterfactual,

> If the glass of water I just drank had been heated to 212 °F. at sea level atmospheric pressure instead, it would have boiled

while (3) does not support the counterfactual,

> If Kareem Abdul Jabbar had attended a university in a tropical climate rather than U.C.L.A., he would have been less than 7′ tall.

As two instances of the same truth-functional forms, however, these statements are syntactically identical and semantically indistinguishable; and as historical generalizations about the world's history, it would be surprising, indeed, if they were distinguishable on non-pragmatical extensional grounds.

The *first* significant reason for rejecting extensional formalizations of lawlike statements, therefore, is that there are no syntactical or semantical criteria for distinguishing those sentences that do from those that do not support counterfactual and subjunctive conditionals; for to explain the support for counterfactuals provided by statements such as (4) and not by statements such as (3) on the grounds that (4) is, while (3) is not, a lawlike generalization, is surely to beg the question. Quine has suggested that

> what could be accomplished by a subjunctive conditional or other non-truth functional modes of statement composition can commonly be accomplished just as well by talking *about* the statements in question, thus using an implication relation or some other strong relation of statements instead of the strong mode of statement composition. Instead of saying:
>
> > If Perth were 400 miles from Omaha then Perth would be in America
>
> one might say:
>
> > 'Perth is 400 miles from Omaha' implies 'Perth is in America',
>
> in some appropriate sense of implication.[8]

But Quine is promising more than he can deliver; for with equal justification one might maintain that

> If Perth were 400 miles from Omaha then Omaha would be in Australia

or, analogously, one might say:

> 'Perth is 400 miles from Omaha' implies 'Omaha is in Australia'

in that same appropriate sense.[9] Quine's way out, therefore, supplies scant comfort for the extensional account: transferring problems from an object- to a meta-language does not automatically provide for their solution. The result would seem to be the choice between a pragmatical distinction and none at all.

The *second* significant reason for rejecting extensional formulations of lawlike statements is that the syntactical and semantical criteria that *are* provided clearly conflate principles (apparently) suitable for *explanation* with principles suitable for *retrodiction*.[10] For if the lawlike statement,

> Whenever a bullet is fired into a piece of pine, it makes a hole

is adequately formalized in extensional language as

(5) $\quad (x)(y)(t)[(Bxt \cdot Pyt \cdot Fxyt) \supset Hxyt']$

– where t' is simultaneous with or subsequent to t – then surely that sentence is *equally* adequately formalized by its extensionally equivalent counterpart,

(6) $\quad (x)(y)(t)[(Pyt \cdot -Hxyt') \supset -(Bxt \cdot Fxyt)]$

which asserts,

> Whenever a piece of pine has no hole in it at t', it is not the case that a bullet was fired into it at t.

Yet while statement (5) would seem to be reliable for explaining why a certain piece of pine displayed a hole at a particular time t', statement (6) could hardly be invoked to *explain why* a bullet was not fired into some whole piece of pine at t. Not the least of the drawbacks to (6) as an adequate expression of the law in question is that it portends the explanation of temporally prior states by reference to temporally subsequent states; and while (6) may serve a perfectly useful and unobjectionable role as a principle of retrodiction, i.e., for the purpose of inferring *what* has happened rather than *why* it has happened, it appears to be both useless and objectionable as a principle of explanation. Since (5) and (6) possess precisely the same syntactical and semantical significance, however, surely (5) is lawlike *if and only if* (6) is lawlike; so if it is an important desideratum that lawlike statements, but not accidental generalizations, *may be invoked to explain their instances*, then (5) may justifiably be held to be lawlike to no greater and no lesser a degree than (6). If (6) is not a lawlike statement, therefore, neither is (5) – on pain of contradiction.

II. CONJECTURES

Considerations such as these, I believe, clearly support the possibility that extensional language itself is not sufficiently strong to capture lawlike properties of the physical world. Indeed, the fundamental difficulty might be explained intuitively as follows: extensional language, in principle, is *historical* and *descriptive*, but lawlike properties are *ontological* and *structural* in character. For it is the ontological structure of the world, so to speak, that generates the world's descriptive history *under the influence of a precipitating set of initial conditions*. As Popper has suggestively remarked,

A statement may be said to be naturally or physically necessary if, and only if, it is deducible from a statement function which is satisfied in all worlds that differ from our world, if at all, only with respect to initial conditions.[11]

The basic inadequacy with the extensional approach, in other words, is that this world could have exhibited many different histories under varying sets of initial conditions; but what *would happen* if those conditions *were different* could never be expressed by extensional language alone. The class of true generalizations of any one such history, as a result, inevitably encompasses a mixture of lawlike and accidental generalizations; for such an approach provides no basis for separating those statements true of all such worlds from those that are true of only one.

Although various theoreticians have acknowledged an intimate connection between the concepts of lawlikeness, dispositions, and counterfactual and subjunctive conditionals, the explication of lawlikeness that will now be advanced makes the relations involved here fully explicit; for the thesis I am proposing is that *a sentence S is lawlike if and only if S is an essentially dispositional statement of logically general form.*[12] The three most important aspects of this intensional explication are: (a) an ontological rather than an epistemological conception of dispositionality; (b) the recognition of dispositional properties of statistical as well as of universal strength; and (c) a distinction between dispositions that are permanent rather than transient properties of things.[13]

(a) The conception of dispositionality invoked here is intended to capture the kind of property dispositions are as features of the physical world wholly *independently of any consideration for the ease with which their presence or absence may be ascertained on the basis of experiential considerations alone*. The class of dispositional predicates is specified as follows:

(7) A predicate is *dispositional* if and only if the property it designates is
   (i) a tendency to exhibit appropriate response behavior under relevant test conditions; and,
   (ii) an actual physical state of some object or collection of objects.

The shape and color of a billiard ball, the malleability and conductivity of a metal bar, and the strength and charge of an electric current are *all* dispositions in this sense. Indeed, as Goodman has remarked,

almost every predicate commonly thought of as describing a lasting objective characteristic of a thing is as much a dispositional predicate as any other. To find non-dispositional, or manifest, predicates of things we must turn to those describing events – predicates like 'bends', 'breaks', 'burns', 'dissolves', 'looks orange', or 'tests square'. To apply such a predicate is to say that something specific actually happens with respect to the thing in question; while to apply a dispositional predicate is to speak only of what can happen.[14]

A dispositional property, therefore, is a tendency to exhibit an appropriate response event when subjected to a relevant test event that is an actual physical state of one object individually or of several objects collectively.

(b) It is important to recognize that dispositions may be either of two distinctive kinds. For the statements, 'This die and tossing device is fair', and, 'That metal bar is magnetic', are both dispositional by virtue of attributing to the object or collection of objects involved tendencies to display appropriate response behavior under relevant test conditions, yet they differ insofar as one property is *statistical*, the other *universal*. While failure to attract small iron objects within the vicinity of a metal bar would constitute *conclusive evidence* that that object was not magnetic (provided, of course, that no interferring conditions were present), for example, failure to show an ace as the outcome of a single toss with a certain die and tossing device would surely *not suffice* to establish that that die and tossing device were not fair (even though, again, the test was a test of the relevant kind). For in the case of the metal bar, the disposition involved is of such a kind that the appropriate outcome is supposed to occur as an *inevitable or invariable* response to every relevant test; while in the case of the die and tossing device, the property involved is of such a kind that an appropriate response, such as showing an ace, is only supposed to occur *more or less often*, i.e., with a characteristic frequency, which is exhibited as the typical result of a long *sequence* of trials rather than as the predictable outcome of any *singular* trial.[15]

What this means, in effect, is that dispositional properties are amenable to *varying degrees of strength*, where, for example, being magnetic is a tendency of universal strength to exhibit an appropriate response under relevant test conditions on *each and every* single trial; and being fair, by contrast, is a tendency of statistical strength to exhibit an appropriate response under relevant test conditions on *each and every* single trial as well. The difference between them, therefore, is that objects or set-ups possessing tendencies of universal strength would exhibit *the very same result* on each and every single trial, even though a trial sequence were infinitely long; while set-ups or objects possessing tendencies of statistical strength would *not* exhibit *the same result* on each and every trial, but would exhibit instead a characteristic distribution of outcomes over a sequence of trials – where the probability that this statistical distribution provides an accurate measure of the relative strengths of the corresponding tendencies for each of the possible outcomes becomes stronger and stronger as the length of the trial sequence increases without bound.[16]

(c) It is significant to note as well that some dispositions are permanent properties of things, while others are properties things have at one time and yet not at another. An ignition temperature of 30 °C and a half-life of 3.05 minutes, for example, are permanent dispositions of lumps of white phosphorous and chunks of polonium^{218}, respectively (although one happens to be a universal disposition and the other is statistical); but the properties of being magnetic and of being painted beige, attributed to a compass needle and to an office wall, are properties those objects might lose (the wall, for example, might be repainted blue or the magnet dropped). Indeed, even though a certain disposition χ may not actually be a permanent property of all objects of a particular kind **K**, nevertheless it might happen to be the case that every member of that class possesses alike that same transient disposition – as it might have been the case that typewriters are always gray or all professors' blond haired, without those properties being entailed by these reference class descriptions. The permanent dispositions of a certain class of things, therefore, may be defined as follows:

(8) A dispositional property χ is a *permanent property of every member of a reference class* **K** if and only if:
   (i) there is no process or procedure – whether natural or contrived – by means of which a member of **K** could lose the property χ without also losing membership in **K**; and,
   (ii) the possession of χ by a member of **K** is not logically entailed by the reference class description of **K**.

The permanent properties of members of the reference class consisting of objects whose molecules have the atomic number 79 and the atomic weight 196.967 include: being a relatively soft, yellow, ductile and very malleable metal, with a melting point of 1063 °C, and so on, for example; but they do *not* include having a certain shape, being sold in a jewelry store, or being referred to by the predicate 'gold'.

On the basis of these preliminary considerations, a more specific characterization of the intensional conception of lawlikeness may be explicitly defined as follows:

(9) A sentence S is *lawlike* if and only if:
   (i) S is logically general, i.e., S is not limited to a finite number of instances on the basis of syntactical or semantical properties alone; and,
   (ii) S is essentially dispositional, i.e., S attributes a permanent dispositional property χ to every member of a reference class **K**.

The class of accidental generalizations, therefore, not only encompasses those sentences that are *not* logically general, e.g., 'All the coins in my pocket are silver' and 'All the members of the Greenbury School Board for 1964 are bald'; but also those that, while logically general, are not *essentially* dispositional, e.g., 'All students who attend universities in tropical climates are less than 7′ tall' and 'All moas die before reaching the age of fifty'.

Since a property χ is a permanent disposition of a member **m** of a reference class **K** if, and only if, although being a member of χ is not logically entailed by the reference class description for **K**, **m** would not be a member of **K** if **m** were not a member of χ, the *basic form* of a lawlike statement is that of a *subjunctive conditional*, which may be represented symbolically by the non-extensional 'fork' operator as

(10) $(x)(\mathbf{K}x \mathrel{\supset\!\!\!-} \chi x)$

which asserts, 'For all x, if x were **K**, then x would be χ'; for example, 'For all x, if x were gold, then x would be malleable' and 'For all x, if x were polonium²¹⁸ then x would have a half-life of 3.05 minutes'. It is also the case, therefore, that subjunctive conditionals are subject to transposition, in the sense that any statement of form (10) is logically equivalent to a statement of form

(11) $(x)(-\chi x \mathrel{\supset\!\!\!-} -\mathbf{K}x)$

which asserts, 'For all x, if x were not χ, then x would not be **K**'; for example, 'For all x, if x were not malleable, then x would not be gold' and 'For all x, if the half-life of x were not 3.05 minutes, then x would not be polonium²¹⁸'.[17]

Since the truth-value of a subjunctive conditional is not determined solely by the truth-values of its atomic constituents when *either* its consequent is true *or* its antecedent is false, the fork is *not* a truth-functional connective. Even if 'This is gold' and 'This is sold in a jewelry store' are *both* true, for example, the truth-value of the compound 'If this were gold, then this would be sold in a jewelry store' is not thereby determined. The fork is an essentially stronger operator than the horseshoe, moreover, since the truth of the corresponding material conditional is a necessary condition for the truth of a subjunctive. Indeed, the fundamental relationship between them may be formalized by the (unrestricted) *principle of subjunctive implication*, namely:

(12)   '$p \mathrel{\supset\!\!\!-} q$'   implies   '$p \supset q$'

but *not* conversely. Nor is the subjunctive conditional the strongest intensional connective this explication requires.

The developments that led Carnap to propose reduction sentences as a method for the partial specification of the meaning of dispositional predicates are exceedingly widely known, e.g., that the explicit definition of 'solubility' by reference to the test event of being submerged in water and response event of dissolving, that is:

(13) x is soluble at $t =_{df}$ if x is submerged in water at t, then x dissolves at t'

logically implies that any object – such as a green leaf, a yellow cow, or a red pencil – is soluble at any time it is not actually being subjected to the relevant test.[18] As Pap has observed, however,

> If Carnap succeeds in deriving from (the concept of solubility) the paradoxical consequence that any substance is soluble in any liquid in which it is never immersed, this is due to his formalizing the definiens, which is ordinarily meant as a *causal* implication, as a material conditional.[19]

Indeed, it seems quite evident that even reduction sentences themselves provide no basis for overcoming the underlying difficulties; for although dispositional attributions are thereby restricted to those cases that actually satisfy test conditions of the relevant kind, these sentences remain purely extensional and therefore provide no support for counterfactual or subjunctive affirmations.

A partial explicit definition for a dispositional predicate χ, however, may be obtained within the framework of an intensional language by introducing one further sentential connective whose logical properties are stronger than those of the subjunctive by virtue of embracing a primitive *brings about* relation as well, i.e., a *causal conditional*. The claim that the occurrence of an event of kind T causes the occurrence of an event of kind 0 – that dropping a fragile glass onto concrete, for example, causes it to break – means not only that such an object *would* break if it *were* dropped but also that dropping it would *bring about* its breaking. Since it is obviously *not* the case that its not breaking would be supposed to *bring about* its not being dropped, the distinguishing property of the causal conditional is that, unlike the subjunctive conditional, it is *not* subject to transposition.[20] Since there are dispositions of statistical as well as of universal strength, the causal conditional is *probabilistic*, i.e., it is applicable with degrees of strength n whose values range through varying statistical strengths p from zero to one, i.e., '$\supset_p$', to universal strength u, i.e., '$\supset_u$' (*not* to be confused with probabilities of one), where the appropriate numerical value is determined by the strength of the disposition that is involved. A dispositional property χ, therefore, may be partially defined as

(14) x is χ at $t =_{df} (T^1 xt \underset{n}{\supset\!\!\!-} 0^1 xt') \cdot (T^2 xt \underset{n}{\supset\!\!\!-} 0^2 xt') \cdots$

where 'T^1', 'T^2', and so on describe relevant test conditions, '0^1', '0^2', and so forth describe appropriate outcome responses. The disposition of fragility, for example, could be partially defined by specifying such test conditions and response outcomes as: dropping x on concrete at t would bring about its breaking at t'; hitting x with a hammer at t would bring about its cracking at t'; and so on. Since there are no *a priori* boundaries on the variety of different kinds of test conditions and outcome responses that constitute manifestations of such a disposition, a partial definition of this kind is *exhaustive* if and only if it captures them all; yet it remains the case that each conjunct of a conjunctive definition simply reflects one necessary condition for the application of such a term and, consequently, the 'open-texture' of reduction chains is, in general, a property of these definitions as well.

Although the basic form of lawlike sentences *is* the subjunctive conditional, therefore, the causal character of lawlike statements only emerges in full force when the dispositional predicate χ in sentences of form (10) is replaced by one of its defining conjuncts, thereby exhibiting the form of

(15) $(x)(t)[\mathbf{K}xt \supset\!\!\!- (T^1 xt \underset{n}{\supset\!\!\!-} 0^1 xt')]$

which asserts, 'For all x and all times t, if x were **K** at t, then the strength of the dispositional tendency for T^1-ing x at t to bring about 0^1-ing x at t' is n'; or, less formally, 'For all x and all t, if x were **K** at t, then T^1-ing x at t would either *invariably* or *probably* bring about x's 0^1-ing at t'.' Among the sentences which reflect the causal significance of the lawlike statement, 'The melting point of gold is 1063 °C', therefore, is one that asserts,

(16) For all x and all t, if x were gold at t, then heating x to 1063 °C at t would invariably bring about its melting at t';

and among those sentences for the lawlike statement, 'The half-life of polonium^{218} is 3.05 minutes', is one that asserts,

(17) For all x and all t, if x were polonium^{218} at t, then a time trial

of 3 minutes duration at t would very probably bring about the loss of nearly half the mass of x by $t+3$.

Although all lawlike statements are expressible *either* as subjunctive conditionals *or* as causal conditionals, therefore, their *explanatory* significance, in general, radiates *exclusively* from their formulation as causal conditionals.[21]

Since the nomological significance of (15) is equally well displayed by

(18) $(x)(t)\,[(\mathbf{K}xt \cdot T^1 xt) \mathrel{\underset{n}{\supset\!\!\!-}} 0^1 xt']$

appropriate principles of commutation and exportation render (18),

(19) $(x)(t)\,[T^1 xt \mathrel{\supset\!\!\!-} (\mathbf{K}xt \mathrel{\underset{n}{\supset\!\!\!-}} 0^1 xt')]$

and (15) logically equivalent. By the principle of subjunctive implication, therefore, lawlike statements logically imply such statements as

(20) $(x)(t)\,[\mathbf{K}xt \supset (T^1 xt \mathrel{\underset{n}{\supset\!\!\!-}} 0^1 xt')]$,

and

(21) $(x)(t)\,[T^1 xt \supset (\mathbf{K}xt \mathrel{\underset{n}{\supset\!\!\!-}} 0^1 xt')]$.

These forms are strikingly relevant for counterfactual attributions; for if an object **m** is a member of **K** that was not subjected to test T^1 at t, for example, it nevertheless follows that if **m** *had been* subjected to T^1 at t, **m** *would have* (invariably or probably) exhibited 0^1 at t'; and so forth.[22] Finally, when n is equal to u (but *not* when n is equal to p), the (restricted) *principle of causal implication* holds, namely:

(22) '$p \mathrel{\underset{u}{\supset\!\!\!-}} q$' implies '$p \supset\!\!\!- q$'

but not conversely. Statistical lawlike statements, however, have no similar purely subjunctive or purely extensional *causal* implications.

The *in principle* inadequacy of extensional language for the formalization of lawlike statements is therefore clearly apparent; for although every (universal) lawlike sentence logically implies the corresponding material conditional, *no set of extensional statements is logically equivalent to any lawlike statement*. And it may be remarked, by way of conclusion, that the paradoxes of confirmation – the *third* important reason for discounting the extensional approach – arise entirely from the extensional implications rather than from the intensional significance of lawlike statements. The benefits to be derived from this account, therefore, are promising, indeed.

University of Kentucky

NOTES

* This essay is dedicated to Linda M. Sartorelli. The author is indebted as well to Donald Nute of the University of Georgia for suggestive criticism.
[1] Nelson Goodman, *Fact, Fiction, and Forecast* (The Bobbs-Merrill Company, Indianapolis, 1965); and Carl G. Hempel, 'Studies in the Logic of Explanation', Sections 6 and 7, and 'Aspects of Scientific Explanation', especially pp. 338–343, *Aspects of Scientific Explanation* (The Free Press, New York, 1965).
[2] A. J. Ayer, 'What is a Law of Nature?', *Revue Internationale de Philosophie* (Brussels, Belgium), No. 36, fasc. 2 (1956); and R. B. Braithwaite, *Scientific Explanation* (Cambridge University Press, Cambridge, 1953). Ayer's article and a relevant selection from Braithwaite's book are reprinted in B. Brody (ed.), *Readings in the Philosophy of Science* (Prentice-Hall, Inc., Englewood Cliffs, N.J., 1970).
[3] The sentences resulting from the instantiation of extensional sentential schemata, in other words, are *eternal* rather than *occasion* sentences in the sense of W. V. O. Quine, *Word and Object* (The M.I.T. Press, Cambridge, 1960), pp. 191–195.
[4] W. V. O. Quine, *Mathematical Logic* (Harper and Row, New York, 1951), p. 11.
[5] For the purposes of chemistry, for example, a classification of elements on the basis of such structural properties as atomic weight and atomic mass has proven useful; for the purposes of biology, a classification of organisms on the basis of phylogenetic and genetic properties has proven useful; and so on.
[6] Not every *named* thing, as Pap appears to think; cf. Arthur Pap, 'Reduction Sentences and Disposition Concepts', in P. A. Schilpp (ed.), *The Philosophy of Rudolf Carnap* (Open Court, La Salle, Ill., 1963), pp. 593–597; and see also Carnap's reply, pp. 948–949.
[7] A finite frequency interpretation is advanced by Bertrand Russell, *Human Knowledge: Its Scope and Limits* (Simon and Schuster, New York, 1948), pp. 350–362. Classical limiting frequency interpretations are Hans Reichenbach, *The Theory of Probability* (University of California Press, Berkeley, 1949); and Richard von Mises, *Mathematical Theory of Probability and Statistics* (Academic Press, New York, 1964).
[8] Quine, *Mathematical Logic*, p. 29.
[9] Cf. Henry Kyburg, *Philosophy of Science: A Formal Approach* (The Macmillan Company, New York, 1968), pp. 313–317.
[10] For an analysis of some significant consequences attending their confusion, see James H. Fetzer, 'Grünbaum's "Defense" of the Symmetry Thesis', *Philosophical Studies* (April, 1974). The present investigation discloses the principles that underlie all such misunderstandings; for although the extensional implications of lawlike statements, for example, may perfectly well serve the purposes of prediction and retrodiction, only *intensional* formulations are sufficiently strong for the purpose of explanation.
[11] Karl Popper, *The Logic of Scientific Discovery* (Harper and Row, New York, 1960), p. 433. Popper's formulation has been subjected to criticism by G. C. Nerlich and W. A.

Suchting, 'Popper on Law and Natural Necessity', *British Journal for the Philosophy of Science* **18** (1967), pp. 233–35. This critique plus Popper's rejoinder and many other related papers are published together in Tom Beauchamp, ed., *Philosophical Problems of Causation* (Dickenson Publishing Company, Belmont, Calif., 1974). With respect to the present explication, Popper's formulation is ambiguous unless it is specified that all permanent and transient dispositional properties of things are kept constant, i.e., they remain permanent and transient properties of things of just the same kinds.

[12] The theory of explanation that attends this explication of lawlikeness is set forth in James H. Fetzer, 'A Single Case Propensity Theory of Explanation', *Synthese* (October, 1974). See also note 21 for a non-trivial clarification.

[13] An intriguing alternative to the present account of dispositions is provided by D. H. Mellor, 'In Defense of Dispositions', *Philosophical Review* (April, 1974).

[14] Goodman, *Fact, Fiction, and Forecast*, p. 41.

[15] James H. Fetzer, 'Dispositional Probabilities', *Boston Studies in the Philosophy of Science*, Vol. VIII, R. Buck and R. Cohen (eds.), D. Reidel Publishing Company, (Dordrecht, Holland, 1971); and James H. Fetzer, 'Statistical Probabilities: Single Case Propensities vs. Long Run Frequencies', *Developments in the Philosophy of Social Science*, W. Leinfellner and E. Köhler (eds.), (D. Reidel Publishing Company, Dordrecht, Holland, 1974).

[16] *Ibid*. The latter includes a discussion of the relevant principles of relevance.

[17] All lawlike conditionals that are true are maximally specific as follows: if a nomically relevant predicate is added to the reference class description (i.e., to the description of K, relative to χ, or of $\mathbf{K} \cdot \mathbf{T}$, relative to 0), then either the resulting statement is a logical truth (by virtue of the fact that its antecedent condition is now self-contradictory) or it is logically equivalent to the original statement (since the additional relevant predicate is actually redundant). Cf. Carl G. Hempel, 'Maximal Specificity and Lawlikeness in Probabilistic Explanation', *Philosophy of Science* (June, 1968), p. 131.

[18] A lucid summary of the most important results involved in these investigations is provided by Carl G. Hempel, *Fundamentals of Concept Formation in Empirical Science* (University of Chicago Press, Chicago, 1952), Part II.

[19] Pap, *op. cit.*, p. 560.

[20] Pap mistakenly believes that 'the distinctive property of causal implication as compared with material implication is just that the falsity of the antecedent is no ground for inferring the truth of the causal implication'; see Arthur Pap, 'Disposition Concepts and Extensional Logic', *Minnesota Studies in the Philosophy of Science*, Vol. II, H. Feigl, M. Scriven, and G. Maxwell (eds.), (University of Minnesota Press, Minneapolis, 1958), p. 212. Davidson purports to demonstrate that the causal connective cannot be a conditional of any kind, essentially on the basis of his observation that, 'My tickling Jones would cause him to laugh, but his not laughing would not cause it to be the case that I didn't tickle him'; 'Causal Relations', *Journal of Philosophy* (November 9, 1967), pp. 691–695.

[21] More precisely, explanations for the occurrence of singular events require the causal formulation of lawlike statements, while the subjunctive formulation of these same statements appears to be appropriate for explaining why particular objects happen to possess certain specified properties.

[22] Counterfactual attributions, in other words, are – implicitly or explicitly – equivalent to conjunctions of the form, '$(T\mathbf{m}t \mathrel{\supset_{\bar{n}}} O\mathbf{m}t') \cdot -T\mathbf{m}t$', e.g., which, in turn, may be accounted for on the basis of the fact that $\mathbf{m}$ is a member of $\mathbf{K}$ and $(x)(t)[\mathbf{K}xt \supset\!\!- (Txt \mathrel{\supset_{\bar{n}}} Oxt')]$. Counterfactuals, therefore, are properly understood as special kinds of subjunctives, i.e., as instantiated subjunctives with historically false antecedents.

PETER KIRSCHENMANN

TWO FORMS OF DETERMINISM

> Die Ereignisse sind von endlicher Ausdehnung in Raum und Zeit, und ihre Individualität ist bestimmt durch die *diskreten Eigenwerte* der Operatoren, die den physikalischen Eigenschaften zugeordnet sind. Dann folgt aber der statistische Charakter der Theorie automatisch. Eine durch Differentialgleichungen ausgedrückte Beziehung zwischen den Ereignissen zu verlangen, würde einen Widerspruch implizieren.
>
> JOACHIM HÖLLING (1936–1968)

I. INTRODUCTION

Philosophers are continually challenged to comprehend chance phenomena and probabilistic relationships. Those cases are especially intriguing in which we speak of chance or probability but can hardly attribute this manner of speaking to our being partially or totally ignorant of the situations in question. Those cases, then, can be taken to possess chance as an objective feature. In its intent, this paper is restricted to a discussion of such cases. Among objective chance relationships there is one kind which seems especially incomprehensible. These are stochastic relationships, i.e., chance relationships between consecutive states of physical systems. The existence of such relationships offends the common conviction that the temporal evolution or succession of states in physical systems is fully deterministic.

It would help our understanding of chance phenomena if one could show that chance is an inevitable feature of the world – that, e.g., some physical systems possess certain quite intelligible properties which imply that the evolution of their states must be determined in a stochastic way. I do not think that one can show quite as much. What one can hope for, however, is to be able to see that, given certain kinds of physical systems with particular properties, it makes sense that there can *at best* be chance relationships holding between their consecutive states. I shall try to provide such a positive view of temporal stochastic determination. The claim is that in such systems we encounter a form of determinism different from, but in no way inferior to, dynamical determinism, i.e., the form we ordinarily refer to as 'determinism'. Judged by the standards provided by dynamical determinism, stochastic determination is often characterized in a merely negative fashion, viz., as a form of indeterminism.

To prepare the ground for this positive view of the form of determinism involving stochasticity, I shall first formulate a general, but precise notion of dynamical determinism and mention some of its conceptual components and presuppositions. It will be compared with a notion of stochastic determination which is the closest possible counterpart of it. Our comparison will include an examination of the compatibility of the two notions and an argument for the non-eliminability of probability statements, which is of some importance for the positive assessment of stochastic determination.

II. DYNAMICAL DETERMINISM

There are many types of deterministic relationships, e.g., deterministic laws of coexistence between two or more properties of a physical system, or between the state of a system and its various properties. The relation mostly referred to by the term 'determinism' is that holding for the dynamical-deterministic evolution of states of certain physical systems. The notion of such a dynamical-deterministic system is well known from theories of classical physics like classical particle mechanics and, though with different meanings of 'physical system' and 'state', classical field theories. This notion, which will serve us as a standard of reference, can be defined as follows (cf. Mackey, 1963, pp. 1f.; van Fraasen, 1972, p. 315).

(DD)   Let S be the state space of a physical system, i.e., the set of all states s that the system is capable of ('admissible' states). Let s_t be the state of the system at time t. The system is said to be *dynamical-deterministic* precisely if there is for each time interval m a mapping U_m of S onto S such that, for any time t, $s_{t+m} = U_m s_t$.

The family of functions $\{U_m\}$ form a one-parameter group of transformations with the properties: (1) $U_0 s = s$; (2) $U_{m+n}s = U_m U_n s$; (3) for each U_m there is a U_m^{-1}, namely U_{-m}, such that $U_m^{-1} U_m s = U_0 s$.

The group is called the dynamical group of the system. It is implicitly defined by the dynamical laws of the system, which, in the theories that the notion (DD) refers to, invariably are expressed by differential equations. Given any state of the system, the dynamical laws thus determine its total past and future history[1]; this idea, when applied to the universe as a whole, amounts to a Laplacean determinism. Any admissible *history*, or trajectory, of the system is a mapping of time into its state space[2], or the set of all states s_{t+m} generated by the group $\{U_m\}$ for varying m from a given possible state s_t.

For the purpose of comparing dynamical determinism with stochastic kinds of determinism, I shall list several general conditions and presuppositions implicit in the notion just defined.

(a) *Uniqueness*. For any state s_t of a dynamical-deterministic system and for every time interval m there is exactly one state s_{t+m}, and conversely. The latter is uniquely determined by the particular state s_t and the time parameter m, and conversely. More precisely, (DD) implies

(DD1) For any state s^i of a dynamical-deterministic system and every time interval m there is one s^j such that $s_{t+m} = s^j$ if and only if $s_t = s^i$.

This uniqueness of determination implied by (DD) can also be expressed as follows (cf. van Fraasen, 1972, p. 315):

(DD2) If any two admissible histories of a dynamical-deterministic system have one state in common then they are identical.

It is especially the 'forward' part of the biconditional in (DD1), viz.,

(DD3) if $s_t = s^i$ then $s_{t+m} = s^j$,

which usually is taken as the essential feature of all deterministic temporal relationships. In general, then, any temporal relationship between an event (occurrence of a state, change of state) and its antecedent conditions is said to be deterministic if the latter (the past) uniquely determines the former (the future), or, in obvious notation,

(DD4) if $C_t = C^i$ then $E_{t+m} = E^j$.

This means that, given conditions C^i at time t, the event E^j will necessarily occur at time $t+m$.

(b) *Temporal determinism*. The dynamical group of a dynamical-deterministic system is a one-parameter group, time being the only parameter. Given the state s_t at some time t, the state $s_{t'}$ at any other time t' depends on the time difference $t'-t$ alone. In this sense, dynamical determinism is a *purely* temporal determinism.

This is only so for the free evolution of isolated, or closed, physical systems. The states of a system open to outside influence do not depend merely on the time differences between them and some given state, but also on other parameters. (It is not clear to me whether the histories of open systems can generally be represented in terms of the operation of some multi-parameter group). One can assume, though, that open systems can be treated as parts of more inclusive closed systems, and thus discussion can be restricted to isolated systems[3].

(c) *'Reversibility'*. Dynamical determinism as defined above is only 'reversible' in the sense that, as has been mentioned, both the future history and the past history of a dynamical-deterministic system are uniquely specified by its dynamical group, given any of its states[4]. Note that this is less than saying that every process that the system can undergo is reversible[5]. What (DD) asserts, then, is not process-reversibility (or, time-reversal symmetry) but the bilateral determinism which is more explicitly brought out by (DD1).

As mentioned, this 'reversible', bilateral, determinism is not regarded as a necessary feature of all deterministic temporal relationships. It is not present in processes that satisfy only (DD3) or (DD4).

(d) *Continuity*. First, the *state space*, the set of all admissible states, of a dynamical-deterministic system, forms a continuum. States, and hence histories, can differ from one another in a continuous fashion. Also, such differences can vary continuously. (This is guaranteed if the state space is assumed to be a vector space.)

Moreover, secondly, any *history* is itself a continuous set of states, viz., a continuous function mapping time into state space. And again, in all classical theories, this function is assumed to be at least twice-differentiable.

(e) *'Homogeneity' of states.* The states of a dynamical-deterministic system are all of the same kind: they are specified in terms of the admissible values of the same state variables. Thus, we can say (comparing, e.g., the general condition (DD4) with the specific condition (DD3)) that the antecedent conditions determining a state are 'homogeneous' with this subsequent state, as the antecedent conditions are nothing else than another state of the system.

(f) *Physical systems and their structure.* The notion of dynamical determinism refers to physical systems with well-defined state spaces. These systems as well as their general structure and properties are simply presupposed as given; dynamical determinism itself concerns only the changes in given systems. Structural peculiarities are restrictions on the state space and specifications of the particular dynamical laws of a given system. The basic structure and properties of the presupposed physical systems, then, are among the factors determining their temporal evolution, but are not affected by it or its laws. (The only exceptions in classical physics are cases of free fields in classical field theories, in which of course there are no pregiven structural constraints or components.)

In dynamical-deterministic theories, one also assumes that the physical systems considered actually are in one state at some time. Given such a state, the dynamical laws determine the subsequent states. But they do not specify, or even restrict, the possible initial states.

Finally, it should at least be mentioned that other general assumptions are made in all dynamical-deterministic theories, assumptions concerning the nature of the systems in question, which define the specific mathematical character of their states and state spaces. In classical particle mechanics, e.g., the systems are considered to be configurations of stable particles.

III. STOCHASTIC DETERMINATION

Just as there are many types of deterministic relationships there is a wealth of different types of probabilistic relationships. There are probabilistic conjunctions of different properties of systems; there are chance coincidences of otherwise independent systems; and there are stochastic relations between events at different times. The type of relationship that comes closest to being a counterpart of dynamical determinism is one which also governs the evolution or succession of states of physical systems (cf. Feller, 1968, p. 420). The notion of such a stochastic determination can be defined as follows.

> (St1)   Let R be the state space of a physical system, i.e., the set of all states r that the system could be in. Let the state r_t of the system at time t be the state r^k, or, $r_t = r^k$. Then, the system is said to be *stochastic* (or, the evolution of its states is said to be stochastically determined) precisely if there is for each time interval m and each state r^l a probability p_m^{kl} (with $0 \leq \leq p_m^{kl} \leq 1$) such that for any time t the probability that the state r_{t+m} is r^l (or, $r_{t+m} = r^l$) equals p_m^{kl}, or briefly[6]: $P(r_{t+m} = r^l / r_t = r^k) = p_m^{kl}$.

The process thus characterized is known as a 'time-homogeneous Markov process' (Feller, 1968, pp. 420f.): 'Markovian' because a given state (the present) alone, and not the past history of the system, stochastically determines the future of the system; and 'time-homogeneous' because the transition probabilities p_m^{kl} do not depend on time itself, but only on the time differences m. Processes of the type (St1), then, share these two features with dynamical-deterministic processes. Moreover, both (DD) and (St1) express lawlike relationships, and just as dynamical determinism has its basis in the particular dynamical laws holding for a given system, so any stochastic determination is (or should be) grounded in a specific stochastic theory of the systems concerned. Further similarities, but also differences, will be discussed later.

The notion of stochastic determination defined by (St1) finds application in the treatment of various processes in physics as well as other fields (see Feller, 1968 and 1966; Prabhu, 1965). Among the physical processes which can be described in this manner are Brownian motion and diffusion processes in general, radioactive decay and other quantum-physical processes[7]. As the notion (DD) captures only some general but essential features of dynamical-deterministic systems and processes, so the notion (St1) captures only certain general features of stochastic systems and processes. In particular, it should be noted that it does not cover all stochastic or probabilistic aspects of quantum mechanics, but only those which have to do with state transitions[8].

While (St1) supplies a quite general notion of stochastic determination, it can still be regarded as a specific case of a more general notion of a lawlike temporal stochastic relationship, which can be characterized as follows:

(St2)   An event (occurrence of a state, change of state) is said to be stochastically determined by certain antecedent conditions precisely if there is a probability p such that whenever these conditions obtain at some time the event will occur at some later time with probability p, or, symbolically, that
if $C_t = C^k$ then $P(E_{t'} = E^l) = p$
for all times t and some time t', where $t' > t$.

The coming up of a certain number of points in a throw of dice is a stochastically determined event E^l in this sense, where C^k comprises the act of the throw and its particular circumstances.

What is characteristic of all temporal and genuine stochastic relationships (i.e., those cases in which none of the particular probabilities have value one) is that the event in question *may* occur but *also may not* occur, or, that it is possible that one or another state occurs in a system. The specification, by a stochastic theory, of the probabilities referred to in (St1) implies a specification of the states which definitely can follow upon a given state: transitions to states which cannot follow will be assigned probability zero[9]. Here 'possible' (or, 'definitely possible' for emphasis) has to be understood in a strong sense: in virtue of the stochastic laws holding for a given system, a possible transition or event is not just one which is not impossible, but rather one which cannot be impossible. For a genuine stochastic system, then, (St1) implies that

(SD1)   if the system is in one state then there are at least two admissible states both of which are possible.

And (St2) implies for a genuine stochastically determined event that

(SD2)   if $C_t = C^k$ then it is at once possible that
$E_{t'} = E^l$ and $E_{t'} \neq E^l$.

Expressed in a merely negative way, this means that, given the conditions C^k at time t, the event E^l is not a necessary consequence.

It is clear that (SD1) or (SD2) do not in turn imply (St1) or (St2), or, that the former can hold without the latter. Specific assumptions about the definitely possible events or states among those which are generally admissible are often made before one turns to a consideration of probabilities. In the treatment of Brownian motion and in that of many other stochastic processes, e.g., one assumes that from a given state the system can pass only to neighboring states. These states, then, are considered possible, while all other admissible states are definitely not.

IV. COMPARISON

In the following sections, I shall examine in what sense stochastic determination is compatible with dynamical determinism, and in what sense it is not. The study of one way in which they can be compatible will supply an argument supporting the view that stochastic theories or objective probability statements cannot be reduced to deterministic theories or statements. In discussing the possible incompatibility of the two forms of determination, we shall also inquire into its general reasons. This will lead to the question of whether stochastic determination is rightly considered as a kind of indeterminism. Taken literally, this view suggests that some of the conceptual components of the notion of dynamical determinism, responsible for the incompatibility, could be weakened so as to yield a notion of stochastic determination. It will become clear that this view is inadequate.

(A) *Compatibility, and Non-Eliminability of Probability Statements*

The notions of dynamical determinism (DD) and stochastic determination of the type (St1) are compatible with each other, even when applied to the same system; and they can be so in at least three ways.

The *first* involves the idea that any strict determinism can be regarded as the limiting case of a stochastic kind of determinism: when one factor (like a state, property, set of conditions) uniquely determines some other factor the former can be said to determine the latter with probability one[10]. Indeed, assuming that both (DD) and (St1) are satisfied by a physical system and refer to the same state space, i.e., S is identical with

R, we can conclude that for any time t and any time interval m

$$P(r_{t+m}=r^l/r_t=r^k)=1 \quad \text{for} \quad r^l=U_m r^k, \quad \text{and 0 otherwise.}$$

The state $U_m r^k$ is the state uniquely determined by the dynamical group of the system, given that its state is r^k at time t. The fact that (DD) and (St1) can both be satisfied by the system shows their compatibility.

Secondly, and more interestingly, stochastic determination (be it a genuine or a limiting case) may have a deterministic underpinning. More precisely, if (St1) is satisfied by a system with regard to state space R there still can be another state space S for the same system with respect to which (DD) holds, and where the states in R and those in S are correlated in some definite way. A simple case in point would be that in which the states r are themselves sets of states s, and thus subsets of S, which exhaust S.[11] In this case, (St1) and (DD) will lead to the following two statements.

(1) The probability that a state in r^k will be followed by a state in r^l at the time $t+m$ is p_m^{kl}.

(2) Any one state s at time t belonging to a particular set $r^{d(l)}$ will with strict deterministic certainty lead to a state in r^l at time $t+m$.

These two statements are compatible; and what is more, they imply[12]:

(3) Given that state s at t is in r^k, the probability that state s is in this particular set $r^{d(l)}$ is p_m^{kl}.

Following common terminology, one could say that the notions of stochastic determination and dynamical determinism are found compatible in this case because the two notions refer to the same state under different 'descriptions': the one refers to it as a member of a subset of S (this subset being an element of R), the other as an element of S. This is quite correct. Only, the term 'description' can be misleading unless it is understood that it does not necessarily indicate some arbitrariness or subjectivity in our reference to the object in question. Different descriptions of things are warranted inasmuch as they all capture objective traits of things through which they bear different lawful relations to other factors. Any description of a certain kind of thing involves some element of abstraction.

Various other views are intimately related to this idea of 'description-relativity'. A rather persistent view is that a more detailed, or more 'complete', description is by itself preferable to, and should eventually replace, any less detailed or incomplete description of a given object. This idea is mostly accompanied by the subjectivist interpretation of probability which takes all talk of 'probability' as a mere expression of our ignorance. Concerning the simple case discussed, one would argue that, knowing certain conditions (that the state is r^k of R), we can predict subsequent states of the system only with some probability; but if the conditions are specified and known in greater detail or 'completely' (switch to S and states $s \in r^k$) then subsequent states can be predicted with certainty. At bottom, the argument is certainly correct as regards the present case, although our making probability statements need not be an expression of ignorance: in the present case they would presumably be based on our knowledge of certain objective features of the system. One would definitely have to object, however, if the argument was meant to show that we can get rid of probability statements in this way and could, at least in principle, always do with deterministic statements alone. As we have seen, this is not so. If the probability statement (1) holds true of a system it can, in this way, only be displaced by another probability statement, namely (3). Probability statements, once admitted, cannot be eliminated.

Analogous points can be made about statistical statements. It is often held that we resort to statistics about certain populations only because we do not or cannot know all the details about its members, which suggests that we would not have to do so and would be better off if we knew them. Poincaré (while still accepting the ignorance interpretation of probability) countered this view with a well-taken practical example: even if an insurance company knew exactly when its clients were to die, this knowledge would have no effect upon the company's dividends (Poincaré, 1912, pp. 3f.; cf. Bunge, 1951, p. 212). To use another example, even if we had 'complete' knowledge concerning any individual throw of dice, the statistical regularity noticed in the results of series of throws could not be forgotten, but would compel us to conclude to a corresponding statistical regularity in the initial and other relevant conditions of throws of dice (cf. Landé, 1973, pp. 112ff.). And statistical regularities can indicate the presence of stochastic relationships; probability hypotheses are capable of accounting for statistical regularities.

Our argument for the non-eliminability of probability statements may appear to resemble von Neumann-type arguments for the completeness of quantum mechanics, which is a stochastic theory, and the impossibility of 'hidden variables' (von Neumann, 1955, pp. 295–328). Of course, our relatively simple argument is not on the same level as those sophisticated arguments, which draw upon many specific quantum-mechanical theorems. Moreover, on the face of it, we seem to have established that an underlying deterministic theory ('hidden-variable' theory) is possible for stochastic theories. A few remarks, then, about the relation between our argument and von Neumann-type arguments are in order[13].

Whereas our argument is concerned with an individual system, those arguments are directly concerned only with ensembles of quantum-mechanical systems. For such ensembles, von Neumann showed that even in the case of 'homogeneous' ensembles not all physical quantities are 'dispersion free'. Insofar as in the 'homogeneous' case all systems are considered to be in the same state, this result applies also to individual systems. It can be taken as re-emphasizing the essentially stochastic character of quantum mechanics.

Von Neumann went on to argue that the laws of quantum mechanics could never be re-derived from a 'hidden-variable' theory, and that these laws could not be retained as correct if 'hidden variables' existed. The second part of this claim is undoubtedly wrong, for 'hidden-variable' reconstructions of quantum mechanics have been worked out and thus it is theoretically possible that 'hidden variables' exist (cf. Bunge, 1967, pp. 291f.). Only, no such theory can do without stochastic (or 'statistical') assumptions. That means, in accordance with the first part of the claim, that quantum mechanics cannot be reconstructed on the basis of a *purely* deterministic theory. This is in line with what our argument is meant to show quite generally, not just for quantum theory, but for any well-founded stochastic theory or probability statement.

One can see this more clearly if our argument above is partly reversed. The probability statement (1) is not derivable from the deterministic statement (2) unless a stochastic premiss to the effect of statement (3) is added (cf. Mellor, 1971, p. 152). This means that a purely deterministic account is incapable of explaining facts captured by probability statements. (By definition, the concept of probability is absent from purely deterministic theories, and will have to be introduced and connected with the theory before it can supply an account of probability statements). Whenever probability statements successfully account for statistical regularities, a switch to a purely deterministic conception of the world would therefore have to leave these regularities unexplained.

As a *third* way in which dynamical determinism can be compatible with stochastic determination I shall merely mention the possibility that dynamical determinism may possess a stochastic underpinning. For instance, stochastic theories may imply dynamical laws as holding for certain averages defined over the state space in question. That this is so is the guiding idea behind the on-going efforts to account for macroscopic and deterministic relationships in terms of microscopic and stochastic (or statistical) theories. (The philosophical significance of the results achieved so far, however, has in many cases not yet been fully clarified; cf. Bunge, 1970). It can of course be questioned whether dynamical laws holding for averages of certain quantities can without further assumption be regarded as constituting a determinism in the strict sense. Yet, this much seems to be clear from our discussion: while stochastic theories – conceptually richer as they are – may provide a foundation for deterministic theories, there can be no hope of reducing stochastic theories to purely deterministic ones.

We have seen three ways in which a stochastic kind of temporal determinism and dynamical determinism can be present in one and the same physical system. Yet, the notions of stochastic determination and dynamical determinism on which our discussion has been based are rather general notions. Thus we can only claim to have shown the general possibility that these two kinds of determinism could hold for the same system. Any more concrete investigation would have to refer to specific physical systems and be based on specific theories proposed for them, which would include an analysis and comparison of the specific structure and presuppositions of such theories.[14]. However, our examination of the general compatibility of stochastic and dynamical determinations is sufficient to bring out the fact that the stochastic elements of a given theory cannot be circumvented: it cannot be recast in a purely deterministic fashion.

(B) *Incompatibility*

The notion of a genuine stochastic determination of the type (St1) clearly is incompatible with that of dynamical determinism (DD) if both refer to the same state space, so that S is identical with R. Let the state of the system at time t be s^i; then, by (DD), there is a s^j such that the state of the system at time $t+m$ is $s^j = U_m s^i$, whence $P(s_{t+m} = s^j/s_t = s^i) = 1$. On the other hand, since we are now concerned with the same state space, we find for genuine stochastic determination, by (St1), that $P(s_{t+m} = s^j/s_t = s^i) = p_m^{ij} \neq 1$, and thus we have a contradiction.

What is the incompatibility in this case due to? In discussing this question we shall refer to the general characteristics of dynamical determinism listed in the beginning. Obviously, the contradiction is directly due to the fact that any strictly deterministic relation, whether dynamical or not, involves the *unique* determination of the determined element, whereas genuine stochastic relations do not and must not. In the case of temporal relations, like the present case, this difference is brought out by the statements (DD4) and (SD2). With the proper identifications of conditions, events, and times made, and with the assumption that the conditions obtain, the two statements contradict each other. Whether other characteristics of dynamical determinism are also (indirectly) incompatible with a genuinely stochastic kind of determinism depends on whether they involve this uniqueness condition or not.

As can be inferred from (St1),[15] *continuity* of the *state space* itself is compatible with stochastic determination. But, as regards the *history* of the system, the situation is different and more complex. It would seem that essential continuity of its history, which is characteristic of a dynamical-deterministic system, cannot be present in the temporal evolution of a stochastic system. Yet, we know about stochastic processes which trace out continuous trajectories. In Brownian motion, e.g., the path of the particle studied is continuous. An examination of the stochastic process describing Brownian motion, however, reveals that the acceleration of the particle does not remain finite, and hence possesses a discontinuity (Prabhu, 1965, pp. 98–102). So, this process is not continuous in all its dependencies on time. If we take 'essentially continuous history' to mean a history which can mathematically be represented by an infinitely often differentiable function, we can safely say that no genuine stochastic process is essentially continuous. For essential continuity of a history would imply that each of its states is, as a function of time, uniquely determined by any given state. Since the uniqueness condition cannot be satisfied by any genuinely stochastic determination, this kind of determinism cannot be essentially continuous, but must involve some discontinuity.

The *purely temporal determinism* inherent in dynamical determinism goes hand in hand with its continuity. The continuity of a dynamical-deterministic process presupposes that of time: it is defined in terms of the continuity of time. But, of course, the continuity of time is not the reason why dynamical determinism is purely temporal. It is because of the specific nature of the laws holding for dynamical-deterministic systems that the evolution of their states is determined in a purely temporal fashion, i.e., any state depends only on some given state and the time difference between them. This 'time-homogeneity', as we saw, can also be found in temporal stochastic determination: it is present in the type defined by (St1). The difference is, however, that in dynamical-deterministic systems the states themselves are determined in a purely temporal way, whereas in a stochastic system of the type (St1) it is the transition probabilities which are so determined. Thus, one can still say that the determination of the evolving *states* in a stochastic system cannot be purely temporal. Subsequent states of a stochastic system depend not solely on the time difference between them and some given antecedent state, but also on other determinants, which is displayed in the facts that there are sets of alternative possible states for each subsequent state, and that the probability of any possible state is specified by probability distributions over such sets.

Quantum mechanics appears to offer a counterexample to the claim that the determination of states in a stochastic system cannot be purely temporal. In a quantum-physical system, the free evolution of what is ordinarily called its 'states' is governed by a one-parameter group of transformations of the kind referred to in (DD), with time as the only independent parameter. However, these 'states' themselves cannot be literally interpreted as physical states; they much rather define probability distributions over sets of possible values of quantitative properties of the system as well as transition probabilities holding for states differing in such values and considered at different times. Thus, here again, it is only probability distributions and transition probabilities which are determined in a purely temporal fashion[16].

Temporal stochastic determination is not compatible with the kind of *'reversibility'* one can have in the case of dynamical determinism, namely, bilateral determinism. Another sort of 'reversibility' may hold for stochastic systems: given its present state, both its future and its past may be specified with certain probabilities. There are good grounds, however, for taking all types of temporal stochastic determination as referring to the future alone: probabilities are assigned to sets of possible states, and there are no objectively possible states in the past besides those that have actually come about.

Finally, the condition of the *homogeneity of states* or, more generally, that of the determining conditions and events determined is clearly compatible with stochastic determination.

(C) *Determinism and Indeterminism*

Is stochastic determination adequately characterized as some kind of indeterminism? As mentioned, this would suggest that a notion of stochastic determination can be obtained by negating or weakening some of the conditions implied by the notion of a strict determinism. As regards a particular notion of determinism like that of dynamical determinism (DD), it is clear that the negation or relaxation of any of the conditions implied by it will mean abandoning this notion. But it need not mean parting with the concept of a strict determinism altogether. As long as the weakened conditions are compatible with some concept of a strictly deterministic relationship they will in general not be sufficient as a specification of any stochastic determination. Now, except for the uniqueness condition and that of the essential continuity of the history, the denial of the characteristics of dynamical determinism in fact is compatible with some concept of strict determinism: there is no contradiction involved in the idea of having heterogeneous elements (which therefore do not form a continuous set), one or more of them being uniquely determined by some others in a nontemporal (and thus certainly not purely temporal) and 'irreversible' way.

The uniqueness condition, or the requirement that it be a necessary relation, which is the earmark of any strictly deterministic relationship, is thus the only one that has to be abandoned in any notion of stochastic determination. (A condition like that of the essential continuity of the system's history in our exemplary case will then follow suit). To this extent one is warranted to call stochastic determination a kind of 'indeterminism'. A mere relaxation or abandonment of this condition, of course, does not lead to a definite concept of stochastic determination, if only because it will not entail any definite probability assignments[17].

The best a specific relaxation can achieve is a specification of the sets of possible alternatives which forms part of any stochastic determination. However, such a positive specification of alternatives is not just a negation or relaxation of the uniqueness condition.

The preceding discussion suggests that stochastic relationships are ill characterized as cases of indeterminism or as deficient forms of determination. Any genuine kind of stochastic determinism amounts to a positive specification of the probabilities or probability distributions over certain sets of alternatives, and thus includes a positive determination of these alternatives themselves. The latter determination is worth separate mention despite the fact that it is implicit in any given probability distribution. It may well be the only determination in some situations. I shall elaborate on these points in the following section.

V. A POSITIVE VIEW

In what follows, I shall try to sketch a positive account of stochastic determination, or chance relationships in time. The discussion will not merely continue the examination of the general notion of stochastic determination defined by (St1). Rather, we shall examine the coherence of this form of determination with other features of physical and, especially, quantum-physical systems, like their stability and the distinctness of their properties. It will be pointed out that the existence of stable systems necessitates a temporal form of determination which is nondynamical. I shall conclude with an analogy between spatial and temporal chance relationships which is meant to help overcome some of the puzzlement about the notion of stochastic determination.

(A) *The 'Necessity' of Chance*

Chance relationships do not only comprise temporal stochastic relations but also any other kind of random, or probabilistic, correlations (cf. Kirschenmann, 1972). If one contrasts dynamical determinism with all possible types of chance relationships and regards the latter as indeterministic, one can say that dynamical determinism itself is only possible at the cost of some indeterminism (cf. Bunge, 1951). Each state of a dynamical-deterministic system is in a lawful way completely determined by the preceding state and consumed in the subsequent state. There cannot be any additional mode of determination through which this process could be lawfully affected or could bear any lawful determining relation to something else. This is not to say, of course, that processes in otherwise dynamical-deterministically evolving systems could not be influenced by extraneous factors or come into contact with and influence other processes. Only, relative to the self-contained and purely temporal mode of evolution of a dynamical-deterministic system, every influence from or on something else cannot be but accidental. For instance, there can merely be chance correlations or chance encounters between the processes in two dynamical-deterministic systems, but no lawful mutual determination.

Moreover, with regard to the lawful, deterministic evolution of states in such a system, the initial state in its history (in general, initial and boundary conditions) also can only be contingent. If there are any statistical regularities found in the correlation of states of different systems they can, as our argument for the non-eliminability of probability statements showed, only be traced back to equipollent accidental regularities holding for the systems' initial states. The very existence of physical systems possessing particular properties must also remain a fact extraneous to the fully determined processes they undergo. In the light of a strict dynamical determinism, they must appear as mere chance arrangements. And if dynamical determinism is considered to hold sway universally, they must be judged to be due to chance arrangements of the initial conditions of the universe, unless one at this point prefers to appeal to some other, supernatural, determining agency (cf. Hawkins, 1964, p. 136; Watanabe, 1966, p. 540). All this is well-known from discussions of the idea of a universal mechanical determinism as advocated by Laplace, who rejected any supernatural explanation. For Newton, on the other hand, the existence of the (for him, almost) deterministic world machine indicated that of a divine maker, and he suggested in an oft-quoted passage that the existence of things with distinct and relatively stable properties was ultimately due to unchangeable particles created in the beginning (Newton, 1952, p. 400). However, it would certainly not be contrary to his scientific spirit to look in this world for the immediate grounds for the existence of relatively stable things and chance relationships.

In dynamical-deterministic systems, given one state as the actual initial state, the dynamical laws mark out one history in the space of admissible states as the one that alone is possible and will become actual. Stochastic determination of the type (St1), as was mentioned, involves a specification of the sets of possible alternative states for any subsequent time once a state is given as actual. Even when the system will actually have assumed a particular state belonging to such a set of alternatives it will still be true that it could have assumed a different one. This is not so with dynamical determinism, which involves only a distinction between generally admissible states and actual ones, since only the actual ones have ever been possible. In stochastic systems of the type considered, by contrast, we have the space of admissible states, sets of possible states, and actual states.

Dynamical determinism, as we have seen, leaves no room for any additional determination. An additional determination would mean over-determination, which generally leads to a contradiction. In particular, we saw previously that a temporal stochastic determination could not be wedded to a dynamical determinism if both referred to the same state space, and that this incompatibility was already evident when one compared (DD4) and (SD2) and identified the conditions and events in question, where (SD2) was the statement asserting the existing of sets of possible alternatives typical of temporal stochastic relationships. Yet, this incompatibility can be construed in two ways.

One way of construing it is to say that, as compared with dynamical determinism, any form of determination which allows for sets of possible alternatives is indeterministic. But one can also argue as follows. The specification of sets of possible alternatives is a positive form of determination. Once such sets are fixed, it is clear that a dynamical determinism concerning the same states cannot hold at the same time, but at most a stochastic kind of temporal determinism, which is able to accommodate and incorporate such sets. Since this determination of possible states is absent from dynamical determinism, one can even say that genuine forms of temporal stochastic determination involve, or at least indicate the presence of, a plus in determination (cf. Hölling, 1971, p. 126).

In view of this incompatibility concerning one and the same state space, a determinist will rather look for a second underlying state space for the same system with respect to which a dynamical determinism held. But, in analogy to the argument for the uneliminability of probability statements, we can say that statements asserting sets of possible states or transitions, once admitted for reason, cannot be eliminated in this way either. The fact that, in the case of stochastic systems, we cannot avoid talking about sets of possible states and their probabilities is, of course, still strange. We would like to see, if not the reasons why such features are present in the world, i.e., the sources of chance and causes of chance relations, so at least the coherence of those features with less intriguing ones.

We found dynamical determinism to be essentially continuous, whereas stochastic determination is not. Discontinuity, then, is a condition necessitating a form of determinism different from dynamical determinism. As long as we are dealing with continuous state spaces there can always be hope that dynamical laws will be found which specify an essentially continuous evolution of the systems concerned (apart from the false hope that they could fully replace genuine stochastic relationships). This is why the particular features of stochastic determination are much more imposing in the case of discrete state spaces. In discrete state spaces, dynamical determinism simply is precluded. The existence of such discreteness, which is a form of discontinuity, can make it comprehensible why there should be stochastic kinds of determinism.

A common objection to this line of argument is to point out that, even in a system with discrete states, the succession of states can be deterministic: each state may lead after some time to a uniquely determined subsequent state. Because of the presupposed discreteness of states, their succession of states cannot be dynamical-deterministic; but, as described, it still would be a strictly deterministic process. For the discussion of this example I shall distinguish two possible cases: (a) the transitions from one state to another occur at determinate time intervals; (b) the transitions occur at irregular intervals. In either case, we would ask for the reason of the respective mode of state succession.

The way the states follow upon each other in case (a), I think, can only be due to an underlying continuous process which determines the time intervals at which the transitions in question occur (unless the common concept of a continuous time were replaced by that of a discrete time). If this is so, we basically have one continuous deterministic process, these transitions being one of its aspects. What would still have to be accounted for is the discreteness of the state space relevant for these transitions. (If one really wanted to maintain that the presence of discrete transitions, even in this case, implies the stochasticity of the process, one could propose to consider it as a limiting, non-genuine form of stochastic process. Of course, eschewing the issue in this way is open to objection).

Case (b), where the state transitions occur at irregular time intervals, clearly involves an element of chance. Even if any given state cannot but lead to uniquely determined subsequent state after some time, the fact that the time of the transition is not uniquely specified suggests that the whole process is a stochastic process (unless there is not even a statistical regularity to be found in the series of time intervals between transitions).

The most impressive examples of stochastic determination involving discrete state spaces are found in quantum mechanics or theories of subatomic systems. The laws of quantum mechanics specify for atoms discrete sets of eigenvalues and eigenstates of certain operators, most notably the energy operator. Such operators are mathematical representations of the properties of atomic systems: their values can be regarded as characterizing the admissible physical states of the system if the operators represent quantitative properties that are conserved in physical processes. Quantum-mechanical laws, furthermore, specify for atoms in excited states or atoms interacting with a radiation field the possible states that an atom, being in a given state, may assume, and thus the possible transitions. Lastly, they also determine the probabilities of these transitions. Any further determination, we saw, would generally involve contradictions (cf. Hölling, 1971, p. 138).

In quantum-physical systems, strict dynamical determinism is not just relaxed, but rather replaced by a different form of determinism, or natural lawfulness. This includes a particular non-temporal determination, which summarily can be called a 'structural determination'[18]. The importance of symmetry considerations in quantum theories is ample evidence for the presence of this new form of determination. It is this structural determination which is expressed in the existence of discrete eigenstates of atomic systems. These in turn account for the amazing definiteness and stability of atoms and their properties. All gold is alike; gold remains gold (for all practical purposes) and does not display degrees of goldness. Derivatively, the relative stability of molecules and macroscopic objects and their properties is accounted for in terms of the stability of atoms and their discrete possibilities of combining with one another, which again follow quantum-mechanical laws[19]. It is these features of the world which in classical physics had to be taken for granted and were left unexplained (Heisenberg, 1958, p. 152). One had to postulate that gold does not change into silver; the notion of continuity prevalent in the conceptual scheme of mechanical determinism would rather have suggested – as it was believed before the rise of mechanics – that a continuous change from gold into silver was possible. One can conclude, then, that an account of the (relative) stability of things and their properties has to renounce the notion of dynamical determinism. This form of determinism has to give way to another which comprises a structural determination of states and a stochastic determination of state transitions. It is one of the greatly appealing features of quantum mechanics that it supplies a unified theoretical basis for the structural (or 'static') and the processual (or 'dynamic') aspects of this form of determinism. (In the history of quantum theory, however, there first was an account of the possible states of atoms, which explained the discrete frequencies of spectral lines, complemented later by theories about probabilistic state transitions, which gave explanations for the intensities of spectral lines).

We can say, then, that the stable definiteness of microphysical and many macrophysical properties is one of the major reasons why there must be chance relationships in our world. This definiteness itself is due to the fact that the corresponding states of a given kind of quantum-physical systems and these systems themselves are all identical. All gold is definitely alike because all gold atoms are identical. It is because of the stability and identity of all gold atoms that we can rely on gold remaining gold. Discreteness is heightened by such identity. A system only capable of discrete states cannot have a continuous history. But the fact that the particular states of an atom or other quantum-physical system are exactly alike when they occur at different times or in different systems entails that they cannot bear any traces of the histories of the systems. This is in marked contrast to dynamical-deterministic systems and their essentially continuous and individual histories, in which each state is unique. In general, the concept of an individual history is applicable only to a system which displays some temporal continuity or connectedness of its states, and also some unique characteristics distinguishing it from other systems, and its present state from its past states. To be self-identical is to have a history; and to be an individual is to possess unique characteristics. The identity of like quantum-physical systems and that of their properties and states means the loss of individuality as well as the loss of individual histories, or self-identity, in cases where such systems interact. (A more technical consequence is that, in treating ensembles of such systems, one has to work with laws of statistics which differ from those employed in classical statistical mechanics).

We have seen that the discreteness of the state space of a physical system is necessarily accompanied by discontinuity, or jumps, in its state transitions, which is an ingredient of stochastic determination. Discontinuity is also introduced into the state space by any 'heterogeneity' of its elements. An example would be the decay of a radioactive atom, where the state and nature of the object in the beginning and those of the consecutive products of disintegration are qualitatively different. One can still think of the process as occurring in one complex and 'heterogeneous' state space. Yet, it seems to be preferable to conceptualize this process in terms of conditions at one time and subsequent events, as in the formulation (St2) of a general notion of stochastic determination. Many macroscopic situations involving stochastic determination or, quite generally, probabilistic relationships, are of this kind. The heterogeneity of conditions and events does not permit a dynamical determinism relating these very conditions and events. A simple example is the throwing of dice. In throwing a die, the antecedent conditions are: the presence of a particular die, the presence of a smooth hard table-top, the fact that the die is thrown; the possible events are the six different outcomes of the throw. Obviously, the conditions are not 'homogeneous' with the events, and they cannot be related in a dynamical-deterministic way, but in a way expressed by appropriate probability statements.

A determinist would point out that the process can also be described on the basis of a homogeneous state space, the space of the detailed mechanical states of the die, with respect to which the process could well be dynamical-deterministic. But even assuming that the process is completely deterministic, i.e., no random factors play a role, from the moment of the throw to the moment the die comes to rest, we know that this will not allow us to do without probability statements. The stochastic character of the whole process would only be traced back to the antecedent conditions: the die was thrown in a random fashion. To throw a die at random means to throw it in a way that is not determined or influenced by, and hence discontinuous with, any personal preference or other imaginable factor. We see here another origin of chance phenomena. Doing something at random can be as much a primary source of chance processes and relationships as are the quantum jumps.

(B) *An Analogy*

For the purpose of these final remarks I shall distinguish three concepts of chance, one of absolute chance, and two of relational, or relative, chance (cf. Nagel, 1961, pp. 324–335; Cahn, 1967).

(1) *Absolute chance* is (or would be) at work in pure chance events; these are considered as events which are completely uncaused, in no way determined by or dependent on anything else in their occurrence.

(2) Relative chance in the sense of a *chance conjunction* is found in the accidental coincidence or correlation of two or more otherwise determinately occurring events or processes.

(3) Relative chance in the sense of a *chance succession* is found in stochastic processes.

Absolute chance, should it exist, would be the most perplexing of the three types. Pure chance events were postulated by Epicurus in antiquity to account for the origination of everything that exists from a stream of uniformly falling particles and, in particular, for the fact that man has free will. Others have followed him in holding that man's free actions must issue from absolute chance (cf. Cahn, 1967). One can object to this view that any free decision worth the name is governed by reasons, thus not due to a pure chance event. This does not exclude that free decisions can exhibit statistical regularities or correlations with other factors. And, as mentioned, free decisions can be a source of chance in those special cases where we decide to be 'absolutely fair' and toss up a coin in an unbiased manner.

Chance coincidences or correlations constitute the type of chance which offers the least offense to our comprehension. We referred to this type when pointing out that even dynamical determinism entailed the existence of chance. Historically, it was especially Chrysippus, Mill and Cournot who proposed to understand all chance phenomena as accidental crossings or encounters of two or more independent deterministic ('causal') processes (Bunge, 1951, pp. 210ff.). But the idea is also found in Aristotle for whom, however, the independent processes in question are all governed by final causes, i.e., ends or goals. We do not find the concept of chance in the sense of chance conjunctions to be very unsettling, I think, because it consistently fits together with the notion of dynamical determinism, which we seem to accept quite readily, at least as far as physical processes are concerned.

Relative chance in the sense of chance successions of events appears to be less mysterious than imaginable pure chance events, but at the same time less comprehensible than chance conjunctions. State transitions in stochastic systems are not events that are completely independent of anything else: they at least depend on the systems in which they occur, and on factors like the available energy of these systems. Still, they are non-deterministic ('uncaused') jumps – which, to minds which are prejudiced in favor of the idea of continuity, seem to be quite unintelligible (despite arguments to the effect that it could not be otherwise). On this count, then, relative chance of type (3) appears to bear little similarity to relative chance of type (2). I intend to point out that the analogy between these two types of chance may be greater than it first appears.

The underlying idea of the following comparison is that of radically analogizing temporal relations with non-temporal and, especially, spatial relations. This approach is not new. N. Goodman (1966, Ch. XI) and R. Taylor (1955; 1963, pp. 71–74), for instance, have pressed such analogies in order to remove some of the mysteriousness traditionally associated with time. A simple example illustrating that temporal and spatial relations are much more alike than we usually think is the following by Taylor. We are wont to point out a basic difference between time and space by saying that, while an object cannot be in two places at once, it can be two or more times at one place. Yet, on reflection, this difference becomes less obvious. When saying that an object can be in one place at two times we tacitly presuppose that the object also exists all the time in between, at the same place or another. But, under the analogous condition, an object can likewise be in two places at one time. We only have to presuppose that it also exists at all places in between, at the same time or another. For instance, a spatially extended earthquake can be and strike in two places at the same time; and it can be considered one 'object' inasmuch as it continuously extends from the one place to the other, whether it strikes at the intermediate places at the same time or not. There are obvious objections to such an analogy, which seems to be a mere philosophical game. Surprisingly, one can go a long way in successfully defending such analogies.[20] To the extent that this kind of comparison can alleviate some of the puzzlement about time, it can help reduce the mysteriousness of chance successions, as compared with chance conjunctions.

I assume then that we admit that there are chance conjunctions. This means, we admit that chance is a mode of coexistence of independent processes. The term 'coexistence' is used to cover both coincidences and other spatial correlations of such processes. 'Coexistence' of two entities means here that they exist at the same time, in a spatial relationship. The processes are independent in the sense that they belong to dynamical-deterministic systems which (except for possible accidental encounters) are causally isolated from one another. Thus the independence of the systems and their processes is due to the self-contained temporal determinism governing the evolution of their states. A process, of course, is something extended in time: the evolution of states occupies time.

To set up the intended analogy we have to translate temporal terms and relations into non-temporal, spatial ones, and conversely. We shall do so, but reverse the account above. In the place of processes, which are extended in time, we introduce something extended in space: spatially extended states or inner configurations of physical systems. They are to be 'independent'; but their 'independence' must be due to some self-contained non-temporal form of determination. What we have called 'structural determination' is a fitting candidate here. Inasmuch as a structural determination specifies discrete states, these will be isolated in the sense that each is fully determined in itself. The counterpart of coexistence of entities is the succession of 'entities', since 'succession' of two 'entities' means that they exist at the same place, bearing a temporal relation to each other. Thus the analogue of the mode of coexistence of (causally) independent processes is: a mode of succession of (structure-deterministically) 'independent' states or configurations. By this analogy, then, we should find nothing particularly puzzling about the concept of chance as a mode of succession of 'independent' states, provided we find nothing especially unintelligible about the concept of chance as a mode of coexistence of independent processes.

In drawing the analogy, I have left the notion of a physical system untouched. For our immediate purpose it was sufficient to compare temporally extended processes in systems to spatially extended states of systems. Some rather fundamental questions, however, will turn up if the analogy is extended to physical systems themselves. I shall mention only one. We consider physical systems to be entities with some spatial extension and, more importantly, some temporal permanence or continuity. Their self-identity is not due to their spatial extensions, but to their temporal continuity which derives from the permanence of the matter they are made of. The analogue of a physical system, then, would be something temporally extended that we consider a 'self-identical entity' because it has some spatial 'permanence' or continuity. This would be something like the earthquake mentioned, but regarded as possessing a 'self-identity' in its own right, which is not based on any temporal permanence of the quake itself or that of some underlying matter. Something having this nature is called a 'field'. The question is whether a consistent account of temporal stochastic determination should not abandon the ordinary notion of a physical or material system and be based on the concept of a field alone. In fact, the field-theoretic approach is the intellectually most satisfying, if not yet fully successful, approach in explaining the diversity of subatomic entities and phenomena.

Clarkson College of Technology
Potsdam, New York

NOTES

[1] If the transformations U_m are defined only for positive m they will form a semi-group, which does not possess property (3); it determines only the future of a system.

[2] In the case of periodical processes, we have to add a phase-variable to make this mapping one-one.

[3] The possibility of embedding any system of classical particle mechanics in an 'isolated' system has been shown (cf. Suppes, 1957, pp. 299ff.). Note that Suppes distinguishes between 'isolated' and 'closed' systems.

[4] See also Note 1.

[5] Whether this is so or not can be decided only on the basis of the particular dynamical theory holding for the system, which specifies the set of admissible states and, in principle, the form of the dynamical group. The admissible states together with the dynamical group define all admissible histories of the system. The system exhibits process-reversibility if there is for any admissible history h an admissible history h' such that, for any m, if $s'_{t'} \in h'$ is the reversed state of $s_t \in h$ then $U_m s'_{t'} = s'_{t'+m}$ is the reversed state of $U_{-m} s_t = s_{t-m}$. As is known,

many theories of classical physics assert process-reversibility for the systems they account for, while the relevant groups can seldom be formulated explicitly. – For more on process-reversibility, especially the notion of a reversed state, see Watanabe (1966, p. 544).

[6] Since R may be a continuum of states, and if we wish to keep the definition as general as possible, we should speak of the probability of r_{t+m} (and possibly r_t also) lying in some subset of R, and indicate this through special notation (cf. Feller, 1966, pp. 205–209, 312; Prabhu, 1965, pp. 65, 87f.). Using the notation for discrete state spaces alone is just a matter of convenience.

[7] The mathematical character of the state space will of course vary with the process studied and the way it is studied. For Brownian motion, the state space is the space of positions of a randomly moving particle (Feller, 1968, p. 354; Prabhu, 1965, p. 102) or, alternatively, the space of its velocities (Prabhu, 1965, p. 98). For radioactive decay, the states can be the successive disintegration states or, if a whole collection of decaying atoms is considered, the number of disintegrated atoms.

[8] Another place in quantum mechanics in which probabilities play an important role is in the relationships between different properties or between properties and states of microsystems.

[9] While impossible transitions or events can always be assigned probability zero, the converse cannot always be asserted according to the mathematical probability theory based on measure theory: in a continuum of admissible states, those making up sets of states of probability measure zero still may happen. In all stochastic theories of physics that I know of, however, the converse is also assumed to hold.

[10] See also Note 9.

[11] I do not make a notational distinction between r as states in R and as sets of states of S.

[12] Although this implication is intuitively clear, here is a sketch of the argument proving statement (3). Let the state with respect to R at time t be r^k. Then, in virtue of (StI),

(1a) $\quad P(r_{t+m}=r^l/r_t=r^k)=p_m^{kl}.$

By assumption, r^k and r^l are sets of states s of S. Thus, r^k obtains at t precisely if $s_t \in r^k$, and r^l obtains at $t+m$ precisely if $s_{t+m} \in r^l$. We can therefore rewrite (1a) in the form

(1b) $\quad P(s_{t+m} \in r^l/s_t \in r^k)=p_m^{kl}.$

On the other hand, in virtue of (DD),

(2a) $\quad s_{t+m}=U_m s_t.$

Quite generally, the mapping U_m, which maps states s_t onto states s_{t+m}, induces a one-one mapping of sets r^y of states at t onto sets r^x of states at $t+m$, where $r^x=\{U_m s_t : s_t \in r^y\}$. Thus the following equivalence holds:

(2b) $\quad s_t \in r^y$ if and only if $s_{t+m} \in r^x.$

This means that r^x must obtain at $t+m$ whenever r^y obtains at t, and conversely, which implies:

(2c) $\quad \begin{array}{l} P(s_{t+m} \in r^x/s_t \in r^y)=1, \\ P(s_{t+m} \in r^x/s_t \notin r^y)=0. \end{array}$

In particular, what has been pointed out holds for r^l, which is a set of states of S taken at time $t+m$. There is then a set $r^{d(l)}$ such that $r^l=\{U_m s_t : s_t \in r^{d(l)}\}$ and

(2d) $\quad \begin{array}{l} P(s_{t+m} \in r^l/s_t \in r^{d(l)})=1, \\ P(s_{t+m} \in r^l/s_t \notin r^{d(l)})=0. \end{array}$

Thus we have a partition of S into the set $\bar{r}^{a(l)}$ of states and its complement $r^{a(l)}$ and, accordingly, a partition of r^k into $r^{a(l)} = r^k \cap r^{a(l)}$ and $r^{b(l)} = r^k \cap \bar{r}^{a(l)}$, with

(2e) $\quad P(s_{t+m} \in r^l / s_t \in r^{a(l)}) = P(s_{t+m} \in r^l / s_t \in r^{a(l)}) = 1,$
$\quad\quad P(s_{t+m} \in r^l / s_t \in r^{b(l)}) = P(s_{t+m} \in r^l / s_t \notin r^{a(l)}) = 0.$

Employing now the Chapman-Kolmogorov identity (see Feller, 1968, p. 445), we have for $m \geq n \geq 0$,

$$P(s_{t+m} \in r^l / s_t \in r^k) = \sum_h P(s_{t+m} \in r^l / s_{t+n} \in r^h) \, P(s_{t+n} \in r^h / s_t \in r^k).$$

(This identity certainly applies here because (DD) could not be satisfied at the same time if the system were to pass from one state to another without going through the intermediate states). The identity holds also for $n=0$. In this case, $P(s_t \in r^h / s_t \in r^k) = 0$ unless $r^h \cap r^k \neq \emptyset$. Further, in our case, we can partition the union of all r^h satisfying this condition, which is r^k, into $r^{a(l)}$ and $r^{b(l)}$, so that only a sum of two terms is left, namely,

$$P(s_{t+m} \in r^l / s_t \in r^k) = P(s_{t+m} \in r^l / s_t \in r^{a(l)}) \, P(s_t \in r^{a(l)} / s_t \in r^k) +$$
$$+ P(s_{t+m} \in r^l / s_t \in r^{b(l)}) \, P(s_t \in r^{b(l)} / s_t \in r^k)$$

Making use of the equations (2e) and the original assumption (1a), we arrive at the final result

(3) $\quad P(s_t \in r^{(l)} / s_t \in r^k) = p_m^{kl}.$

Note that the left side of (3) no longer depends on m. This is as it has to be: if a probability is assigned to a state s_t, then, because of (2a), the same probability must be assigned to all states s_{t+m} which are part of the same history, regardless of the value of m.

An analogous, yet much briefer, argument can be given in terms of ordinary conditional probabilities (which, however, are not suitable for describing transition probabilities as long as they are supposed to refer to an 'event' space taken at one time).

(1)   Let the probability of E given C be p, i.e., $P(E/C) = p$.
(2)   Assume there is an additional factor D such that C and D together necessitate the occurrence of E, while E, given C, cannot occur in the absence of D. This implies
$\quad\quad P(E/C \cdot D) = 1,$
$\quad\quad P(E/C \cdot \bar{D}) = 0.$

Elementary probability calculus suffices to establish

(3) $\quad P(D/C) = P(E/C) = p.$

For similar arguments see Hawkins (1964, p. 146); Mellor (1971, pp. 151f.).
[13] For a recent discussion of the case against 'hidden variables' see Bub (1973).
[14] See, e.g., Hooker (1973) for a thoroughgoing analysis and comparison of classical particle mechanics, classical field theories, and quantum mechanics.
[15] See also Note 6.
[16] While more would have to be said about quantum-mechanical 'states', the points made, I think, are enough to show that quantum mechanics is no exception in this respect. I will discuss some further aspects of quantum-physical states in Section V(A).
[17] The ease with which one can relax, or generalize, a statement like $P(A/B) = 1$ (which can be taken to be implied by an appropriate strictly deterministic relationship) by formally passing over to the statement $P(A/B) = p \neq 1$ tends to obscure this obvious fact.

[18] This term is used by Hawkins (1964, p. 173). Other adjectives have been proposed to characterize this new kind of determination apparent in quantum theory: 'morphic' (cf. Hawkins, 1964, p. 178, footnote; Weisskopf, 1972, p. 295); 'global' (Hooker, 1973, pp. 212f.).
[19] Even the transition probabilities between quantum-mechanical states serve to account for relatively stable macroscopic properties. On this point and those mentioned in the text, cf. Wigner (1973, p. 373).
[20] For their limitations see Schlesinger (1975).

BIBLIOGRAPHY

Bub, J.: 1973, 'On the Completeness of Quantum Mechanics', in C. A. Hooker (ed.), *Contemporary Research in the Foundations and Philosophy of Quantum Theory*, D. Reidel, Dordrecht, 1973, pp. 1–65.
Bunge, M.: 1951, 'What is Chance?', *Science and Society* **15**, 209–231.
Bunge, M.: 1967, *Foundations of Physics*, Springer, New York.
Bunge, M.: 1970, 'Problems Concerning Intertheory Relations', in P. Weingartner and G. Zecha (eds.), *Induction, Physics, and Ethics*, D. Reidel, Dordrecht, 1970, pp. 285–325.
Cahn, S. M.: 1967, 'Chance', in P. Edwards (ed.), *The Encyclopedia of Philosophy*, Macmillan, New York, 1967, Vol. 2, pp. 73–75.
Feller, W.: 1966, *An Introduction to Probability Theory and Its Applications*, Vol. II, John Wiley, New York.
Feller, W.: 1968, *An Introduction to Probability Theory and Its Applications*, Vol. I, 3rd ed., John Wiley, New York.
Fraassen, B. C. van: 1972, 'A Formal Approach to the Philosophy of Science', in R. G. Colodny (ed.), *Paradigm and Paradoxes*, University of Pittsburgh Press, 1972, pp. 303–366.
Goodman, N.: 1966, *The Structure of Appearance*, 2nd ed., Bobbs-Merrill, Indianapolis.
Hawkins, D.: 1964, *The Language of Nature*, W. H. Freeman, San Francisco.
Heisenberg, W.: 1958, *Physics and Philosophy*, Harper & Row, New York.
Hölling, J.: 1971, 'Zur Kategorialanalyse des physikalischen Feldbegriffs', in J. Hölling, *Realismus und Relativität*, W. Fink, Munich, 1971, pp. 126–139. First published in *Philosophia Naturalis* **10** (1968), 343–356.
Hooker, C. A.: 1973, 'Metaphysics and Modern Physics', in C. A. Hooker (ed.), *Contemporary Research in the Foundations and Philosophy of Quantum Theory*, D. Reidel, Dordrecht, 1973, pp. 174–304.
Kirschenmann, P.: 1972, 'Concepts of Randomness', *Journal of Philosophical Logic* **1**, 395–414.
Landé, A.: 1973, *Quantum Mechanics in a New Key*, Exposition Press, New York.
Mackey, G. W.: 1963, *The Mathematical Foundations of Quantum Mechanics*, W. A. Benjamin, New York.
Mellor, D. H.: 1971, *The Matter of Chance*, Cambridge University Press, Cambridge.
Nagel, E.: 1961, *The Structure of Science*, Harcourt, Brace & World, New York.
Neumann, J. von: 1955, *Mathematical Foundations of Quantum Mechanics*, Princeton University Press, Princeton.
Newton, I.: 1952, *Opticks*, Dover, New York.
Poincaré, H.: 1912, *Calcul des probabilités*, Gauthier-Villars, Paris.
Prabhu, N. U.: 1965, *Stochastic Processes*, Macmillan, New York.
Schlesinger, G.: 1975, 'The Similarities Between Space and Time', *Mind* **84**, 161–176.

Suppes, P.: 1957, *Introduction to Logic*, D. van Nostrand, Princeton, New Jersey.
Taylor, R.: 1955, 'Spatial and Temporal Analogies and the Concept of Identity', *Journal of Philosophy* **52**, 599–612. Repr. in J. J. C. Smart (ed.), *Problems of Space and Time*, Macmillan, New York, 1964, pp. 381–396.
Taylor, R.: 1963, *Metaphysics*, Prentice-Hall, Englewood Cliffs, New Jersey.
Watanabe, S.: 1966, 'Time and the Probabilistic View of the World', in J. T. Fraser (ed.), *The Voices of Time*, G. Braziller, New York, 1966, pp. 527, 563.
Weisskopf, V. F.: 1972, *Physics in the Twentieth Century*, MIT Press, Cambridge, Massachusetts.
Wigner, E. P.: 1973, 'Epistemological Perspective on Quantum Theory', in C. A. Hooker (ed.), *Contemporary Research in the Foundations and Philosophy of Quantum Theory*, D. Reidel, Dordrecht, 1973, 369–385.

PETER A. BOWMAN

THE CONVENTIONALITY OF SLOW-TRANSPORT SYNCHRONY*

I. INTRODUCTION

The conventionality of distant simultaneity, as maintained by Hans Reichenbach and Adolf Grünbaum, is by now so widely known that it can be stated very briefly. Let us consider two points A and B which are separated from one another in an inertial frame K. For a light signal emitted from A and reflected at B back to A, we compare the time interval for the out-going trip to that for the round trip. This ratio is called 'epsilon' (ϵ). In formulating the special theory of relativity, Einstein effectively took ϵ to be $\frac{1}{2}$; thus, we may use $\epsilon = \frac{1}{2}$ in defining what is now called 'standard signal synchrony'. Reichenbach views ϵ as being restricted *only* by the causal relations involved in the signaling process. That is, the reflection of the light ray at B must take place after the ray's emission at A but before its return to A. These considerations require us to restrict ϵ between zero and one, but Reichenbach (1958, p. 127) insists that within these limits values of $\epsilon \neq \frac{1}{2}$ "could not be called false". He claims that there are no facts which would mediate against using these values in definitions which are now called 'nonstandard signal synchrony'. This allegedly physical possibility of choosing ϵ between zero and one is the conventionality of distant simultaneity as determined by signals. Grünbaum (1973, p. 353) also argues for this thesis, making clear that it obtains within a single inertial frame.[1] In this paper I will not be concerned with the situation in more than one inertial frame.[2] Rather, I will consider a nonsignaling definition of synchrony and ask what sort of conventionality it manifests in a single inertial frame.

What characterizes the Reichenbach-Grünbaum approach is establishing the basic concept of special relativity, distant simultaneity, through the transmission of signals such as electromagnetic or gravitational waves or particles. The approach I wish to take is characterized by the transport of actual clocks; e.g., mechanical, light, or atomic clocks. Since a clock is defined as any physical system which passes through the same process periodically, and since on some construal a signal might possess this property, it would not be possible to distinguish sharply between a clock and a signal in that sense.[3] However, for the purpose of my characterization of a signal, it has the salient feature of being an infinitesimal disturbance or a point mass which, outside of its own presence and absence, carries no information.

The nonsignaling definition of synchrony on which I will rely is a modification of the following: two clocks U_A and U_B, at rest at the places A and B in any inertial system K, are synchronous if a clock V first synchronized locally with U_A at A and then transported infinitely slowly (along a line) to B is found to be synchronous locally with U_B at B. To this definition might be lodged the objection that, in determining the limit as the transport velocity v approaches zero, one must determine v; and, in order to do so, one must have a pre-established synchrony. However, instead of v, one can use a quantity v', called an 'intervening "speed"', which is determinable without vicious circularity or regress and which tends to zero as v tends to zero. Henceforth, I will assume this definition to be without difficulty[4] and call it 'standard slow-transport synchrony'.[5] Ellis and Bowman (1967, pp. 128, 130) have shown that clocks with this synchrony in K will also be found synchronous with clocks set using the standard signaling definition.

II. ATTEMPTED CONSTRUALS OF A 'NONSTANDARD' DEFINITION

Still, it may be argued that there is no more reason to accept standard slow-transport synchrony than to accept any other such definition – call it 'nonstandard slow-transport synchrony'. If one is to pursue such an argument, of course he must say how he construes the latter. I want to consider three possible ways of construing it, but to make clear the differences among these ways it is necessary to introduce a distinction. In some theories, among them special relativity, it is assumed that the temporal interval on a clock at a particular position or instant is equal to other intervals on that clock when it is at any other position or instant. These assumptions are called 'standard temporal congruence'. The denial of one or the other or both of these assumptions, i.e. the assumptions that the temporal intervals on a clock may vary with position or time, are called 'nonstandard temporal congruence'. When this distinction is applied to spatial intervals in the same way, 'standard spatial congruence' and 'nonstandard spatial congruence' are defined. In physical terms standard spatial and temporal congruence are given by 'rigid rods' and 'isochronous clocks', respectively, i.e. rods and clocks which are at rest in an inertial system and which have been corrected for differential perturbing influences. Following the traditional and most common definition of special relativity, I will take standard congruence to be a necessary (but not sufficient) condition for the theory. But, in so doing, I am only adding to the usual separately necessary conditions, the flatness of space-time and the restriction to inertial frames, the condition that rigid rods be independent of orientation.[6] In using this reportive (or lexical) definition of special relativity rather than some stipulative one which is incompatible with it, I do *not* imply that such a stipulative definition cannot be used in some other context. I do imply that the resulting theory would not be special relativity as it is commonly conceived.

Now, the first possible construal of a 'nonstandard slow-transport' definition assumes standard congruence, but maintains that the conventionality of ϵ is as characteristic of slow-transport synchrony as it is of signal synchrony. In other words, slow-transport synchrony 'contains' epsilon in some sense. John Winnie (1970, §2) sees this containment as occurring in the relative speed of transport, when he claims to have

...shown that the determination of relative speeds depends on the choice of ϵ, and thus the *conventionality of relative speeds* follows from the conventionality of simultaneity....

It follows from these results then, that the conventionality of the one-way speed of light results in the analogous conventionality of *all* one-way relative speeds.

But Winnie cannot say '*all*' (his emphasis) categorically. For he has not shown that infinitely slow relative speeds are conventional; and, indeed, according to his results they are not. He derives the following formula for the one-way, out-going relative speed:

$$\vec{v}_\epsilon = \frac{cv}{c+v(2\epsilon-1)},$$

where v is the constant speed with $\epsilon = \frac{1}{2}$ of a point moving with respect to a frame K. But, according to this formula, when $v=0$, $\vec{v}_\epsilon=0$, and vice versa. Thus, at most Winnie's conclusion applies to relative speeds which are not zero; i.e. speeds which are not calculated from times read off clocks synchronized by slow transport. And the first possible construal of 'nonstandard slow-transport synchrony' fails to get started.

The second possible construal of a 'nonstandard slow-transport' definition continues to assume standard congruence, yet it holds that clocks synchronized by infinitely slow transport can get out of adjustment. Thus, one might allow that clocks so synchronized will always be found to give the same readings when they are juxtaposed, but deny that they continue to do so when they are separated. To be sure this would show that distant synchrony is conventional in the same sense as any other quantitative equality at a distance which depends for its determination on local coincidences (e.g., 'distant mass' or 'distant temperature'). For we can always modify our quantitative scales so long as we preserve those coincidences. Therefore, this is one sense of conventionality which we have to admit.[7]

But the more interesting question is, what else is implied by such a modification? If distant synchrony is to be conventional in a second stronger sense, it has to be established that no factual assumptions at all would have to be made or given up in modifying the slow-transport definition of synchrony. And this, in turn, requires that the controversial notion of a factual assumption be explicated. Such an explication lies beyond the scope of my own work, though I can indicate the sort of account I favor; namely, a 'system-based' one in which the smallest unit of meaning is a theory – a set of inseparable statements – as opposed to a single statement or concept. In this short paper I cannot give any general argument designed to show that such an approach is preferable. Instead I simply rely on what many perceive as the general drift of research in the philosophy of science: it seems to show that the physicist presupposes some theory or other; not that he regards certain facts as secured. In short, facts are theory-dependent. Accordingly, I would insist that the theoretical context of all factual assumptions be specified. This requires a qualification in the second kind of conventionalism: if distant synchrony is to be conventional in that strong sense, it has to be established that, *within a given theoretical context*, no factual difference is entailed by a modification of the slow-transport definition. I will now determine whether the latter conventionalism obtains in special relativity or in a closely related theoretical context.

III. FORCES IN SPECIAL RELATIVITY AND ITS PREDECESSORS

Let us first try to say in special relativity that, even though two clocks have been synchronized through infinitely slow transport to read the same values when they are juxtaposed, they may read different values when they are separated. To do so one would have to account for a deviation from what special relativity dictates is the natural behavior of clocks through the invocation of some force. According to an analysis of force given by Ellis (1963, pp. 178f), a force is acting on a given system if and only if the system persists in an unnatural state. An object is in a natural state if and only if no causal explanation is necessary for its persistence in that state. And by a causal explanation Ellis means any explanation which invokes the existence of perturbing agencies. Thus, an object is in an unnatural state if and only if its persistence in that state requires some causal explanation. In the light of this explication of the concept of force, the difficulty facing the second kind of conventionalism is that special relativity allows the existence of *no* (net or unbalanced) forces on clocks transported with uniform, rectilinear motion. It does admit the rate-retardation effect in such cases, but as the natural behavior of clocks in motion – not as the result of the action of the sort of perturbing agent, the aether, hypothesized by the electron-theoretic predecessors of special relativity. For clocks transported infinitely slowly the rate effect does not alter their synchrony values; thus, if they change anyway, this effect cannot be attributed to their natural behavior, which is to remain in agreement once they have been locally synchronized. Instead it must be that they are in an unnatural state, the existence of which requires causal explanation. The perturbing agency thereby invoked is a force in Ellis' sense. Therefore, we see that modifying the slow-transport definition does result in a difference, namely the presence of a force. Now, if we can establish that this is a physical or empirical assumption, we can conclude that in special relativity distant synchrony is not conventional in the second sense.

In a subsequent paper elaborating his account of forces, Ellis (1965) seems to vacillate between giving them factual and conventional interpretations. While he may not regard the two as being incompatible, at least two pairs of commentators, Hunt and Suchting (1969, p. 233) and Earman and Friedman (1973, p. 347), see him as coming down on the side of the conventional to the exclusion of the factual. However, both note that he allows for different 'theoretical commitments' in two otherwise equivalent force-descriptions (cf. p. 253 and p. 348, respectively). And, in a more recent paper (which neither pair of commentators has taken into account), Ellis (1971, pp. 186f) shows that the choice between them "is not one that can be made on grounds of descriptive simplicity". Rather, "choices of this kind... determine the future pattern of scientific development...": made in accord with other theoretical changes, they tell us what need and what need not be provided with physical or empirical explanations.

A philosopher disposed to the second kind of conventionalism,[8] while admitting the foregoing with regard to the entirety of special relativity, could perhaps uphold his contention with regard to a minimum core of the theory, say just the separately necessary conditions of the theory. Thus, within that most basic context, a change from standard to nonstandard slow-transport synchrony should not change the facts of special relativity.[9] This claim represents a considerable weakening of the theoretical context, since the electron theories of H. A. Lorentz are also compatible with the necessary conditions named: the flatness of space-time, the restriction to inertial frames, and the orientation-independence of rigid rods. In Lorentz' 1895 theory, which contains the length-contraction hypothesis but no rate-retardation effect, two clocks synchronized by infinitely slow transport would have different readings when they were separated in a frame other than the aether frame.[10] The aether frame's being a physically preferred frame distinguished the 1895 theory from special relativity. However, once Henri Poincaré fully implemented his principle of relativity in Lorentz' 1904 version of the electron theory, there was no possibility of giving experimental evidence for the absolute motion of the Earth,[11] and it would seem that the electron theory had become factually equivalent to special relativity.

But what, then, is to be made of an hypothesis like 'Poincaré's pressure'? He adduced it as a causal explanation of the contraction of electrons. And, as such, it played a necessary role in the electron theory developed by him, Lorentz, and their followers: it was the perturbing effect of the aether that accounted for the flattening of the naturally spherical electron as it moved with respect to the aether. By Ellis' account of forces, that motion produced a force which distorted the constituent parts of material objects and caused them likewise to contract. Further, if Poincaré were to be consistent, he would have to hypothesize a force produced by the motion which would account for the retardation of clocks, both on the electronic and macroscopic level. Hence, the electron theory has to tender the existence of forces in a very literal sense.

When that theory was rejected as a factual hypothesis regarding microparticles, the existence of those entities had to be denied. Thus, it turns out that special relativity and the electron theory are empirically equivalent *only if* the electron theory of matter can be divorced from the electron-theoretic modification of Newtonian mechanics, a spatial and temporal theory. But this is physically (if not logically) impossible. The electron theory is the very *raison d'être* of the latter theory: the electromagnetic equations as applied to systems of moving electric charges must be transformed to a moving frame of reference; if this is to be done without alteration of their mathematical form, the transformation equations of Newtonian mechanics must be modified. Hence, the separation of the two does not make physical sense. Yet this distinction can be and is made between special relativity and the atomic theories that have been developed subsequent to the rejection of the electron theory of matter.[12] And, empirically, special relativity and the electron theory are not equivalent, since the electron theory makes a (false) ontological claim regarding forces, which has no counterpart in special relativity.

IV. UNIVERSAL FORCES IN RELATIVITY

A philosopher of a strongly empiricist persuation is not apt to be swayed by a consideration of such matters, however; and Reichenbach (1958) marshalls a distinction to overcome it: universal forces "affect all materials in the same way" and are such that no materials can be used as insulation against them; "all other forces are differential forces". Now universal forces make for a certain difficulty in determining which geometry characterizes physical space. This determination depends on knowing whether or not two lengths are equal, but if universal forces are admitted, this can never be decided (pp. 13f). On analysis, Reichenbach then finds the equality to be "not a matter of cognition but of [coordinative] definition". Thus, "the determination of the geometry of a certain structure depends on the definition of congruence" (p. 18). Arguing that universal forces are not demonstrable, he concludes "that they have to be set equal to zero by definition if the question concerning the structure of space is to be meaningful" (p. 27).

Nonetheless, this conclusion does not support the conventionalist's position. On the one hand, the empirical status of Poincaré's forces is undercut if they are a matter of definition. This would return the conventionalist to the situation in which the electron theory and special relativity are empirically equivalent, if there are no other factual differences between them. On the other hand, setting the universal forces on the slowly transported clock to zero eliminates the possibility of the clock's reading differently, which was after all what we were trying to make sense of as the second construal. Following Grünbaum (1973, p. 49) one might attempt to come to the conventionalist's rescue by interpreting such a setting as a metaphor for what is literally a change in the congruence definition – a change which could just as easily involve a nonzero setting. However, this third construal of 'nonstandard slow-transport synchrony' is of no avail, since in assuming the nonstandard congruence of temporal intervals, it leaves the second context considered. A congruence can be in that context only if it is compatible with the congruences allowable within inertial frames ($g_{\mu\nu}=0$ for $\mu\neq\nu$). For this is one of the defining conditions of that context. Yet, nonstandard temporal congruences (in flat space-time) are equivalent to the congruences of certain non-inertial frames, as is shown in the Appendix, and are thus incompatible with even the minimum theoretical context.

In other theories of relativity, the assumptions that space-time is flat and that inertial frames of reference comprise a preferred class are given up. Thus, those theories deal with curved space-times or non-inertial frames, and nonstandard temporal congruences are properly considered in those contexts.

The University of Tennessee, Knoxville

APPENDIX[13]

Consider the x, t-plane as being subdivided by innumerable curvilinear lines so as to create layers of quadrilaterial figures which are roughly parallelograms. The figure represents just one of these, in which coordinate values have been indicated.

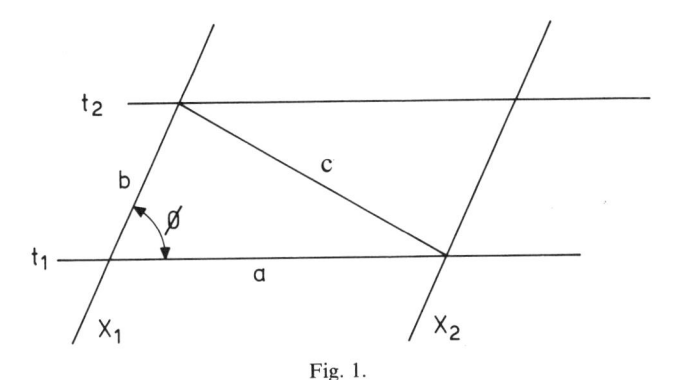

Fig. 1.

We wish to determine the interval or line element of this surface, i.e. $ds = c$. For the orthogonal case it would be given by $c^2 = a^2 + b^2$, but this expression will have to be generalized for the nonorthogonal case: $c^2 = a^2 + b^2 - 2ab\cos\phi$. These two expressions are approximations, of course, which become exact in the limit as $a, b \rightarrow 0$. The relations in which a and b stand to the infinitesimal coordinate differences dx and dt can be written as

$$a = \alpha\, dx \quad \text{and} \quad b = \beta\, dt,$$

where α and β are factors. Substituting the latter into the formula for the interval, we obtain

$$ds^2 = \alpha^2 dx^2 + \beta^2 dt^2 - 2\alpha\beta\, dx\, dt \cos\phi,$$

in which α, β, and $\cos\phi$ are numbers characterizing the particular parallelogram we have considered. In the general case, other parallelograms would be characterized by different numbers (though the form of the expression would be invariant), and we would have to think of them as functions of position and time. But, being concerned with the flat space-time of special relativity, we will think of them as constant 'functions' characterizing all other parallelograms in the x, t-plane.

Now, to test the compatibility of nonstandard temporal congruence with special relativity, we make the following associations with the components of the metric tensor: $\alpha^2 = g_{11}$, $\beta^2 = g_{44}$, $-\alpha\beta\cos\phi = g_{14}$. For flat space-time, these components are constants (see n. 7):

$$g_{11} = \pm k, \quad g_{44} = \pm k_4, \quad g_{14} = \pm k_{14}.$$

Merely using x and t coordinates assures that α and β must be nonzero. Thus g_{44} must be a nonzero constant in special relativity. However, if the temporal interval is to vary with position or time, as it does for nonstandard temporal congruence, β and hence g_{44} cannot be constant. This is all that is required to show such congruences are incompatible with special relativity in the orthogonal case ($\phi = 90°$ and $\cos\phi = 0$). In the nonorthogonal case, g_{14} can be a nonzero constant if α and β are required to be nonzero constants and ϕ is constant. Similarly, using y and z coordinates with the t coordinate in flat space-time means g_{24} and g_{34} can be nonzero constants. Thus, in this case, nonstandard temporal congruences (in flat space-time) are the congruences $g_{\mu 4} = k_{\mu 4} \neq 0$ where $\mu = 1$, 2, or 3. However, the latter obviously constitute a subclass of the congruences associated with non-inertial frames; namely, $g_{\mu\nu} \neq 0$ for $\mu \neq \nu$, where μ, $\nu = 1$, 2, 3, and 4. Therefore, the nonorthogonal case is also incompatible with special relativity.

NOTES

[*] This paper derives from section F of Bowman (1972): it summarizes subsection 2 and follows subsection 5 with only minor changes in the exposition.

[1] Ellis and Bowman (1967, §1) derive an expression for one kind of nonstandard signal synchronization of the clocks of any inertial frame.

[2] Ellis and Bowman (1967, §2) show that the conditions of linearity and reciprocity of relative velocities can both be satisfied only if the clocks of all inertial systems are in standard signal synchrony. Grünbaum (1969, pp. 13–18) disputes the interpretation they give of this result, but he is answered by Ellis (1971, pp. 194–196) and by Bowman (1972, §H.3; 1976, §2).

[3] The construal mentioned above occurs in Grünbaum (1969, p. 21), who uses a clock as a signal. This conflation produces a certain difficulty, as is pointed out in Bowman (1972), subsection J.1, part a, and in Bowman (1976), subsection 2.1.

[4] An attempt to obviate a difficulty raised by Winnie (1970, pp. 227f) can be found in Bowman (1972, §F.3), but it is unsuccessful.

[5] A detailed statement of the slow-transport definition is to be found in Ellis and Bowman (1967, §4).

[6] In general the components of the metric tensor $g_{\mu\nu}$ (where μ, $\nu = 1$, 2, 3, 4) are functions of x, y, z, t. By implementing the necessary conditions of flat space-time, we require (1) $g_{\mu\nu} = \pm k_{\mu\nu} = 16$ constants. The condition that we are dealing only with inertial frames yields (2) $g_{\mu\nu} = 0$ for $\mu \neq \nu$. Combining (1) and (2) we get $g_{\mu\mu} = \pm k_\mu = 4$ constants. Adding

the condition on rigid rods yields $g_{\mu\mu} = \pm k$ for $\mu = 1, 2, 3$ and $g_{44} = \pm k_4$. Obviously, we cannot specify the relationship between $g_{\mu\mu}$ and g_{44} without presupposing the speed of light and hence some value of epsilon. In other words, there has been deleted what Reichenbach (1969, p. 89) calls 'Axiom X': "The time unit of natural clocks is always of such a kind that the time $\overline{ABA}$ of a light signal measured by these clocks is the same everywhere if AB is the same when measured by rigid rods."

[7] As Ellis (1971, pp. 171f) points out, Grünbaum (1969, §3) mistakenly claims that Ellis and Bowman (1967, §6) deny this sense of conventionality, which Grünbaum calls 'Riemann-conventionality'. A remaining point of contention is whether all quantitative equalities other than spatio-temporal ones are Riemann-conventional: Ellis can adduce his book (1966) in support of the claim that they are; Grünbaum's latest reasons for denying this claim are to be found in his 1970 article, though it is unclear there (cf. esp. pp. 471, 475f) whether they are in the published Part A or the unpublished Part B.

[8] Reichenbach's discussion of transported clocks (1958, p. 133) has been interpreted along the electron-theoretic lines of the above paragraph; cf. part b of subsection G.2 in Bowman (1972). Bridgman (1962, pp. 20–22) also does not espouse this position explicitly, but he may be committing himself to it when he talks about "spreading time over space". He prefaces those remarks by assuming clocks at different places have the same rates (p. 19) and says that the remarks apply to *"any method whatever"*, mentioning one which involves "a suitable correction to the reading of our local clock" (p. 21). Bridgman's theoretical situation further shares with Lorentz' 1895 theory an ambiguity as to the possibility of using arbitrarily fast causal chains to determine synchrony. If he rules out such a possibility, his and Reichenbach's positions become the same.

[9] This is a variation of the second construal since standard congruence is entailed by the conjunction of those necessary conditions.

[10] This simultaneity effect can easily be calculated from equation (34) of §31 in Lorentz (1937) by letting $v_y = v_z = 0$ and $v_x = v$ (so that the coordinate frames have their usual positioning) and by retaining terms of the order v^2/c^2; cf. subsection B.1 of Bowman (1972) or section II.A of Bowman (1976a).

[11] Poincaré made these claims in his two papers entitled 'Sur la dynamique de l'électron', the 1905 *Comptes rendues* paper and the 1906 *Rendiconti* paper; cf. Poincaré (1954), pp. 489 and 495. A detailed discussion of these claims is to be found in part c of subsection B.4 in Bowman (1972) or in Part III of Bowman (1976b).

[12] Schaffner (1969, p. 513) makes this same point.

[13] This demonstration is adapted from Reichenbach (1958, pp. 242f).

BIBLIOGRAPHY

Bowman, P. A.: 1972, 'Conventionality in Distant Simultaneity: Its History and Its Philosophy', Ph.D. dissertation, Indiana University.

Bowman, P. A.: 1976, 'On Conventionality and Simultaneity – Another Reply', *Minnesota Studies in the Philosophy of Science*, Vol. VIII (ed. by J. Earman and C. Glymour), forthcoming.

Bowman, P. A.: 1976a, 'The Rejection of Newtonian Simultaneity by Lorentz and Larmor', forthcoming.

Bowman, P. A.: 1976b, 'Henri Poincaré and the Light Principle', forthcoming.

Bridgman, P. W.: 1962, *A Sophisticate's Primer of Relativity*, Wesleyan Univ. Press, Middletown, Conn.

Earman, J. and Friedman, M.: 1973, 'The Meaning and Status of Newton's Law of Inertia and the Nature of Gravitational Forces', *Philosophy of Science* **40**, 329–359.

Ellis, B. D.: 1963, 'Universal and Differential Forces', *The British Journal for the Philosophy of Science* **14**, 177–194.

Ellis, B. D.: 1965, 'The Origin and Nature of Newton's Laws of Motion', in R. G. Colodny (ed.), *Beyond the Edge of Certainty*, Univ. of Pittsburgh Press, Pittsburgh.

Ellis, B. D.: 1966, *Basic Concepts of Measurement*, Cambridge University Press, Cambridge.

Ellis, B. D.: 1971, 'On Conventionality and Simultaneity – A Reply', *Australasian Journal of Philosophy* **49**, 177–203.

Ellis, B. D. and Bowman, P. A.: 1967, 'Conventionality in Distant Simultaneity', *Philosophy of Science* **34**, 116–136.

Grünbaum, A.: 1969, 'Simultaneity by Slow Clock Transport in the Special Theory of Relativity', *Philosophy of Science* **36**, 5–43; reprinted in Grünbaum (1973, pp. 670–708).

Grünbaum, A.: 1970, 'Space, Time and Falsifiability: Introduction and Part A', *Philosophy of Science* **37**, 469–588; reprinted in Grünbaum (1973, pp. 449–568).

Grünbaum, A.: 1973, *Philosophical Problems of Space and Time*, 2nd ed. (Synthese Library), D. Reidel Publishing Company, Dordrecht and Boston; 1st ed. published 1963.

Hunt, I. E. and Suchting, W. A.: 1969, 'Force and "Natural" Motion', *Philosophy of Science* **36**, 233–251.

Lorentz, H. A.: 1937, 'Versuch einer Theorie der elektrischen und optischen Erscheinungen in bewegten Körper', in *Collected Papers*, Vol. V (ed. by P. Zeeman and A. D. Fokker), Martinus Nijhoff, The Hague.

Poincaré, H.: 1954, *Oeuvres de Henri Poincaré*, Vol. IX, Gauthier-Villars, Paris.

Reichenbach, H.: 1958, *The Philosophy of Space and Time* (translated by M. Reichenbach and J. Freund), Dover Publications, New York; German ed. published 1928.

Reichenbach, H.: 1969, *Axiomatization of the Theory of Relativity* (translated by M. Reichenbach), University of California Press, Berkeley and Los Angeles; German ed. published 1924.

Schaffner, K. F.: 1969, 'The Lorentz Electron Theory [and] Relativity', *American Journal of Physics* **37**, 498–513.

Winnie, J.: 1970, 'Special Relativity Without One-Way Velocity Assumptions: Parts I and II', *Philosophy of Science* **37**, 81–99 and 223–238.

SYMPOSIUM

TECHNOLOGY ASSESSMENT

TOM SETTLE

THE BICENTENARY OF TECHNOLOGY ASSESSMENT

1. INTRODUCTION

My thesis in this paper is that because technology assessment has an irreducible moral component it challenges some rather common assumptions among scientists, moral philosophers, democratic theorists and philosophers of science. Quite apart from my sympathy both with technology assessment in general and with its moral component in particular, I think the challenge to established modes of thought is very wholesome. Philosophers of science would be failing in their ordinary jobs, let alone in their moral responsibilities, if, having noticed the social phenomenon of a resurgence of interest in technology assessment, they refrained from inspecting its repercussions upon the ontological, epistemological and axiological presuppositions of science, especially of political science and of economics.

It has sometimes been suggested that technology assessment is a child of the 1960's, born in North America, weaned in 1972 with the enactment of the Technology Assessment Act by the congress of the U.S.A. I do not want to quarrel with giving credit to Philip Yeager, counsel for the House Committee on Science and Astronautics for coining the term 'technology assessment' – but I do quarrel with the notion that technology assessment is a late arrival on the world stage. In my view, it is a child of the 1760's and was weaned as early as 1776 with the publication of Adam Smith's *Wealth of Nations*. It was born and raised in Great Britain.

Now this is neither a mere historical quibble, nor a piece of nationalistic pleading by a British native, but a point of some consequence in philosophy of science, especially the philosophy of the social sciences. The very idea of technology assessment as it is presently generally understood poses with fresh force and urgency questions which have hitherto been answered wrongly – as was the case with Smith, who was already aware of the deleterious effects of the technology on the social fabric of

Great Britain and misjudged them as a fair price to pay for what he judged to be economic benefits – or, if not answered wrongly, ignored or suppressed, but which challenge theories or beliefs or prejudices in political economy which are widely held and deeply cherished in the Western world.

In the light of this challenge to established ways of thought, some thinkers who are opposed to the possibility of an objective theory of morals or who are opposed to metaphysics of any kind and to metaphysics of morals in particular, may urge giving up technology assessment, arguing (against its proponents) that the distribution of the effects of technological innovation is properly left to the normal hazards of the interplay of power in some so-called marketplace. Other thinkers may welcome, with me, the possibility of assessments of technology being undertaken with the ethical presuppositions out in the open, and thus open to criticism; and may welcome the opportunity the renewal of interest in technology assessment affords to take a quite fresh look at such matters as the metaphysics of morals and the presuppositions of political economy.

The argument of the paper proceeds as follows: I shall say what I think is currently meant by the term 'technology assessment', and on the basis of this explain the significance of my point that we are at or near the bicentenary of technology assessment. Even trying to say what is meant by 'technology assessment' will bring into the discussion what my fuller explanation of the point about the bicentenary will emphasize, namely, the moral component of technology assessment. This component will show up the contrast between two different modes of analysis in political economy: (1) what I call the power struggle analysis, in which society is analysed as a power struggle between insatiable utility maximizers – this is the amoral degenerate form of Adam Smith's morally complacent natural law analysis; and (2) what I call the public good analysis, which is the alternative morally principled analysis with which technology assessment challenges the currently popular view in political economy. Finally I shall stress some implications for philosophy of science.

II. 'TECHNOLOGY ASSESSMENT'

The first step of the argument then is for me to try to say what is usually meant by the term 'technology assessment'. In their background study of technology assessment for the Science Council of Canada, Gibbons and Voyer define it as

the activity to provide information about, and systematic analyses of, the internal and external consequences (short, medium and long term) for a society of the application and diffusion of a technological capability into its physical, social, economic and political systems. This information and systematic analysis is to be so structured and presented as to aid the decision makers charged with the responsibility of operating those systems. (Gibbons and Voyer, 1974, pp. 26–7)

This is pretty well in agreement with the definition arrived at by Marstrand and Pavitt in a survey of the literature for the Science Policy Research Unit at Sussex, England, except that these latter are ambivalent about the responsibility of the decision makers. For them, technology includes developing methods "for comparing different technologies as means to stated ends". The argument in the law establishing the Office of Technology Assessment in the U.S.A. is similarly ambivalent:

It is essential that, to the fullest extent possible, the consequences of technological applications be anticipated, understood, and considered in determination of public policy on existing and emerging national problems.

By the choice of language "... means to stated ends" or "essential" for "determination of public policy" the specification of responsibility is avoided. This is not a small point, since the crux both of the rationale and of the effectiveness of technology assessment is that it involves a (moral) responsibility. Even if the mere collection of information "to the fullest extent possible" on the consequences of implementing a technological innovation were to be construed as a scientific task, and thus, supposedly, free of value judgment (but see arguments in my 1974 that not even science is free of value judgments) the use of this information for responsible government cannot be so construed. The very idea of technology assessment as a type of assessment oriented towards decision making in government, or at least towards decision making by those who share government's responsibilities, forces us to ask questions about what is responsible government. People wishing to avoid such questions may view technology assessment as the mere collection of data that will aid in "comparing different technologies as means to stated ends", as Marstrand and Pavitt put it, without discussing the ends or who did the stating of the ends, and on what criteria, hence by-passing the moral question of which ends to pursue and the political question of responsibility both for stating them and for pursuing them. But things are not so simple as that. The social sub-system which does the assessing is not some value-free observer who, without charge, collects and arranges fact, predictions, trends, imaginative guesses and the like into comparative tables. On the contrary, assessing technology is very difficult and very expensive; and much hangs on the outcomes of the judgments to implement or not to implement a particular technology. In real life, the social sub-system that does the assessing – what Gibbons and Voyer call "the technology assessment system" (p. 14) will include those interested parties both who will be affected by the implementation and who have heard about it. Those predicates are conjunctive: the technology assessment system will be unlikely to include persons whose real interests are at stake, but who do not know what is being planned. It is a genuine moral problem for people planning to implement a technological novelty to decide whom to inform in time for their reaction to be significant. Gibbons and Voyer go so far as to say that "in order to have a *true* technology assessment it is essential that those groups which *should be* involved, but which for some reason... are not, become part of the technology assessment system" (p. 15) (Italics in the original).

It is also a judgment with a moral component to decide how far is "the fullest extent possible" to which a technology assessment system may go in collecting information. Undoubtedly, in practice, the Herculean task of providing the decision-makers with all conceivably relevant information that could with effort and cost be assembled is cut down to a manageable size by a series of judgments about what kind of information is not only relevant but important enough to warrant the cost of assembling it. Importance, in this situation, has an irreducible moral dimension, in my view.

Gibbons and Voyer may or may not be right about what constitutes what they call a *"true"* technology assessment but I think they are right that judgments involved in doing a technology assessment – whom to inform and involve, how much information to make public, what kind of information to seek – have an irreducible moral component connected with the responsibility of governments, in guiding some of the social systems in our various nations, to promote public good. The point that technology assessment involves morals turns on whether promoting public good involves morals. A very common view is that a government has simply to reflect the views of the citizens – the theory of consumer sovereignty in the marketplace of public policy, perhaps – and this certainly does not involve the government's being moral beyond its reflecting the morality of its citizens. In my opinion, this view is importantly wrong. Trying to tell you why I think it wrong will take us into the question of the bicentenary of technology assessment, and with this we enter the second step of the argument.

III. THE BICENTENARY

Let me begin this step by characterizing a little more fully the competing analytical modes to which I referred. First, the power struggle mode of analysis. Commonly it is thought that individual citizens are motivated primarily by desires (wants, needs, preferences, etc.) and that when rational, they maximize the satisfaction of these desires. Moreover, it is thought that the peaceful and orderly interplay of many rational agents resolves conflicts of interest in an equitable manner. For Adam Smith 'equitable' had moral overtones: nowadays in discussions in economics and in theory of rational decision making, it often seems as though equity is identified with Pareto-optimality. The role of government on this analysis is to keep the peace and administer the law, members of government being motivated, it is commonly thought, by a desire to stay in office. Thus society is analysed as a balance of power between the citizens, individually or in groups, including the government as a group of citizens elected to keep order. In this way, what happens in a society is what the people will, as expressed in their mutual accommodations (market prices etc.) and by the actions of their democratically elected representatives. In this way what was for Smith a matter of providence has become democratized.

This power-play model is incompatible with the rival model that I have called the public good model. On the public good model, it is not presupposed that the interplay of power between interested citizens and groups of citizens necessarily results in equity. In other words, to put it more crudely, it is not presupposed that the will of the stronger, (which obviously prevails where it is asserted,) is the same as justice. Both models agree that there is justice where the will of the stronger is just, but the public good model has other criteria of justice than merely that there has been a more or less peaceful resolution of conflict.

It is a matter of common knowledge among philosophers that the incompatibility of these models was discussed in Western philosophy at least as long ago as Plato. It is perhaps less well known that there have been similar discussions in classical Hebrew and classical Chinese literature of even greater antiquity. It has been the remarkable achievement of students of political economy of the last 200 years that these models have become thoroughly confused in Western thinking, especially, I might remark, in North American thinking. The history of the confusion and the reasons for its power especially over American minds would make an interesting topic for discussion, but it would take us far too deeply both into history and into comparative democratic theory for our purposes here today. I shall content myself with a few remarks about the origins of this mischievous confusion in Great Britain in the seventeenth century. Hopefully, in that way, we shall arrive at the central issue: what impact technology might have on established modes of thought.

Adam Smith, much influenced both by the physiocrats and by natural theology conceived the laws of operation of the free market, the laws for the allocation both of the factors and of the products of industry, as being natural laws, that is: God-given laws. Thus liberal capitalism appeared to enjoy the double blessing of following scientifically sound principles (economics) and of doing what God wanted. There is little doubt about the role of religion in the rise of capitalism. Liberal capitalism appeared to enjoy both rational and moral sanction. Both these sanctions were of importance to Smith and to many contemporaries. In my view, Smith was wrong on both counts, but I shall concentrate here only on the moral sanction. Given Smith's view, clearly the role of governments in such a political economy was to leave well alone.

Now let us turn to my point about our being at or near the bicentenary of technology assessment. The industrial revolution in Great Britain is not merely a matter of technology. It followed hard on the heels of, and compounded the social effects of, the agrarian revolution of enclosure of common pasture – which created instant riches for the landowner and instant poverty for many peasants, who perforce became the cheap labour force for the new industries, and the new citizens for rapidly expanding urban centres like Manchester (my home city). The industrial revolution impinged on the whole fabric of social life. Smith, knowing about all this, surveyed the scene, and pronounced it good. He could have pronounced it good even without surveying the scene, since for him the social consequences of the two revolutions were the outcome of the play of natural laws. Hence they conformed to God's will, and were good.

Only recently has the complacency of the received wisdom, the complacency that goes with the view that the natural consequences of the free interplay of economic and political forces will be good, been widely challenged in North America, though it has not gone without earlier challenge in other parts of the world. The first well known reaction against the social impact of technological advances was among English workers, for example the followers of "Ned Ludd", who reacted violently in 1811–13. It is noteworthy that the Luddites did not destroy machinery indiscriminately: they destroyed only those machines that belonged to employers who refused to pay certain kinds of compensation to workers. The first well known thoroughgoing theoretical rejection of Adam Smith's views of inevitability and goodness was Karl Marx's. Marx had his own theory both of what was good and of what was inevitable, which we need not here digress to discuss. Recent interest in technology assessment in North America is a child of this continent's belated discovery that some of the consequences of technology in the hands of our style of capitalism are neither good nor inevitable.

My point about the bicentenary is this: we miss the philosophically interesting challenge in the freshly awakened demand for technology assessment if we conclude that it is merely something in the nature of contemporary technology or of contemporary society which is called in question by technology assessment, something perhaps to do with the scale of some recent novelties or with the congestion in some contemporary cities.

I think Smith's theologically-based theory that public good is fostered naturally by the more or less peaceful interplay of more or less self-interested power groups within an economy, is being challenged by the discovery that not all consequences of technological innovations are good, especially if one accepts the commonplace that a technological novelty considered in isolation from the society in which it is being implemented seems to be neither good nor bad in itself. Nowadays economists and political scientists in the West do not usually share Smith's theological assumptions, though they do share his optimistic conclusions. In an attempt to press the challenge against contemporary social scientists who commonly suppress the moral component in political economy, let us turn to the third step in the argument.

IV. THE PROBLEM OF PROMOTING PUBLIC GOOD

Let us now draw out the contrast between the power struggle model and the public good model, and thus challenge what I take to be popular views in economics and political science. I shall argue that there can only be public good, including social justice, if some powerful groups in a society, especially governments, transcend the game of merely balancing conflicting interests. Tullock has recently made this point very neatly with respect to judging (1971). He has shown both that it is not in the interest of judges to judge wisely and fairly and that if they acted in their own interests justice would be in jeopardy. There is no escape from this difficulty by arguing that legislators could make it worth a judge's while to judge justly, since that merely shifts the problem onto legislators, whose private interests do not coincide with the public interest and who would not, therefore, be motivated to make it worth the judges while to be just unless they themselves transcended the game of power play of conflicting interests. Somewhere in the analysis we need individuals to act differently from the norm laid down by Smith, and widely accepted since, if public good is to emerge otherwise than by accident. I think we may view the extent to which public good is actively being promoted, as opposed to merely happening by accident, as a measure of how far individuals do as a matter of fact act indifferently to, or even against, their own interests in favour of the interests of others, and even of the public interest. Furthermore, we may take this measure as a refutation of the thesis that people are insatiable utility maximizers engaged in a power struggle. In short, it is difficult to explain the actions of the public-spirited public servant, on Smith's assumptions.

Another strategy of the believers in the power struggle model of political economy is to explain the pressure on elected representatives to act in the public interest as being a result of citizen pressures conveyed at voting time. But this strategy is futile. The idea, that the will of the people determines policy, lacks both rational and moral coherence. First, and crucially, the criticisms of Schumpeter (1942), Arrow (1950, 1951) and others seem to imply the impossibility of a rationally coherent social will; secondly, matters of morals are hardly to be settled by popular vote, unless, of course, we analyse morals as mere mores or customs.

This brings us to the heart of the matter: technology assessment raises fundamental questions. For example, in moral theory: Is an objective theory of moral obligation possible?; in political theory: Does the existence of democratic institutions, for the election of a small number of persons as rulers, guarantee government in the public interest? If not, can there be any institutional safeguards that would offer such a guarantee and if, as I suspect, not: what might bring about government in the public interest and has it anything to do with morality? Psychology and economics are challenged to take account of the human being as a morally autonomous agent rather than simply a desire-gratifier of insatiable greed, otherwise we cannot account for the publicly good actions of public spirited judges, legislators, and administrators, except as signs of irrationality or sickness. I shall not pursue any of these questions further here, though I have discussed some of them at greater length in my work for the Science Council of Canada (Settle, 1975). My purpose here is to show how the demand to assess technology raises them. Let me close by drawing attention to a challenge of special interest to philosophers of science.

V. THE CHALLENGE TO PHILOSOPHY OF SCIENCE

For too long, the fetish for objectivity among scientists, encouraged by most philosophers of science, has led to the ritual rejection from the domain of scientific inquiry and even from rational discussion of any matter suspected of being subjective or of being a matter of social convention. This in its turn has led to the suppression of what I think is true, namely: that there is an irreducible subjective component in all pure sensation and in all imaging of the real world; and to the suppression or at least to a brushing aside of the idea there is an irreducible social-conventional component in all reports of experience and in all laws or theories purporting to explain the reports. The suppression of subjectivity has led to the identification of scientific inquiry with empirical inquiry and has raised all sorts of problems in accounting for theoretical knowledge, which seems to be beyond the reach of the justification experience was supposed to give to observation reports.

Not surprisingly, matters of morals, which are not obviously empirical matters, were judged subjective, arbitrary, matters of taste, meaningless, or otherwise beyond the pale of rational discourse. If there is an irreducible moral component in technology assessment as commonly understood, then the social phenomenon of a growing demand for it challenges philosophers of science to re-examine a number of epistemological and ontological matters, in particular those related to the idea of the human being as morally autonomous rational agent. Not even the methodological rule of parsimony should be allowed to excuse the suppression any longer.

Obviously what a particular person regards as morally obliged will have a subjective component; and what a community permits or requires will have a social or conventional component. But in view of the irreducible subjective and social components in all scientific knowledge, this should not excuse excluding normative theories of moral obligation (theories that is, of what is universally obligatory or of what is obligatory on governments, and so on) from the arena of rational discussion. Clearly empirical proofs or refutations can hardly be made to bite directly on theories of moral obligation; but this property of remoteness from test is shared with abstract theoretical knowledge in science: we should not expect the remote paradigms of scientists to be easily overthrown any more than we should expect ultimate value judgments to be easily overthrown. Perhaps we have a sufficient safeguard for intellectual integrity if we require, for the purposes of rational discussion, that a person's views in moral matters, including matters related to the distribution of costs and benefits resulting from applying technological capability, be held open to criticism. Perhaps not. Either way, this point is worth full treatment. (See my 1971b, 1974, and Settle, Agassi and Jarvie, 1974, on the rationality of openness to criticism).

In sum: technology assessment challenges the *ontology* presupposed by science: whether man is morally autonomous, whether he has moral obligations and so on; it challenges its *epistemology*: how far objectivity can be achieved and how much that matters for knowledge; and its *axiology*: the status of judgments of value and the content of any universal moral obligation. For myself, I welcome the challenge; and I hope philosophers of science will find fruitful ways of responding to it.

Philosophy Department,
University of Guelph

BIBLIOGRAPHY

Arrow, K. J.: 1950, 'A Difficulty in the Concept of Social Welfare', *Journal of Political Economy* **58**, 328–46.
Arrow, K. J.: 1951, *Social Choice and Individual Values*, Yale University Press, New Haven.
Gibbons, M. and R. Voyer: 1974, *A Technology Assessment System*, Information Canada, Ottawa.
Marstrand, P. K. and K. Pavitt: 1974, 'An Approach to Technology A Assessment' *MS*.
Schumpeter, J.: 1942, *Capitalism, Socialism and Democracy*, Allen and Unwin, London.
Settle, T. W.: 1971a, 'The Relevance of Philosophy to Physics', in M. Bunge (ed.), *Problems in the Foundations of Physics*, Springer-Verlag, New York, pp. 145–162.
Settle, T. W.: 1971b, 'The Rationality of Science versus the Rationality of Magic', *Philosophy of the Social Sciences* **1**, 173–194.
Settle, T. W.: 1974, 'Induction and Probability Unfused', in P. A. Schilpp (ed.), *Philosophy of Karl R. Popper*, Open Court, Illinois.
Settle, T. W.: 1975, *In Search of a Third Way: Is a morally principled political economy possible?* McClelland and Stewart, Toronto (forthcoming).
Settle, T. W.: I. C. Jarvie and J. Agassi, 1974, 'Towards a Theory of Openness to Criticism', *Philosophy of the Social Sciences* **4**, 83–90.
Smith, A.: 1776, *The Wealth of Nations*.
Tullock, G.: 1971, 'Public Decisions as Public Goods', *Journal of Political Economy* **79**, 913–918.

JOSEPH AGASSI

ASSURANCE AND AGNOSTICISM

The scientific ethos of the modern Western world contains two conflicting moods, of utter lack of assurance and of utter assurance. The question posed here is, how do the two go together? The answer I offer is, all assurance is contingent on the supposition that our system as a whole still survives, but there is no assurance for the survival of the system as a whole.

I. THE COMPLEAT AGNOSTIC

The mood of agnosticism, expressed in the music of Schoenberg and Webern (even though the latter, at least, was profoundly religious), in the art of Jackson Pollack and Franz Kline, in Buñuel's 'Los Olvidados', and in ever so many other manifestations of the modern world, is philosophically best expressed in Bertrand Russell's 'Free Man's Worship'.

This work, Russell's 'Free Man's Worship' is mainly a manifesto. (The general idea in it he inherited from T. H. Huxley and H. G. Wells.) Russell himself said he did not like, later in life, its pseudo-Miltonian prose; but he never withdrew its content, and his biographer Wood testified that he felt strongly, later in life, that the view and mood expressed there is correct.

It is a true sleeper: there are few references to it in the philosophic literature, none of any import. And perhaps there is, indeed, no need to refer to it. Yet, any survey, even a superficial one, will show that it is immensely popular – students of philosophy read it and fall under its spell, people unfamiliar with it will assent to anyone who quotes to them crucial passages from it. It truly expresses both the view and the mood of most Western intellectuals who profess agnosticism and even of some of those who profess some religious denomination or other.

Free Man's Worship – and Free Man is both a species and an individual, so that the reader gets a whiff of a sense of eternity from the mere style of the work – is the erect pose with which he walks over a narrow bridge, not knowing where that bridge ends, if anywhere, not having any assurance that it leads anywhere: the bridge is only seen for a brief distance and then disappears in the fog, or, still better, in the void. The meaninglessness and cruelty we see around is inescapable, yet Free Man does not succumb. He goes on regardless, hardly even hoping for the best, but acting as if he does.

II. THE IMAGE OF INDUCTIVE SCIENCE

The traditional religious agnosticism is usually coupled with a traditional scientific assurance. Again, I am making a broad and sweeping empirical observation. Let me, then, describe scientific assurance. I wish to state at once, however, that whereas the image of religious agnosticism described above is congenial to me, the image of scientific assurance I wish to draw now is personally most disagressable to me. But bitter experiences led me to observe repeatedly that I was, on this matter, the odd man out; bitter, since often my being intellectually and temperamentally out on this issue was driven home to me by those who felt they could not easily associate with such a rare bird. What helped me overcome the bitterness was my otherwise successful social life, my professional success regardless of its enormous obstacles, and my effort to understand, rather than complain about, the mood I wish now to describe.

The mood, the inductive mood for short, is that of a self-sufficient intelligent agent. Self-sufficiency is a terrific feel, one which is described in many stories and ballads and plays and movies, from Robinson Crusoe to Citizen Kane. It has diverse components, it has diversity; but usually it contains the idea of self-assurance, of self-reliance, of self-confidence and, above all, of confidence and optimism. When C. P. Snow – now Lord Snow – described the scientific temperament, meaning the temperament given to the inductive mood, he described it as optimistic, progressive, future-oriented. For Snow this will do; for us here it will not.

The specific thing to the inductive mood is the rational grounding of the self-reliant, optimistic mood. Anyone can be self-reliant and optimistic, say, by living in a fool's paradise, by refusing to look far enough into the future to be able to forecast disasters and willing to try to take proper measure against them. Not so the optimist in his inductive mood: he is rational and his optimism is grounded in his rationality.

It does not matter much who should serve as an example, but I wish to take, for the sake of convenience, as my example, none other than Bertrand Russell, the author of 'Free Man's Worship'. Russell's grim vision of Man was no mere abstraction. In his late *Has Man A Future?* Russell seriously considers the possibility that we shall destroy the earth and all its inhabitants. He does not express any optimism or hope – he pleads with us not to destroy. Even in his most inductivist and celebrated *Human Knowledge, Its Scope and Limits* he starts by the image science has of man: trapped in a very narrow strip of barely livable and barely stable corner of the universe, most of which is utterly uninhabitable for Man; in a hostile universe we have to make do as best we can. Yet this making do, which is largely science and scientific technology, is based on induction, on the idea that science validates its theories and assures us – not absolutely but quite very probably – that our expectations based on our scientific theories will indeed come true.

The idea that science has no validity at all, that it offers no assurance whatsoever, is dismissed by Russell off-hand as too defeatist. In his preface to the new edition of Nicod's classic book on induction Russell has one line on Popper's view of science: it is defeatist and our duty is to try and do better than he.

Are Russell's pessimism and optimism reconcilable? I think not. Before we examine that we can ask, do we really need assurance? The answer is, alas, yes.

III. EMPIRICAL FACTS ABOUT ASSURANCE

It is, I fear, an empirical fact that we need assurance constantly. If any one idea of the whole of the Freudian corpus has become commonplace in the modern world, it is that a child needs its mother's assurance, that cuddling and cooing to one's baby, that physical proximity to it, are of extreme importance to the baby, who would soon panic without it. Whether we call this infant-sexuality or the need for security matters here not at all: no one denies that the mother offers her child assurance at every step and that this is essential for the child's well-being.

Experience shows that industrial workers are in need for assurance too, that this need can be easily undermined by minor but noticeable modifications of their surroundings which lead to repeated disappointments of their expectations.

It is an empirical fact that by regularly disappointing an individual's expectations we can render him a nervous wreck: people who blow hot and cold, who seemingly erratically offer good or bad will to their associates and dependants are known as holy terrors. Inquisitors and Brainwashers of all sorts tighten their screws on their victims by making their violent shifts more systematic and damaging. These cruel instances show how much stability is important for normal human beings if they are to lead normal smooth lives.

Stability is not here any inflexibility. The disappointed expectations that fall into patter and elicit well prepared responses are in no way unnerving since a higher-level pattern of satisfaction and disappointment can help. And patterns can also be disappointed and altered, but not so rapidly as to shatter the process of readjustment. That is to say, a most important aspect of stability is not only that of satisfied expectation, but also of rational adjustment to unsatisfied ones. Hence both stability and rationality are important for human well-being, as far as experience can tell us. All this, to repeat, is widely accepted as empirical facts.

Hence, the inductive mood seems to be of the essence. And it seems to be there, to this degree or that, in stable modern enclaves. Hence, the question, can we reconcile this stability with the utter lack of expectation on the large scale, the utter agnosticism with which we have started?

IV. THE NON-JUSTIFICATIONIST MOOD

There are quite a few components in our complex picture, yet one is still missing. We have the agnostic mood which is defiant of its own desperation; the inductive mood which is the optimistic, progressive and assured; and the need for assurance which is a matter of empirical fact, psychological, sociological, and common sense. The missing component is that of moral and psychological independence or autonomy or self-sufficiency. I have already mentioned it. I also noted that its importance lies in the fact that one can have stability and flexibility, a readiness to cope with counterexpectation: this very readiness is autonomy. The inability to cope pushes one towards heteronomy, namely towards emotional and moral and intellectual dependence.

Again the drastic instance, however rare or common it may be, is easy to grasp and it helps make the point sharply. The candidate for conversion, as William Sargant tells us in his *Battle for the Mind*, is ruthlessly broken down by his guide: the guide shatters the candidate's expectations; watches new ones arise and shatters these; helps the candidate build still newer ones and them shatters them as well. The candidate is bewildered, lost, helpless, destitute. In other words, his sense of dependence is hightened to a peak. Then the guide offers love and solace and a new set of expectations to go with all of these.

The crux of the present study may indeed be the contrast of the mood of the brainwashed with the mood of an independent person – politician, philosopher, scientist – who makes a great and decisive switch. For Sargant all switch is brain-wash. This is (unintended) relativism at its rather ludicrous end. For me the opposite of brainwash is the simply deliberate and autonomous decision of the well-composed flexible individual who faces certain alternative routes from the junction of his own understanding of his previous mistake, refuted expectation, or misguided aspirations. There is no hysteria or disturbed spirit here, simply the opening up of new possibilities hitherto unnoticed.

But the two extremes, the brainwashed and the serene deliberator, are not such that middle positions are mixtures of both. We have, admittedly, measures of emotional dependence and intellectual dependence; we have degrees of responsibility, and I shall contrast it with the mood of justificationism, to use Bartley's apt term.

When in civil society I stand before a judge properly accused, I am obliged to justify myself. Otherwise I am obliged not to. Justifying myself to my peers I in fact appoint them as my judges; and so long as I do so without their expressed consent I have imposed on them; I have made myself dependent on them.

The fact that people are all too often guilty of such immorality is the central theme of all of Kafka's works. Kafka also showed that this guilty act is felt by the guilty party as nothing short of a supreme effort of expiation of guilt and/or of proof of innocence or uprightness or guiltlessness – at least by comparison. But, of course, the act is self-defeating.

Most common people in common situations are, again, not so extreme. They explain themselves, they show that their actions, their positions, their views, are eminently reasonable; that indeed their peers before whom they expound their own reasonableness cannot but endorse their views, justify their actions, share their positions. They are not behaving quite like the accused at the dock as Kafka's hero does; but they mention witnesses, they describe significant facts and experiences. They appeal to the supreme authority of religion, of social norms, or of science.

Here the inductive mood and brainwash come extremely close. Many thinkers, I suppose, have felt uneasy about it and tried to throw a wedge between the two: brainwashed people rely on other people's say so, scientists rely on their own experiences. But this means that I, as a scientist, rely on the experiences of other scientists – science becomes my social norm. If and when the scientific world happens to have a dogma, then I am an easy victim of it, with no redemption!

The only way out is to realize that my peers are not my judges: my actions, my expectations, my views, are not such that I need justify them except under very specific and legally well-specified circumstances. And then I have the right for legal advice.

V. CONVERSION TO AUTONOMISM

There is a final and cruel twist to all this. The justificationist mood, to repeat, is extremely common, and it fosters one's dependence under the guise of one's acceptance of the authority of science; the justificationist mood is disguised as the inductive mood. This permits the freakish cruel – if unintended – twist I now wish to illustrate by way of a real empirical example.

In recent years the philosophy of Sir Karl Popper and his followers has gained increasing popularity. It is no secret that I am a product of that school, disciple or apostate, I cannot say. I have observed a process akin to brainwash there. When a philosopher, especially a novice, comes to meet with members of the Popper camp, usually from the inductivist camp, he tries the Popperian idiom for size and checks his expressions with the adept – as is quite common. He is used to the inductive terminology which refers to validation of scientific theories, positive evidence, supporting evidence or argument, confirmation, etc. etc. Here he finds a blank wall. In desperation he tries to translate some of what he thinks are truisms of the philosophy of science to the Popperian jargon; he learns not to use the word 'confirmation' but uses the word 'corroboration' to the same end (in line with one of the Master's most regrettable moves. I say 'regrettable' because it is not words but their imports that matter, as Popper has repeatedly declared, yet he has opened the door to a change that is merely terminological: I wish to report that in a draft of one of my papers I used the word 'verification' with the Popperian import of positive evidence, yet my Popperian colleagues were shocked by my use of a word so branded as belonging to the justificationist camp – as if words can be monopolized. Yet words are monopolized though they cannot be, and 'corroboration' is today a Popperian monopoly. This is regrettable.)

If our novice's interlocutors in the Popperian camp are any good they are not fooled by his switch from the word 'confirmation' to its cognate 'corroboration'. Indeed they are expecting him to do just this and are waiting around the next corner to trip him. At this point, unwittingly, they act like our brainwasher from a previous section. No matter how our novice tries to express a justificationist view, his guide sees that it is justificationist and so not autonomous and so otiose. Our novice does not thereby learn to become autonomous; he attempts, no doubt, to express quasi-autonomy and his expectation of sympathy and understanding and at least partial approval is cruelly shattered. Hence his measure of dependence is heightened, and his conversion may amount to nothing short of brainwash, mild or severe.

Of course, once he enters the circle of the elect – and Popperianism is still not yet so vulgar as to be devoid of this booster – the moment he is boosted, our novice is told that of course their is positive evidence, support, or corroboration, that his need to have some measure of stability of recognized. That the Master's social philosophy is, indeed and alas!, quite conservative!

Does this help our not very autonomous novice become critical and independent or does he become a fanatic upholder of the dogma of the new critical philosophy? Truth to tell I have observed it happen both ways. Now, apart from the educational aspects of our novice's autonomy, he has a genuine theoretical question, that his guide is not quite frank about: how do we combine agnosticism and assurance?

VI. THE ASSURED AGNOSTIC

I do not know the true answer. I have devised an answer that satisfies me for the time being, and I shall repeat it here in the briefest outline.

But I wish to impress upon my reader my impression that this is a problem that has troubled many in the past and does trouble many at present. Many thinkers have felt that a total global agnosticism, a total skepticism on the large scale, precludes all assurance on the small scale. That, conversely, the existence of any assurance on the small scale is evidence enough (perhaps a transcendental proof, even) that somewhere, somehow, assurance exists also on the large scale. And I wish to deny just this.

I think the starting point is Kant's synthetic a priori, which is now better known by such words as the conceptual framework, an overview, a general view, even a welt-anschauung or world view. We have to perform two operations on the conceptual framework to de-Kantianize it. It has to be socialized: it has to be identified with what the sociologist says a member of a given society has internalized in order to share his peers' outlook. And it has to be completely deprived of all privileged status of validity, knowledge, or whatever else. The second step has been taken already by Solomon Maimon; the first is new in the philosophy of science though it has Marxian origins. (Marx himself blew it: he was a relativist.)

Once we deprive our conceptual framework of its finality, we can add to it incentives for change in the form of canons for internal criticism. Among these we should count not only those listed by diverse authors from Plato to Popper but also the canons of commensurability. Let me explain.

Some frameworks are, or at least are meant to be, incommensurable: they include the rule: prefer me over all others. Traditionally, I suppose, Judaism is the paradigm here. Other canons differ. For example, both Newtonians and Einsteinians declare Einsteinism better: the Newtonians had to convert to Einsteinism while following their own canons. Oh, yes, some philosophers have declared that no two conceptual frameworks are commensurable. They illustrate their view with historical facts that conflict with their views. But at least their conceptual frameworks contain no canons of criticism, so that they can stick to their view: the doctrine of incommensurability (of Duhem, Evans-Pritchard, Polyani, etc. etc.) is, indeed, incommensurable, very much like Judaism (though the leading thinkers just mentioned were Catholics of diverse sorts).

So much for the alterability of the framework. But, as it is an institution, it cannot alter so fast as to deprive citizens of their need for assurance. And within any conceptual framework there are all sorts of standards of assurance. There are commercial assurances, including banks and banking insurance, and insurance companies and federal reserves; there are social assurances, there are government agencies to test patent medicines and all sorts of inventions, including, in particular, safety gadgets; travel safety, especially by air; etc. etc. Hence, Popper's theory of corroboration is patently false: it does not operate within any specific conceptual or legal framework but in a social and scientific vacuum.

Finally there is the great old wisdom: diversify; do not put all your eggs in one basket. It means that if my system collapses at a point and say, my electric system goes bust, I can depend on yours; but we may go down together electricity-wise and have a total blackout. In such a case we need a substitute which should be available right now. We thus construct our system so that each detail is connected to other detail and with the framework too. This does insure that the detail is kept afloat, but at the cost of making the whole framework heavy and capable of sinking all in all.

This was realized recently with the series of ecological crises which have led to mass hysteria and to attempts to go back to Mother Nature.

But, as I have explained, this runs counter to Free Man's Worship, which is still the modern enlightened man's manifesto. We can try to criticize and improve the conceptual framework; we can only seek assurance within it while realizing that there is nothing by which to justify our enterprise as a whole.

Boston University and
Tel Aviv University

HENRYK SKOLIMOWSKI

TECHNOLOGY ASSESSMENT AS A CRITIQUE OF A CIVILIZATION

My concern with Technology Assessment is born out of my larger concern with Philosophy of Technology. What is Philosophy of Technology? It is a systematic reflection on the nature of contemporary technology, its role and function in society and civilization at large. It may be said without exaggeration that technology is the major force shaping the destiny of the present western civilization: thus shaping the destiny of Society, and therefore, to a large degree, shaping the destiny of its individuals. It is quite obvious that technology is not a collection of tools, but a vital social and cultural force determining our future. It is not an assembly of gadgets, but a part of our world view, indeed an intrinsic part of the western mentality: whenever we westerners think technology, we invariably think 'manipulation' and 'control.' The primary locus of Philosophy of Technology is not a simple accumulation of insights, a merely analytical and dispassionate inquiry, but an attempt to find some answers to the most urgent social, moral and human dilemmas of our times – the dilemmas which have largely been caused by the relentless unfolding of technological progress. In short, Philosophy of Technology is the philosophy for our times.

The relation between Technology Assessment and Philosophy of Technology is quite clear: the former is a subclass of the latter. Genuine Technology Assessment is, and must be, a form of a socio-moral (therefore philosophical) reflection on the large scale unintended consequences of technology at large. As such it is inextricably bound to a larger corpus of philosophy which I call Philosophy of Technology.

Moreover, unless and until Technology Assessment is seen in a broader philosophical framework, it is bound to be a curious conceptual hybrid, belonging neither to technology, nor to the field of morality. Most of the attempts to define or describe Technology Assessment are ill-conceived because they do not take seriously enough the very idea of *Assessment*. Which is to say, real assessments of Technology must be essentially *critical* not apologetical with regard to Technology. We have had enough apologias for Technology. The idea of Technology Assessment was born at the time when all these apologias were perceived as fundamentally misleading us as people, as a society, as a civilization. Hence sprang the very conception of assessing Technology in social and human terms. Such an assessment must be, by definition, essentially critical.

We are now ready to suggest a new comprehensive characterization of Technology Assessment: *Technology Assessment is a critique of the technological civilization, its foundation and its mythology, from a perspective which must, in some fundamental way, transcend this foundation and this mythology.* Though ambitious and broad, this characterization is one which goes beyond self-serving and circular definitions. For if such a perspective is not accomplished, we are bound to be subservient to the entity we wish to evaluate: we are then servants to the edifice of Technology, rather than its critics.

It suffices to look at various existing 'technology assessments' in order to be at once persuaded that present technology assessment is unfortunately, more often than not, an adjunct to technology itself, a set of technical procedures generated by technology and ultimately serving its purposes. Ironically, many a time the same people who had developed a given technology or a given process, later themselves are assessing this technology, usually applying predominantly technical criteria. It will perhaps be more adequate than harsh to say that quite often technology assessments are fraudulant from a social and human point of view, for while paying lipservice to 'social aspects' the overall tenor, methodology and conclusions are technical: a technical exercise performed by technicians.

Technology Assessment is not *another* branch of technology. Technology Assessment is not an insignificant evaluation of some aspects of Technology. Technology Assessment is a social critique of Technology at large. This critique may be vital to the survival of the technological society; or should we say: may be vital for the preservation of society and its evolution towards a post-technological society. Therefore, this critique must not be left in the hands of those who are themselves responsible for creating powerful but sometimes lethal tools. We must therefore be aware that Technology Assessment does not degenerate into a servile adjunct of Technology. We must also be aware of another danger, of a much subtler nature. We are all *technicians*. Our attitudes and our mentality are profoundly effected by the ideals of technology, and by the assumptions on which technology and the whole civilization are based. We must be therefore aware that even when we genuinely try to assess Technology from a social or a moral point of view 'the process,' as one astute critic has put it, "may be totally slanted in favor of the assumptions underlying the technological civilization." [1]

In the first issue of the periodical *Technology Assessment* there is a comprehensive survey of the field by Genevieve J. Knezo – 'Technology Assessment: A Bibliographic Review' – and in my opinion it shows a distinctive bias towards the treatment of Technology Assessment as an adjunct of Technology. It is slanted in a subtle and often an explicit way in favor of the assumptions underlying the technological civilization, of which it is supposed to be an assessment. Knezo writes, for instance: "The growing literature on technology assessment takes a variety of forms. Some of it is of an emotional, neo-luddite, and polemic nature designed to arouse and mold mass public opinion to the 'uncontrollable' hazards of technology, such as Muller's *The Children of Frankenstein: A Primer of Modern Technology and Values*, [66] or Schwartz's *Overskill: The Decline of Technology in Modern Civilization*, [72] or Douglas' *The Technological Threat*. [71] Much of it, on the other hand, usefully serves to inform the public, through a responsible press, of the pros and cons of a public issue of national importance,"[2] It would thus appear from Knezo's statement that no literature on Technology Assessment should be of a 'polemic nature,' that none should arose public opinion of the *uncontrollable* (she puts the term in quotes, why?) hazards of technology, and that the literature that does so is not a 'responsible press.' Perhaps just the reverse is true: that only critical literature is a 'responsible press.'

It is quite clear that Technology Assessment has not yet found its identity. However it is too important a field to be left in the hands of the apologists of Technology. I would therefore like to propose what I call Skolimowski's laws of Technology Assessment as a set of guidelines for a genuine process of assessing Technology at large. Here are the laws:

(1)   No system can adequately assess itself.
(2)   The more satisfactory is the assessment from the quantita-

tive point of view, the less valuable it is from the social point of view.
(3)   All genuine assessment must terminate in value terms.
(4)   The 'real expertise' in Technology Assessment is social and moral not technical.

These laws stem from the general characterization of Technology Assessment as "a critique of the technological civilization from a perspective which must in some fundamental way transcend the foundation and mythology of the technological civilization." In this context the relevance of philosophy of technology for Technology Assessment is undoubted and overwhelming. One could go even further and suggest that each technology assessment, even performed by the most crass of technocrats, is a consequence, and an expression of a philosophy of technology. Let me now comment on Skolimowski's laws.

Regarding 1. "No system can adequately assess itself" is equivalent to saying that systems must be assessed by agencies outside the system. The very essence of assessment is just that: it is not something that follows from the system, but something that X-rays the system from outside. Can an X-ray machine X-ray itself? No. In the same sense no system can *adequately assess* itself. There is one exception however. And this exception is the human being. If we think about the human being as a system, then it is the only system capable of adequate self-assessment. (Some people would doubt even that.) Because of his ability to judge himself on the variety of meta-levels simultaneously, because of his self-consciousness, and because of his peculiarity to self-judge himself, the human being may be the only system capable of adequate self-assessment. The history of philosophy is an impressive record of the human being self-assessing himself as a species being, as a social being, as an individual being. Returning to the first law: the criteria for assessing a given technological system cannot be derived from this system and cannot be limited to the system, because then there is an apology for the system, not an assessment of it.

Regarding 2. "The more satisfactory is the assessment from the quantitative point of view, the less valuable it is from the social point of view." This law will raise many a hair and objection. I am not saying that it is analytically true, that is to say, that there is a logical relationship between the two. Logically speaking, there is no reason for quantitative perfection to be achieved at the expense of qualitative judgements. But I am suggesting that this is a contingent law, based on actual observations. It *just* happens to be the case that the more successful we are in quantifying, the more successful we are in eliminating qualitative judgments. One immediate reason for this state of affairs is: within the present technological system the process of quantification has almost invariably served the cause of instrumental values, so that quantification came to mean an exclusion of qualitative, or intrinsic values. Unless it is demonstrably *shown* that quantification can serve the cause of intrinsic values, the second law holds.

Regarding 3. "All genuine assessment must terminate in value terms." This law is, of course, a consequence of law 2. But it must be clearly stated because very often instrumental values usurp for themselves the place of ultimate values. It is often argued that the maximization of the economic value, or raising of the standard of living for all is the ultimate value which justifies all our pursuits. This is a favorite strategy for subverting genuine assessments by instrumental or economic ones. Another strategy is to argue that "A strong capacity in applying adequate techniques of technology assessment is necessary in both the executive and the legislative branch." Then methodology takes precedence over value and we gently ride on the high horse of quantitative techniques towards the instrumental paradise. Genuine assessments must be moral, human and social assessments, related to some intrinsic values in which ultimate ends of man's life are expressed, and not merely means leading to these ends. Here again the importance of a larger philosophical basis, whether we call it philosophy of technology or not is irrelevant, is of vital importance.

Regarding 4. "The real expertise in Technology Assessment is social and moral not technical." Such a formulation may strike one as a contradiction in terms, for expertise is something technical, while morality and sensitivity are considered the qualities of the mind and the heart clearly lying outside 'spheres of expertise.' This is a real dilemma which must be met face to face. We must be able to evolve a set of rules and criteria based on moral and social sensitivity which alone will make sense of technology assessment in the long run, and on the scale of the whole society and the entire civilization. Perhaps what is needed is a new social science as a guardian and executor of Technology Assessment. Such a social science does not exist; and it undoubtedly will take a while before we work it out. Such a social science will be directly opposed to Skinner's behaviorism.[3]

Technology Assessment is sometimes abbreviated as T.A. On the other hand, we have another movement which is called Alternative Technology, and it is abbreviated as A.T. Looking at the symbols we see that one is the reversal of the other. These symbols correspond to their respective realities. For while Technology Assessment has, by and large, been an arm and justification of the *status quo*, Alternative Technology has, by and large, been pursued by the opponents of the *status quo*. Ideological controversies apart, I wish to bring to the focus one important thing, namely that Alternative Technology is a form of Technology Assessment; whoever does not see this, does not realize important consequences of either. Alternative Technology is not a mere extravaganza of the hippies, confused drop-outs, half-baked revolutionaries (although it is sometimes associated with those). Alternative Technology is a profound critique of the entire existing technology. It is profound because it not only offers a critique of the existing system, but also suggests a thorough-going alternative. Alternative Technology is Technology Assessment through and through. The sooner we realize that the more we can learn from this movement, whose individuals often possess keen and sharp insights into the nature of present reality.

The end of the thing is often close to its beginning. So let me finish where I started. Technology Assessment is a branch of Philosophy of Technology. Philosophy of Technology is, in a sense, an extension of philosophy of science. But Philosophy of Technology has been lamentably neglected by philosophers of science. In concentrating on exclusively technical aspects of their disciplines, philosophers of science have lost touch with new epistemological and moral realities of the scientific-technological world. The result is: philosophy of technology has already superseded philosophy of science, but philosophers of science do not know it yet.

University of Michigan,
Ann Arbor

NOTES

[1] Dr Frances Svensson, in a private communication.
[2] Genevieve J. Knezo, 'Technology Assessment: A Bibliographical Review', *Technology Assessment* **1**, 67.
[3] The distinguished historian Lynn White, who is perhaps more accomplished in interpreting technology in cultural contexts than any living historian seems to be quite in accord with my views. He writes: "Systems analysts are caught in Descartes's dualism between the measurable *res extensa* and the incommensurable *res cogitans*, but they lack his pineal gland to connect what he thought were two sorts of reality. In the long run the entire Cartesian assumption must be abandoned for recognition *that quantity is only one of the qualities and that all decisions, including the quantitative, are inherently qualitative.* [italics, mine: H.S.] That such a statement to some ears has an ominously Aristotelian ring does not automatically refute it."

"There is a second present defect in the art of technology assessment: the lack of a sense of depth in time; this may be called the Hudson Institute syndrome. It is understandable not only because most social sciences that normally take a flat contemporary view of phenomena, but also because the concrete problems set before systems analysts for solution look toward future action and discourage probing the genesis of things." (Lynn White, 'Technology Assessment from the Stance of a Medieval Historian', *The American Historian Review* **74** (1974), 5.)

SYMPOSIUM

VELIKOVSKY AND THE POLITICS OF SCIENCE

LYNN E. ROSE

THE DOMINATION OF ASTRONOMY OVER OTHER DISCIPLINES

Nature is *one*. Science *should be* one. But we see that contemporary 'science' is fragmented into disciplines, departments, and specialties in an arbitrary and artificial manner. Our 'science' has fallen to these present depths under the sheer weight of its accumulated data and literature. Since no one of us can any longer hope to sift through all of the paper that has been accumulated, we face a choice: we can admit how little we know about nature as a whole – or we can restrict our areas of knowledgeability to fit our capacities, and thus continue to appear knowledgeable. Most scholars choose the course that is more self-flattering: since we cannot be experts on the panorama of nature, we instead become experts on solar prominences, or on Homeric epic, or on Bristlecone pines, or on Mayan sculpture, just so that there remains at least *something* on which we can be experts.

This surrender to a short term temptation has become a long term tragedy. For departmentalized 'science' is pseudo-science: instead of being geared for discovery, it is programmed to self-destruction – for it is guaranteed *not* to discover the truth. *Only* an interdisciplinary approach, seeking one unified and coherent theory to describe our one universe, has much prospect of finding the truth. The mere summation of isolated special theories will be a disconnected jumble, inevitably lacking coherence and probably lacking mutual consistency as well.[1] And where the theories developed within the various isolated disciplines or departments do lack mutual consistency, the resolution of the inconsistencies is essentially a political rather than a scientific process, with the more passive departments designing their own theories in such a way as to conform to the theories that have been favored within the stronger departments.

The phrase, 'politics of science', has many levels of applicability. It could refer to the manoeverings of an *individual* seeking increased status and success within the scientific community (such as the way astronomers need to be 'team players' in order to gain access to the telescopes or other instruments required for their research). Or it could refer to the manoeverings of one *group* against another. Often these groups are *nationalistic*[2] (such as when the Germans claim that Copernicus was German rather than Polish; or when the English write histories of science crediting Harvey with the discovery of the circulation of the blood and ignoring the prior work of da Vinci, Columbo, and Cesalpino – or stressing the role of Dalton in modern atomism while neglecting Avogadro). But the political arrangements that will be emphasized here are those in which the units are the *departments* or *disciplines* themselves. The political form that we find here is doctrinal *monarchy*: Urania, or astronomy, has long been enthroned as 'the queen of the sciences'. This monarchical rule by astronomy is not restricted to the so-called 'natural sciences', but extends also to such departments of knowledge as folk lore, linguistics, art history, Egyptology, psychoanalysis, and even philosophy. (All of these, for purposes of this paper, will arbitrarily be called 'sciences', since they are included in the domain over which astronomy is said to be queen.) Let us take a closer look at the background of this monarchical structure within science, this long-established pecking order of the disciplines.

Once upon a time there was a young Macedonian named Aristotle who believed that Earth and the objects on it were radically different both in composition and in behavior from the objects in the heavens. Aristotle worked out a very complex, purely speculative theory that objects on Earth are composed of one sort of material and obey one set of laws of motion and that objects in the heavens are composed of an entirely different sort of material and obey an entirely different set of laws of motion. Furthermore, objects in the terrestrial realm are subject to all sorts of changes – such as coming-into-being and passing-away, or generation and corruption – whereas in the celestial realm nothing ever changes: the heavenly bodies move along forever in perfect circles on perfect spheres, with perfectly uniform speeds. By their very nature, substance, and laws of motion, the unchanging heavens are completely divorced from the terrestrial world of change and corruption. The precise and uniform speeds and geometrically perfect paths make it possible to calculate the past and future positions of the heavenly bodies with absolute precision, for none of the variation or indefiniteness of the terrestrial realm interferes with such calculations.

The Aristotelian view is that the heavens are made of a nobler substance than are the humble objects on Earth, and that the science that studies the heavens is therefore a nobler science than the mundane sciences of biology, geology, chemistry, physics, history, and psychology. Thus we have the view that astronomy is 'the queen of the sciences'. This view that astronomy is the queen of the sciences has been with us for more than two thousand years; its popularity remains high, especially among astronomers.

Aristotle's pupil, Alexander the Great, conquered a large part of the known geographical world, but his conquest endured for only a few years. On the other hand, Aristotle's circle-happy view of the unchanging heavens has conquered the intellectual world, and has endured right down to the present day. It is no wonder that Aristotle has been called – it is usually intended as a compliment – 'the master of those who know'.

The most vigorous and most successful of the critics of this circle-happy view was Giordano Bruno. It was Bruno who singlehandedly demolished the astronomy of the circle and the sphere, and who thoroughly undermined the Aristotelian distinction between the celestial and the terrestrial.[3] After reading Bruno, no rationally-thinking person could continue to defend the Aristotelian position. But there are problems here: (1) not enough people have read Bruno; and (2) not all of our behavior is rational. The result is that even today the Aristotelian view of an unchanging heaven is very difficult to eradicate from our minds and our thinking, its untenability notwithstanding.

But has not the Aristotelian viewpoint been abandoned by modern astronomy? Not really. Astronomy has changed its tools and some of its terms, but not its stripes. It has only superficially emancipated itself from the circle-happy thinking of Aristotle, Ptolemy, and Copernicus. The circular has been replaced by the cyclical and the periodic, but the effect is the same; just as a combination of perfect and uniform circular motions once permitted backwards and forwards calculation of planetary positions without effective limit, so today it is claimed that a combination of precise cyclical or periodic components (with some secular ones taken into account also) permits calculation of planetary positions for any time in the future or in the past.[4]

The sort of argument used by the astronomers is this. If a body such as Earth is on a well-defined orbit around a central body such as the Sun, we can extrapolate from the present movements of that orbiting body and determine its orbital position for some given time in either the past or the future. Thus the orbital period of Earth is 365.256 days, and that means that every 365.256 days Earth will come back to the sector of its orbit in which it is located today. Exactly five, or fifty, or five hundred, or even five thousand sidereal years ago today, Earth must have been in the same sector of its orbit that it is passing through today. The same applies to all the other planets. To many astronomers, all this seems to be merely a matter of definition.[5] And yet any calculation of the past or future positions of an orbiting body from the characteristics of the orbit being followed at present is fallacious, unless that argument is supplemented by an extra assumption that begs the question that is at issue between Velikovsky and most of the astronomers: namely, whether catastrophic events caused such planets as Earth, Venus, and Mars to be thrown onto their present orbits as recently as within historical times. If you know that there were no such events, then retrocalculation may be one way to determine the approximate positions of the planets at some point in time. But if you do not make that extra assumption or have that extra knowledge, then there is no way to tell if the calculation is correct. *If nothing changed the orbits, then we can integrate the orbits back far enough to show that Velikovsky is wrong and that there were no near-collisions.* But the integration is informative only if no substantial orbital changes are involved.

The considerable success that modern astronomers have enjoyed in predicting planetary positions years in advance has eroded their caution. They forget that such predictions work out only if the present factors affecting the orbits are unchanged. If a black hole or other massive body passes through the solar system and near Earth, then – to say the very least – those calculations and predictions will have to be done over. And while *retro*calculation *may* be effective back some twenty-six centuries, whether it is effective farther back depends upon whether it is or is not the case that Earth's orbit was changed in the year 687 before the present era, which is the very point at issue. Whether there was or was not an Earth-Mars near-collision in 687 cannot be settled by retrocalculation. It must be settled on the basis of historical evidence. In such cases as this, history should set guidelines for astronomical theory, rather than having astronomical theory dictate what shall be admissible as historical fact.

As many have already pointed out, one does not have to be an astronomer to refute the argument of the astronomers. Retrocalculation of the position of some artificial satellite now on a highly stable orbit would show where that satellite was located three decades ago. But of course the artificial satellites did not even exist three decades ago! Calculating where a planet was three millennia ago is no more valid than calculating where an artificial satellite was three decades ago. In order to make such arguments valid, you have to know that the orbits were not subject to any disturbing or perturbing factors other than those operative at present. In order to use such a calculation to prove that there were no near-collisions of planets within historical times, you must assume as a premise of your argument that there were no near-collisions of planets within historical times; to prove Velikovsky wrong, you must assume Velikovsky wrong – a classic case of *petitio principii*.

This procedure has repeatedly been shown to be fallacious, but it remains extremely popular among astronomers. One of its attractions is that it permits astronomy to continue to be the queen of the sciences, and to exercise a rule over other disciplines, such as history. Thus the retrocalculation of past planetary and lunar positions permits astronomers to engage in arm-chair historiography – the writing of history without having to bother with data, records, documents, archeological finds, or other irrelevancies. Modern historiography concerning the ancient world has been *constructed*[6] on a framework of astronomical calculation, and is not a mere summary of the recorded facts. The result is that Velikovsky finds that modern writings about antiquity have left those ancient Ages in Chaos,[7] and he insists that both the history of the solar system and the history of ancient civilizations need to be *reconstructed* on the basis of historical evidence rather than on the basis of astronomical question-begging.

Representatives of those disciplines upon which Velikovsky's work has had repercussions frequently say that they need not look into Velikovsky's radical proposals because the astronomers have already shown that Velikovsky cannot be correct in claiming the occurrence of near-collisions of planets within historical times. Thus one of the reasons that Velikovsky has been denied an adequate hearing within the various relevant disciplines is that those disciplines have accepted the question-begging arguments offered by the astronomers. For example, certain historians have refused to look at Velikovsky's historical evidence because the astronomers have said that the historical evidence cannot show the near-collisions suggested by Velikovsky, since those near-collisions cannot have taken place.

Thus it is that historians and others have allowed their disciplines to remain subalternate and subordinate to queen astronomy. But monarchies, like totalitarian relationships generally, are unlikely to endure for as long as two thousand years unless there is a passivity or a willingness to be ruled and to be led on the part of the subjects. And that is the case here. The geologists, the biologists, and especially the historians have passively accepted the thesis that the astronomers can tell them what has been going on on this Earth, and they have then sketched in those details permitted by the astronomers' general outline. The astronomers inform these other disciplines that Earth has been substantially undisturbed on its orbit for billions of years, and that whatever changes may have taken place on Earth are traceable exclusively to causes of the sort that we see operating today. This is uniformitarianism: the unproved assumption that the only factors that could have operated in the past are the factors that we see operating today. The other disciplines have meekly accepted this uniformitarian thesis, and have done so without seriously examining the evidence or the lack of evidence. It has become standard operating procedure to divert one's critical eye from everything but one's own specialty. Thus it is that even though the students of the other sciences have spent several millennia kneeling at the feet of queen astronomy, and accepting retrocalculated 'history', the habit of diverting the eyes (and the mind) has caused most of those students to remain unaware that these feet at which they have been kneeling are nothing but clay; and the uniformitarian pronouncements of queen astronomy thus remain unchallenged by the other sciences.

Item: Since we do not see new species suddenly appearing all around us today, biologists assume that the evolution that occurred in the past must have occurred so slowly that an observer at some point in the past would have been unaware – just as we today are unaware – of the new species slowly evolving all around.

Item: Since we do not see new mountain ranges or new oceans suddenly coming into being today, geologists assume that the formation of mountains and oceans that occurred in the past must have occurred so slowly that observers back then would have been unaware that they were watching new mountains and new oceans develop, just as we today are unaware of these developments.

Thus neither the biologists nor the geologists have given adequate consideration to the possibility of cataclysmic evolution of new biological and geological forms and cataclysmic extinction of older forms. The astronomers have prohibited the biologists and geologists from invoking a solar system that has intermittently had its Worlds in Collision and that has at times put Earth in Upheaval; and yet these are the very prospects that make cataclysmic evolution a workable and a successful theory.[8]

The historians, too – whether they are historians of art, or historians of languages, or historians of politics and wars – are guided by the astronomers. For the astronomers tell us that our safe haven, Earth, has been moving on its same basic orbit for billions of years, accompanied by its faithful and steady Moon, and that it is therefore possible to calculate, with pencil and paper, or with a computer (the fallacy is the same), all the past and future positions of Earth and Moon, including the dates of all past and future eclipses of the Sun and eclipses of the Moon, without effective limit.[9] If the historians find an ancient record that seems to describe an eclipse, the astronomers have laid out the possible dates to which that report might be assigned. The historians, without putting up any argument, meekly agree that they must either reject their ancient record as inaccurate or else rig their sequences of historical events in such a way that the ancient record of the eclipse comes out at one of the dates that the astronomers will allow.

This subalternation or subordination of the various disciplines to the discipline of astronomy has by now become sacrosanct. To quarrel with this state of affairs means rejecting some of the most sacred beliefs and most cherished habits of at least a score of different disciplines at once. It means telling the established leaders in those disciplines that the labor of decades that they have invested and the professional status that they have acquired may all have to be set aside because of the claims of an outsider. It means tackling both the astronomical queen and her geological, biological, and historical handmaidens. It takes a person of exceptional courage and of exceptional breadth and competence to dare to do this.

Such a person is Immanuel Velikovsky....

State University of New York
at Buffalo

NOTES

[1] See Rose (1972).
[2] For discussion of nationalism in science, see Paterson (1973).
[3] See Paterson (1970, especially pp. 9–49).
[4] Examples are legion: see Oppolzer (1887 and 1962); Lockyer (1894); van den Bergh (1954); Tuckerman (1962); Hawkins (1965); and Hawkins and Rosenthal (1967). For criticism of such procedures, see Bass (1974a and 1974b).
[5] It should be pointed out that those astronomers who make the most sweeping claims for retrocalculation are usually not the leading specialists in "mathematically *rigorous* celestial mechanics" and that the loose sort of retrocalculation that is so widely used is "more an art, based upon 'hope,' than a strict science based upon strict logic". Bass (1974b, pp. 22, 26).
[6] See Velikovsky (1973).
[7] See Velikovsky (1945) and Velikovsky (1952).
[8] See Velikovsky (1955).
[9] See, in particular, Oppolzer (1887 and 1962) and van den Bergh (1954). Lockyer (1894, p. 129) speaks of carrying retrocalculations "as far back as we choose".

BIBLIOGRAPHY

Bass, Robert W.: 1974a, 'Did Worlds Collide?', *Pensée* (Summer, 1974), 8–20.
Bass, Robert W.: 1974b, '"Proofs" of the Stability of the Solar System', *Pensée* (Summer, 1974), 21–26.
Hawkins, Gerald S.: 1965, *Stonehenge Decoded*, Doubleday, Garden City, New York.
Hawkins, Gerald S. and Rosenthal, Shoshana K.: 1967, *5,000- and 10,000-Year Star Catalogs, Smithsonian Contributions to Astrophysics*, Volume 10, Number 2, Smithsonian Institution, Washington, D.C.
Lockyer, J. Norman: 1894, *The Dawn of Astronomy*, Cassell, London.
Oppolzer, Theodor Ritter von: 1887, *Canon der Finsternisse*, Karl Gerold's Sohn, Wien.
Oppolzer, Theodor Ritter von: 1962, *Canon of Eclipses (Canon der Finsternisse)*, translated by Owen Gingerich, Dover, New York.
Paterson, Antoinette Mann: 1970, *The Infinite Worlds of Giordano Bruno*, Charles C. Thomas, Publisher, Springfield, Illinois.
Paterson, Antoinette Mann: 1973, *Francis Bacon and Socialized Science*, Charles C. Thomas, Publisher, Springfield, Illinois.
Rose, Lynn E.: 1972, 'The Censorship of Velikovsky's Interdisciplinary Synthesis', *Pensée* (May, 1972), 29–31.
Tuckerman, Bryant: 1962, *Planetary, Lunar, and Solar Positions 601 B.C. to 1 A.D. at Five-Day and Ten-Day Intervals*, American Philosophical Society, Philadelphia.
van den Bergh, G.: 1954, *Eclipses in the Second Millennium B.C. (−1600 to −1207) and How to Compute Them in a Few Minutes*, H. D. Tjeenk Willink & Zoon N.V., Haarlem.
Velikovsky, Immanuel: 1945, *Theses for the Reconstruction of Ancient History From the End of the Middle Kingdom in Egypt to the Advent of Alexander the Great, Scripta Academica Hierosolymitana*, Simon Velikovsky Foundation, New York.
Velikovsky, Immanuel: 1950, *Worlds in Collision*, Macmillan, New York.
Velikovsky, Immanuel: 1952, *Ages in Chaos*, Doubleday, Garden City, New York.
Velikovsky, Immanuel: 1955, *Earth in Upheaval*, Doubleday, Garden City, New York.
Velikovsky, Immanuel: 1973, 'Astronomy and Chronology', *Pensée* (Spring-Summer, 1973), 38–49.

M. W. FRIEDLANDER

SOME COMMENTS ON VELIKOVSKY'S METHODOLOGY

I. INTRODUCTION

For the 25 years since *Worlds in Collision* was published, Immanuel Velikovsky has awaited the recognition and acceptance that he feels he justly deserves. Whatever recognition he has achieved has come from outside the scientific community; the vast majority of professional scientists has either not examined his work, or has rejected it with varying degrees of vehemence. Those scientists who have reviewed his earlier works (*Worlds in Collision* and *Earth in Upheaval*) have been sufficiently negative in their assessments as to deter the scientific community at large from any extended consideration. During these 25 years, the scientific community has been under intermittent attack for its alleged failure to give Velikovsky a fair hearing and his theories a fair test. As early as 1950, some scientists brought pressure on his publishers, the Macmillan Company, that led to Macmillan transferring the publication of Velikovsky's books to other publishers who did not have text-book divisions through which they were laid open to threats of boycott in the large academic market. This particular episode has coloured all later discussions of Velikovsky and his theories; it has and probably will continue to re-emerge and haunt the scientific community and while it can possibly be understood in some ways, it can or should in no way be condoned. But, no matter how reprehensible that occurrence was, it cannot be taken as lending any strength to the intrinsic merits of Velikovsky's theories, even though some of his supporters have attempted to draw this kind of inference.

The title of this symposium, 'Velikovsky and the Politics of Science,' focussed on the relations between Velikovsky and professional science. The record of the past 25 years is public, but what has not always been adequately spelled out are some of the major reasons why scientists accord him so little attention. It is my purpose in the present paper to restrict my discussion to a few of these points. It would be an easy task to point to the many serious errors in Velikovsky's work and to show quantitatively why his theories fail; alternatively, for their success, his theories would require the capricious relaxation of physical laws that have served us well and which we do not yet have any adequate reason to abandon. Some of this has been taken up in the past. The printed debate, covering many journals and years, provides an interesting study in itself, but the main topic that I wish to deal with involves Velikovsky's methodology. One brief qualification is required: I shall deal only with Velikovsky's writings, for it would be unfair to burden him with the errors of his supporters. At the same time, nothing in this paper should be read as a defence either of any errors on the part of other professional scientists or of their less temperate comments.

My discussion will focus first on tests that Velikovsky has proposed, and then on some of his claims for the validation of his thesis. In all of this, I shall confine my comments to the relevant physics and astronomy, and shall not be concerned with the origins of Velikovsky's theory nor with the other branches of learning such as anthropology and geology, that provide similar arenas of controversy. Finally, I have some brief comments on the deference given by scientists to the scientific opinions of other scientists.

II. PREDICTIONS

By now, the physical sciences have advanced to a degree of sophistication that places a heavy burden on any new theory. There is now an enormous body of experimental data that a successful new theory must encompass. The quantitative demands are severe, and there are few remaining areas where qualitative theories and predictions will be of any use.

A successful theory should incorporate most of the known facts that are considered important and relevant, and should not be based on the systematic selection of just those facts that can be made to support some preconceived idea. Further, when it comes to testing a theory, its predictions should not only be preferably quantitative but should also follow logically and unambiguously from the theory. Where a 'prediction' is seen to have only a tenuous relation to the theory, then even if the 'prediction' is later verified, little or no support accrues to the theory.

It is useful to start with a discussion of some points that Velikovsky has himself selected. In his article in the *Yale Scientific Magazine*, Velikovsky wrote (1967a):

For the thesis that Venus erupted from Jupiter in historical times and went through a series of stormy events before settling in an orbit, the natural sciences must possess independent evidence.... To show that they have happened, I have formulated the following claims as critical tests of the concept:
(1) Venus must be very hot....
(2) Venus must be enveloped in hydrocarbon (and possibly carbohydrate) dust and gases,
(3) The motion of Venus has been disrupted in near collisions with other celestial bodies and in capture by the sun.... And therefore its motion may well be retrograde.

(From this quotation, I have omitted only the references that Velikovsky gave.)

These tests will be discussed in turn.

(1) "Venus must be very hot." A very wide range of temperature is encountered in astronomy: around 1000° in some infra-red objects, around 5000° for the outer surface of the sun, around 30,000° on the surface of hot stars, to millions of degrees within stars. "Hot" needs qualification, and in other places Velikovsky has referred to "petroleum fires" if oxygen is present in the Venusian atmosphere. Nowhere does Velikovsky give a clear indication of how hot Venus should *now* be, nor is there any calculation to show how much Venus might have cooled between its suggested earlier violent existence and today. This lack of precision then makes it possible to claim to have accurately predicted the approximately 600° that was detected more recently via long-range space probes. The clouds that we see are opaque to visible light; the temperatures of around $-40°$, (deduced from observations at visible and near infra-red wavelengths) are well known to refer to the topmost cloud levels. The surface of Venus is indeed hotter that the long-known temperatures of the upper clouds, but the ambiguity of Velikovsky's predictions precludes attaching any weight to it as a predictor of the surface temperature of that planet.

(2) The presence of hydrocarbons in Venus' atmosphere is essential to Velikovsky's theory. To date, there is no evidence at all for the presence of hydrocarbons even at the trace level of 1 part per million. An earlier claim that hydrocarbons had been detected during the Mariner II mission in 1962, was based on a casual verbal comment during the first press conference and has never been substantiated in a scientific publication. From Velikovsky's writings, one would infer that hydrocarbons should be an important constituent of the atmosphere of Venus, but the range of

expected concentrations has never been spelled out. Non-detection, at the present levels, would seem to be critical to his theory, unless the hydrocarbons are later detected at presumably some lower concentration than $1:10^6$ and which can then be shown to be consistent with his theory. But this very ambiguity in prediction has the interesting effect of inverting the burden of proof: it would now seem that a burden of disproof is being charged to those who doubt the theory, rather than the initial burden of proof being carried by Velikovsky as he proposes a radically new theory. A novel theory requires, at the least, a degree of plausibility, before it will receive serious consideration; this plausibility will depend upon the extent to which it covers most already-known facts. Only after an initial and tentative appearance of plausibility, does part of the burden shift to one of disproof.

(3) The rotation of Venus. We classify the directions of rotation of the planets and their satellites as prograde or retrograde. With only two possibilities, of what predictive value is the statement "its motion may well be retrogate"? In the form of the original prediction, it is made to appear that the retrograde rotation would be the result of the motion of Venus being "disrupted". Even an elementary consideration of the collisions of rigid bodies, however, shows how the results are sensitive to their moments of inertia and angular momenta, and these important parameters are conspicuously absent from Velikovsky's discussion, vitiating any meaningful predictions. So, when the retrograde rotation of Venus was later discovered, it could not be taken as in any way confirming Velikovsky's theory, for the relation of the direction of rotation to the planet's earlier history has never been defined.

Another "prediction" of Velikovsky has received much attention and deserves discussion, and that is the radio emission from Jupiter. In *Earth in Upheaval*, (Velikovsky, 1955) there appears the following:

In Jupiter and its moons we have a system not unlike the solar family. The planet is cold yet its gases are in motion. It appears probable to me that it sends out radio noises as do the sun and the stars. I suggest that this be investigated.

At the time of that writing, radio astronomy was in its infancy, and a reasonable reading of that paragraph, at that time, would seem to imply a hot atmosphere for Jupiter with corresponding thermal radio emission. Only a few years later, very strong radio emission was indeed detected from Jupiter, but we now recognise that this arises from a situation similar to the Van Allen regions of trapped particles around our Earth, and the radio emission mechanism involves what we now term the "synchroton" process, which is quite different from thermal emission. Here again there is considerable room for ambiguity as to the meaning of Velikovsky's prediction. Did he predict radio emission? Yes. Did he predict the right kind? No. Does this strengthen the support for his theory? No, in my view. (For a comparison, the reader is referred to the account given by Shklovsky (1960) of the development of a quantitative theory of synchrotron radiation in astrophysics.)

It might seem that Velikovsky's predictions are being held up to unreasonable scrutiny, and that a more open-minded review would grant him the credit for several successful predictions of phenomena that had been quite unanticipated by others. It is therefore useful to compare this situation with the accuracy and specificity of the predictions derivable from Einstein's special theory of relativity, or later from the quantum theory, to see just how precise a good theory can be. Despite representing radical departures from hitherto accepted theories, these soon gained wide acceptance, as their specific predictions were subjected to searching and quantitative testing.

III. CLAIMS FOR VERIFICATION

A scientific theory or hypothesis gains support when its predictions are verified. Verifications may involve the use of experimental data or the application of other already-accepted theories. However, verification depends critically on the *accurate* assessment of experimental data and the *accurate* perception of theories that are being claimed as supportive. Velikovsky's methodology in this area falls far short of what is considered acceptable in the physical sciences today, and specific instances will be cited below.

(a) Origin of Venus: Velikovsky has suggested that Venus originated in an expulsion from Jupiter, and he has cited Lyttleton's work in support of this thesis: Velikovsky quotes Lyttleton, "the so-called terrestrial planets, Venus included, must have erupted from the giant planets, actually from Jupiter" (Velikovsky, 1967) and, again, "Lyttleton showed why the terrestrial planets. Venus included, must have originated from the giant planets, notably Jupiter, by disruption." (Velikovsky, 1963). A careful examination of Lyttleton's writing, (Lyttleton, 1961) has failed to produce an instance of the use of "must" in this context. Lyttleton has written extensively on the origin of the planets, both in professional journals an in semi-popular books and articles. I can find carefully qualified statements such as "If... then... could...." (Lyttleton, 1961) but nothing as strong as Velikovsky's "must."

(b) Thermal history of the moon: Velikovsky has drawn on results from the recent analyses of samples brought back from the moon, to claim support for his thesis that the moon's surface was very hot in relatively recent (historical) times, and he has written (Velikovsky, 1972a):

> The thermoluminescence study by R. Walker and his collaborators at Washington University, St. Louis was made on Apollo 12 cores. They reported tersely "The TL (thermoluminescence) emitted above 225°C by samples between 4 and 13 cm show anomalies resulting from disturbances 10,000 years ago." The "disturbances" referred to were of a thermal nature.

Because this thermoluminescence study was carried out by some of my colleagues, I have had the opportunity to discuss this work and Velikovsky's use of it, in detail. The statement that "the disturbances referred to were of a thermal nature" is Velikovsky's, and is quite unsupported by the paper to which he refers (Hoyt et al., 1971). In fact, the overwhelming conclusion of that long paper is that the surface of the moon was *not* molten in recent times; in as complex a problem as the history of the lunar surface, there are (not surprisingly) some data that cannot be fitted exactly into a simple picture but which can still be generally understood within the developing framework. A general heating of the lunar surface, to the degree that Velikovsky seems to consider these few data points can be generalised, would have obliterated the record that has been so painstakingly reported in the bulk of the paper.

(c) The role of electrical forces in the determination of planetary orbits. Over the years, Velikovsky has laid great emphasis on what he considers to be the unjustified neglect of electromagnetic forces by those who are concerned with the mechanics of large celestial objects. Recently he had drawn on the professional literature for support of his ideas, and has written (Velikovsky, 1972b):

> ... concerning the additional ease with which Venus could achieve a circular orbit in an electromagnetic solar system, I would refer the reader to a 1970 paper by D. K. Sarvajna in *Astrophysical Space Science* **6**, 258 and to a 1971 paper by I. P. Williams on 'Planetary

Formation from Charged Bodies' in the same publication (Vol. 12, 165–171) showing how "a charged body ejected from the Sun can be captured in orbit because of electromagnetic effects." (Williams' model requires a much smaller charge than Sarvajna's.)...

From Velikovsky's comment, one would infer (i) that Venus could easily enter a circular orbit through the action of electromagnetic forces, (ii) that Sarvajna's paper support this, and (iii) Williams' later paper adds support, even requiring a smaller charge. A reading of Williams' paper quickly shows that neither (ii) nor (iii) of these inferences should have been drawn. The two papers in question [Sarvajna (1970), Williams (1971)] start by dealing with the *assumed* ejection of matter from the sun, and Sarvajna explicitly disclaims "it is therefore outside the scope of the problem to discuss whether it is reasonable or not that a body be projected from the sun with a velocity approaching the escape velocity." The assumption is that the ejected body is electrically charged, and the subsequent analysis bears on the point of whether this body can go into a circular orbit or must re-enter the sun. Since Velikovsky has not yet postulated Venus' origin in the Sun, the relevance of these calculations to its origin in Jupiter is unclear. However, Williams points out that "the charge assumed by Sarvajna is too high by many orders of magnitude" and then shows that a feasible model can be constructed, but with a constraint that the mass of an ejected body (from the sun) must be less than 3×10^{14} gm. As this is substantially less than Venus' mass (by a factor of around 10^{13}) the relevance to Venus' origin is again unclear.

(d) Rotation of the Earth: Velikovsky has written (Velikovsky, 1963b):

Assertions that the Earth's axis could not have changed its geographical or astronomical position constituted one of the main arguments against *Worlds in Collision*. They gave place to the theory of wandering poles. Th. Gold (1955) shows the error in the view of G. Darwin and Lord Kelvin, and stresses the comparative ease with which the globe could – and did – change its axis, even with no external force applied.

Reference to Gold's paper (Gold, 1955) fails to reveal the use of anything resembling Velikovsky's quotation "and did". Gold's paper is titled 'The Instability of the Earth's Axis of Rotation'; the closest support one can find is a comment qualified by "it is thus tempting to suggest" and pointing to a "time scale of the order of 10^5 to 10^6 years" for a swinging of the Earth's axis. This is a far longer time than Velikovsky has suggested, and, overall, Gold's paper cannot be taken as providing the strong support that Velikovsky claims.

Other instances can be cited, and all bear on the same point. Velikovsky has repeatedly claimed to find support for his theories in papers published in a wide variety of professional journals. However, these claims are not sustained when one examines those papers: as shown by the cases that I have cited, Velikovsky's use of the primary literature is, at best, unacceptable.

IV. THE ROLE OF AUTHORITY IN SCIENCE

A persistent complaint against scientists has been their apparent willingness to accept the adverse judgements of Velikovsky's works, as have appeared in many book reviews. At issue here is the role of authority: why should one scientist's view be so passively accepted? I have discussed this in more detail elsewhere (Friedlander, 1972), but feel that it deserves mention here, for it also brings up a point more closely related to those discussed above.

By publication in the regular professional journals, involving the refereeing of the manuscripts, scientists gradually establish their scientific credibility. One's scientific reputation is partly built upon this record: the consistency of one's results with other results that are currently accepted, the precision of one's data, etc. When an established scientist then gives it as his opinion that some new theory is quite contrary to accepted views, and has no basis for acceptance when judged in the usual ways, then it is likely that most scientists will accept this judgement. Most of us have enough to do without following up all of the very many crank proposals that emerge each year, and, if one of our colleagues has performed the necessary judgement, we accept it. In the same way, we normally will accept most results in the professional journals: we simply *cannot* check for ourselves everything that everyone else publishes – life is too short, and not all of us have available the skills and equipment. It is a regular procedure, therefore, to accept the view of another scientist on a scientific matter.

Occasionally, results are published which cannot be immediately shown to be in error: they report quite unexpected results for which no confirmation then exists, but neither does there exist any basis for disbelief. Judgement must be reserved, and often it takes years before support arrives, through a repeated measurement or the observation of related phenomena. On the other hand, sometimes support never seems to arrive, and the journals contain many papers whose contents may never be refuted directly, but which we feel must be wrong, for they imply occurrences that should have been seen again by now, in other experiments. Occasionally, also, journals publish papers that are wrong: not all of these are later withdrawn, and unless the paper is attracting major attention, it is rare for anyone to publish a rebuttal. Its errors are often obvious, and those who work in that field know which papers to ignore and which are reliable.

V. SUMMARY AND CONCLUSIONS

Velikovsky's publications appeared first in book form, not in journals where the papers have been subjected to professional refereeing; *Worlds in Collision* was preceded by extensive coverage in *Harpers* and the *Readers Digest*. This style of publication is certainly unorthodox for science today, but it should not have been taken as necessarily reflecting on the scientific validity of the contents. The reviews by scientists were highly unfavorable, and, 25 years later, there seems to be no reason to reverse those early judgements, nor do the more recent claims of verification withstand scrutiny.

There are severe methodological shortcomings in Velikovsky's works: his 'predictions' and his use of the professional literature have been reviewed (in the present paper) and shown to be deficient.

This whole case, however, still repays study by the scientific community, and there are many other aspects not covered in this present review. One major item relates to the reasons for some scientists initiating the pressure on the Macmillan Company in 1950. A second topic would deal with the mutual misunderstandings and antagonisms between some scientists and non-scientists, as revealed by the guerrilla-style debate that has dragged on for far too long. It is important that both sides gain a far clearer understanding of how the other actually operates, and it is certainly important for scientists to understand the roots of the suspicion and even hostility with which we are sometimes viewed. Too often, the arguments of scientists have been too dogmatic; too many of the non-scientists have revealed their view of science and its methods as so grotequely distorted that any reality is hard to recognise. Perhaps the best conclusion that we should draw is that there is still an urgent need for us to work hard at improving science education, if the continuing debate is any indication of past successes.

Washington University

BIBLIOGRAPHY

Friedlander, M. W.: 1972, *The Conduct of Science*, Prentice-Hall Inc., Englewood Cliffs, N.J.
Gold, T.: 1955, *Nature* **175**, 526.
Hoyt, H. P., Miyajima, M., Walker, R. M., Zimmerman, D. W., Zimmerman, J., Britton, D., and Kardos, J. L.: 1971, *Proc. Second Lunar Science Conf.* **3**, 2245.
Lyttleton, R. A.: 1961, *Man's View of the Universe*, Little, Brown & Co., Boston, Mass., p. 36.
Sarvajna, D. K.: 1970, *Astrophys. Space Sci.* **6**, 258.
Shklovsky, I. S.: 1960, *Cosmic Radio Waves*, Harvard Univ. Press, Cambridge, Mass, p. 191.
Velikovsky, Immanuel: 1955, *Earth in Upheaval*, Doubleday & Co., N.Y. (Dell edition, 1968; p. 276).
Velikovsky, Immanuel: 1963a, *American Behavioral Scientist* **7**, 39.
Velikovsky, Immanuel: 1963b, *American Behavioral Scientist* **7**, 40.
Velikovsky, Immanuel: 1967a, *Yale Scientific Magazine* **41**, 8.
Velikovsky, Immanuel: 1967b, *Yale Scientific Magazine* **41**, 9.
Velikovsky, Immanuel: 1972a, *Pensée* **2**, 21.
Velikovsky, Immanuel: 1972b, *Pensée* **2**, 43.
Williams, I. P.: 1971, *Astrophys. Space Sci.* **12**, 165.

A. M. PATERSON

VELIKOVSKY VERSUS ACADEMIC LAG
(THE PROBLEM OF HYPOTHESIS)

Professor E. A. Burtt [1] suggested that we might add "tertiary" qualities to our primary and secondary qualities when we undertake an analysis of reality. These tertiary qualities are the ones embodied in the human institutions of the world. He suggests that we may be arranging nature with our methodologies due to main conditions in us rather than due to main conditions in nature. Nothing can provide the satisfactory generalized concept of the world that does not entail *extensive historical analysis of the major factors that have conditioned us*, according to Burtt. The new cosmology would hardly be worth the effort, if it were merely the synthesis of scientific data or the logical criticism of the basic assumptions of that data, for Burtt. Sound insight must *supplement* scientific data and its assumptions, through the reasoned expression of the intellectual insight of all the ages. He goes on to point out that we see readily the role that wishful thinking has played in older methodologies, but we do not understand the role that wishful thinking plays in our own methodologies. The mechano-morphic period was absorbed in the mathematical nature of physical motion: they ignored the ultimate suppositions that they used in order to frame their laws and their hypotheses. Clarification is necessary and cannot proceed without the historical studies that can expose the fundamental motives and human factors in each of the characteristic analyses that have been adopted. In choosing between hypotheses of two kinds, the tertiary qualities are at work in us. The rise and fall of scholarly interests is conditioned by factors as yet totally unexplored. Burtt goes on to state that though science may reject final cause, it does harbor in its fundamental categories *functioning values* or tertiary qualities that remain completely unseen. A comparative study of the different stages of the growth of scientific thinking would throw light on the nature of our models of the universe and the tertiary qualities of the structure of contemporary scientific procedure.

As Bridgman points out: In "pure physics" the problem of the "observer" must eventually deal with the observer as thinking about what he observes.

Historically, and also in our time, the logical foundation of the natural sciences has been founded and grounded for the sake of usefulness and wealth for man. Problems of controlling natural bodies have been studied primarily for the purpose of gaining technology that would yield fruitful patents and political power. Analysis of the course that mechanics carved out demonstrates this graphically. The story of mechanics is a record of rapid appropriation of experimental results into technology and the scrapping of leftover implications if they held no immediate payoff for industrial giantism or national political edge.

We can observe that serious and critical scientific *insights* have been systematically shelved and laid to rest for 50-100-200 years, until the technology that proceeded *without* comprehensive theory became blocked, unwieldy, and chaotic due to lack of understanding of the primary processes that were involved.

Then and only then, in desperation for further progress in profit, a neglected generalization or analytical clue would be retrieved from the junk heap to serve the technology. In this way, science has limped along with ad hoc hypotheses and the narrow ranges of validity that these hypotheses yield. This is in service to ongoing industrial wealth and national power. The myth of purity in the sciences has arisen in the scientific community as a compensational device. As disinterested pursuit of truth lost its virtue early in the game, the purity that had been rejected from scientific *goals* was claimed to be housed in scientific *ways and means*. It is clearly understood, as one analyzes the record, that from 1650 to 1850, science tinkered technologically and mathematically with the primary processes of physical theory, which were already well known, but these primary processes were unwanted because they involved true or scientific hypothesis rather than ad hoc hypothesis. In scientific hypothesis, a scientist remains disinterested insofar as he stands willing to go wherever the data might lead. During these years, science was mechanomorphic, and these models *suppressed* all unwanted data. As Bruno taught, "we only look for the fruit of the trees that we ourselves plant." Or as Planck observed, we tend to suppress the common nature of all phenomena.

Expectancy, then, determined their hypotheses, and their hypotheses determined their results. Results that were outside of their expectancy and were due to nature were ignored. No mathematics was developed for them during this time.

Bruno noticed that "Each man takes enough fruit to fit his own bowl." In this way, the comprehensive development of natural science has experienced heartbreaking "academic lag".

Philosophers follow rich men, according to Bacon, and the Socratic maxim "know thyself" has been scrapped. The acquiescence of critical philosophy and the flight of philosophers into Romanticism in the 19th century speaks for itself. As Bacon would say: "We have not had a whole man working." Today, things have changed. We have a philosophy of science that holds the promise of becoming a critical philosophy. The danger is present, though, that our philosophy of science will tend to become a mere handmaiden to mechano-morphic science. We have played the handmaiden role before. We must never play it again. There is a middle way; but it does not call for the intimidation of the philosophical search. The middle way for philosophy of science calls for firm and rigorous watchdogging. We are the watchdogs of intellectual integrity and intellectual freedoms.

In 1888, the American Academy of Science reminded us that Young had been evaluated by his colleagues like this: "the absurdity of this writer's 'law of interference' as it pleases him to call one of the most incomprehensible suppositions that we remember to have met with.... This paper contains nothing which deserves the name of experiment or discovery; and it is, in fact, destitute of every species of merit."[2]

The scientific lords also wrote on Ohm's book: "He who looks on the world with the eye of reverence must turn aside from this book as the result of an incurable delusion, whose sole effort is to detract from the dignity of nature." And again on Ohm: "a physicist who professed such heresies was unworthy to teach science." Ohm was penniless for six years because he was "unworthy to teach". And again at the trial of Deforest: "Deforest has said in many newspapers and over his signature that it would be possible to transmit the human voice across the Atlantic before many years. Based on these absurd and deliberately misleading statements, the misguided public, Your Honor, has been persuaded to purchase stock in his company."

And Crookes wrote: "We have actually touched the borderland where matter and force seem to merge into one another. I venture to think that the greatest scientific problems of the future will find their solution in this border land". The orthodoxy immediately said of his work, "How beautiful and how useless", and did everything they could to keep young men from being exposed to him, because he was dangerous.

Let us not forget that the principle of conservation of energy had to be explicated by five men, none of whom was a physicist. Helmholtz, a physiologist working in sound; Carnot, an engineer; Joule, an industrialist and brewer; Colding, an engineer; and last but hardly least, Mayer, a physician who calculated the mechanical equivalent of heat and who was labeled by scientists a freak or a madman and driven to a suicide attempt by the isolation and ridicule of the keepers of orthodoxy.

We have here a record that is a disgrace. Cultural lag is one thing, but academic lag is inexcusable. The man in the street makes no advertisement or commitment to full and open inquiry. The scientist does. So do the philosophers.

It is the fallacy of composition, the fallacy of the "partial view" mistaken for the "whole". This is a special problem in physics and astronomy today. The idea that all of the special sciences are reducible to physics is ridiculous. Especially when we see that what passes for physics is natural philosophy developed in special areas of interest for various phenomena. The phenomena (man included) do not belong to physics. The phenomena constitute the great book of nature, and, as deSebonde insisted over 500 years ago, "The great book of nature can be intelligible to every man."

We see that technology uses the *accumulated everyday* experience of the human race *along with* scientific theories. The interplay here is crucial. We also see that mechanics was developed only after insemination of physics by chemistry, electrical engineering, optical and auditory experiments, and medicine. There is no disciplinary purity to be seen in the grasping of successful scientific understanding of natural phenomena. On the contrary, until the relationships between laws of light, sound, chemistry, electricity, heat, and mechanical energy were understood, researchers explained mechanical energy in terms of the power of the horse. After insemination of mechanics by the heterogeneous phenomena, mechanics was explained in mathematical terms.

The wedding of astronomy to mechanics was a camouflage necessary at that time of Newton. Much of the pictorial modeling in science has been empirical camouflage. As Bacon taught, if one would transplant knowledge from one century to the other, he must quietly and in an unnoticed way graft onto the knowledge-plant at its root – so that the change would seem to occur naturally. For the change would start out as a tiny helpless green shoot, and the keepers of knowledge-plants would get used to it gradually and even believe that they brought it about, somehow.

It has taken 250 years for us to get used to the idea that terrestrial phenomena can be explained without the models of practical mechanics. The same job remains to be done now for astronomy.

The accummulated record of man in his physical environment includes his experience with stars, moon, planets, and the sun, but early in the game these physical bodies were tied to a rich emotional milieu. The time has arrived to sever the intellectual analysis of these physical bodies from old cultural hysterias. To hold that ancient reports of human experience (set aside in holy books) are the sole basis of scientific knowledge is, of course, ridiculous. *But* to hold, *because* some records of ancient human experience have been previously set aside *as* holy books, that they cannot *now* be used to demonstrate the *source* of careless metaphysical assumptions by the human race – is equally ridiculous. The law of uniformity is one such metaphysical assumption.

Analysis can demonstrate that human life devised this principle as a means of coping with unpredictable global physical catastrophe. The basic laws of both analytical psychology and behavioristic psychology demonstrate the avoidance phenomenon that humans develop in order to cope with massive anxiety. The hypothesis of racial amnesia is stated by Dr. Velikovsky:

> The memory of the cataclysms was erased, not because of lack of written traditions, but because of some characteristic process that later caused entire nations, together with their literate men, to read into these traditions allegories or metaphor where actually cosmic disturbances were clearly described.

This hypothesis is actually a scientific or site[3] hypothesis, as opposed to technological hypotheses, such as a working hypothesis or an *ad hoc* hypothesis. The site hypothesis is a statement of *admittedly unproven assumptions* that are adopted as a basis for inference.

Ancient written experience is used in all the natural sciences. It is sorted out and sifted through analytically. Only relevant parts of the material are used for the scientific perspective. It is not a case of all or nothing at all.

We have always ignored much of Newton. He spent over fifteen years writing material not relevant to scientific perspective. We must ignore a lot of Kepler, Bruno, Descartes, Cusanus, Occam, and Archimedes, etc., because not every idea they left behind is relevant to the scientific perspective. And yet, as we sift through their works carefully, we find the clues relevant to our scientific perspective. It is the same for *all* of the ancient records that report human experience. Scientists decide to accept Chinese records of the sudden appearance of the Crab Nebula, because it fits their expectations. They reject some of the historical evidence cited by Velikovsky, because it does *not* fit their expectations.

But to arbitrarily take the position that some books are automatically off limits to analysis, because of the *way* past cultures have used them, is a case of political expediency for the scientific orthodoxy. As philosophers of science, should we support academic lag on that basis?

Let us look at Dr. Velikovsky's hypotheses in terms of their predictive successes. The record is overwhelmingly a success. Yet the astronomer, like the 19th century "physicist", prefers to wait, and *after* the actual fact to look around for an *ad hoc* hypothesis to cover it. This will not do. The immature need to retain the pictorial image – e.g., the "greenhouse effect" – must be given up. Physics had to give up its "pictures", and astronomy must also yield.

Actually, the battle is over. Dr. Velikovsky has emerged the victor because his scientific hypothesis that there have been physical planetary catastrophes in historical times has been proven to have enormous predictive power. For example, a few from very, very many may be listed: Radio noise from Jupiter, strong charge on Jupiter (1953); Earth's extensive magnetosphere (1950); An extensive magnetic field in the solar system, turning with the Sun (1950); The Sun is charged (1950); Venus is hot, has a heavy atmosphere, and an anomalous rotation (1950); Mars' atmosphere contains quantities of argon and neon (1945); Mars is moon-like, battered, and geologically active (1950); There have been many reversals of Earth's magnetic poles (1950); Some of Earth's petroleum was deposited only a few thousand years ago (1950); Earth's rotation varies (very slightly) due to electromagnetic factors such as solar flares (1950).

And successful deduction about the Moon: Hydrocarbons, carbides, and carbonates will be found (July 2 and July 21, 1969); Strong remanent magnetism in rocks (May 19, 1969); Pockets of radioactivity (March 14, 1967); Excessive argon and neon (leading to incorrect age estimates) (July 23, 1969); Thermal gradient in from the surface (July 2, 1969); Many bubbles, some burst (1950).

A scientific or site hypothesis is a statement of admittedly unproven assumptions that are adopted theoretically as a basis for inference of a predictive nature. In this site hypothesis, inference is pointed in a certain direction but is not limited in that direction. The parameters of this fundamental or site hypothesis may be very broad or very narrow in scope. If this site hypothesis is very narrow, the range of this hypothesis is broadened by re-adjustment of the site itself through deductive inference from working hypotheses in combination. This is a logical fanning out from an analytic position gained through reflection upon a combination of working hypotheses.

A working or techno-hypothesis is a statement that attempts to explain a particular physical *process* that takes place during a particular physical *experiment*, and so it involves making provisional assumptions with vested interest in the experiment. Another type of techno-hypothesis is the *ad hoc* hypothesis that provides a reasonable description of the physical experiment, and so it *suggests that there is a physical explanation* for that particular experiment, and the job that the experiment does.

The site hypothesis provides the fundamental parameters of a scientific theory that usually has been demonstrated with various models. These models are actually constituted of three levels of hypotheses, ad hoc, working and site or scientific. With the site hypothesis in view, reflexion fans out the collapsed position of deep analysis that is an attained working hypothesis. When the reflexive flow that is directed toward the site hypothesis indicates any chance of a "fit" into the site model, analysis takes over and picks up the scent, to pursue the possibilities. The coupling is accomplished through analytical deductive inference. This entails the collapsing of the reflexive flow into the mould of the analytical deductive inference. A simpler working hypothesis is now achieved.

From this newly achieved position, and still in view of the site hypothesis, reflexion fans out the new material. This dialectical process of reflexive fanning out and analytical collapsing of the material under study can proceed without limit. However, the site hypothesis demands parameters. It is crucial that the site hypothesis be constantly dilated to accommodate all the families of ad hoc hypotheses that are supporting the numerous working hypotheses that eventually combine to form ground for remodeling the site hypothesis.

Since a working or techno-hypothesis attempts to explain a particular physical process that takes place during physical experiment, it involves making provisional assumptions of a general and peculiar nature. These working hypotheses are written in the highly technical language of mathematics. Using the precision of the mathematical language and the mathematical inference, the working hypothesis establishes an objective frame of reference within which site or scientific hypothesis can be expressed in terms of the ad hoc hypothesis that first generated it.

The working hypothesis sits between the site or scientific hypothesis on the theoretical side and the ad hoc hypothesis on common sense side. The working hypothesis integrates these two logical thresholds by demonstrating in mathematical terms and operations what kinds of inferences are being used in order to thread the ad hoc experiment into the theoretical logic of the site or scientific hypothesis being used.

The working hypothesis is the co-incident of the ad hoc hypothesis and the site hypothesis (a global methodological perspective). It represents the conceptual lift-off from the ad hoc operation, in co-incidence with the basic assumptions of the site hypothesis or global methodological perspective that is involved. This formulation involves an analytical collapse or deductive inference that provides a theorem to the basic axioms or assumptions of the site hypothesis or global methodological perspective. The richer and more structural this mathematical formulation is, the more families of ad hoc hypotheses it will include. There will be entire families of ad hoc hypothesis covered by a deeply reflexive working hypothesis. If the site hypothesis has been well chosen, it will be able to support myriads of these working hypotheses and their innumerable families of supportive ad hoc hypotheses.

We see here a regular pattern of intermittent fanning out reflexively and collapsing analytically as the three levels of hypothesis are accomplished. These complementary logical operations move the reasoning forward from the ad hoc to the site level as surely as the caterpillar is able to move forward by this style. And in human gait, the same phenomenon can be understood. Ambulation is achieved by the dialectic of expansion and contraction operations. Human logic focused on conceptual problems operates by the same physical law.

In the techno- or working hypothesis, the statement is understood ahead of time to be inadequate to explain or control the facts that have appeared. However, because a limited part of the new facts can be handled through such an hypothesis that it can control an interesting fragment of the new facts, this type of hypothesis can be used. It is a function of special interest in a very fragmentary portion of a phenomenon. The chain of higher level reasoning that follows the adoption of this techno-hypothesis eventually will support deductions that will help form the statement of a scientific or site hypothesis. Or this techno-hypothesis can lead to simple formulations that can logically fan out, and at that level this fanning can suggest a deduction that fits into the site hypothesis.

The second kind of techno-hypothesis is the *ad hoc* hypothesis. This type I shall call the inductive hypothesis, because it does not catalyze deductive fall out. Its range of validity is not only narrow, but this range is unlike the narrow range that might be involved in either the working hypothesis or the site hypothesis. This narrow range of validity is tied to the common sense by a direct relationship. The success of this *ad hoc* hypothesis is tied to simple observation. The one redeeming feature of the ad hoc hypothesis is that the human agent present in the demonstration of this hypothesis can (IF HIS LOGICAL OBSERVATION RUNS AHEAD OF WHAT THE *AD HOC* HYPOTHESIS DEMONSTRATES) extend the results of the ad hoc hypothesis. These inductive extensions usually involve analogy and memory in a very active and trained mind totally immersed in the relevant problem. The ad hoc hypothesis is the one that the technicians who live by government subsidy will prefer. Here, they take a very small calculated risk, and results are pretty well guaranteed. It is slow, and it is safe. However, it has nothing to do with the disinterested pursuit of truth and is instead the father of technology and applied science.

It is the scientific or site hypothesis that the truly disinterested curiosity yearns to produce. The record demonstrates that the ability to abandon the senses at the correct stage in reflexion is the only way in which the objects under analysis can appear under the discipline of the reflexive reason. Here, the objects under analysis are no longer pictorial images. Instead, they are structural relationships that the reflexive reason can readily arrange in structures that the intellect suggests.

Perhaps the most crucial characteristic involved in these three levels of hypothesis is flexibility and contingency. Whereas, the procedures *within* each hypothetical step may be certain and necessary, the steps of hypotheses themselves must be capable of expansion as the staircase of theory matures.

Speculative and descriptive method are integrated to produce what is commonly called the empirical method. To proceed merely by speculative method is also impossible. The rub comes in when it cannot be decided what can be included as descriptive data. For Dr. Velikovsky and a host of other scientists, ancient written records can be submitted to critical scrutiny in order that distinct descriptions of cosmic disturbances experienced and recorded by the human race can be recognized and used. Tertiary qualities generated by socialization patterns must be set aside by the scientist who scrutinizes ancient written records.

Analysis is merely the function of reason. Reason faces both the senses and the intellect. The descriptive method uses the sensual side of reason and its function, analysis. The reflexive method is based in the reason but faces the intellect or structural reasoning (mathematics).

The reflexive method entails the stripping of descriptive pictorial images into structural relationships. This is accomplished by logically restructuring them with the active suggestions of the intellect, which stands over the objects under reflexion much like a lamp. In this way, the reflexive capacity can yield structures from deductive relationships (working hypotheses). These reflexive structures (models) themselves can be taken under the reflexive study, and the denuding of these structures can yield sophisticated re-modeling.

Sophisticated re-modeling of the scientific or site hypothesis may also be the result of direct, severe reflexion on and analysis of, raw descriptive data. In this case, the re-modeling *appears* as idio-syncratic or novel. These re-modelings must never be rejected automatically because they call for a re-organization of working and ad hoc hypotheses and a re-modeled site hypothesis.

Scientific or site hypothesis, then, is the tuber or the pod for reflexive deductions. It can be historically demonstrated that sociological conditioning has greatly impaired this insight. Undue fantasy has caused the faculty to come into disgrace. Undue social control has caused the faculty to hold tight to the ad hoc hypothesis. However, the site hypothesis is the result of a critical and courageous, well-trained mind controlled by a mature personality. Because we have the constant emergence of mature personalities despite the sociological threats, true hypothesis will not ever be completely suppressed. Because every man can form crude hypothesis from practical experience, men of scientific reflexion are often intimidated by the resemblance. All one can say to this is that philosophers from Plato to Broad have understood this problem. It is treated by the demand for reflexive, critically sufficient reasons. One man will have them and another man will not.

Dr. Velikovsky has the proper scientific or site hypotheses rather than the techno-hypotheses of the science that is merely a public utility. When the site or scientific hypothesis remains in the shadow of petrified tertiary material fed to it through working and ad hoc hypotheses that are themselves limited to the tertiary qualities – those tertiary qualities had better be severely reflexive and analytical.

State University of New York
College at Buffalo

NOTES

[1] E. A. Burtt, *The Metaphysical Foundations of Modern Physical Science*.
[2] The quoted reactions of the scientific community to new ideas in the past have been taken from Lloyd Taylor's book, *Physics: the Pioneer Science*. This book was chosen because it is 33 years old and is an introductory physics text. Yet, its recorded lessons have seemingly escaped the attention of the academic towers.
[3] The term "site" is used here in an attempt slowly to drop the term "scientific". The term "scientific" has been found to be misleading and ambiguous, with no clear meaning. The concept "hypothesis" is more logically broken down into three types, site, working, and ad hoc.

BIBLIOGRAPHY

Benjamin, A. Cornelius: 1965, *Science, Technology, and Human Values*, University of Missouri Press, Columbia, Missouri.
Bridgman, P. W.: 1959, *The Way Things Are*, Harvard University Press, Cambridge.
Burtt, E. A.: 1932, *The Metaphysical Foundations of Modern Physical Science*, Doubleday, Garden City, New York.

Campbell, Norman Robert: 1957, *Foundations of Science*, Dover, New York.
Dampier-Whetham, William Cecil Dampier: 1929, *A History of Science*, Macmillan, New York.
Evans, Herbert M. (ed.): 1959, *Men and Moments in the History of Science*, University of Washington Press, Seattle.
Farber, Marvin: 1959, *Naturalism and Subjectivism*, Charles C. Thomas, Publisher, Springfield, Illinois.
Friedlander, Ernst: 1965, *Psychology in Scientific Thinking*, Philosophical Library, New York.
Haldane, J. B. S.: 1951, *Everything Has a History*, George Allen and Unwin, Ltd., London.
Helmholtz, Hermann von: 1904, *Popular Lectures on Scientific Subjects*, Longmans, Green, and Co., New York.
Humphreys, Willard C.: 1968, *Anomalies and Scientific Theories*, Freeman, Cooper & Company, San Francisco.
Jordan, Pascual: 1955, *Science and the Course of History*, translated by Ralph Manheim, Yale University Press, New Haven.
Kemeny, John G.: 1959, *A Philosopher Looks at Science*, D. van Nostrand, Princeton, New Jersey.
Maslow, Abraham H.: 1966, *The Psychology of Science*, Harper & Row, New York.
Muller, Herbert J.: 1943, *Science and Criticism*, Yale University Press, New Haven.
Ornstein, Martha: 1928, *The Role of Scientific Societies in the Seventeenth Century*, University of Chicago Press, Chicago.
Scheffler, Israel: 1967, *The Anatomy of Inquiry: Philosophical Studies in the Theory of Science*, Alfred A. Knopf, New York.
Struik, Dirk J.: 1957, *The Origins of American Science*, Cameron Associates, New York. Original title: *Yankee Science in the Making*, 1948.
Struik, Dirk J.: 1962, *Yankee Science in the Making*, new, revised edition, Collier Books, New York.
Taylor, Lloyd W.: 1941, *Physics: the Pioneer Science*, Houghton Mifflin, Boston.
Wiesner, Jerome B.: 1961, *Where Science and Politics Meet*, McGraw-Hill, New York.
Williams, Henry Smith: 1904, *A History of Science*, assisted by Edward H. Williams, Vols. 1–5, Harper & Brothers, New York.

SYMPOSIUM

QUANTUM LOGIC

(See also the contributed paper by William Demopoulos
at the end of this volume)

PETER MITTELSTAEDT

QUANTUM LOGIC

I. INTRODUCTION

It has been shown by Birkhoff and v. Neumann (1936) and by Jauch and Piron (1963, 1964, 1968) that the subspaces of Hilbert space constitute an orthocomplemented quasi-modular lattice L_q, if one considers between two subspaces (elements) a, b the relation $a \subseteq b$ and the operations $a \cap b$, $a \cup b$, $a^\perp$. Furthermore, since the subspaces can be interpreted as quantum mechanical propositions, and since the operations $\cap$, $\cup$, $\perp$ have some similarity with the logical operations $\wedge$ (and), $\vee$ (or) and $\neg$ (not), the question has been raised already by Birkhoff and v. Neumann, whether the lattice of subspaces of Hilbert space can be interpreted as a propositional calculus, sometimes called *quantum logic*.

There are many kinds of lattices which can be interpreted as a propositional or logical calculus. A Boolean lattice L_B of propositions corresponds to the calculus of classical logic and an implicative (Birkhoff, 1961) lattice L_i has as a model the calculus of effective (intuitionistic) logic. Therefore one could ask the question, whether also the lattice L_q of quantum mechanical propositions can be interpreted in some sense as the calculus of quantum logic. A formal condition which must be satisfied, if one wants to interpret a lattice as a propositional calculus, is that the partial ordering relation $\leq$ can also be defined as an operation. This condition results from the requirement that the lattice must be the Lindenbaum Tarski algebra of the logical calculus considered. For the lattices L_B, L_i and L_q it is well known that this condition is fulfilled (Mittelstaedt, 1972).

More important is the question of the semantic interpretation of the lattice L_q. A Boolean lattice L_B can be interpreted by a two-valued truth function. It has been shown by Gleason (1957), Kamber (1965), Kochem and Specker (1967), that neither a two-valued truth function nor a conveniently generalized truth function does exist on the lattice L_q. On the other hand, it is well known that also the intuitionistic logic, i.e. the lattice

L_i cannot be interpreted by truth values. However, an implicative lattice can be considered as a logical calculus if one takes into account a more general method – the operational method which makes use of the dialogic technique (Lorenzen, 1962; Lorenz, 1968; Kamlah and Lorenzen, 1967). Although this interpretation cannot directly be applied to the lattice L_q it turns out that a generalisation of the dialogic method can be used as an interpretation of a lattice L_{eq} which is isomorphic to L_q if one adds to its axioms the 'tertium non datur' $I \leqslant a \vee \neg a$ (Mittelstaedt, 1972a; Mittelstaedt and Stachow, 1974; Stachow, 1973).

In order to demonstrate the possibility of a logical interpretation of the lattice L_q by means of the dialogic technique we proceed in the following way: (Figure 1). We start with the discussion of some obvious modification of the operational method, which come from the incommensurability of the quantum mechanical propositions, i.e. from the theory of quantum mechanical measurement. With the aid of the modified dialogic method we derive a propositional calculus Q_{eff}, which is similar to the calculus of the intuitionistic logic but contains a few restrictions which come from the incommensurability of quantum mechanical propositions. Therefore, we call this calculus the calculus of the effective quantum logic. Taking account of the definiteness of truth values for quantum mechanical propositions – which property follows again from the theory of measuring process – we arrive at the calculus Q of full quantum logic, which also contains the 'tertium non datur'. This calculus represents a model for the lattice L_q and can therefore be used as a logical interpretation of this lattice.

II. THE DIALOGIC INTERPRETATION OF LOGIC

We consider elementary statements about a quantum mechanical system, which we will denote by a, b, c. We will assume that it has been established how these elementary propositions can be proved, i.e. that for each proposition a method is known which is to be regarded as evidence for the respective proposition. On the set S_e of elementary propositions we define connected propositions $c(a, b)$, where $a, b \in S_e$. In particular we use the connectives:

Conjunction: $\quad a \wedge b \quad$ (a and b)
Adjunction: $\quad a \vee b \quad$ (a or b)

Subjunction: $a \to b$ (a then b)
Negation: $\neg a$ (not a)

Apart from the colloquial denotation 'a and b', 'a or b', 'a then b' and 'not a', we explain these connectives as follows by a dialog (Lorenzen, 1962; Lorenz, 1968; Kamlah and Lorenzen, 1967): One of the participants in the discussion (Proponent P) asserts the connected proposition $c(a, b)$, the other (Opponent O) attempts to refute this assertion. If the proponent asserts, for instance, the proposition $a \to b$, then he assumes the obligation, in the case that a can be proved by the opponent, of justifying b. Hence a dialog in which P successfully defends the proposition $a \to b$ reads:

P	O
$a \to b$	proof of a. Why b
proof of b	

On the basis of explanations of this kind, which are described in detail in the literature (Lorenzen, 1962; Lorenz, 1968; Kamlah and Lorenzen, 1967; Mittelstaedt, 1972a; Mittelstaedt and Stachow, 1974), it is defined how a connected proposition $c(a, b)$, formed by the connectives $\wedge$, $\vee$, $\to$ and $\neg$ can be proved, if a and b are elementary propositions.

It is obvious that the connectives $\wedge$, $\vee$, $\to$ and $\neg$ can be iterated and that the concept of dialogic proof can be generalized to iterated connected propositions. A simple example of such a dialog is

P	O
1. $b \to (a \to b)$	1. proof of b. Why $a \to b$
2. $a \to b$	2. proof of a. Why b
3. proof of b.	

The set of all arbitrarily connected propositions – including the elementary propositions – will be denoted here as S_P. The connectives $\wedge$, $\vee$, $\to$, $\neg$ are then *operations* on this set.

If a proposition $c \in S_P$ can be proved, i.e. defended in a dialog, we write $\vdash c$. It is useful in addition to the *operations* $\wedge$, $\vee$, $\to$, $\neg$, on the set S_P of propositions to define a *relation* $a \leqslant b$ with $a, b \in S_P$ by

$$a \leqslant b \rightleftharpoons \vdash a \to b.$$

This relation is called 'implication' and must be distinguished from the operation $a \to b$, denoted here as 'subjunction'. According to the definition the relation $a \leqslant b$ between the propositions a and b holds if and only if the proposition $a \to b$ can be defended in a dialog. If for two propositions the implications $a \leqslant b$ and $b \leqslant a$ are valid we call them equivalent and write $a = b$.

A proposition which can always be successfully defended in discussion irrespective of the elementary propositions it contains, is denoted as a logical statement or, more precisely, as an *effective logical* statement. A logical statement is one for which there is always a certain strategy of success within the scope of dialog, which does not depend on the content of the elementary propositions contained in it. A certain strategy of success can only exist if the proponent can never be committed by the opponent to prove a proposition explicitely – since the proposition in question could be false and in that case the proponent would lose the dialog. As an illustrative example we use the dialog mentioned above. In the framework of the possibilities of defending a proposition in a dialog so far considered the proponent is committed to prove proposition b in $P3$. However this proposition has been proved previously by the opponent in $O1$. If the proposition b proved in one instance is still valid at a later stage of the discussion, when b is challenged again, then the proponent should be allowed to refer to proposition b in $O1$ instead of proving it again in $P3$.

P	O
1. $b \to (a \to b)$	1. proof of b; why $a \to b$
2. $a \to b$	2. proof of a; why b
3. b (refer to $O1$)	

The assumption that a proposition once proved in the dialog remains valid throughout the discussion, is usually accepted in ordinary logic. Therefore the proponent should always be allowed to cite a proposition which has been proved previously by the opponent. By means of this additional possibility some connected propositions (i.e. logical statements) can be defended in a dialog independent of the particular propositions contained in it.

It is obvious that the totality of logical statements can only be presented if the rules for the dialog are formulated in detail. This has been done extensively in the literature (Lorenzen, 1962; Lorenz, 1968; Kamlah and Lorenzen, 1967) and should therefore not be repeated here. Using these rules for a dialog the calculus of effective logic can easily be derived.

III. THE CALCULUS OF EFFECTIVE QUANTUM LOGIC

If one considers a connected proposition about a quantum mechanical system, for the dialogic proof of this proposition it is of importance whether the elementary propositions contained in it are commensurable or not. The reason is that a proposition b, proved in one instance, may not simply be cited at a later stage of the discussion, when b is challenged once again. For in general in the course of the dialog other propositions not commensurable with b will have been proved and thus proposition b is no longer available. This situation can again be illustrated by the proposition $b \to (a \to b)$. If a and b are not commensurable, b may not simply be cited in the dialog mentioned above, and therefore $b \to (a \to b)$ cannot (in general) be successfully defended.

In order to incorporate this point of view into the detailed rules for a dialog, it must be clarified whether two propositions can be considered as commensurable. Generally one denotes two propositions a and b as commensurable, if the corresponding observables can be measured in arbitrary sequence on the investigated system without thereby influencing the result of the measurement. From this definition it follows that two propositions a and b, between which one of the implications $a \leqslant b$ or $b \leqslant a$ is valid, are always commensurable. Furthermore, it can be shown, for instance, that a proposition a is commensurable with $a \to b$, independently of b, and with $\neg a$.

A reformulation of the rules of dialog taking account of the restrictions important for quantum mechanical propositions has the consequence that not all dialogically provable implications can still be successfully defended in dialog. Propositions which, even after the inclusion of these restrictions, can still successfully be defended, will be denoted as *quantum-dialogically* provable. The detailed formulation of the rules of dialog incorporating the restrictions, which come from the incommensurability of quantum mechanical propositions, has been given elsewhere (Mittelstaedt, 1972a; Stachow, 1973) and shall therefore not be repeated here.

The totality of statements, which are formed with the connectives $\wedge$, $\vee$, $\rightarrow$, $\neg$ and can always be successfully defended quantum dialogically, i.e. independent of the content of the elementary propositions contained in it, will be called *effective quantum logic*.

For a formulation of the effective logic, which is most convenient for a comparison with lattice theory, it is useful to introduce the two special propositions $\bigvee$ (truth) and $\bigwedge$ (falsity). The use of both of these propositions within the framework of the dialogic method shall be established in such a way that $\bigvee$ cannot be questioned by either participant of the dialog and that whoever maintains $\bigwedge$ shall have lost the dialog.

From this definition it follows that the propositions $a \rightarrow \bigvee$ and $\bigwedge \rightarrow a$ can be proved in dialog for all propositions a. Therefore we have the general validity of the implications

$$a \leqslant \bigvee \qquad \bigwedge \leqslant a$$

from which in particular it follows that $\bigvee$ and $\bigwedge$ are commensurable with all propositions. According to the dialogic definition of $\bigvee$ we find that for an arbitrary proposition a the statement $\vdash a$ is equivalent to $\vdash \bigvee \rightarrow a$. On the other hand $\vdash \bigvee \rightarrow a$ is equivalent to the relation $\bigvee \leqslant a$. Therefore a proposition a can be defended in a dialog, i.e. $\vdash a$, if and only if the relation $\bigvee \leqslant a$ holds. Furthermore, since for any proposition a we have $\vdash a \rightarrow \bigvee$ the relation $\bigvee \leqslant a \rightarrow \bigvee$ holds for an arbitrary a. Correspondingly since for any a we have $\vdash \bigwedge \rightarrow a$ the relation $\bigvee \leqslant \bigwedge \rightarrow a$ and consequently the relation $a \leqslant \bigwedge \rightarrow a$ is valid for any a.

We wish to represent the effective quantum logic in the form of a calculus, that is, we will present a system of rules with the aid of which all propositions of effective quantum logic can be derived from a few statements which we will include in the rules as the point of departure of the calculus. For the formulae of the calculus we use combinations of the symbols $\bigvee$, $\bigwedge$, $a, b, c, \ldots$ with $\wedge$, $\vee$, $\rightarrow$, $\neg$ and $\leqslant$ and the bracket symbols. With these formulae we establish the rules for the derivation of implications $a \leqslant b$. For the designation of the rules we use the double arrow $\Rightarrow$ and the double comma ,,. With the aid of the rules of dialog which take account of the restrictions for quantum mechanical propositions we arrived at the following calculus, which will be denoted by Q_{eff}. ($b * c$ in 4.4 means any one of the connected propositions $b \wedge c$, $b \vee c$, $b \rightarrow c$, $\neg b$).

THE CALCULUS Q_{eff}

1.1	$a \leqslant a$
2	$a \leqslant b \,,\, b \leqslant c \Rightarrow a \leqslant c$
2.1	$a \wedge b \leqslant a$
2	$a \wedge b \leqslant b$
3	$c \leqslant a \,,\, c \leqslant b \Rightarrow c \leqslant a \wedge b$
3.1	$a \leqslant a \vee b$
2	$b \leqslant a \vee b$
3	$a \leqslant c \,,\, b \leqslant c \Rightarrow a \vee b \leqslant c$
4.1	$a \wedge (a \to b) \leqslant b$
2	$a \wedge c \leqslant b \Rightarrow a \to c \leqslant a \to b$
3	$a \leqslant b \to a \Rightarrow b \leqslant a \to b$
4	$a \leqslant b \to a \,,\, a \leqslant c \to a \Rightarrow a \leqslant (b \ast c \to a)$
5.1	$a \wedge \neg a \leqslant \bigwedge$
2	$a \wedge b \leqslant \bigwedge \Rightarrow a \to b \leqslant \neg a$
3	$a \leqslant \neg a \to a.$

The rule $\alpha \Rightarrow \beta$ states that if the implication α can be proved quantum dialogically, then the implication β can also be justified quantum dialogically. For the proof of a rule $\alpha \Rightarrow \beta$ the implication α will therefore be presupposed by the opponent as a *hypothesis* before the dialog. The proponent has then to defend the implication β (quantum dialogically) whereby he may refer to the hypothesis α which has been accepted by the opponent. It is obvious that the referability of the hypothesis α is not restricted in any way.

In order to illustrate the meaning of the calculus Q_{eff} a few important consequences, which can be derived from Q_{eff}, will be given here:

THEOREM I. The following implications can be proved in Q_{eff}:

(1)	$a \wedge b \leqslant a \to b$
(2)	$\neg a \leqslant a \to b$
(3)	$\neg a = a \to \bigwedge$
(4)	$\bigvee \leqslant a \to a$
(5)	$a \leqslant \neg \neg a.$

Note: (3) is well known as the intuitionistic definition of the negation;

(4) shows that the special proposition $\bigvee$ is equivalent to $a \to a$ for any arbitrary a; the inverse of (5) i.e. $\neg\neg a \leq a$ cannot be proved within the calculus Q_{eff}, just as in the ordinary effective logic.

THEOREM II. The following rules can be proved in Q_{eff}

(1) $\quad a \leq b \Leftrightarrow \bigvee \leq a \to b$
(2) $\quad a \leq \bigwedge \Leftrightarrow \bigvee \leq \neg a$
(3) $\quad a \leq b \Rightarrow \neg b \leq \neg a$
(4) $\quad b \leq a \,,\, c \leq \neg a \Rightarrow a \wedge (b \vee c) \leq (a \wedge b) \vee (a \wedge c)$
(5) $\quad \bigvee \leq a \vee \neg a \Rightarrow \neg\neg a \leq a$.

Note: (1) reconfirms in terms of the calculus Q_{eff} the dialogic definition of the relation $a \leq b$ by $\vdash a \to b$; the inverse of the rule (3) cannot be proved within the calculus Q_{eff}, just as in the ordinary effective logic. The rule (4) will be called 'weak quasimodularity', since in the framework of full quantum logic II.4 expresses the quasimodularity of the lattice of propositions. (5) is important for the transition from the effective quantum logic to the full quantum logic.

Furthermore, the connected propositions $a \wedge b$, $a \vee b$, $a \to b$ of two propositions a and b and the negation of $\neg a$ of a proposition a, which had been originally explained by a dialog, are now uniquely defined by the calculus Q_{eff}, i.e. we have the

THEOREM III. For two given propositions a and b it follows from Q_{eff} that
 (1) the conjunction $a \wedge b$ is uniquely defined by 2.1, 2.2, 2.3.
 (2) the disjunction $a \vee b$ is uniquely defined by 3.1, 3.2, 3.3.
 (3) the subjunction $a \to b$ is uniquely defined by 4.1, 4.2, 4.3, 4.4.
 (4) the negation $\neg a$ is uniquely defined by 5.1, 5.2, 5.3.

The meaning of these uniqueness theorems is that the calculus Q_{eff} of effective quantum logic can be used without any further reference to the dialogic definitions of the connected propositions. Whereas the uniqueness of $a \wedge b$ and $a \vee b$ is well known from ordinary logic, the uniqueness of $a \to b$ and $\neg a$ is by no means trivial, since the 4.1–4.4 and 5.1–5.3 of Q_{eff} are quite different from the corresponding rules in the calculus L_{eff} of effective logic. In the calculus L_{eff} of effective logic, one has instead of 4.1–4.4 and 5.1–5.3 the rules

4.1* $\quad a \wedge (a \to b) \leqslant b$
4.2* $\quad a \wedge c \leqslant b \Rightarrow c \leqslant a \to b$
5.1* $\quad a \wedge \neg a \leqslant \bigwedge$
5.2* $\quad a \wedge b \leqslant \bigwedge \Rightarrow b \leqslant \neg a$

from which the uniqueness of the propositions $a \to b$ and $\neg a$ follows immediately. The proofs of Theorem I–III have been given elsewhere and should therefore not be repeated here. (Mittelstaedt, 1972a; Mittelstaedt and Stachow, 1974; Stachow, 1973).

IV. CONSISTENCY AND COMPLETENESS

The calculus Q_{eff} of effective quantum logic is *consistent* with regard to the class of quantum dialogically provable implications, i.e. every implication derivable from Q_{eff} can be proved quantum dialogically. The proof of this consistence property is straightforward. However, since we have not formulated in this paper the detailed rules of a quantum dialog, we cannot present this proof here and must refer to the literature (Mittelstaedt, 1972a; Mittelstaedt and Stachow, 1974; Stachow, 1973). Furthermore, the calculus Q_{eff} is also *complete* in respect to the class of quantum dialogically provable implications, i.e. every quantum dialogically provable implication can be derived from the calculus Q_{eff}. Consistency and completeness of Q_{eff} guarantee that the dialogic technique can indeed be completely replaced by the calculus Q_{eff}.

The proof of the completeness property is rather complicated. It is well known from ordinary logic that in order to prove the completeness of a logical calculus one first has to transform the rules for a dialog into an equivalent calculus for so called tableaux (Lorenzen, 1962; Lorenz, 1968). This tableau calculus can be shown to be equivalent to a calculus of sequences which was first introduced by Genzen. In a last step one shows the equivalence of this calculus of sequences with the Brower calculus of effective logic. The calculus Q_{eff} of effective quantum logic, which we have presented here, has the form of a Brower calculus. Therefore, for the completeness proof we have to proceed in the same way just mentioned. At first, one has to construct a tableau calculus T_{eff} of effective quantum logic which is equivalent to the rules of quantum dialog. In a second step, one must transform this calculus T_{eff} in an equivalent cal-

510   PETER MITTELSTAEDT

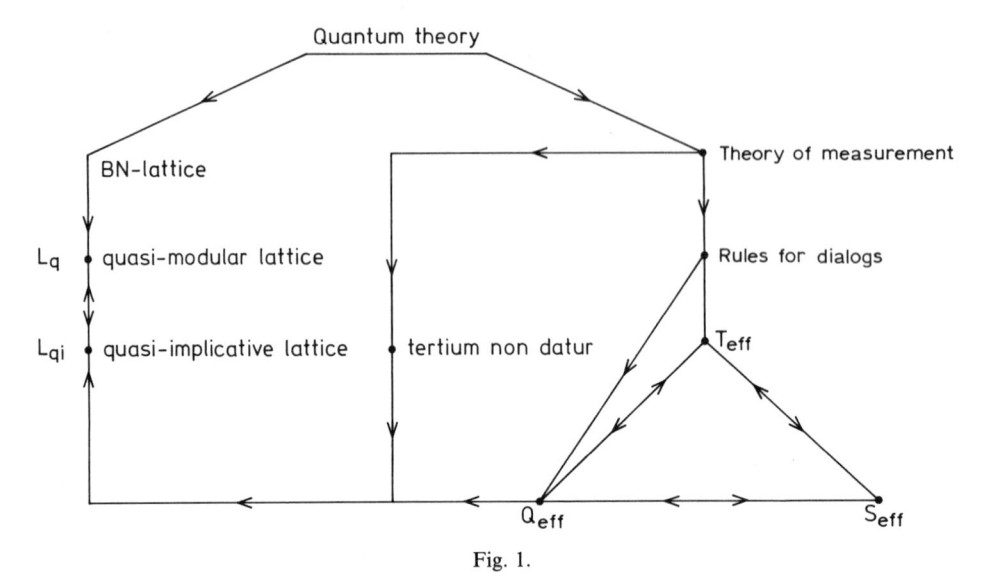

Fig. 1.

culus S_{eff} of quantum logical sequences. Finally, in a third step one has to show the equivalence of this calculus S_{eff} with the Brower like calculus Q_{eff} of effective quantum logic. Alternatively, one can also show directly the equivalence of T_{eff} and Q_{eff} (Figure 1). All these completeness proofs have now been finished by E. W. Stachow and will be published in a forthcoming paper (1975).

V. THE CALCULUS OF FULL QUANTUM LOGIC

In order to construct a quantum logical calculus which can be shown to be equivalent to the lattice L_q of quantum mechanical propositions mentioned above, we have to incorporate into the logical calculus the fact that the propositions considered have well defined truth values, that is, it can be decided whether they are true or false. For quantum mechanical propositions this assumption is always true, since for every quantum mechanical proposition one can decide between the two possibilities by an experiment. This weak assumption, that every proposition can be decided to be true or false, must not be confused with the much stronger postulate of the existence of a two-valued truth value function. It has already been mentioned above, that on the lattice L_q of quantum mechanical propositions such a two-valued truth function does not exist (Kamber, 1965; Gleason, 1957; Kochen and Specker, 1967), and that this fact is one of the essential reasons for the difficulty to interpret the lattice L_q as a logical calculus.

The assumption that the propositions considered have well defined truth values has the important consequence that for any arbitrary proposition a the 'tertium non datur' can successfully be defended in dialog, since either a or $\neg a$ can actually be proved (Kamlah and Lorenzen, 1967). Furthermore, from $\vdash a \vee \neg a$ it follows $\bigvee \leqslant a \vee \neg a$. Consequently, for a quantum mechanical proposition, the relation $\bigvee \leqslant a \vee \neg a$ is always valid. In a dialog which contains the elementary quantum mechanical propositions $a, b, c, \ldots$ the relations

$$\bigvee \leqslant a \vee \neg a, \quad \bigvee \leqslant b \vee \neg b, \quad \bigvee \leqslant c \vee \neg c, \ldots$$

respectively can therefore be presupposed as hypothesis.

In order to incorporate the 'tertium non datur' into the quantum logical calculus one has to add the implication

6.1 $\qquad \bigvee \leqslant a \vee \neg a$

to the rules of the calculus Q_{eff}. The resulting system of rules will be called the calculus Q of *full quantum logic*. This calculus differs essentially from Q_{eff}. The important consequences of the incorporation of the 'tertium non datur' into the quantum logical calculus can be demonstrated by

THEOREM IV. *The following rules can be proved in the calculus Q but not in Q_{eff}*

(1) $\quad \neg \neg a \leqslant a$
(2) $\quad \neg b \leqslant \neg a \Rightarrow a \leqslant b$
(3) $\quad a \to b = \neg a \vee (a \wedge b)$.

Theorem IV.1 is the inverse of Theorem I.5 and follows from the 'tertium non datur' according to Theorem II.5. Theorem IV.2 is the inverse of Theorem II.3. Theorem IV.1 and IV.2 is important for the orthocomplementarity of the lattice of propositions. A further consequence which can be derived in the calculus Q of full quantum logic is the fact that the subjunction $a \to b$ of the two propositions a and b can be expressed by the other operations (4) due to Theorem IV.3. In the effective quantum logic the subjunction $a \rightarrow b$ is also uniquely defined by a and b (Theorem III.3), but in the framework of the calculus Q_{eff} the subjunction cannot be replaced by the other operations.

With the aid of the rules of the calculus of full quantum logic Q, the set of all quantum mechanical propositions can be considered as a lattice. Since we have already defined the relation $\leqslant$ between two elements, it is not necessary here to construct first the Lindenbaum-Tarski algebra of the propositional calculus. The calculus Q of full quantum logic can immediately be expressed in terms of lattice theory. However, for the lattice theoretical characterisation of Q it is not useful to translate the calculus Q directly into a lattice, but first to reformulate the rules of the calculus by means of Theorems I–IV in a system of more convenient but equivalent rules.

Because of the rules 1.1–1.2 the set of propositions forms with respect to the relation $\leqslant$ a partly ordered set. For every two elements a and b the rules 2.1–2.3 cause the existence of a greatest lower bound $a \wedge b$ and 3.1–3.3 provide the existence of a least upper bound $a \vee b$, so that the structure investigated is a lattice. The assumption that a proposition $\bigwedge$ and $\bigvee$ exists means that the lattice has a null and a unit element respectively. The rules 5.1, 6.1 together with Theorems I.5, II.3, III.4, IV.1, IV.2 imply that the lattice considered is orthocomplemented and that the orthocomplement of an element a is the uniquely defined negation $\neg a$. The rules 4.1, 4.2 and Theorem IV.3 means that the orthocomplemented lattice considered is also quasi-implicative, that is, for any two elements a and b there exists one and only one element $a \rightarrow b$, the quasi-implication which satisfies the rules 4.1, 4.2 and Theorem IV.3. The lattice of quantum mechanical propositions is therefore an orthocomplemented quasi-implicative lattice, which we will denote by L_{qi}. It can be characterized by the following axioms.

L_{qi} (1) $a \leqslant a$
L_{qi} (2) $a \leqslant b, \ b \leqslant c \Rightarrow a \leqslant c$
L_{qi} (3) $a \wedge b \leqslant a$
L_{qi} (4) $a \wedge b \leqslant b$
L_{qi} (5) $c \leqslant a, \ c \leqslant b \Rightarrow c \leqslant a \wedge b$
L_{qi} (6) $a \leqslant a \vee b$
L_{qi} (7) $b \leqslant a \vee b$

L_{qi} (8) $a \leqslant c,\ b \leqslant c \Rightarrow a \vee b \leqslant c$
L_{qi} (9) $a \leqslant \bigvee,\ \bigwedge \leqslant a$ lattice with null and unit element
L_{qi} (10) $a \wedge \neg a \leqslant \bigwedge$
L_{qi} (11) $\bigvee \leqslant a \vee \neg a$ ⎫
L_{qi} (12) $a = \neg \neg a$ ⎬ orthocomplemented lattice
L_{qi} (13) $a \leqslant b \Leftrightarrow \neg b \leqslant \neg a$ ⎭
L_{qi} (14) $a \wedge (a \to b) \leqslant b$ ⎫ quasi-implicative lattice
L_{qi} (15) $a \wedge c \leqslant b \Rightarrow \neg a \vee (a \wedge c) \leqslant a \to b$ ⎭

It can be shown (4) that the axioms $L_{qi}^{(14)}$ and $L_{qi}^{(15)}$ are equivalent to the axiom

L_{qi} (16) $b \leqslant a,\ c \leqslant \neg a \Rightarrow a \wedge (b \vee c) \leqslant (a \wedge b) \vee (a \wedge c)$

i.e. the lattice L_{qi} is quasimodular. Therefore we arrive at the lattice L_q, which we have already mentioned in connection with the subspaces of Hilbert space.

In conclusion, we find that the lattice of subspace of Hilbert space, which has been shown to be an orthocomplemented quasimodular lattice, can be considered as a model for that lattice L_q, which has been obtained on the other hand from the investigation of the quantum mechanical propositional calculus. This result shows that the lattice of subspaces of Hilbert space can indeed be interpreted as a propositional calculus. Our conclusion is that this interpretation can be based on the operational foundation of logic which makes use of the dialogic technique.

Institut für Theoretische Physik
Universität zu Köln

BIBLIOGRAPHY

Birkhoff, G.: 1961, 'Lattice Theory', *Ann. Math. Soc. Coll. Publ.* XXV, Rev. Ed., p. 147, 195.
Birkhoff, G. and Neumann, J. v.: 1936, *Ann. of Math.* **37**, 823.
Gleason, A. M.: 1957, *J. of Math. and Mech.* **6**, 885.
Jauch, J. M.: 1968, *Foundations of Quantum Mechanics*, Addison-Wesley Publishing Co., Reading, Mass.
Jauch, J. M. and Piron, C.: 1963, *Helv. Phys. Acta* **36**, 827.
Kamber, F.: 1965, *Math. Ann.* **158**, 158.
Kamlah, W. and Lorenzen, P.: 1967, *Logische Propädeutik*, Bibliographisches Institut, Mannheim.
Kochen, S. and Specker, E. P.: 1967, *J. of Math. and Mech.* **17**, 59.

Lorenz, K.: 1968, *Arch. f. Math. Logik und Grundlagenforschung.*
Lorenzen, P.: 1962, *Metamathematik*, Bibliographisches Institut, Mannheim.
Mittelstaedt, P.: 1972, *Z. Naturforsch.* **27a**, 1358.
Mittelstaedt, P.: 1972a, *Philosophische Probleme der modernen Physik*, Bibliographisches Institut, Mannheim, English ed.: *Philosophical Problems of Modern Physics*, D. Reidel Publishing Company, Dordrecht, Holland, 1976.
Mittelstaedt, P. and Stachow, E. W.: 1974, *Found. of Physics* **4**, 335.
Piron, C.: 1964, *Helv. Phys. Acta* **37**, 439.
Stachow, E. W.: 1973, Diplomarbeit, The University of Cologne.
Stachow, E. W.: 1975, Dissertation, The University of Cologne (to be published).

JOHN STACHEL

THE 'LOGIC' OF 'QUANTUM LOGIC'*

One can and often does use the word 'logic' in a variety of metaphorical ways, speaking of "the logic of the situation", "the logic of events", "the logic of history", and so on. No harm is done, so long as one realizes that these usages are metaphorical, and should not be given any deeper significance – at least without extensive discussion of the meaning of the word 'logic' in these phrases. In this metaphorical sense, one may speak of "the logic of quantum mechanics", just as one may speak of "the logic of modern art", for example.

But if we want to use the term 'logic' in a non-metaphorical sense – and I take it that this is the intent in all serious discussions of 'quantum logic' – we are under some obligation to give an account of the significance to be given to the word 'logic', and to demonstrate that the proposed usage does not do violence to the several-thousand-year-old tradition in the field. Otherwise, the claim that there is a new quantum logic risks asserting no more than that quantum theory is different from classical theories – a claim few would want to challenge.

The usage of the word 'logic' implicit in discussions of the various proposed quantum logics seems to be restricted to 'formal logic'; more especially, the propositional calculus, and perhaps the first order predicate calculus in some cases. Thus, we are asked to consider some sort of extension of that science of logic which consists in drawing inferences from premises, by virtue of the form of these premises, to give a very brief – and admittedly in many ways problematical – characterization of this subject.

Perhaps the most problematic issue, and one most relevant to any assessment of the claims made for various quantum logics, concerns the nature of, and the relationship between, these premises and the conclusion(s) to be drawn from them. In other words, what is the object of the various 'laws of logic?'

Although discussions of quantum logic are often carried out at a high level of abstraction, it is often useful, and indeed for my purposes indispensable, to remind ourselves of some of the basic facts about logic. I cannot say that everyone is going to agree with me that these *are* the basic facts about logic; but at least I will have laid my cards on the table and said what I think the issues are.[1]

People, organized into the various social groups which humanize them in the first place, act upon the world in many ways (as well as being acted upon by the world, of course). They form a part of the world that they act on, and in speaking of their activities in the world, I mean to include their actions upon themselves and each other. Many of these activities are, or involve, direct, unmediated interactions with the outside world – such as biological metabolism, production and exchange of material goods and services (or social metabolism, as Marx would say), etc. Other of these activities are, or involve, mediated interactions, mediated through thought processes, which are embodied in linguistic or other symbolic systems. These include all the non-material modes of appropriation of the world, including various aesthetic and intellectual activities. The non-mediated activities are often called practices these days (or *praxis*, if we like Greek); the mediated activities, however, are also human practices. Especially important for our discussion is theoretical practice and in particular scientific practice.[2]

The subject matter of formal logic, it seems to me, is certain aspects of the linguistic practice of humanity. This linguistic practice, it must never be forgotten, has as its aim to aid in the mastery of the world through aiding in the cooperative activity of human beings in the material, unmediated aspects, as well as through helping them to understand the world through their mediated activities.[3]

The various practices, it cannot be too strongly emphasized, are interwoven in the most varied and complicated ways, and their singling out for separate conceptual discussion in no way implies that a deeper comprehension of them can avoid the problem of understanding the specific ways that they fit together into the totality of human practice.

Now, it seems evident (if one accepts my premises) that there is no way to think about the world except through the mediation of language or some other symbolic system. This means that our theoretical practice in general, and thus, our scientific practice in particular, cannot put us in unmediated contact with 'the real world', but only with the world as appropriated and worked over by the human brain into concepts, theories, etc. It also implies that there is no such entity in the world as a 'pure thought', unembodied in a language or other symbolic system. We cannot think, and certainly not in the communicable, socialized form that scientific practice demands, without making use of such a system.

But these symbolic systems can themselves become the subject of study and further elaboration. Formal logic, I believe, is just such a study and elaboration of the inferential structure of natural language, to begin with, and then of artificial languages. It has the normative function of ensuring that certain types of linguistic usages, which may in themselves violate no grammatical rules, do not lose or distort the meaningful content originally contained in more elementary linguistic units as a result of their further elaboration and combination into more elaborate linguistic structures.

For example, (assertoric) propositional logic treats of such questions as whether one may simultaneously assert and deny a certain premise (which is certainly not forbidden by the rules of linguistics) in the course of an argument; and what implication for various types of combinations of premises follow from the possible truth or falsity of these premises.

If we conceptually isolate certain aspects of this linguistic practice in order to study it in a separate science of logic, this must never lead us to forget that this abstraction is made from a context of activities, mediated and unmediated, which has as its aim the understanding of and changing of the world. To forget this is to run the risk of reification of logical concepts – a risk that is always present when we are dealing with concepts abstracted from a concrete totality. These abstractions are necessary to the understanding of that totality (at least, if they are correctly abstracted from it), and their correct interweaving in thought is our most powerful method of grasping the concrete totality intellectually; but we run the risk, on the one hand, of isolating and reifying them into independent entities, presumed to have significance totally apart from the context from which they were abstracted; and on the other hand, of identifying the process by which they are combined and built up in thought to give us an intellectual appropriation of the concrete entity, with the real ontological process of formation of the concrete totally; so that the more abstract concepts are taken for the more fundamental in an ontological sense. To come back to the case at hand, we run the risk of taking logical abstractions as fundamental ontologically, a sin that has been called pan-logism.[4]

The view that the laws of logic are about the inferential use of language emphasizes why there cannot be any logic without human beings, or other creatures using symbolic systems as a major instrument in their attempts to cope with the world. It also firmly maintains the important distinction between the real object in the external world, and the object as conceptualized via ordinary language or other symbolic systems such as mathematics. It also enables us to understand why logic may be developed as an independent, objective science; and why alternative logical systems are possible, as I will discuss elsewhere.[5]

There is an often-made (at least by some quantum logicians) analogy between physical geometry and logic: Geometry was once thought to be unique and therefore immutable in its application to the physical world. Then, it was realized that there are many possible mathematical geometries, so the question of which one of them describes the physical geometry of the real world became an empirical one. Analogously, so runs the claim, it was once thought that logic was unique. Now we know that it is not, so the question of which logic applies to the physical world becomes an empirical one, too. I think, on the basis of the view of logic I have taken, we can see where this analogy breaks down. Language deals with the world (or at least *can* be used to deal with the world); while logic deals with language – using language to do so of course (when this point becomes important we have to distinguish between language and meta-language). Different physical geometries, once the appropriate coordinative definitions have been adopted, deal with different worlds – or at least with different aspects of the world, or different possible worlds. Different logics deal with different ways that languages can handle the world – and these different ways may – or may not – reflect differences in the world being handled.[6]

Thus, there is no intention in my comments to deny the possible usefulness of alternative logics. Indeed, I think it has been established that even natural language contains elements of alternative logical systems, which can be built upon and formalized if the need arises; to say nothing of the possibilities for artificial languages to be constructed which embody non-standard logical systems. Yet these alternative logical systems may be used to deal with the same world – just as one logical system may by used to deal with different worlds. Of course, I would be more accurate, if less dramatic, if I spoke of different conceptions of or theories about the world, and I ask that my references to the world in the last few sentences be so understood. We should really never lose sight of the distinction between the world and our theories about the world: between the real object and the conceptual object. At any rate, unless we believe that the world is of the nature of someone's or something's language, we see that the analogy between physical geometry and logic breaks down at a crucial point for the argument of those who wish to argue that logic is empirical in the same sense that geometry is.

A language, of course, unites both syntactical and semantical aspects: it utilizes a formal structure to convey meanings, to use plainer language. These aspects themselves may be formalized to a greater or lesser extent, depending on how formal the language is to start with; or how much we choose to formalize certain aspects of it, as when we formalize certain subsections of natural speech; or how much we choose to formalize the way in which we discuss the language itself (metalanguage). Formalization of the syntactic aspects of a language leads to the formulation of certain criteria for an expression to be well-formed, certain rules for the compounding of well-formed expressions, certain rules for the deduction of new well-formed expressions from others, etc. Various mathematical structures are needed for the formalization of these syntactical aspects of language such as Boolean and partially Boolean algebras, lattices, etc.; just as they arise in the formalization of many other subjects.

Mathematical structures, precisely because they represent abstractions from so much of the content of entities to which they can be related, are capable of a wide variety of applications and interpretations. Thus, the fact that a mathematical structure, such as a Boolean algebra, or a lattice, occurs in the syntactical study of logical systems by no means implies that everywhere else that we may meet with this identical mathematical structure we automatically are dealing with a logic.

One cannot even claim that if one does utilize such a mathematical structure logically, that the logical interpretation follows uniquely from the mathematical structure ("reading the logic from the lattice", as has sometimes been suggested in discussions on quantum logic). The same mathematical structure may have more applications to different logics. Elsewhere, I will discuss a simple example of a Boolean algebra, which of course can be used for a standard logic; but also be given a non-standard application in a logic of color mixing, with negation interpreted as color complementation. (See Note 5).

In the development of a physical theory, the central aspect in my view is the singling out and elaboration of the basic concepts of the theory, and the development of a network of relations between these concepts which enables us to grasp in thought some elements of that particular aspect of reality with which the theory proposes to deal. This involves a constant process of testing and development of the theory, involving both theoretical elaboration and the confrontation of the theory with experimental situations. (See Note 2).

A major aspect of the theoretical elaboration of a mature physical theory, at any rate, will involve the correlation of various aspects of the conceptual relationships within the theory with various mathematical structures. Elaboration of the resulting mathematical relationships will then lead, on the one hand, to a deepened understanding of the concepts and their interrelationships – the development of the theory; to the discovery of novel consequences of the assumed relations – the deduction of testable consequences of the theory; and perhaps, ultimately, to the realization of the limits to the usefulness of the correlation – the reaching of the limits of applicability of the theory, in its existing form.

Alternative mathematical structures may be developed and found useful in the course of this formal elaboration of a theory. But the mathematical formalism itself cannot dictate the level of interpretation at which these structures are elaborated into meaningful expressions, which have logical significance. This will depend on the semantical interpretation given to the concepts of the theory and to their interrelationships. Of course, formal elaboration may suggest shifts in the semantical interpretation of concepts and their interrelation. But the formalism does not create its own interpretation – the interpretation calls for the formalism.[7]

Now, after this long and indirect approach, let me finally turn directly to the problem of quantum logic. Whenever, in the course of dealing with a physical theory, we formulate statements about the properties of some system (or a subsystem of some larger physical system), which properties cannot be completely specified except by reference to features outside the system (or subsystem) in question (what I like to call 'open systems', in at least one acceptable usage of this term), it will be found that such statements do not necessarily have a truth value, if they don't include a full description of those aspects of the wider context which allow us to completely specify the property in question. Various classically valid laws for the composition of propositions will also be found to be invalid.

In particular, it will be meaningless to combine assertions about two properties of the system, if the conditions for the appearance of one property preclude the appearance of the other. For example, one cannot combine assertions about the hardness and viscosity of a certain sample of H_2O in contact with a temperature bath, since the temperature range for which the concept of hardness is applicable (below $0°C$) does not overlap with the temperature range ($0°C-100°C$) for which the concept of viscosity is meaningful.[8] The possibility of the introduction of various non-classical logics to treat such situations does indeed arise – not however at the level of assertoric logic, but by means of a modal treatment of the propositions. It will then be found that, for example, the distributive law fails regularly under these circumstances.[9]

The basic question, as far as contemporary microphysics goes, is whether the quantum theory treats microsystems as completely specified by the attribution to them of internal properties, such that meaningful propositions about them may be formulated; or whether properties of such microsystems are only completely specified in the context of a macroscopic preparation, interaction and registration cycle which constitute essential elements of the complete description of almost all quantum-mechanical properties of a system.[10]

This complete cycle is often loosely, and poorly, described as a 'measurement'. I say poorly, because the problem being posed here is not whether measurement *creates* the result or not – the (in) famous 'uncontrollable interaction of the microsystem with the measuring apparatus'. Rather, the question is the fundamental one: whether the very formulation of the relationship between microsystem and macrosystem in quantum theory can be fitted into the traditional (mechanistic, analytic) framework: system *A* with its properties; system *B* with its properties; plus some interaction between *A* and *B* characterized by some sort of interrelation between the variables characterizing the properties of *A* and those of *B*?

The claim being made here is that such a formulation, which lies at the basis of all formulations of the 'measurement problem' since the classic work of von Neuman, is a remaining vestige of mechanistic thinking in physics; and that the relationship between microsystems and macrosystems within the quantum theory is much more integral, more 'structural', calling for new concepts of structural causality falling outside the system A plus system B framework altogether. Certainly, if we are to make sense of the Einstein-Podolsky-Rosen type of experiment, with its inextricable linking of two systems once they interact, no matter what their future fates may be, we seem forced to take some such step.[11]

I am not here claiming to have fully justified this approach to the quantum theory, and certainly not that we have reached any sort of ultimate stage in our understanding of the relationship between macrosystems and microsystems with the existing formulation of quantum mechanics. Nor that macrosystems must forever be introduced into theory as unanalyzed 'wholes', to be treated purely phenomenologically. I am just trying here to argue that there is an important problem of this relationship, and suggest that existing quantum theory may represent the first way station along the road of such an investigation. In my view, quantum mechanics may not be so much the final answer, as an invaluable way of posing the problem.

One may attempt to justify the claim that only statements about microsystems which include reference to a macroscopic background should be admitted as meaningful propositions about microsystems, as a corollary of a phenomenological viewpoint in philosophy; indeed it tended to be so justified even by Bohr, for example, on some occasions. But I have been trying to show that it may also be put forward on quite different grounds, fully committed to a materialist approach to microphysics.

The problem I have been discussing can be posed in a linguistic mode, of course. If one believes that it *is* meaningful to give a specification of the properties of a microsystem apart from reference to any macrosystem of which it is an element, then one will naturally want to give a semantic interpretation to assertions at this level, and attach truth values to such statements as: 'The momentum of the electron is p', etc. One will then try to develop a logic to handle the relationship between such statements. Such a logic will certainly have to be nonclassical; and depending on exactly which type of elementary statements one chooses to admit as meaningful, and what types of combinations of such statements one chooses to allow, one can develop a variety of nonclassical logics to do the job. Two, three or infinite-valued logics; partial Boolean or non-Boolean propositional systems, etc. have been proposed as quantum logics. Their variety is only restricted by the need to reproduce within the logic some aspect or aspects of the mathematical structure of the theory: its Hilbert space, or convex set, or C^*-algebraic aspects; depending on which aspect(s) of the various possible mathematical formalisms associated with the quantum theory is chosen for interpretation in a logical mode. I shall not in this paper enter into a criticism of this approach (or better, these approaches), leaving that task for another time (Reference 5).

If, on the other hand, one believes that microsystems are only completely specified within quantum mechanics as aspects of some macrosystem, one will regard phrases such as 'The momentum of the electron is p', as incomplete sentence fragments not warranting semantical interpretation as true or false. One will only assert as meaningful propositions sentences such as: 'The momentum of the electron issuing from accelerator A, passing through magnetic field M and detected by grating G is p'. Then one will find that classical logic, used with the usual cautions, will do quite well for handling such statements.[12]

Let me emphasize again that what is at issue here is the meaning of the quantum theory as it now exists, not what might be true of some future theory; nor whether one should be satisfied or dissatisfied with the existing quantum theory. Interesting as these questions may be, they seem to require an understanding of the significance of the present theory before any useful discussion of them becomes possible.[13]

What also needs to be emphasized, it seems to me, is that whether or not one formulates problems about the meaning of the quantum theory in a logical or a non-logical mode, they are the same problems. They do not go away by a logical reformulation. The basic mathematical structures of the theory are used in ways that express the non-classical nature of the theory. [The same structures could also be used in classical theory in ways that would express that classical nature: that one can do classical mechanics using Hilbert spaces has been known for over forty years. Does this mean classical mechanics requires a non-classical logic?] If this were the only claim of the quantum logicians, who could object? But if, because assertions about the non-classical nature of the theory are put into the logical mode, some greater profundity is thereby claimed for them, it seems to me that either:

(1) A claim is being subtly made that the basic nature of the world is logical; and thereby a pan-logistic and therefore an implicitly idealistic account of the world is being advocated. One should at least be aware of what is being done, if one chooses to do this.

(2) It is being claimed that the switch from talking about the world to talking about statements about the world deepens our understanding of it. One is invited to feel that one has understood the two slit experiment better, for example, by understanding the non-applicability of the distributive law. This reminds me of the explanation given by the quack doctor in Molière that opium puts one to sleep because of its dormative powers.

(3) It is being claimed that use of a non-classical logic proves that the properties of a quantum system can be interpreted in the same way as the properties of a (closed) classical system. This approach is based on a reversal of the proper relationship between conceptual structure and formalism in a theory, as well as on a misapprehension of the role of logic.

Acknowledgement is gratefully made to Robert S. Cohen and Donald Hockney for many useful comments on earlier versions of this paper.

Boston University

NOTES

* Research partially supported by the National Science Foundation.
[1] In reading some of the articles on quantum logic, I cannot help wishing that their authors had done the same, and that one didn't have to try to infer their positions on these basic questions about logic from their discussions of quantum theory.
[2] I have elsewhere written 'A Note on the Concept of Scientific Practice', in *For Dirk Struik* (*Boston Studies in the Philosophy of Science*, Volume XV, D. Reidel Publ. Co.; Dordrecht, Holland and Boston, 1974).
[3] I leave aside here the important ideological functions of language, which may aid in the mastery of one group of men by another, as not directly relevant to the topic of quantum logic; although on other occasions I would be prepared to discuss the question of whether certain misinterpretations of quantum theory do not – in highly mediated fashion, of course – play a role, however modest, in certain general ideological misinterpretations of scientific practice. See note 2 for reference to a paper containing the beginnings of such a discussion.

⁴ Those who have followed recent discussions on the methodology used by Marx in his economic studies will recognize my debt to them. My formulation here is essentially an attempt to apply the methodological criteria outlined by Marx in his introduction to the *Grundrisse*. For an excellent translation of this introduction, with extensive commentary and other materials, see Terrell Carver, ed., *Karl Marx: Texts on Method* (Basil Blackwell, Oxford, 1975). For an excellent discussion of Marx's views on science, with extensive quotations, see the biographical article by Robert S. Cohen, 'Karl Marx', to appear in *The Dictionary of Scientific Biography*.

⁵ 'How Logical is Quantum Logic', to appear in the *University of Pittsburgh Series in the Philosophy of Science*. Let me add here, dogmatically again in the interests of brevity, but to avoid possible misunderstandings of my position, that I do not accept the view of the sciences which sees logic as the foundation of mathematics, and mathematics as the foundation for the so-called empirical sciences. The relations between the development of the various sciences are too complex and many-sided to be fruitfully encompassed in such a linear geneology. I think the by now traditional distinction between the formal and the empirical sciences must be re-examined and replaced by a more adequate image, which will avoid the pseudo-problems that arise when we seek to link up the 'formal' and the 'empirical' — which should never have been sundered in the first place.

⁶ Hilary Putnam, one of the chief advocates of this analogy, gives us a nice example of a shift in the analogy when he first writes:

$$\text{``}\frac{\text{GEOMETRY}}{\text{GENERAL RELATIVITY}} = \frac{\text{LOGIC}}{\text{QUANTUM MECHANICS}}\text{''},$$

and then states:

"...just as it is impossible to understand the true nature of space and time as long as it is assumed that 'space' obeys the laws of Euclidean geometry, so it is impossible to understand the true nature of microprocesses as long as it is assumed that physical propositions obey the laws of Aristotle's logic".

After the antecedent: "nature of space and time... 'space' obeys...", we should have expected the consequent: "nature of microprocesses" ...*microprocesses* "obey" – but Putnam is too good a logician for that! So we get "...propositions obey...". Of course he could have restored the balance of the phrase by writing "so it is impossible to understand the true nature of *propositions about* microprocesses..." – but there would have gone the point of his argument! Hilary Putnam, 'How to Think Quantum-Logically', *Synthese* **29**, 55 (1974).

⁷ "Physicists know what it means to interpret physical experience in terms of mathematical concepts; but to imagine that they might have to interpret somehow pre-existing mathematical symbols in physical terms is an idealistic construction alien to the spirit of science". Leon Rosenfeld, 'Misunderstandings about the Foundations of Quantum Theory', in R. S. Cohen and J. Stachel (eds.), *Leon Rosenfeld, Selected Papers* (*Boston Studies in the Philosophy of Science*, Volume XXI, D. Reidel Publ. Co.: Dordrecht, Holland and Boston, forthcoming).

⁸ Martin Strauss had long ago pointed this out for quantum theory. See the translations of his papers on complementarity dating from the 1930's, in Martin Strauss, *Modern Physics and its Philosophy* (D. Reidel Publ. Co., Dordrecht, Holland and Boston, 1972).

Peter Mittelstaedt has also emphasized this for quantum theory, with his concepts of 'restricted availability', versus 'unrestricted availability' for propositions. (*Boston Studies in the Philosophy of Science*, Vol. XVIII, D. Reidel, 1975). But it is just as true for properties of open systems in classical physics, as I will discuss in more detail elsewhere. (See Note 5).

[9] A detailed discussion of this question will appear in Reference 5. A brief treatment will be found in my 'Comments on 'Logical Problems Suggested by Empirical Theories'', in *Boston Studies in the Philosophy of Science*, Volume XXXI, to appear shortly.

[10] Those few properties which serve to characterize the type of microsystem in question, such as charge, mass, spin, etc., do indeed behave like classical properties of a closed system. We may assert: 'The charge of the electron is e', without further qualification. But whether the contingent properties of a microsystem, such as its position, momentum, etc., which serve to characterize its particular physical interactions, can be so asserted is the fundamental issue.

[11] I will discuss these issues more fully in Reference 5. Brief discussions of various aspects will be found in Reference 2, pp. 425–430; and in my 'The Rise and Fall of Geometrodynamics', *Boston Studies in the Philosophy of Science*, Volume XX (D. Reidel Publ. Co., Dordrecht, Holland and Boston, 1974), especially pp. 40–41.

[12] One may also choose to allow compound statements, referring to compatible properties of a quantum system (such as 'The x-component of the momentum of the electron is p_x, and the y-component of its position is y_o'), while forbidding compounds referring to incompatible properties (such as 'The momentum of the electron is p and the position is x'), without attaching a truth value to such combinations apart from a specification of the macroscopic context. This approach, leading to the use of a partial Boolean algebra for quantum logic, has been advocated by Martin Strauss (See Note 7 for reference). It appears that one is even able to allow combinations of all statements about properties of microsystems, modally qualified, in such a way as to utilize a lattice for quantum logic, without thereby being committed to a non-standard assertoric logic. See Note 9 for reference to a preliminary discussion of this problem, and Note 5 for reference to a fuller discussion.

[13] Perhaps I should state that I am fully aware that many interesting, and even important, studies of various mathematical structures in quantum theory, which have shed new light on the nature and implications of the theory, have been carried out under the label of 'quantum logic', and indeed that various attempts at the generalization of the existing theory have been similarly labelled. But these studies must stand on their own merits, and do not help to resolve the properly logical issues. It is only the labelling which is open to question here, not the contents of the package.

CONTRIBUTED PAPERS

SESSION IV

IAN I. MITROFF

INTEGRATING THE PHILOSOPHY AND THE SOCIAL PSYCHOLOGY OF SCIENCE OR A PLAGUE ON TWO HOUSES DIVIDED

I. INTRODUCTION[1]

A few years ago at the conclusion to one of his papers, Norwood Russell Hanson wrote:

> ... scientific observation and scientific interpretation need neither be joined nor separated. They are never apart, so they need not be joined. They cannot, not even in principle, be separated, and it is conceptually idle to make the attempt. Observation and interpretation are related symbiotically so that each conceptually sustains the other, while separation kills both. This will not be news to any practicing scientist, but it may seem heretical indeed to certain philosophers of science for whom *Analysis* has, alas, become indistinguishable from *Division* (1967, p. 99).

While not everyone would agree with Hanson regarding the role of observation and interpretation, I have often wondered what would have happened had Hanson attempted to make the same argument regarding the philosophy and the social psychology of research, if indeed he would have been inclined to attempt such an argument at all. I suspect that if Hanson had made such an argument the reaction to it would have been intense. First of all, it is a fact of academic life that we have separated the philosophy and the social psychology of science. Second, many, if perhaps not most, of those who practice the philosophy and the social psychology of science see no need in joining them whatsoever, let alone that each "conceptually sustains the other." Yet this is the very heretical position I would like to argue. To be more specific, I would like to argue that the philosophy and the social psychology of science not only *need* to be joined but that they *ought* to be joined. They ought to be joined because not only does each conceptually *sustain* the other but that each conceptually *presupposes* the other.

In arguing my case, I am perhaps more aware than anyone else as to how much stronger I wish the evidence was in favor of such a merger. Rather than pretend otherwise, I think it is better to acknowledge outright that the situation has a certain vicious quality to it. That is, before one would invest the time and the energy into bringing about a greater union between the philosophy and the social psychology of science, one would like to have some evidence that its benefits would outweigh its costs, and yet prior to the establishment of such a union one is hampered in producing some of the strongest evidence that could be mustered in favor of such a merger. But then this is to assume that unions and mergers are only brought about when the "evidence" is in favor of them.

The argument of the paper proceeds in three main parts. The first part involves a very brief assessment of the arguments and devices that philosophers of science have used to separate themselves and their concerns from the sociology of science. A more exacting and proper assessment would unfortunately take us too far afield, if not constitute a whole study of its own. I must therefore confine myself to a few brief observations. The second part is concerned with an equally brief assessment of the arguments and devices that sociologists of science have used to separate themselves and their concerns from the philosophy of science. The paper is thus not meant to be a one-sided attack on the philosophy or the sociology of science but rather a challenge to both. If the sociology of science has been received with less than enthusiastic acceptance by philosophers of science, sociologists of science have not exactly been overenthusiastic in their pursuit of the philosophy of science. I realize of course that both of the partners to the union I am proposing are less enthusiastic about it than I am. Such is undoubtedly always the fate of those who arrange alliances for the "benefit" of others. There is nothing worse than the benefits imposed upon us by others.

The third part of the paper concerns the presentation of a model of science that is equally grounded in philosophical and social psychological considerations. The model is in large part based on this author's four year study of the lunar scientists over the course of the Apollo missions (Mitroff, 1974a, b, c).

II. HOW THE PHILOSOPHY OF SCIENCE HAS ATTEMPTED TO SHIELD ITSELF FROM THE SOCIAL PSYCHOLOGY OF SCIENCE

I would like to raise briefly a few of the many devices that the philosophy of science has used to shield itself from the sociology of science. The devices that I am especially concerned with in this paper are: (1) the issues of psychologism and sociologism, (2) the Reichenbach (1968) distinction between the context of discovery and the context of justification, (3) the charge that a social psychologically based theory of science inevitably leads, à la Kuhn (1962) and Feyerabend (1970) to relativism, and (4) Lakatos' (1970) charge that Kuhn's view leads to science ultimately being governed by a "mob psychology." For reasons that will become apparent, the latter two issues are better dealt with in the concluding parts of the paper.

The issue of psychologism is an interesting one. If by psychologism one means that psychology enjoys some privileged status as a science and that everything is nothing but a matter of psychology, i.e., everything can fundamentally be reduced to psychology and to psychology alone, there I quite agree with the charge of Popper (1964, 1965) and others (Morgenbesser, 1967) that this conception of psychology is psychologism, and as such, is a dubious, if not dangerous philosophical doctrine. The problems of philosophy are no more reducible to psychology than are the problems of any field. But if on the other hand the charge (or better yet the *label*) of psychologism is meant to refute any and every application of psychology or to deny that psychology is a factor which is present in every human situation, then I take strong exception. If psychologism and sociologism are the occupational diseases that psychologists and sociologists are prone to, then the strength with which logicians of science seem to oppose any application of psychology (or with which they demean it by relegating it to second place behind *their* privileged science, i.e., logic) makes them liable to the charge of "logicism" (see Toulmin, 1972, on this point).

The best way to meet the charge of psychologism is to show that try as one might one can never avoid all references to psychology and therefore that it is futile to try and do so.[2] Indeed it would be a great deal easier to swallow the arguments of those who are most vehement in their opposition to psychologism if they were not guilty of perpetuating an even worse psychologism of their own, i.e., their own homespun, implicit, highly idealized psychology of which they are, of course, unaware.

The following observation may help to put the matter in perspective: the distinguished sociologist Robert K. Merton has noted with more than just a touch of irony that whenever a scientist declares in print that he or she is not concerned with being anticipated in his or her discoveries, then we can be sure that not so many pages later in their writings we may be sure to find an extreme concern being expressed about being scooped (Merton, 1963, 1969). I would venture to speculate that the same holds true of those who are most vehement in their opposition to psychologism and sociologism.

If, as I believe it to be the case, that every theory or view of science makes certain assumptions about the nature of scientific observables and their relation to a human observer (see Mitroff, 1974b, c), then every view of science presupposes some form of psychology and sociology whether it is aware of it or not. It is futile then to resist joining what has already been presupposed. Indeed, the question is not whether to join or not to join the philosophy and the social psychology of science but whether we can join the best theories that the philosophy of science has to offer with the best theories and data that psychology and sociology have to offer.

If psychologism and sociologism are repugnant, it is because that down deep we feel that it is futile to try and explain the whole of reality in terms of one set of variables and one set of variables alone. Views of reality which have as their aim the promotion of a narrow and insular isolationism or reductionism should always be suspect. It is for this very reason that I have always regarded the Reichenbach[3] distinction between the contexts of discovery and that of testing as a poor or naive one. Even stronger, I regard it as an exceedingly harmful distinction.

I regard the distinction between the contexts of discovery and of justification as a poor and naive one because I do not believe that the process of science can be neatly partitioned into discovery versus testing phases. To me the most creative part of science requiring explanation and study is the problem of how a scientist *discovers* the *tests* of his or her ideas (see Mitroff, 1974a, b, c). The context of testing is discoveryridden and laden no less than the supposedly distinct and pure context of discovery (Mitroff, 1974c). This is why I disagree so strongly with Popper (1964) when he tells us that not only is the context of justification all that the logician of science need concern himself with, but stronger still, that the context of discovery is actually irrelevant to our understanding of science. To me none of the aspects of science are irrelevant. They are all important. I do not see where the process of testing can be explicated *even in theory* independently of the process of testing and vice versa.

I also regard the distinction as harmful because it promotes and perpetuates a separatist, piecemeal view of science. It discourages, if not actively prohibits, the kind of broad-based, interdisciplinary studies (see Churchman, 1971) which could uncover some of the most interesting evidence which could effectively challenge the basis of the distinction. That is, the distinction takes as unproblematic the very thing requiring justification – the distinction itself. If social psychology is relegated to studying the discovery aspects of science and philosophy to the testing aspects, then pray tell which discipline or "science" is it that studies the whole process? Better yet, which discipline is it that tests the distinction itself? It begins to be clear that the distinction has the distressing side-effect of making itself immune to testing. This is perhaps one of the most distressing things about strict disciplinary thinking. The disciplines erect such formidable barriers to the arguments, methods, and data of other disciplines that they insulate themselves from the strongest possible challenges that could be mounted against their most basic concepts. As Paul Feyerabend has put it:

Scientific education as we know it today... has the purpose of carrying out a rationalistic simplification of the process "science" by simplifying its participants. One proceeds as follows. First, a domain of research is defined. Next, the domain is separated from the remainder of history... and receives a "logic" of its own. A thorough training in such a logic then conditions those looking in the domain so that they may not unwittingly disturb the purity (read: the sterility) that has already been achieved. An essential part of the training is the inhibition of intuitions that might lead to a blurring of boundaries. A person's religion, for example, or his metaphysics, or his sense of humor must not have the slightest connection with his scientific activity. His imagination is restrained and even his language will cease to be his own.

It is obvious that such an education, such a cutting up of domains and of consciousness, cannot be easily reconciled with a humanitarian attitude. It is in conflict "with the cultivation of individuality which [alone] produces, or can produce well developed human beings"; it "maim[s] by compression, like a Chinese lady's foot, every part of human nature which stands out prominently, and tends to make a person markedly dissimilar in outline" from the ideal of rationality that happens to be fashionable with the methodologists (1970, pp. 20–21).

III. HOW THE SOCIAL PSYCHOLOGY OF SCIENCE HAS ATTEMPTED TO SHIELD ITSELF FROM THE PHILOSOPHY OF SCIENCE

If the philosophy of science has shielded itself from the sociology of science by arguing that not only can the epistemic structure of science be studied independently of its social structure but that the philosophy of science only need concern itself with the epistemic structure, then it is not surprising to find that the sociology of science has shielded itself from the philosophy of science by arguing just the reverse. Notwithstanding Kuhn and Feyerabend there is one notable difference, however, between the philosophy and the sociology of science. While the philosophy and the sociology of science have both attempted to shield themselves from one another, in my opinion the sociology of science has been far more critical of the division than has the philosophy of science. That is, within contemporary philosophy and sociology of science, I find that sociologists of science have been much more critical of the division of labor between fields than have philosophers of science (see Barnes and Dolby, 1971; King, 1971; Mulkay, 1969). In philosophical terms, I find that sociologists of science have been much more *reflective* (i.e., self-critical and self-analytical) with regard to some of the basic assumptions underlying their field than have philosophers of science. Lest I be accused of lumping historians of science with sociologists of science, let me put it this way. I find that those who have approached the Reichenbach distinction from either the sociological *or* historical side have been much more critical of the distinction than those who have approached it from the philosophical *or* logical side. Consider, for example, the following passages from M. D. King's essay, 'Reason, Tradition, and the Progressiveness of Science:'

> The majority of sociologists of science have... [drawn] a sharp distinction between science as a "system of ideas" governed by an "inner logic" and science as a "social system" shaped by non-logical forces, and by arguing that though sociological analysis can add little or nothing to our appreciation of the former, it is the obvious means of understanding the latter. They have, in other words, accepted a clearly drawn division of labor. Science as a system of knowledge is, they accept, simply not their business; it is the province of the history or perhaps the philosophy of science... sociology, from this view, seeks to explain the behavior of scientists... chiefly in terms of the values and norms to which scientists *qua* scientists are committed.
>
> This division between the history of scientific ideas and the sociology of scientific conduct, between the study of science as "a particular sort of knowledge" and as "a particular sort of behavior" has met with the ready consent of historians and sociologists alike. One can see why such a division of labor should appear attractive to both sides – whatever its intellectual justification. It saves intellectual historians from the indignity of being told that the "real" causes of scientific growth lie beyond their professional comprehension; and it relieves sociologists of the necessity of understanding scientific ideas.
>
> However, at least one figure has spoken out against this "divorce of convenience." T. S. Kuhn in his book *The Structure of Scientific Revolutions* moves freely backward and forward across the boundary between the history of ideas and the sociology of scientific behavior. Whatever the merits of the particular account he gives of scientific change, his work forces us to reopen the question of whether its intellectual and social dimension can be properly understood in isolation from one another (1971, pp. 3–47).

It is unfortunately beyond the scope of this paper to show how the preceding points can be supported in detail. Suffice it to say that my own study of the Apollo moon scientists lends strong theoretical as well as empirical support to the thesis that the social norms of science cannot be explicated independently of the cognitive structure of science as well as vice versa (Mitroff, 1974a, b, c). Contrary to some of Merton's early ideas regarding the sociology of science (1949), the norms of science neither arise nor function in a substantive vacuum (King, 1971). As Michael Mulkay, for one, has pointed out scientific theories themselves function as norms (1969). That is, the norms of science not only function within the context of scientific theories but competing theories also serve as competing normative guides for scientists. As such, neither the norms of science nor the theories can be understood independently of one another. Further, my own work also points to the existence of sets of competing norms of science. Instead of science being governed by a single set of unambiguous norms, it can be shown that science gives rise to a dialectical opposition between different sets of norms (Mitroff, 1974b). For every one of Merton's (1949) norms of science, it can be shown that there exists an opposing counter-norm. But then this means that there is another sense in which the sociology of science is not independent of the philosophy of science, namely, the understanding of dialectical systems, or at the very least, the understanding of how incommensurable thought patterns function in science.

At this point let me give a brief example of what I have been talking about; i.e., I would like to give an illustration of the direct consequences that follow from the acceptance of a division of labor between the philosophy and the sociology of science. Harriet Zuckerman (1967, 1968) has recently completed a highly interesting as well as important study of over 40 Nobel prize winners. Zuckerman intensively interviewed each Nobel prize winner with regard to (a) who collaborated with whom in the doing of research and writing papers, and (b) how the Nobelists assigned credit to themselves and to their lesser known colleagues in publications. Zuckerman found a number of markedly distinct patterns of collaboration. Perhaps the most interesting finding was that the age at which a man or a woman received the Nobel prize had great bearing on his or her scientific social behavior with regard to the sharing of credit with other colleagues. As interesting as I find what Zuckerman did, I find what she did *not* do even more interesting. To the best of my knowledge, Zuckerman missed a golden opportunity to survey what Nobel prize winners thought about science; that is, what is their methodology, their philosophy of science? Do Nobel prize winners have a concept of science that is distinctly different from that of lesser scientists? To systematically interview elite scientists and not raise these kinds of issues is a disappointment. It is a poignant and vivid illustration of the encapsulation of disciplines.

IV. TOWARDS AN INTEGRATED PHILOSOPHY AND SOCIAL PSYCHOLOGY OF SCIENCE

On July 20, 1969, the night men first walked on the moon, I had the idea for a study that was to occupy me for the next four years. Looking at the events that night, the significance of what was unfolding struck me. Rocks were going to be returned from the moon in order to test earth-based scientists' ideas about the moon. Since I knew that some of the scientists had held strong prior views about the nature of the moon and that the views of different scientists were in sharp opposition, one would have the rare and golden opportunity to study the participants in on-going scientific controversy. Further, since some of the scientists had formed and thus held their views over a long period of time, one would also have the opportunity to study how easily scientists in reality parted with their views or like most of us held on to them with their dear life.

Almost three months to the day of Apollo 11's landing, I began a series of interviews with 42 of the most eminent scientists who participated in the lunar missions.[4] Each scientist was intensively interviewed on four separate occasions in the time period just after the completion of one Apollo mission and before the start of another. The interviews thus covered the time period between Apollo 11 and 16. The general purpose of the interviews was to assess how the scientists' ideas were or were not changing in response to the Apollo data.

If there was a central question with which the study began and which occupied it throughout it was the following: Would it be possible to identify and to study those scientists, if any, who exhibited a high degree of *prior* commitment (see Kiesler, 1971) (i.e., before Apollo 11) to certain pet hypotheses or theories regarding the nature and/or origin of the moon and who as a result showed a high degree of reluctance to give up their pet hypotheses in spite of the data returned subsequently from the moon? A major part of the study thus concerned itself with showing empirically the possible normative functionality of the intense commitment of scientists to their ideas instead of assuming it to be inevitably dysfunctional, unscientific, or irrational. In terms of the interviews with the 42 scientists, a small number of scientists were overwhelmingly and consistently nominated by their peers as the ones "most likely to hang onto their pet hypotheses 'til death do them part." The perceptions of these few key scientists by their peers (the 42 scientists) were repeatedly studied over each interview round for their combined implications for the sociology, psychology, and philosophy of science. The general issues and subsequent topics that were investigated were far too numerous just to be able to list them, let alone to report the results here. However, in terms of the central issue of commitment it was shown that it was possible to measure systematically and precisely the differences in psychology between (1) those scientists who were judged most likely to become committed to pet hypotheses and to take strong stands on scientific issues and (2) those who were judged to avoid taking strong stands or not to develop intense commitments. Not only were the attitudinal differences between these two "types" of scientists immensely striking, but they were also, in statistical terms, highly significant. Furthermore, they emerged continually. No matter what was used to measure them, the same differences were obtained repeatedly. In short the study revealed that there are indeed definitive and very strong systematic differences between different kinds of scientists. Most of all, it is possible to capture and to measure these differences in psychology systematically and precisely. This paper attempts to draw the significance of these results for the philosophy of science.

Another major part of the study was concerned with the precise measurement, in information theoretic terms (see Mitroff and Mason, 1974a) of the change in scientific beliefs of the scientists with respect to certain hypotheses proposed for the origin of the moon as well as for its detailed properties. To this end, repeated measurements were taken with respect to how probable each scientist thought various scientific hypotheses were over the conduct of the Apollo missions. The beliefs and attitudes of the scientists were also studied and measured with respect to certain basic issues in the philosophy of science. Considerations of space prevent me from reporting here on the information theoretic analysis of the change in beliefs with respect to scientific hypotheses. Of necessity, the reader is referred to a recently published book, *The Subjective Side of Science: A Philosophical Inquiry Into the Psychology of the Apollo Moon Scientists* (1974c), for a full description of the study, i.e., a detailed description of the sample of scientists, the study's methodology, its results, and most of all, its full implications.

Very early during the first round of interviews, a number of the scientists, independently of one another and of their own accord, suggested a typology of different kinds of scientists. In their opinion, their experience suggested that there were very sharp differences between a relatively small number of fundamentally differing kinds of scientists. On presenting the typology to the rest of the sample for their reaction and evaluation, a strong degree of consensus was obtained as to, first, the face validity of the typology, and then, second, the names of those specific scientists who "best typified" each of the various "types." The precision of the typology was continually refined on subsequent interview rounds. In addition, the perceptions of those few scientists who were most frequently nominated as best typifying the various types were systematically measured. Each of the 42 scientists in the study reported their perceptions of these few scientists in terms of a variety of attitudinal scales (Mitroff, 1974b, c).

Although the differences between the various "types" cannot be reduced to a single underlying dimension, the most fundamental dimension and the one which originally suggested the typology was that of "speculativeness" or "willingness to extrapolate beyond the available data." In any social grouping, there are always those who prefer to stay close to "the facts" and those who prefer to venture out beyond them even at the risk of ignoring them. Type I scientists, as they were referred to, were distinguished by the essential defining quality that they excelled at extrapolating from data. Although they were often fine, detailed experimenters themselves and even at times enjoyed experimental results or numbers for their own sake, theorizing was obviously their most pleasurable and exalted task. *Type I scientists were willing and even relished the bold intuitive and theoretical leaps always required in making inferences from incomplete data to a comprehensive and encompassing theory.* Type III scientists, as they were referred to, represented the other end of the spectrum. Here, numbers are often relished primarily for their own sake. There is a preoccupation, even an obsession, with data gathering. There is often an extreme disdain of theorists who deal with highly inferential and abstract concepts. Speculation or extrapolation from data is valued little and *only* engaged in when the data clearly warrants such extrapolation and then only with extreme cautiousness. Where type I's more readily tended to see the positive advantages of speculation in science and to speak glowingly of it, type III's tended to disparage it. They tended to equate speculation with wild theorizing and refer to it disparagingly as "finger-painting in the sky."

Type III scientists are often seen as brilliant but extremely narrow and specialized experimenters. In some instances, they are regarded as nothing more than "super-technicians with a Ph.D." (In fairness, it was noted that theorists can often be just as narrow. In general though, it was the consensus opinion of the sample that it was much more difficult for a theorist to be a narrow specialist than it was for an experimenter. Theorizing on something so broad as the origin of the moon required, by its very nature, that a scientist be familiar with, if not competent in, several diverse scientific fields.) Type II scientists represented something in between. Here were to be found scientists who were equally capable of doing competent experimental work as well as engaging in modest theorizing and extrapolating activities. From time to time, they could even rise to bold feats of theorizing and extrapolation, but in general, they represented the middle ground, running to neither of the extremes represented by types I and III.

In response to the question. "What scientists are, in your opinion, most committed to their pet hypotheses or theories and as a result least likely to shift their commitment as a consequence of the Apollo data?", three scientists were consistently mentioned time and again. These three scientists were also nominated as the most outstanding examples of type I scientists. More than any other group of scientists, these scientists excited the jealous envy and hostility of their peers. They also excited the most positive superlatives. They not only dazzled their peers with their spectacular feats of speculative theorizing but they also offended them with the abrasive, provocative, and often aggressive manner with which they presented and defended their theories. These reasons alone warranted comparing these three particular type I scientists among themselves as well as against the other types. In addition it was felt that it was important enough to get a collective portrait of those scientists who were perceived as most committed to their ideas to warrant the study of three particular type I's in detail.

Even more revealing were the differences between the type I scientists and the type III scientists who were studied in detail. In order to appreciate the significance of these differences it is necessary to point out that throughout the study the type III scientists were judged most representative of the *average* or *typical scientists* in the lunar program. (It should be emphasized in this regard that throughout the study the scientists talked extremely freely about one another. There were as a result many opportunities to obtain consistent portraits of the various scientists. That is, the descriptions and inferences with respect to the types are not based on isolated, infrequent observations.) The three particular type I scientists on the other hand were the ones who were most frequently judged the outstanding, extraordinary scientists of the program, the scientists who most consistently stood at the creative apex of the profession. It is thus both interesting and important to compare the differences in adjectives that were used to differentiate between these two extremes.

If the type I scientists most nearly correspond to Kuhn's Extraordinary scientist and if the type III scientists most nearly correspond to the Normal scientist, then the results of the study unambiguously imply that the Extraordinary scientist is more likely to be more creative in the production of bold and speculative ideas. They are also most likely to be the kinds of scientists who become most rigidly committed to their ideas. That is, the three type I's were perceived as extremely creative in the sense of their being able to produce and having produced many original innovative ideas over a long period of time and in this sense they were regarded as extremely flexible. They possessed the requisite mental agility and nimbleness of mind to see old problems in a new light and to perceive (literally invent) highly imaginative patterns in a complex set of data; thus, the judgements of "brilliant," "theoretical," "generalist," "speculative," and the tendency towards "vagueness." On the other hand, they were perceived as extremely attached to their ideas once they were produced; thus, the strong judgements of "bias," "aggressiveness," and "rigidity." Indeed, independent exercises (Mitroff, 1974c) establish that over a span of three years, there is virtually *no* perceived shift (according to the 42 scientists) in the positions of each of the three type I scientists with respect to certain key scientific hypotheses with which they have long been associated. This point cannot be overemphasized. Although the psychological differences are interesting and important for their own sake, they are even more important for what they imply for the understanding of the growth and change of scientific ideas. While it is beyond the scope of this paper to demonstrate how these psychological differences operate in detail, it can be shown (Mitroff, 1974c) what kinds of rationalizations *and* rational arguments that different types use to hang onto, preserve, as well as, change their ideas. In other words, under certain circumstances, it is possible to relate explicitly differences in psychology to the growth of a scientist's ideas.

The implications of these results for the philosophy of science, in particular the views of Kuhn (1962) and Feyerabend (1971), are as follows. It is well known that both Kuhn and Feyerabend have been accused of overemphasizing the discontinuities, the disagreements, between successive phases of the course of scientific development. They have been accused of placing a greater emphasis on the conflict between competing theories, paradigms, and scientists than actually exists or can be shown to exist. They have been accused of viewing the successive phases between competing theories or paradigms as an all or nothing phenomenon, i.e., as either perfect agreement or perfect disagreement. As Shapere has put it:

... Kuhn has committed the mistake of thinking that there are only two alternatives: absolute identity or absolute difference. But the data at hand are the similarities and differences; and why should these not be enough to enable us to talk about more, and less, similar views and, for certain purposes, to classify sufficiently similar viewpoints together as, e.g., being in the same tradition? After all, disagreements, proliferation of competing alternatives, debate over fundamentals, both substantive and methodological, are all more or less present throughout the development of science; and there are always guiding elements which are more or less common, even among what are classified as different 'traditions.' By hardening the notion of a 'scientific tradition' into a hidden unit, Kuhn is thus forced *by a purely conceptual point* to ignore many important differences between scientific activities classified as being of the same tradition, as well as important continuities between successive traditions. This is the same type of excess into which Feyerabend forced himself through his conception of 'theory' and 'meaning.' Everything that is of positive value in the viewpoint of these writers, and much that is excluded by the logic of their errors, can be kept if we take account of these points (1966, p. 71).

While basically in strong sympathy with the views of Kuhn and Feyerabend, I recognize the validity of many of the arguments of their critics. Even more the data from my own study allows me to support a number of their critics' points with empirical evidence as well as with theoretical arguments. And yet their critics are not all right either. There is a sense in which Kuhn and Feyerabend are right and a sense in which they are wrong, or better yet, incomplete. If, as their critics have pointed out it is not an "all or nothing" choice between competing paradigms or theories, then it is not an "all or nothing" choice between Kuhn and Feyerabend. There is a point in between.

The central issue is that of "agreement." Kuhn is right when he emphasizes the tremendous conflicts and disagreements that exist between competing paradigms, theories, and scientists. The data of my study more than supports him on this point. He is even right when he emphasizes that there are occasions for which there are theories, paradigms, and scientists which, for all practical purposes, may be regarded as in total disagreement. However, Kuhn is wrong when he either fails or neglects to point out that what is true on the *micro* level is not necessarily true on the *macro* level. That is, there are *individual* scientists who for all practical purposes may be regarded as in extreme disagreement with one another, and indeed the results of my study would make it seem that there are issues on which all scientists disagree some and maybe even all of the time. But likewise there are also *individual* scientists who for all practical purposes may be regarded as in complete agreement with one another. Since both of these forces or tendencies operate in every system, it would seem incorrect to characterize a collectivity of scientists as existing in either a state of total or near total disagreement or agreement.

There is a more telling way to put this point. It seems to me that the data of the study suggests that in every social system there are those kinds of individuals (type I's) who have a compulsive need to make revolutions, to disagree *as strongly as possible* with established ways of thinking – paradigms if one prefers to call them. These individuals have an almost consuming need to produce radical counter ideas and theories to those currently in existence if not in vogue. They seem to need to go out of their way to produce extremely novel ways of looking at old phenomena. However the data also suggests that there are also those kinds of individuals (type III's) who have a compulsive (security) need to preserve continuity with old established ways of thinking, to differ as little as possible from the tried and the true. Rather than complete disagreement or agreement being an actual state of affairs, they are instead "states of mind," attitudinal ideals (Churchman, 1948, 1971; Mitroff, 1973), or in Toulmin's terms (1972) the divergent "disciplinary aims" of radically distinct types of men. There are, in other words, scientists who act to bring about "revolution in perpetuity" but there are also those who act to preserve the status quo, to bring about "continuity in perpetuity." Even more so, there is a large body of scientists in between (type II's) who are blends of these two extremes and much more besides. To say then that after a scientific revolution, there is complete lack of communication between the two sides is to distort the situation. It is a partial truth at best in the sense that there are at any one point in time – before, during, and after a revolution – those scientists who are unable to communicate across any one of a hundred different gaps – theories, attitudes, issues. But there are also those kinds of individual scientists (type II's) or forces acting within the system whose ideal is to discover links between the past and the present. It is too much then to say that after a revolution there is a complete breakdown between the two sides. For one this assumes a far greater degree of cohesiveness and homogeneity between sides, groups, or schools of thought than is warranted. I see as much diversity within the system or Game of Science at any point in time – even within supposedly tight common factions – as between successive time periods. As much as there are strong forces ("revolutionaries") acting to produce drastic changes in The Game at any one point in time, there are also just as strong forces acting to keep The Game the same as it has always been ("reactionaries"), and further yet, there are those ("moderates") acting to steer a course in between, to usher in the new while acting to preserve its links with the past.

In sum, I think Kuhn was right in his basic appeal to social psychology and history as one way of doing philosophy of science. Where Kuhn went wrong in my opinion was in his detailed application of social psychology. In fact it can be said that Kuhn wasn't social psychological enough to be able to give a good accounting of the workings of scientists. For instance, Kuhn accords far too passive a role to individual scientists in bringing about a revolution. As I have stressed, there are certain very specific kinds of scientists (type I's) who actually seek to promote revolution as well as those (type III's) who seek to suppress it. Even more important, there are those kinds of scientists in between (type II's) who serve the extremely important function of providing a bridge between the other two and of thereby helping to lend continuity to the system of science. In other words, whatever "continuity" there is in the system of science is put there by scientists themselves. To paraphrase Kant, we would not have found continuity, or its absence, in science had we not put it there ourselves in the first place.

V. CONCLUDING REMARKS: TOWARDS A PHILOSOPHY AND A SOCIOLOGY OF KNOWLEDGE

The situation I have been describing as well as bemoaning – the deliberately enforced separation between the philosophy and the social psychology of science – is only a symptom of a much larger problem or disease if you will. The larger problem is that of the general separation and isolation of disciplines from one another (Churchman, 1971). For too long we have uncritically accepted – even worse, actively perpetuated – the notion of the autonomy of disciplines, the notion that knowledge is best gained by splitting the world into separate knowledge domains (read: fiefdoms). The result has not only been the extreme compartmentalization of knowledge but something far worse – reductionism on a scale far grander than psychologism and sociologism ever dared imagined in their wildest dreams. Operating on the autonomy principle, each discipline not only believed in its view of the world but it reinforced it. In effect each discipline came to believe that it could explain the world – or at the very least its own exclusive part of it – in terms of its primitive explanatory variables and its variables alone.

It is no random accident that the most pointed examples of such thinking are to be found in the long history of repeated attempts to find simple irreducible elements of knowledge. This holds true whether the irreducible elements were those of empiricism or of rationalism. In each case, the intent was the same, i.e., to found an autonomous, ground or "bottom-up" theory of knowledge. If the analysis and arguments of this paper have substance and merit, then I believe they argue for the serious consideration of "top-down" theories of knowledge. If each discipline is in reality dependent upon each of the others, i.e., if the irreducible elements (the givens) of one discipline are the theoretically and infinitely expandable topics of empirical inquiry of another discipline, then we are always in a top-down universe of inquiry. That is, the problems of knowledge are not those of a single isolated discipline but those of all disciplines. Reductionism does not consist in the adoption of a particular discipline with which you or I happen not to like but in the adoption of *a* single discipline in *any* given inquiry (Churchman, 1948, 1971).

Were the point of view I have been outlining to become more prevalent, several consequences might follow. For one, I think we would view distinctions in a new light. We would be highly more critical and suspicious of any distinction which knowingly or unknowingly threatened to further divide and isolate disciplines from one another. Is the Reichenbach distinction between the contexts of discovery and of justification, for example, really necessary? Are our imaginations so impoverished that we cannot create a philosophy and a sociology of science that could not only exist but get on with another without such a distinction? What I want from the logicians and the philosophers of science is a theory of science that is grounded in the social psychological system of science; I do not want theories that ignore these important elements for the sake of some silly protective notion of what "the proper function of philosophy is." By the same token, I want from the sociologists and psychologists of science a theory of science that is well grounded in the philosophy of science.

Were such a greater sense of partnership to come about, it might help to dissolve some bothersome problems. For example, to anyone with the least bit of sociological knowledge, Lakatos' charge that Kuhn's views necessarily lead to "mob psychology" is laughable. Lakatos' charge only shows how naive he was in sociological matters. It does not follow that because "groups" of people are involved in a decision process that the outcome of the process can be characterized as governed by a "mob psychology." Lakatos was naive because he revealed that he was unable to appreciate the crucial difference between a "group" and a "mob," a difference that has preoccupied psychologists and sociologists to no end (Ackoff and Emery, 1973).

A second issue that might not be so bothersome is that of relativism. If Kuhn's views do lead to relativism, then part of the cause is for the reasons referred to earlier, i.e., his failure to distinguish between the individual and the collective behavior of scientists. To repeat: if there are particular individual scientists seeking to promote radical divergence between competing views, then there are also scientists continually seeking convergence. In this sense, if complete relativism is the mark of a chaotic state within a particular science, then its complete absence is the hallmark of stagnation, i.e., that there is nothing exciting enough going on in the science worth arguing and disagreeing about in the most violent of terms. The point is that what is a sign of extreme disease from one point of view or discipline can be a benefit and sign of health from another (Churchman, 1948). A "moderate amount" of relativism can be a good thing even in science.

We are never in a bottom-up world. The human mind, let alone scientific inquiry, does not start on the road to knowledge with unproblematic "bits" of knowledge. We start with whole patterns, with "chunks." The problem of knowledge is the problem of how the disciplines relate to one another (Churchman, 1971). For this reason, as much as we have needed a sociology of knowledge, we need even more a philosophy of knowledge. We need a philosophy of inquiry that is free from the disease of extreme disciplinary thinking. Thorsten Veblen described it well when he pointed that learning how to think within the confines – the rules – of a single discipline promotes "trained incapacity."

Philosophy is unfortunately no exception in this regard. Indeed, given its extreme preoccupation with identifying the essense of rationality with hard-bound, fixed rules, it is perhaps only inevitable as well as natural that there should have arisen in opposition anarchistic theories of knowledge (Feyerabend, 1971). One can not help but wonder: is anarchism inevitable or is it instead the result of our having bought initial clarity and order by means of an enforced separation of the disciplines? In other words, is anarchism the price we pay in the end for our decision to separate the disciplines from one another in the beginning? What *is* worse – to face anarchism in the end or the difficult problem of relating the disciplines to one another in the beginning?

University of Pittsburgh

NOTES

[1] This paper is an expanded version of a talk presented at the 1974 Philosophy of Science Association meetings at Notre Dame. The present paper was presented in its entirety at the fourteenth series of the University of Pittsburgh invited lectures in the History and Philosophy of Science.

² I refer the reader to an earlier paper (Mitroff, 1972) where I attempt to show this with regard to certain of Popper's views (1964, 1965).
³ See also Popper (1965).
⁴ If the general population of Apollo scientists represented an elite body of scientists, then the sample of 42 scientists is best described as an elite of elites (see Zuckerman, 1972). The sample contained some of the most distinguished scientific analysts of the Apollo missions. Two of the 42 scientists have the Nobel prize; six are members of the National Academy of Scientists. Nearly all are at prestigious universities or top-ranked government research labs.

BIBLIOGRAPHY

Ackoff, R. L. and Emery, F.: 1973, *On Purposeful Systems*, Aldine-Atherton, Chicago.
Barnes, S. B. and Dolby, R. G.: 1971, 'The Scientific Ethos: A Deviant Viewpoint', *Archives Européens de Sociologie* 11, 3–24.
Churchman, C. W.: 1948, *Theory of Experimental Inference*, Macmillan, New York.
Churchman, C. W.: 1971, *The Design of Inquiring Systems*, Basic Books, New York.
Eiduson, B. T.: 1962, *Scientists: Their Psychological World*, Basic Books, New York.
Feyerabend, P. K.: 1970, 'Against Method: Outline of an Anarchistic Theory of Knowledge', in M. Radner and S. Winokur (eds.), *Analyses of Theories and Methods of Physics and Psychology*, University of Minnesota Press, Minneapolis.
Hanson, N. R.: 1967, 'Observation and Interpretation', in S. Morgenbesser (ed.), *Philosophy of Science Today*, Basic Books, New York, pp. 89–99.
Kiesler, C. A.: 1971, *The Psychology of Commitment*, Academic Press, New York.
King, M. D.: 1971, 'Reason, Tradition, and the Progressiveness of Science', *History and Theory* 10, 3–32.
Kuhn, T. S.: 1962, *The Structure of Scientific Revolutions*, University of Chicago Press.
Lakatos, I.: 1970, 'Falsification and the Methodology of Scientific Research Programmes', in I. Lakatos and A. Musgrave (eds.), *Criticism and the Growth of Knowledge*, Cambridge, University Press.
McClelland, D. C.: 1970, 'On the Psychodynamics of Creative Physical Scientists', in L. Hudson (ed.), *The Ecology of Human Intelligence*, Penguin, London.
Merton, R. K.: 1949, 'Science and Democratic Structure', in R. K. Merton, *Social Theory and Social Structure*, The Free Press, New York.
Merton, R. K.: 1963, 'The Ambivalence of Scientists', *Bulletin of the Johns Hopkins Hospital* 112, 77–97.
Merton, R. K.: 1969, 'Behavior Patterns of Scientists', *American Scientist* 58, 1–23.
Mitroff, I. I.: 1972, 'The Mythology of Methodology: An Essay on the Nature of a Feeling Science', *Theory and Decision* 2, 274–290.
Mitroff, I. I.: 1973, 'Systems, Inquiry, and the Meanings of Falsification', *Philosophy of Science* 40, 255–276.
Mitroff, I. I. and Mason, R. O.: 1974a, 'On Evaluating the Scientific Contribution of the Apollo Moon Missions Via Information Theory: A Study of the Scientist-Scientist Relationship', *Management Science*, 20, 1501–1513.
Mitroff, I. I.: 1974b, 'Norms and Counter-Norms in a Select Group of the Apollo Moon Scientists: A Case Study of the Ambivalence of Scientists', *American Sociological Review* 39, 579–595.
Mitroff, I. I.: 1974c, *The Subjective Side of Science: A Philosophical Inquiry Into the Psychology of the Apollo Moon Scientists*, Elsevier, Amsterdam.

Morgenbesser, S.: 1967, 'Psychologism and Methodological Individualism', in S. Morgenbesser (ed.), *Philosophy of Science Today*, Basic Books, New York, pp. 160–174.
Mulkay, M.: 1969, 'Some Aspects of Cultural Growth in the Natural Sciences', *Social Research* **36**, 22–52.
Popper, K. R.: 1964, *The Poverty of Historicism*, Harper, New York.
Popper, K. R.: 1965, *The Logic of Scientific Discovery*, Harper, New York.
Popper, K. R.: 1970, 'Normal Science and Its Dangers' in I. Lakatos and A. Musgrave (eds.), *Criticism and the Growth of Knowledge*, Cambridge University Press, pp. 51–58.
Reichenbach, H.: 1968, *The Rise of Scientific Philosophy*, University of California Press.
Roe, A.: 1961, 'The Psychology of the Scientist', *Science* **134**, 456–459.
Shapere, D.: 1966, 'Meaning and Scientific Change', in R. G. Colodny (ed.) *Mind and Cosmos*, University of Pittsburgh Press, pp. 41–85.
Toulmin, S.: 1972, *Human Understanding*, Princeton University Press.
Zuckerman, H.: 1967, 'Nobel Laureates in Science: Patterns of Productivity, Collaboration and Authorship', *American Sociological Review* **32**, 391–403.
Zuckerman, H.: 1968, 'Patterns of Name-ordering among Authors of Scientific Papers: A Study of Social Symbolism and Its Ambiguity', *American Journal of Sociology* **74**, 276–291.
Zuckerman, H.: 1972, 'Interviewing an Ultra-Elite', *Public Opinion Quarterly* **36**, 159–175.

ROBERT M. ANDERSON, JR.

THE ILLUSIONS OF EXPERIENCE

ABSTRACT. On reading the grain argument as advanced by Meehl and Sellars, I find that there is not one but two grain arguments. According to one argument, mental events cannot be the same as neural events because mental events have a continuity that neural events do not have. The other argues for the same conclusion from the simplicity of experienced quality. I answer these arguments by claiming that these properties of experience are illusory. I detail a dual threshold theory of visual experience and show that given this model the mind-brain identity theory predicts the existence of these illusions.

The grain argument (labeled by Sellars) has been expounded most eloquently by Sellars and Meehl. In Sellars' paper 'Philosophy and the Scientific Image of Man,' he uses an example involving a pink semi-transparent cube.

It does not seem plausible to say that for a system of particles to be a pink cube is for them to have such and such imperceptible qualities and to be so related to one another as to make up an approximate cube. *Pink* does not seem to be made up of imperceptible qualities in the way in which being a ladder is made up of being cylindrical (the rungs), rectangular (the frame), wooden, etc. The manifest ice cube presents itself to us as something which is pink through and through, as a pink continuum, all the regions of which, however small, are pink. It presents itself to us as ultimately homogeneous; and an ice cube variegated in color is, though not homogeneous in its specific colour, "ultimately homogeneous," in the sense to which I am calling attention with respect to the generic trait of being colored (Sellars, 1963, p. 26).

Another way to put this is that there are certain continuities in perceptual space which do not exist in the neurophysiological processes which are the perceptual space. In our spatial experiences, color expanses consist of regions which are color expanses, which in turn consist of regions which are color expanses, and so on. Sometimes there are color patches in the visual field without texture, which are smooth and continuous, and which reveal no internal distinctions over the extent of their expanses. This fact in itself, of course, is not problematic. It is, however, a difficulty for the identity theory. According to the identity theorist, experiences of space are one and the same with neurophysiological processes. The microstructure of these processes, which are asserted to *be* the experiences, is much finer than the structure of our experiences. The problem then is: how can these things (brain processes and experience) which appear to differ radically in their structure (sometimes, as in the case of the continuous patch, there is even no structure in the experience and yet some in the process) be one and the same thing? The grain argument against the identity theory then is that since experiences and neural processes differ from one another in their structural properties (grain), by Leibniz's principle, which states that for two objects to be the same they must have the same properties, they must be different. Therefore, the identity theory must be false.

A similar statement of the grain argument may be made using the absence of perceptual change over time instead of space. Meehl states the perceptual fact as follows:

The most careful and sophisticated introspection will fail to refute the following argument: "There is a finite subregion ΔR of the raw-feel red patch, ψr, and a finite time interval Δt, such that during Δt no property of ΔR changes" (Meehl, 1966, p. 167).

The grain argument continues by substituting the physical process state Φr for the psychological state. One obtains:

There is a finite region ΔR of the brian state Φr and a finite time interval Δt such that during Δt no property of ΔR changes (Meehl, 1966, p. 168).

Suppose then, says Meehl, that we take the time to be 500 m sec. and ΔR to be a 5° circle from center of the phenomenal circle. The above statement with the blanks filled in this way will be true of Ψr but not of Φr. Thus Ψr and Φr cannot be the same.

There seems to be two arguments in Sellars' presentation of the grain argument. Sellars talks of the continuity of the cube, but he also speaks of the pink color itself. "Pink does not seem to be made up of imperceptible qualities in the way in which being a ladder is made up of being cylindrical." Here it is not continuousness in space of which Sellars is talking. It is rather the simplicity or intrinsicness[2] of pink. This second type of grain argument also can be found in Meehl in his statement of the grain objection in terms of properties.

> Sim (P)
> Sim (Φ)
> $P \neq \Phi$
>
> Where 'Sim' is a second-type one place predicate designating the property *simple* (a property of first-type properties) (Meehl, 1966, p. 177).

Thus we have another grain argument which begins with the premise that the colors we perceive are simple, first-order (see Maxwell, 1968), or intrinsic properties. Colors are simple in that they have no parts. Red can make up an expanse, but red itself or what makes red different from green does not depend on a further difference or is not spatially represented (representable). Red is somehow ultimately simple. Colors thus cannot be identical with neural processes which are themselves not simple. The problem again for the identity theorist is how can the neurophysiological process and experience be the same thing when they appear to differ in their properties, i.e., simplicity.

My answer to these grain arguments is that the continuity or smoothness, and simplicity of our experiences are mere illusions. Thus I call these seeming qualities of our experience "the illusions of experience." These experiential illusions are peculiar in that they cannot be accounted for in the usual way, by retreating to a representation Y that is said to really have the property that is mistakenly perceived to belong to X. For example, in Köhler and Wallach's (1944) experiments with figural aftereffects, first the retina is stimulated with an inspection Figure 1, then this figure is removed and the subject fixates on X in the test figure T. The

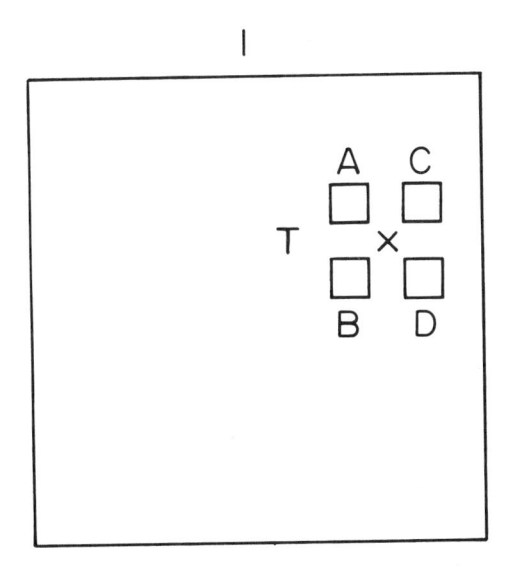

Fig. 1. Figures for Köhler and Wallach's Experiment.

squares *C, D* are seen to be farther apart than *A, B*, even though both the squares in the sets are equally distant. Köhler hypothesized that this illusion resulted from the development of increased resistance between parts of area 17 of the brain due to the exposure of the retina to the inspection figure (current flowing through the neural tissue was thought to increase resistance). In other words, Köhler explained the illusion by reference to an appearance (the spatially separate brain activity) which had, at least structurally, the property that the object (the squares on paper) seemed to have. C. D. Broad uses sensa to the same end. To explain the bent stick illusion, for example, he postulates a sensum with the property of bentness separate from the stick in water.

If a light is flickered at a rate of 60 cps, the human eye will not distinguish it from a light that is burning continuously. At a frequency of less than 60 cps, a flicker will be detected. Sixty cps is called the critical fusion frequency (CFF) – the frequency of subjective fusion (Poggio, 1968). This illusion can be explained in terms of the properties of single visual neurons. After 60 cps the neurons fail to fire at each pulse – their firing cannot be differentiated from the firing that occurs when the retina is continuously illuminated. Again we have the "object" (the flickering source of light) and the "appearance" (the record of the pulses in the visual neurons).

The illusions of experience cannot be explained this way, however. To see this consider a perception of red, X. On our schema X will be the object. X seems smooth but, by the grain argument, it is not. So we posit Y, an appearance of X which has the smoothness which X cannot have. Y in turn, however, is grounded in processes that are not smooth so again we are forced to posit Z, *ad infinitum*. The twine of experiences must come to an end – at the end is the world knot.

As a start on understanding how we are mistaken about our experiences, I will ask a question. Why do not we perceive a grain rather than a smooth expanse or a simple quality? Why do not we perceive, why are not we conscious of, each electron, proton, etc., as distinct? We do not because it takes more than an electron or a proton to be a consciousness. It takes *at least* a neural impulse,[3] and an impulse requires numerous elementary particles interacting in a complex way. Awareness of every particle in the brain or of every particle involved in a certain experience is a physical impossibility.

The living process bears a similar relation to its substrate. The individual atoms which make up the protoplasm are not alive. It is only by virtue of the elementary particles which make up a cell's protoplasm being related in a dynamical structure that the protoplasm is alive. A certain minimal amount of organization is necessary for life to exist. As I shall show, it is due to the fact that there are such minima of biological organization that the illusions of experience exist.

Since animals with many more than one neuron are incapable of consciousness, it is highly unlikely that even a single neuron would have the capacity. Consciousness in humans probably requires a coordinated activity of the entire brian. Let us assume for a moment, however, that a certain minimal cluster of neural firings is required for the existence of visual consciousness. A visual field of consciousness could then be constructed from such elements à la Bertrand Russell's method for constructing space out of events (Russell, 1954, p. 290).

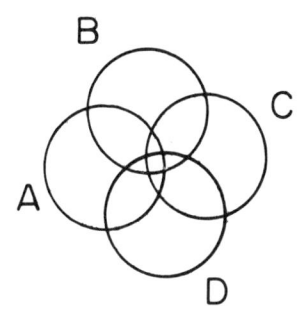

Fig. 2. Elements of visual consciousness overlapping to form point in visual space.

This schema is born out by neurophysiology. Area 17 (the primary projection area for vision in the brain) is organized in a columnar manner and is composed of what might be termed basic computational units.[4] These units are basically vertical in operation but overlap enough horizontally to allow for the construction of a continuous visual space à la Russell. Area 17 is wired to area 18 and 19 in a complex many-many way thus allowing for knowledge of relations. Rather than centers of consciousness, we might think of the basic units of 17 (or some collection thereof) as minimal elements of perceptual spatial consciousness; consciousness itself would be a more widespread phenomenon. The minimal elements of spatial consciousness would be the minimal centers of neural activity which when in combination with the processes of consciousness would be a visual sensation.

Thus we see how extensity comes about in the visual field, and how we can be mistaken about the smoothness of visual consciousness. We suffer this illusion because visual consciousness requires a certain minimal

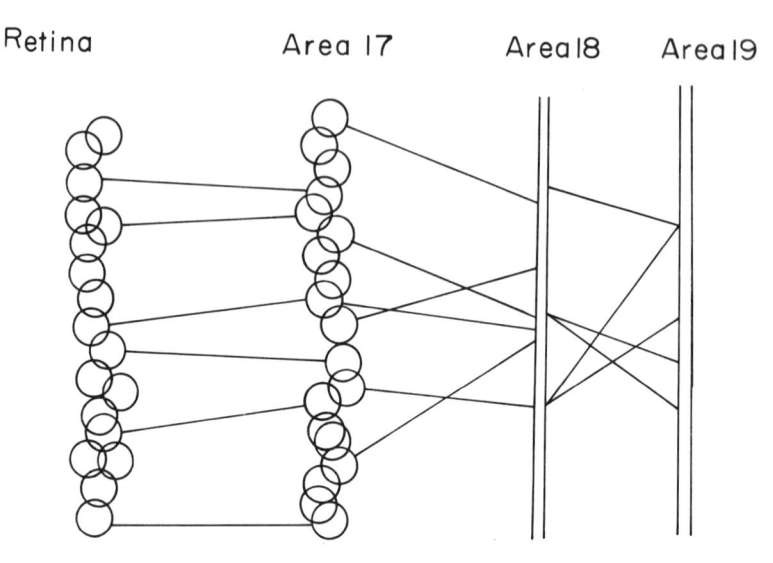

Fig. 3. Topographical mapping.

amount of organizational complexity. Consciousness is not sensitive to processes or elements which lack the requisite activity. Processes which exceed the minimum can become part of the global process of consciousness involving all modalities, motor activity, etc., and can overlap with one another in such a way that there are no discontinuities (pick any place on the cortex and there is always the requisite amount of organization of neural firings) thus forming the experienced visual field.

If this rendering of how brain activity is consciousness is correct, the grain argument loses its punch since it would be predicted by the identity theory in conjunction with this hypothesis that we should *not* be aware of all the microprocesses that interact to form a parcel of visual consciousness, and that indeed we should experience smoothness in the visual field when there is in fact a grain.

Thus far I have only dealt with bidimensional extensity in the visual field and the problem of the smoothness of visual experience. I shall now consider the problem of the simplicity of quality. When we perceive a color, a certain pattern of firing occurs in the visual cortex. This pattern, however, does not exceed a certain threshold value and, therefore, is not experienced as a pattern. Differences in such patterns, however, although not experienced as differences in pattern, are experienced as differences in quality. These patterns are local affairs basically involving a particular area in the visual cortex. To understand more clearly what is meant by this, a brief excursion into the physiology of the visual system is helpful.

The retina has been found to be related to the cortex by a topographic mapping (Talbot and Marshall, 1941). The mapping cannot, however, be described accurately as point to point. It is more precisely an area-to-area relation. To explain what is meant by this, we must describe what has become known as the columnar organization of the cortex, first discovered by Mountcastle (1957) in the somatosensory area. In the visual cortex, Hubel and Wiesel (1963) have found the neurons to be arranged in columns according to their receptive field orientation. (The *receptive field* of a cell is defined as the receptor area of the retina within which a light stimulus has an effect upon the firing of the cell.) Cells with horizontally oriented receptive fields are found in a column together, while obliqueness detectors of various degrees are found in other nearby columns. It seems that each point on the retina is represented in the cortex by about fifteen columns of cells, the cells of each column covering a particular stimulus orientation. Also, within a column the receptive fields of the cells vary a bit in their position on the retina. Corresponding to each column there is a cluster of overlapping receptive fields. Thus the topographic projection from the retina is better described as a mapping from an area on the retina (cluster of overlapping receptive fields) to an area in the cortex (cluster of columns).

Just as cells sensitive to orientation have been found in the visual cortex, so have cells sensitive to color (DeValois and Jacobs, 1968). These cells have been hypothesized to be color detectors and, as such, responsible for the experience of color. A more plausible conjecture, however, is that patterns of firing of cells form color experiences (Anderson, 1975). According to this hypothesis the cortex is composed of clusters of columns. However, within these columns are neural groups whose cells are interconnected so that when stimulated a spatiotemporal pattern of firing occurs which is a particular quality of experience. Each column cluster has the potential to fire in patterns differentially to different inputs and to produce most of the range of qualities that can be experienced. Since certain patterns of firing in a cluster of columns will give rise to experiences of lineness, tilt, etc., the intensive dimensions of experience must be expanded to include experiences of orientation and shape.

Extensity is experienced on this model when enough cells of overlapping clusters of columns are firing. Thus we have a dual threshold view of visual consciousness. The first threshold is one of organizational complexity. If a pattern of firing exceeds this threshold, a quality becomes part of consciousness. This is not enough, however, for the experience of extensity. A spreadout is experienced only when a minimal amount of the cortex is firing with the quality activity, when a minimum number of clusters of columns are firing with patterns of activity sufficient to exceed the quality threshold. The illusion of the simplicity of qualia is explained by the existence of this threshold. We do not experience the parts of redness because although the red process is above the threshold for coming into consciousness it is not of sufficient magnitude to come into consciousness as extended and differentiated in its parts. Thus the second thrust of the grain argument is parried.

The model requires only the concept of elements of the striate cortex that are capable of sustaining the patterns of firing that are the quality experiences. In Figure 4, for example, the element is sustaining a greenness pattern of firing. Neither extension nor differential parts are perceived in the firing of a lone quality element in the cortex. Only when such units band together in a system of overlapping firing can extension and spatial

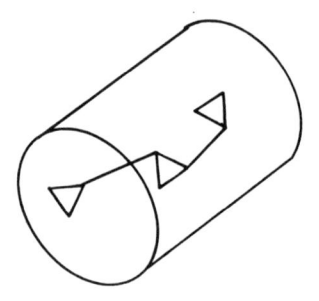

Fig. 4. Quality element firing in a greenness pattern.

THE ILLUSIONS OF EXPERIENCE 557

differentiation come into consciousness. One might object here that this cannot be so because unextended colors are never experienced. I reply that this is not due to their being no quality element – extension element dichotomy in the visual cortex. It is due to the fact that quality elements never perform solos. The firing of one never occurs without a great accompanying volley. The fact that there is a delay of at least one-half second upon electrical stimulation of the brain before a conscious experience is evoked implies that a mass of firing is usually going on by the time something is experienced (Eccles, 1970).

Figure 5 contains two views of a cluster of quality units overlapping to

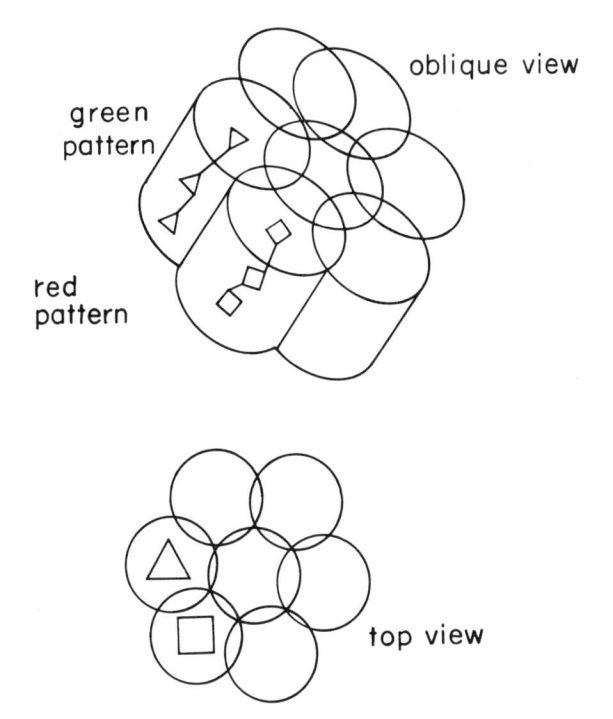

Fig. 5. Quality elements overlapping to form a minimal element of visual extension.

form a minimal element of visual extension. One element is firing in a green pattern, the other in a red pattern. Figure 6 shows a number of minimal elements of visual extension overlapping, in turn, to form part of the field of visual experience. It should be noted that the boundaries of the elements are somewhat artificial in that, especially in the case of the elements of visual extension, there are no corresponding hard and fast anatomical units in the cortex. Analogically, there are not hard and fast minimal units of life in the protoplasm of a cell. Yet in the cell, there is a certain minimal amount of organization that is necessary for a part of its protoplasm to be living. Similarly, in the visual cortex, there is a certain

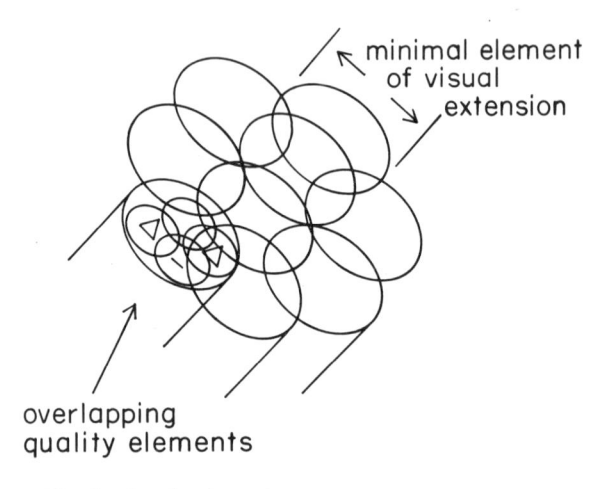

Fig. 6. Overlapping units of visual consciousness.

minimal amount of firing of conjoint quality elements necessary for visual extension to exist. You may cut the cake where you like, but the pieces may only be so small.

In Figure 7 we have the cortical firing in response to a green bar on a red background. The qualities of the line are reflected in the firing of patterns of lineness and greenness in the quality elements; the line is experienced as extended, both in width and in length, because the quality elements are strung out over the cortex and in width cover about two diameters of minimal elements of visual extension.

This model allows a response to Professor Meehl's qualms arising from the grain problem. Meehl says:

It strikes me as very odd that I could fully understand the intension of the phenomenal predicate "red-hued," this predicate designating a configural physical property and could also fully understand the intension of the physical predicate "r"; and yet not "understand why" they designate the same *property* (Meehl, 1966, p. 172).

THE ILLUSIONS OF EXPERIENCE 559

This perplexity may be resolved by seeing that although we may fully understand the intension of the phenomenal predicate "red-hued," we misapply it to our own experience. The smoothness and simplicity of our perceptions is an illusion. We do not know the microstructure of our

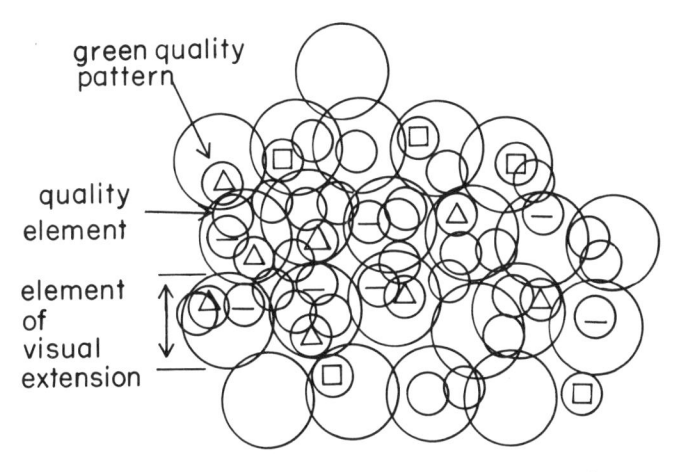

Fig. 7. Experience of green bar on red ground.

experiences, and in this we are deceived. If red-raw feels are defined as having the requisite properties of smoothness and simplicity, then they do not exist. If they are not so defined, then we are simply mistaken about raw feels being smooth and simple, about their having these properties.[5]

Department of Psychiatry,
Stanford University Medical School, Stanford

NOTES

[1] This paper has been written first while being supported by a grant from the Carnegie Corporation as a research assistant at the Minnesota Center for the Philosophy of Science and second while receiving support as a Biological Sciences Training Fellow on a grant from the National Institute of Mental Health: MH8304-09. I have profited from dis-

cussions with Dr. Grover Maxwell, Dr. Herbert Feigl, Gary Gunderson, Dr. Karl Pribram, and Dr. D. N. Spinelli.
[2] The term "intrinsic" is explicated in Grover Maxwell's (1968).
[3] It may be that the neural impulse plays a relatively insignificant role in experience, since most information processing in the brain takes place at the synaptic junctures (Pribram, 1971). But even if this is so, the quantal nature of most synaptic events (Eccles, 1970, p. 13) would provide a basis for the grain argument.
[4] This has been demonstrated in, for example, Hubel and Wiesel's experiments (1962 and 1963). Kabrisky (1966), in his *An Information Processing Theory of Vision*, speaks of basic computational units.
[5] Sellars (1966) has put forth a solution in some of his writings. Stephen Pepper in his *Concept and Quality* has used the concept of critical thresholds of fusion in his attempt to deal with the grain problem. Carnap has also had similar thoughts (Feigl, 1967, p. 141).

BIBLIOGRAPHY

Anderson, Jr., R. M.: 1975, 'Wholistic and Particulate Approaches in Neuropsychology', in W. B. Weimer and D. S. Palermo (eds.), *Cognition and the Symbolic Processes*, V. H. Winston, Washington, D.C., 1975.
De Valois, R. L. and Jacobs, G. H.: 1968, 'Primate Color Vision', *Science* **162**, 533–540.
Eccles, J. C.: 1970, *Facing Reality*, Springer-Verlag, Berlin.
Feigl, H.: 1967, The *'Mental' and the 'Physical'*, University of Minnesota Press, Minneapolis.
Hubel, D. H. and T. N. Wiesel: 1962, 'Receptive Fields, Binocular Interaction and Functional Architecture in the Cat's Visual Cortex', *Journal of Physiology* **160**, 106–154.
Hubel, D. H. and T. N. Wiesel: 1963, 'Shape and Arrangement of Columns in Cats' Striate Cortex', *Journal of Physiology* **165**, 559–568.
Kabrisky, M.: 1966, *A Proposed Model for Visual Information Processing in the Human Brain*, University of Illinois Press, Urbana.
Köhler, W. and H. Wallach: 1944, 'Figural After-Effects: An Investigation of Visual Processes', *Proceedings of American Philosophical Society* **88**, 296–357.
Maxwell, G.: 1968, 'Scientific Methodology and the Casual Theory of Perception', in I. Lakatos and A. Musgrave (eds.), *Problems in the Philosophy of Science*, North-Holland, Amsterdam, 1968.
Meehl, P. E.: 1966, 'The Complete Autocerebroscopist: A Thought-Experiment on Professor Feigl's Mind-Body Identity Thesis', in P. K. Feyerabend and G. Maxwell (eds.), *Mind, Matter, and Method: Essays in Philosophy and Science in Honor of Herbert Feigl*, University of Minnesota Press, Minneapolis, 1966.
Mountcastle, V. B.: 1957, 'Modalitity and Topographic Properties of Single Neurons of Cat's Somatic Sensory Cortex', *Journal of Neurophysiology* **20**, 408–434.
Pepper, S.: 1967, *Concept and Quality*, University of California Press, Berkeley.
Poggio, G. F.: 1968, 'Central Neural Mechanisms in Vision', in V. B. Mountcastle (ed.), *Medical Physiology*, C. V. Mosby, St. Louis, Missouri, 1968.
Pribram, K. H.: 1971, *Languages of the Brain: Experimental Paradoxes and Principles in Neuropsychology*, Prentice-Hall, Englewood Cliff, New Jersey.
Russell, B.: 1954, *The Analysis of Matter*, Dover, New York.
Sellars, W.: 1963, *Science, Perception, and Reality*, Humanities Press, New York.
Sellars, W.: 1966, 'The Refutation of Phenomenalism: Prolegomena to a Defense of Scientific Realism', in P. K. Feyerabend and G. Maxwell (eds.), *Mind, Matter, and*

Method: Essay in Philosophy and Science in Honor of Herbert Feigl, University of Minnesota Press, Minneapolis, 1966.

Talbot, S. A. and Marshall, W. H.: 1941, 'Physiological Studies on Neural Mechanisms of Visual Localization and Discrimination', *American Journal of Ophtthalmology* **24**, 1255–1263.

SYMPOSIUM

DEVELOPMENT OF THE PHILOSOPHY OF SCIENCE

GROVER MAXWELL

SOME CURRENT TRENDS IN PHILOSOPHY OF SCIENCE: WITH SPECIAL ATTENTION TO CONFIRMATION, THEORETICAL ENTITIES, AND MIND-BODY

This discussion will be limited to two main topics, the existence of and our knowledge of unobservable entities, on the one hand, and problems about the confirmation of our knowledge claims (scientific and common sense ones), on the other. Unfortunately, this isn't much of a limitation; for the topics are so large, and their ramifications and implications are so numerous and run so deep that the most that a brief treatment such as this can accomplish is to hit a few of the high spots and, I hope, to generate a little more interest and activity in the areas in question.

It is interesting to me that none of the three of us in this symposium knew much of anything about what subjects the others had selected for their talks and nothing at all, except by way of speculation, about what they were going to say. This, of course, was not as it should have been, but my point is not to moralize about the dilatory practices of McMullin, Putnam, and Maxwell. It is, rather, to suggest that the selection of pretty much the same topics by all of us was not accidental but resulted from recognition of the centrality of the matters considered, the recent and current interest in them, and the importance of their numerous philosophical *and* scientific implications.

Problems about unobservables and those associated with confirmation are, of course, by no means independent of each other. There are numerous implications in both directions from one area to the other. Perhaps the most crucial ones, however, are those of the latter for the former. As we all know, Hume revealed to us in agonizing detail and with awesome clarity the forlorn predicament of the empiricist when he faces, as he *must*, the problem of confirming even the most mundane knowledge claim that goes beyond observation of the moment. I believe that Hume was the first and, perhaps, the last consistent empiricist. He could not bring himself to *abandon* empiricism so he forthrightly grasped the other horn of the dilemma and embraced skepticism. My contention is that attempts by subsequent empiricists to slide down between the horns have all ended in failure. I have argued this in considerable detail elsewhere;[1] here I can only summarize briefly and, I fear, rather dogmatically what I take to be some of the most important results and implications of recent work in confirmation theory.

Put bluntly, the most important result is this: *no kind of inductive inference can be justified in any sense of 'justified' that is strong enough to save empiricism.* In other words, Hume's dilemma is inescapable. I am using the phrase 'inductive inference' to include any kind of procedures that, on the basis of premises or *evidence* at hand, select certain knowledge claims as preferable to others – no matter how tentative or qualified such selection and such preferability may be. This is broad enough to include even Popper's falsificationist or "critical" approach. He would, no doubt, object to the term 'inductive inference', but let us not quarrel about words. I stated the matter in such a fashion not out of any desire to "beat up" on empiricism – like most of us, I was brought up, philosophically, on empiricism, and I feel that it is a shame that it would not work. I did it, rather, because the statement seems not only succinct but also quite suggestive of some of its most important implications, for example: *observational evidence is not sufficient, in general, for deciding between conflicting knowledge claims.*

This last statement holds when it is qualified (vacuously?) by saying, "observational evidence plus logic (including inductive logic – if only it existed) are not sufficient..." Russell emphasized these results at least as early as 1948,[2] and they are strongly supported by vast amounts of recent results (especially the preponderance of negative ones) from research in confirmation theory and foundations of statistical inference, as well as by the fact that so many of the researchers are settling for some variety of Bayesian subjectivism or "personalism."

But this socio-academic evidence, interesting, important, and (to me) encouraging as it may be, is not required to establish the point, for Russell's contention just cited that *simple inductive inferences* that are not bolstered by "common sense" *(contingent, but unconfirmed)* presuppositions lead, in principle, infinitely more often to erroneous results than to correct ones – this contention can easily be demonstrated (deductively!), again as noted by Russell *(op. cit.)*. (What I have called 'simple inductive inferences' include not only induction by simple enumeration and its extension to the "straight rule" but also Mill's canons and virtually all simple statistical inferences that one would encounter, say, in a course in elementary statistics.) Virtually the same thing can be demonstrated to hold for the only other serious candidate for methods of confirmation or selection among competing knowledge claims, namely the hypothetico-deductive (better: *the hypothetico-inferential*) method. (The result holds with full force when this method is interpreted liberally enough to include Popper's approach. [Popper's method also fails because of the fact that the more interesting and important knowledge claims *(theories)*, especially in scientific inquiry, are not falsifiable (see, e.g., Maxwell (1974a) and (1975)).]) For, again, it is easy to demonstrate that, given *any* amount of observational evidence, there always exist an infinite number of mutually incompatible theories that explain (entail *or*, in cases of statistical explanation, imply statistically) the evidence at hand. In other words, as far as purely logical considerations, indeed, as far as *any* non-contingent considerations are concerned, observational evidence will always be neutral as to which knowledge claim from among an infinite number of mutually incompatible competitors we would be well-advised, at any moment and however tentatively, to select.

If evidence, no matter how abundant, plus logical and linguistic or conceptual factors are neutral towards this bewildering multiplicity of mutually incompatible possibilities, what hope is there for the existence of a significant amount of knowledge, or, even, of true belief? It seems obvious that there is *no* hope unless at least the two following conditions obtain: (1) the knowing subject must have the *innate* capacity to acquire, in appropriate circumstances, the ability to make *good guesses* in a significant portion of trials – or, what is the same, the ability to conjecture, fairly often, good knowledge claims or good *theories*, and (2) the subject must have the ability to make not too hopeless *estimates* of the *prior probabilities* of the knowledge claims that *are* proposed and considered.

That *something like* the first condition must hold has usually been granted, even by empiricists, but it has not been recognized, I believe, how remarkable and, in a sense, how extremely unlikely this is. Moreover, it has usually been dismissed with statements such as, "Yes it *is* rather surprising, but it has to do only with the *context of discovery*. Little is known yet about the *psychology* of discovery and creativity, but this is, after all, a matter *for* the psychologist. As inductive logicians, or as epistemologists, or as philosophers in general, it is none of our affair. We are concerned only with the *context of justification*." I think that there are many things that count against such statements, including the nonexistence of a sharp distinction between "context of discovery" and "context of justification," or, even, between science and philosophy. However, at this point, I only want to remark that *justification* in the sense meant, or, at least, *aimed at* by empiricists, that is, on the basis of nothing other than evidence and logico-conceptual factors – that justification, in this, *operative* sense, is impossible.

In view of this fact, it would seem that the fundamental question of epistemology is *not*, "How are our knowledge claims to be justified?" It is, rather, I have contended,[3] "What are the conditions that must hold in order for us to have a significant amount of knowledge (or true belief[4])?". The answer, I believe, is given by the statements of conditions (1) and (2) above.

Condition (1) may be stated alternatively: The frequency with which we propose good hypotheses must be surprisingly high, given the number of possibilities for selecting "bad" hypotheses that nevertheless account perfectly for all of the evidence at hand. And this, in turn, may be put slightly differently: The *probability* of our proposing good hypotheses must be surprisingly high, given.... Now, relative to appropriate reference classes (of hypotheses, statements, or propositions, etc.) we can equate the *probability* that a given hypothesis is true (or fairly close to the truth, etc.) with the *relative frequency* of true (or close to true, etc.) hypotheses among hypotheses that resemble it in certain respects (the reference class). A basis thus emerges for interpreting the probabilities of hypotheses (and, indeed, of all propositions) in terms of truth frequencies.[5] By appropriately selecting different kinds of reference classes, we may specify both prior probabilities and posterior probabilities of hypotheses as well as prior probabilities of statements reporting the evidence at hand. Naturally, the next step is to apply the division theorem of the calculus of probability (a form of Bayes's Theorem) to obtain the (posterior) probability of the hypotheses (theory, knowledge claim, etc.) on the basis of the evidence at hand. (I apologize for the impossibility of giving even a sketch of the details here. For them, see Maxwell (1974a) and (1975)).

Many, of course, have objected to talking at all about the *probability* of a theory or hypothesis. But this interpretation seems quite straightforward and innocuous to me, and I am convinced that it provides a means of either meeting or viably circumventing all of these objections with which I am acquainted. If the interpretation *is* allowed, there can be no objection to using Bayes's Theorem in computing probabilities (or "degrees of confirmation"), for in it, as in *all* frequency interpretations, all theorems of the calculus of probability are truths of arithmetic (or, rather, of set theory).

We should note here that, if condition (1) holds, this is sufficient for providing that a significant portion of our beliefs are true (or close to true, etc.). This portion may be great enough to be of considerable survival value[6] and to foster the illusions that we are pretty knowledgeable creatures. But the "surprisingly high" portion of true (or close to true) beliefs required by condition (1) may be, indeed *is*, I should venture, quite low in absolute value. Condition (2) is needed both to enable us to discard the less preferable (less *probable*, in the sense just explained) of the competing hypotheses that are being considered and to give us an (admittedly rather crude) estimate of the probabilities of those hypotheses (theories, etc.) that we tentatively accept and use. And, let us face it, the only viable way of doing this is to make (subjective) estimates (guesses) of the prior probability of the hypothesis or theory under consideration and of the prior probability of the evidence at hand.[7] (See again Maxwell (1974) and (1975), where I argue at length that there just is no way of getting around "all of this subjectivity".)

Although the process of estimating or guessing at the prior probabilities is subjective, the prior probabilities themselves are objectively existing (contingent) relative frequencies, as explained above. Analogously, you may subjectively guess that I am six feet and five inches tall, plus or minus one inch, but this does not make my true height a subjective matter. It remains objectively (and contingently) at six feet and two inches plus or minus one inch. And, although your *guessing* is subjective, the truth or falsity of your *guess* is, again an objective matter. (This was emphasized with great clarity by Salmon (1965), and I have tried to develop some of its consequences in Maxwell (1974) and (1975).)

This objectivity of probabilities coupled with the necessity of subjectively estimating them provides one component of the basis for a *rapprochement* between *personalists* or *subjectivists* such as L. J. Savage,

Richard C. Jeffrey, and Abner Shimony, on the one hand, and *frequentists*, or at any rate *Bayesian frequentists*, such as Wesley Salmon and myself, on the other. The other component is the fact that it is perfectly "meaningful," indeed crucially necessary to talk of, theorize about, and confirm hypotheses about relative frequencies that never have been or, perhaps, never will be *determined directly* by, say, counting the number of individuals in a given sample that have the attribute of interest. Assertions that the relative frequency of a certain attribute is so much in a certain population should be thought of, in general, as theoretical assertions. Often they can be confirmed or disconfirmed, if at all, only very indirectly by testing the consequences of their conjunction with other theoretical assertions and/or with "background knowledge." For example, when a meteorologist asserts that the probability of rain during the next ten hours is 0.65, he should, I maintain, be interpreted as meaning something like: the relative frequency of occasions when rain occurs during the subsequent ten hours among occasions when the current meteorological conditions are about as they are now is 0.65. Now while it *may* be true, although it usually is not, that the *grounds* that he has or *would* give for the assertion consist of actually observed relative frequencies of rain in similar circumstances, the probability statement neither asserts nor entails anything about such observations. In most cases, the grounds for such a probability statement (statement of relative frequency) are much more complex and are related to the statement in indirect and devious ways. For example, they may consist mainly of certain meteorological and other physical theories some of which are statistical. And in some cases the grounds may consist almost entirely of the weatherman's educated "hunch." He might say, "Well, I just cannot say what evidence, if any, or what theoretical considerations, if any, make me feel this way, but I am just pretty sure that, when conditions are like they are now, it will rain within then hours 65 per cent of the time."

The personalist or subjectivist can, thus, have as much subjectivism as he deems necessary or desirable in the crucially important *process* of *estimating* prior probabilities. And the *probabilities* themselves can be unproblematically interpreted as (*objectively* existing) relative frequencies. The objection that talk about such frequencies is useless because they cannot, in general, be determined or measured is mistaken, and it loses its sting entirely as soon as hangovers from positivism, operationism, and other empiricist dogmas[8] are discarded. In other words, the objection is an instance of the *fallacy of epistemologism*; it confuses the *meaning* or *what is asserted* by a probability (relative frequency) statement with *(one kind)* of grounds or evidence that can be used in its support. Relative frequency statements about unobserved (and, indeed, unobservable) frequencies are not only perfectly meaningful, but, as noted above, they can often be confirmed or disconfirmed (indirectly) and/or given theoretical support by theories which, in their turn, may be confirmed or disconfirmed. (It should be remembered, of course, that I am using here the Bayesian-frequentist notion of *confirmation*, which always involves use of *some* statements that are *not* confirmed or disconfirmed [although they are contingent], such as the estimated values of the requisite prior probabilities.)

One important consequence of these developments in confirmation theory is the necessity to reject or drastically revise currently popular views about the nature of *philosophical inquiry*, how it is supposed to be sharply distinguished from science, etc. (Again I must be brief and dogmatic here and refer to Maxwell (1970a) and (1975), where I have discussed the matter in considerable detail.) If all statements are either analytic or synthetic and if all (and only all) synthetic statements are *empirical* in the sense of 'empirical' used by empiricists, then it is easy to provide a convincing defense of the views that I am calling into question. For surely, the defense would run, *empirical* statements are decided, or confirmed, or disconfirmed by collecting relevant observational evidence, and proposing and testing empirical statements (empirical theories, empirical hypotheses, etc.) in such a manner is exactly what constitutes *scientific* inquiry – no more and no less. And, of course (argument continues), it is no business of the philosopher to stick his nose into the business of the scientist! What *does* remain for the philosopher? Obviously, in the realm of *statements*, only the analytic ones, although proponents of this view permit him to engage in the *activities* of "logical analysis", "linguistic analysis," "conceptual analysis," etc., the results of which are, I suppose, expressed in some metalanguage or other by *statements* which are analytic, or at any rate, true by virtue of the meaning of the language used and which are nonsynthetic, nonempirical, and *non-contingent*.

Let me pause to acknowledge that by no means all proponents of the main core of the view in question (philosophy is essentially an "analytic" discipline) would subscribe to the line of defense of it that I have been suggesting. But this does not matter, for I only meant it to *be* suggestive. Nor is it necessary to be concerned with the many vexing complications that would arise if we pursued the line and examined it critically, such as various meanings (or lack of meaning) of 'analytic', 'analysis', 'synthetic', 'meaning' itself, etc., or the question whether the analytic-synthetic distinction is a clear one or a viable one at all (or whether, as our good friend and fellow symposiast, Hilary Putnam (1960) has held, there is a clear distinction, all right, but it is a trivial, uninteresting, and unimportant one). These items do not matter much and much of what I have been saying is merely suggestive because I intend to attack the premise that I believe is at the very heart of any plausible defense of the view under consideration and that, I believe, provides the philosophical motivation for holding it. I am going to attack the assumption that science is concerned entirely or, even, mainly with "empirical" statements, in the required sense of 'empirical'. The basis of my attack is my contention that, except for certain degenerate cases, there *are* no empirical statements in this sense. All that I have in mind here is something that I have stressed already: experience (observational evidence) plus logic (of any kind worthy of the name) is insufficient to confirm or disconfirm even the most mundane knowledge claim that goes beyond observation of the moment [9] *unless* it is bolstered by powerful (contingent) assumptions that are neither known by experience nor confirmed by knowledge by experience. It follows that empirical statements, in the required sense, do not exist save for those that report observations of the moment (the degenerate cases) – as was clearly recognized and elucidated by Hume; again we are back where we started.

"But surely," I will be told, "one does not have to be a radical empiricist like your straw men to hold that statements in the realm of science must be such that observational evidence of the appropriate kind can either count for them or against them and that this is not true of statements that are in the province of philosophy." This popular retort sounds quite reasonable and even, perhaps, rather mild, but it fails, I believe, for reasons to be given in a moment.

However, let me pause to emphasize that I am all in favor of "analysis," *in its place*, whether it be "logical," "linguistic," or "conceptual." In philosophy, as in chemistry or political science, we should try to make our ideas clear and to have some degree of selfconsciousness about the meaning of the language in which we express them. And I grant and insist that, in philosophy, as in other disciplines, contemporary logical and mathematical resources can provide powerful aids in many avenues of inquiry. I might even grant that all of this holds with more force and in a larger portion of cases in philosophy than in most other areas and, even, that *some* traditional philosophical problems that were once thought to be substantive and/or contingent can be solved or "dissolved" entirely by logical or linguistic analysis. (Note, however, that Einstein also solved or "dissolved" a few problems in *physics* through linguistic or conceptual analysis of the notion of *simultaneity*.)

What I deny – and deny vigorously and, I fear, sometimes tiresomely and self-righteously – is the contention that *analysis* (or something very much like analysis or something very closely related to and involved with analysis) is the *only* legitimate tool available to the philosopher and that all problems that can legitimately be classified as philosophical can be solved only by analysis (or something very much like it, etc.).

Professor Putnam, whose talk came after mine in the symposium, remarked that he did not care or did not think it important whether philosophy was similar to or quite different and distinct from science – or something to that effect. Since I am rewriting this paper after *his* symposium talk is over, I cannot resist saying that I find this attitude, especially coming from *him*, puzzling. He made it quite explicit in Putnam (1960) – and the symposium indicated that he still feels the same way – that he does not believe that there is any more than a rather trivial and unimportant kind of distinction between analytic and synthetic statements and, moreover, that great philosophical mischief results from the view that a profoundly important distinction exists here and that deep and important philosophical truths result from unearthing deep and important analytic truths that are tacit in our system of concepts, beliefs, scientific postulates, or what have you. I now agree, in large part, with the spirit of Putnam's position, although I still have considerable disagreement with its letter. However, what I want to ask now is: does not the Quine-Putnam view on *analytic-synthetic* (and related issues) support, indeed, *entail*, the conclusion that I draw about the nature of philosophy and its relation to science *and* undermine, indeed *refute*, the position that I am attacking? It seems undeniable that it does; however, as much as I admire the work of Professor Putnam, as well as large portions of that of Quine, I am not inclined to count them as allies in this intance. For it seems to me that they win the battle too easily (or, rather, Quine does; I am still puzzled as to why Putnam refrains from drawing the obvious conclusion from his premises). I do not want to join them in drawing the correct conclusion from what I believe to be the wrong reasons. And if Quine and Putnam *were* right about *analytic-synthetic* (which, literally, I still believe they are not [10]), the case against the "popular view" would be "overdetermined," and, in my opinion, by far the more important (and conclusive) reason agnist it would be the one from confirmation theory discussed earlier. But before returning to this, Quine's own views about confirmation theory merit attention here. Let us consider, for a moment, the "other" "dogma of empiricism" (Quine, 1951).

Quine sees, quite correctly I believe, that what I called above 'simple inductive inferences' do not get us very far towards accretion of interesting and important knowledge and he agrees with Hume that they are logically defective, anyway, as far as confirmation is concerned. He recognizes that the fundamental tool actually used is the hypothetico-inferential one, and he sees, moreover, that, as a confirmation device in and of itself, this too is defective. He appreciates the importance of Duhem's insight that theories (and other nondegenerate knowledge claims) are, in general, not falsifiable by experience – that it is only a conjunction of a theory or hypothesis with numerous other items from our store of putative knowledge that can be contradicted by an observation statement. And when this does happen (when "trouble strikes"), he correctly insists, there is a *multiplicity* of ways in which our "conceptual system" (I should say, "our system of theories, hypotheses, other beliefs, etc.") can be modified to accommodate the result. I am not sure, however, that he has seen just how great and how bewildering such "multiplicities" are and that they are, in principle, just as great and just as troublesome when we observe "positive results" as they are when "trouble strikes," and, finally, that the only way to get out of the trouble (if there *is* a way) is to rely on our hoped-for creative ability to dream up theories and hypotheses that have surprisingly high prior probabilities and our ability to estimate, with not-too-hopeless results the prior probabilities of the competing theories that may be proposed.

At any rate, it is here, I believe, that Quine makes a serious mistake. If I have misread him, I apologize, but if I have not, I think that, this time, he draws a mistaken conclusion from correct premises. He reflects on the fact that, given any amount of observational evidence that we may happen to have at a given time, there will always be a number of *apparently* mutually incompatible theories that, as far as logical considerations are concerned, account for the evidence equally well and that, moreover, attempts to collect more evidence – attempts to conduct "crucial experiments" to decide among competing theories – are doomed, for familiar Duhemian reasons. He apparently concludes that, since evidence (plus logic) is insufficient to produce a decision, we must select from among the competing theories on "pragmatic" grounds (whatever this may mean; fortunately, for our purposes here, it does not matter what it means). Others, as we know, conclude that the decision to accept one competitor over the others is a matter of *convention*. I believe that both conclusions are gravely mistaken; I believe that, in general, we select (or we *ought* to select) the theory that we do because we estimate its *prior probability* to be (contingently) higher than that of any of its competitors (even though we do not have at hand, and may never have, any way of testing this estimate).

But let us leave aside my last contention for the moment and suppose that the only grounds we have for deciding between two given statements *are* pragmatic ones or conventional ones – or both. *It would by no means follow that there was only a pragmatic or a conventional difference between the contents of the statements – between what was asserted by one statement and what was asserted by the other; it would not even follow that the statements could not differ in truth values.* To suppose otherwise, as so many seem to suppose, is another instance of the *fallacy of epistemologism,* confusing the grounds for accepting a statement (or deciding between statements, etc.) with what the statement asserts. I do not contend that Quine makes exactly the mistake warned against in the italicized sentence above; I do not know whether he does or not. However, he certainly does seem to hold that when experience plus logic is insufficient to decide between two statements (which he correctly holds is the rule rather than the exception) – when this predicament obtains, no distinction exists between their having different (contingent) contents, the meaning of their common terms remaining the same, on the one hand, and, on the other hand, having the same content but with some terms having meanings in one of the statements differing from their meanings in the other; in other words, no distinction exists between their differing *synthetically* and their differing *analytically*.[11] This may seem more plausible, but I believe that it is just as grave a commission of the fallacy of confusing *grounds* with *content* as the others just described. Like most of us raised in the tradition, Quine seems not completely immune to seduction by some (at least one) of the dogmas of empiricism.

Let me return to the charge that I have made things easy for myself by using too stringent a version of empiricism as a straw man. For (the charge continued), a sharp distinction between philosophy and, say, science can be drawn by relegating to the domain of the latter those statements (and questions, etc.) for which observational knowledge would have some relevance as evidence and to the realm of philosophy the statements (and questions, etc.) for which no such relevance obtains. Concerning this, the discussion of Quine's views was not as much of a digression as it might have seemed, especially the part that was intended as a sympathetic exposition and an endorsement of his position concerning the defects of all extant empiricist theories of confirmation. The existence of these defects together with others that I have cited make it apparent that notions about *what* is relevant for *what* need to be critically re-examined. I shall argue presently that this will reveal that observation, experimental results (as well as scientific [and common sense] theories) are relevant for a number of important issues that are generally acknowledged to be philosophical. With this, I shall rest the case against what I have termed 'the popular view' about the nature of philosophy.

At this point someone might be inclined to remark, "Well, even so, does not Putnam still have a point? Should not what you (claim to) have established reinforce our conviction that both scientists and philosophers (and others, for that matter) should be more concerned with the substantive problems that face them than with the nature of their methodology? Should not each use those tools that seem necessary or helpful without getting into a tizzy about whether they should be labeled 'scientific', or 'philosophical', or 'analytic', or 'common sensical', or etc., etc.?" If this is what Putnam had in mind, I am with him, or, rather, close to him. Aside from the considerable intrinsic interest of the matter,[12] my main motivation for trying to establish that there is no one special method peculiar to philosophy (and none for science either) has been to remove the debilitating restrictions on both philosophical and scientific activities that convictions to the contrary generate. Even here, however, a word of caution is in order. As others have noted, contempt for methodological considerations is usually either a rationalization for ignorance or, perhaps more often, a cover for out-dated, half-baked, and disastrously restrictive methodology. (Consider, for example, the situation in large segments of psychology and the social sciences today.) Methodological self-consciousness is similar to certain kinds of medicines; too much can be fatal, but, sometimes, none or too little can be fatal, too.

Let us turn now to a brief consideration of some of the problems about unobservable entities (during the course of which I shall try to make good my promise to show the relevance of observational knowledge for some [at least one] "philosophical" problems). I have tried to show in some detail elsewhere (Maxwell (1970a) and (1972)) that it is a contingent matter for which observational knowledge *is* relevant whether realism or phenomenalism is true and that the same holds, regarding scientific theories, for the realism-instrumentalism issue. Consider the question: *if* our theories *are* interpreted realistically so that we understand them to be about unobservable entities, are they reasonably well-confirmed by the observational evidence that we have at hand? It seems to me that the answer is *yes* for *some* or our extant scientific theories. Now I admit and insist that *as far as observational evidence (plus logic) alone is concerned*, an instrumentalistically interpreted theory will be just as well supported as the corresponding realistically interpreted one. [13] But, let us remember, that there exist an infinite number of other (meta-) theories that will explain how it is that the words we utter or write to do what we call 'expressing' the theory in question – how it is that these sounds or these inscriptions produce the observation statements that they do. If this last sentence is confusing, just forget it and recall that we have seen, ad nauseam, that observational evidence (plus logic) alone is impotent as far as effecting confirmation or disconfirmation is concerned and must always be bolstered by *estimates of prior probabilities*. I, for one, estimate the prior probability of realism to be *much greater* than the prior probability of instrumentalism.

Is it appalling that I stake the case for realism on my subjective guess that its prior probability is greater than that of instrumentalism (and greater than that of its other infinitude of competitors, virtually all of which I have never heard and which I have not been bright enough to formulate myself)? Dear friends, what is the alternative? As Hume clearly saw, we make the same kind of reckless wager when we (tacitly) predict that the next piece of bread we eat will nourish rather than poison us. Of course, more can be said. Indeed, I *have* said much *much* more elsewhere. However, no matter how far we go, we must stop somewhere; and, when we do stop, we will still be faced with the necessity of making prior probability estimates to eliminate competitors.

When I (and others) say that both instrumentalism and realism – or both phenomenalism and realism – can account for all of our observational evidence and that none of them can be refuted by any conceivable observation, it should always be remembered, but is often forgotten, that this holds *only* when each, in its turn, is conjoined with appropriate auxiliary theories, "background knowledge," and assumptions about *initial conditions* some of which are unobservable. Indeed, the theories cannot account for *any* of the evidence unless such conjuncts are added. *But the same is also true of almost any "scientific" theory.* In general, the set of auxiliary theories and other conjuncts will vary, more or less, from theory to theory even when the evidence is the same. Now, in general, almost any theory can be made to yield almost any observation statement as a consequence if it is conjoined with appropriate auxiliary theories, etc., even if it is stipulated that the theory of interest must function nonvacuously in the derivation of the observation statement. Moreover, we know from Duhem as well as from subsequent developments that almost any (consistent) theory of much interest and importance can be saved from refutation by any (consistent) set of observation statements by appropriate conjuncts of auxiliary theories, etc. (This, indeed, follows from the previously cited principle.)

It might seem to follow from this that almost any observation statement is evidence for almost any theory. It *does* follow, I believe, that our notions of relevance (of what *is* evidence for what must be re-examined and "explicated." As a beginning, let us return to realism vs. phenomenalism. I have contended (Maxwell (1970a) and (1975)) that realism accounts for our experience (or accounts for the "observational evidence") in a simpler, less convoluted, more "natural," more plausible manner than does phenomenalism and that, therefore, the evidence favors realism over phenomenalism. I should now like to add some additional considerations in support of this contention. In the Bayesian confirmation procedure that I have advocated it is necessary to take into account the prior probability not only of the theory of interest (in this case realism – or phenomenalism) but also the probabilities of all of the auxiliary theories, assumptions about initial conditions, etc. This can be done in a number of equivalent ways, but for our purposes here it will be convenient to take the prior-probability factor of the numerator (what we, so far, have taken to express just the prior probability of the theory of interest) to be the probability of the conjunction of the theory of interest with all of the auxiliary theories, etc. Obviously, we would then estimate the probability of a complex, convoluted, unnatural, implausible conjunction to be lower than that of a simple,[14] natural, plausible one, even in cases where we would estimate the prior probabilities of the various theories of interest involved from case to case to be equal. I, therefore, calculate the degree of confirmation (the posterior probability), on the basis of the observational evidence, of realism to be higher than that of phenomenalism not only because of the higher prior probability that I estimate for realism itself but also for the higher probability that I estimate for (most of) the other members of the conjunction that includes it and that is used in deriving *(explaining)* the evidence.

These considerations, I maintain, are sufficient to establish the relevance of observational evidence for the hypothesis that realism is true. Very tentatively and with misgivings, I suggest that it may be possible to define, in a fruitful manner, a *degree* of relevance of a putative item of evidence as a function of the probability of the conjunction of all of the auxiliary theories, bits of background knowledge, initial conditions, etc., that are conjoined with the theory of interest in order to yield (entail, or imply statistically, as the case may be) this item of evidence (or its contradictory or one of its contraries).[15]

It follows from this, I believe, that deciding between realism and phenomenalism differs, if at all, only in degree from deciding between two theories that are generally acknowledged to be "scientific", such as Newtonian mechanics and relativity. The difference of degree would inhere in the somewhat greater degree of relevance (in the sense just suggested) of the available observational data for the latter case than for the former one.

Now it must be granted that, in order to contend plausibly that realism (vs. instrumentalism) is a (contingently) true theory, one must make a good case for the possibility of meaningfully formulating sentences that express assertions that are about or that *refer* to unobservable entities and also for the possibility of confirming or disconfirming such assertions. Again, I have devoted many pages to these tasks (see, e.g., Maxwell (1970b) for the former and Maxwell (1975) for the latter), and, this time, I really *will* be brief.

Russell has elucidated the details of how we can refer (in one important sense of 'refer') to individuals with which we are not acquainted. His theory of descriptions requires no explaining or defending here. Using existentially quantified variables and terms whose *direct* referents are observable, we can refer *indirectly* to items that we never have or, in some cases, never will observe. There is no reason why we cannot do the same with unobserv*ables* provided they have specifiable connections (usually causal ones) with observables. Using existentially quantified predicate variables we can refer to unobservable properties, properties of properties, sets, sets of sets, etc. One special result of this technique is, of course, the Ramsey sentence of a theory. This eliminates all of the endless vexations about the meanings of theoretical terms,[16] simply because no theoretical terms (no theoretical *words* – no theoretical *constants*) remain. But it by no means eliminates reference to theoretical entities (unobservables). The Ramsey sentence can and *should* be interpreted just as realistically as the original theory. By referring to unobservables *indirectly* by means of existentially quantified variables it "keeps us honest" by making explicit our ignorance of just *what* the unobservables *are* although we do know *somethings* about them such as their (causal) relations with each other and with observables. This explicates, I believe, Russell's ((1927) and (1948)) assertion that we have knowledge only of the *structural* properties [of unobservables] and are ignorant about their *intrinsic* properties.

We have already seen in some detail how *any* theory including theories about unobservables are to be confirmed – by means of their observational consequences and estimation of appropriate prior probabilities. (Incidentally, showing how knowledge claims about unobservables can be expressed and confirmed overcomes the two main traditional arguments against representative realism and views similar to it such as those of the later Russell – see Maxwell (1970a) and (1972).)

When we take seriously the results of contemporary physics, physiology, and psychophysiology (or neuropsychology) and conjoin them with the considerations that have been occupying us here, they have rather conclusive implications, I believe, for some of the traditional problems about perception, "knowledge of the external world," and the mind-body problem. (For detailed discussion see, e.g., Russell (1948) and Maxwell (1970a), (1970b), (1972), and (1974).) First of all, the scientific account of perception implies, I am convinced, that we should count as observables only the entities (the properties, individuals, etc.) that are exemplified in our private experience. (The scientific account, moreover, sanctions the use of word 'private' here.) If this is done and if we then "Ramsify" (replace with existentially quantified predicate variables of appropriate type) all of the other descriptive (nonlogical) terms in our knowledge claims, the result is an explication or, perhaps, an *admission* of Russell's claim that *all* of our knowledge of the "external" world is knowledge of structure only and that we are ignorant of *its* intrinsic nature. If this is true, then, *a fortiori*, we are ignorant of the intrinsic properties of our brains. This leaves open the possibility that these properties *just are* the qualities that we know in experience, indeed, that *constitute* our private experience. Not only are almost all of the common objections to an identity theory of mind-body thus removed, but other strong arguments in its favor follow (see Russell (1949) and Maxwell (1972) and [forthcoming]). Obviously, however, it is not the identity theory of the traditional materialist. The mental does not become any less mental by virtue of being identified with (one component or one set of constituents of) the brain.

The theoretical and systematic importance of the matters discussed in this brief essay is, I believe obvious. I apologize for the fact that most of the references are to my own work. My excuse is that they are the ones at my finger tips and with which I am most familiar. Readers who follow them up will find a more detailed account of – and a more adequate acknowledgment of indebtedness to – the work of others than can be given in this brief account.

Finally, I should like to remark that I think that these current trends in philosophy of science also have considerable practical import. Their implications for the actual practice of scientific inquiry are rather obvious (see Maxwell (1970a) and (1975) for brief discussions and for other references). One of them, I trust, is the desirability of close cooperation between scientists and philosophers. This is vital, for example, in psychophysiology (or neuropsychology) where scientists are working away with renewed vigor on the old mind-body problem (for example, note the work of the neurophysiologists Eccles, Pribram, Sperry and others in Globus et al. (forthcoming). There is also, I think, direct and important relevance for the problems facing society and its individuals in our current era of crises, especially from the developments in confirmation theory. For, if the policy decisions that are aimed at solving the problems that face us are to take into account the best of our current scientific knowledge (as well, of course, as the values that we want to perpetuate), surely the application of *decision theory* is imperative. In order to effect this we need to understand as fully as possible both the resources and the limitations of the best available confirmation theory for determining the *probabilities* that must be used in decision-theoretic calculations.

University of Minnesota

NOTES

[1] See Maxwell (1970a), (1972), (1974a), (1974b), and (1975).

[2] For example, one of Russell's many beautiful statements on the matter: "It used to be supposed by empiricists that the justification of [nondeductive] inference rests upon induction. Unfortunately, it can be proved that induction by simple enumeration, if conducted without regard to common sense, leads very much more often to error than to truth. And if a principle needs common sense before it can be safely used, it is not the sort of principle that can satisfy a logician. We must, therefore, look for a principle other than induction if we are to accept the broad outlines of science, and of common sense in so far as it is not refutable. This is a very large problem" (Russell, 1948).

[3] I have discussed this in detail in Maxwell (1975).

[4] I am *not*, of course, equating knowledge and "mere" true belief. However, the considerations at hand make it clear that we cannot possess knowledge in the sense of 'justified true belief' if 'justified' is to have the force desired by empiricists and others. If we are not willing to settle for hoped-for true belief, we shall have to liberalize our conception of *knowledge*. I have made tentative efforts towards this in Maxwell (1975).

[5] I am indebted to the work of Wesley Salmon (1965) for the central ideas in this notion of the probability of an hypothesis. I have discussed possibilities of overcoming the difficulties involved in specifying reference classes of hypotheses that are not hopelessly inadequate, as well as the possibility that we can make do without specifying, even crudely, a reference class but by merely entertaining the pious hope that a pretty good reference class exists and that we can use it and even refer to it (indirectly or *by description à la Russell*). See Maxwell (1974a) and (1975). (For those who prefer, the probabilities can be defined, of course, in terms of the propensities to realize such and such relative frequencies.)

[6] There is, of course, no guarantee that it will not operate, occasionally and, perhaps, finally, in the opposite direction. A little, or perhaps a lot, of knowledge may be a dangerous thing either for individuals or for the race.

⁷ These two estimates are sufficient in the hypothetico-*deductive* case, for, in such cases, the second factor of the numerator of the right-hand side of Bayes's formula (i.e., the "likelihood") is unity. In hypothetico-inferential cases when the inference from theory to evidence is statistical this, of course, does not hold. In such cases, however, the likelihood factor will, in general, be determined by the rules of statistical inference that have been adopted (either explicitly or tacitly). Such rules, in turn, I have argued (Maxwell, 1975), rest on *contingent but unconfirmed* assumptions about the distributions of relative frequencies from subset to subset of the general population. Thus, another contingent but unconfirmed (and in a sense, unempirical) postulate must be included in the foundations of confirmation theory and of general epistemology.

⁸ No allusion to Quine intended at *this* point, although his "other" dogma (now largely forgotten) is not entirely unrelated, as will soon be noted.

⁹ I am writing as if knowledge deriving directly from observation of the moment were entirely unproblematic. Of course it is not, and this only worsens the plight of empiricism.

¹⁰ See Maxwell (1960 and 1961).

¹¹ I apologize for this tortured sentence. It seemed, at the time, the most succinct way of stating what seems to be Quine's most plausible argument for rejecting the analytic-synthetic distinction.

¹² Other things being equal, it is nice to have *some* idea of what we are doing.

¹³ Strictly speaking, this is stated incorrectly. An "instrumentalistically interpreted" theory is not really interpreted, or, rather it is not completely interpreted and, therefore, is neither true nor false and, thus, cannot be confirmed or disconfirmed. Strictly speaking, the question we are considering is about which of two mutually incompatible, *contingent* metatheories is better confirmed by the observational evidence. One is the metatheory that the theories in question, interpreted realistically are true; the other is the metatheory that the theories in question are mere calculating devices (etc., etc.) which for some (strange, unexplained) reason grind out correct observational predictions, etc.

¹⁴ However, I have argued at length (Maxwell, 1975) that attempts to rescue empiricism by stipulating that "the simplest" theory be selected when other things are equal are doomed to failure.

¹⁵ Since there will always be many such conjunctions, we might stipulate that the one with the highest probability should be sought. For an excellent discussion of the importance of relevance considerations, see Salmon (1975).

¹⁶ The charge that use of the Ramsey sentence would entail that every time we change a theory, be it ever so slightly, the meanings of all the theoretical terms would change and that this is very naughty – this charge is mistaken. For the use of the Ramsey sentence *eliminates* theoretical terms. It is true that its use amounts to a "reconstruction" and does not faithfully reflect all of our ordinary usage of theoretical terms. But may it not be that some facets of such usage is wrong, even though it may not be wrong enough to cause much, if any, trouble in ordinary cases?

BIBLIOGRAPHY

Globus, G., Maxwell, G., and Savodnik, I. (eds.): forthcoming. *Consciousness and the Brain: A Scientific and Philosophical Inquiry*, Plenum Press, New York.

Maxwell, G.: 1960, 'The Necessary and the Contingent', in H. Feigl and G. Maxwell (eds.), *Minnesota Studies in the Philosophy of Science*, Vol. III, University of Minnesota Press, Minneapolis.

Maxwell, G.: 1961, 'Meaning Postulates in Scientific Theories', in H. Feigl and G. Maxwell,

Current Issues in Philosophy of Science, Holt, Rinehart, and Winston, New York.

Maxwell, G.: 1970a, 'Theories, Perception, and Structural Realism', in R. Colodny (ed.), *Pittsburgh Studies in the Philosophy of Science*, Vol. IV, University of Pittsburgh Press, Pittsburgh.

Maxwell, G.: 1970b, 'Structural Realism and the Meaning of Theoretical Terms', in M. Radner and S. Winokur (eds.), *Minnesota Studies in the Philosophy of Science*, Vol. IV, University of Minnesota Press, Minneapolis.

Maxwell, G.: 1972, 'Russell on Perception: A Study in Philosophical Method', in D. Pears (ed.), *Bertrand Russell: A Collection of Critical Essays*, Doubleday, New York.

Maxwell, G.: 1974a, 'Corroboration without Demarcation', in P. S. Schilpp (ed.), *The Philosophy of Karl Popper*, Open Court Publishing Co., LaSalle, Illinois.

Maxwell, G.: 1974b, 'The Later Russell: Philosophical Revolutionary', in G. Nahknikian (ed.), *Russell's Philosophy*, Duckworth, London.

Maxwell, G.: 1975, 'Induction and Empiricism: A Bayesian-Frequentist Alternative', in G. Maxwell and R. M. Anderson, Jr. (eds.), *Minnesota Studies in the Philosophy of Science*, Vol. VI, University of Minnesota Press, Minneapolis.

Maxwell, G. and Anderson, Jr., R. M. (eds.): 1975, *Minnesota Studies in the Philosophy of Science*, Vol. VI: *Induction, Probability, and Confirmation*, University of Minnesota Press, Minneapolis.

Maxwell, G.: (forthcoming), 'The Relevance of Scientific Results for the Mind-Brain Problem', in G. Globus, G. Maxwell, and I. Savodnik (eds.), *Consciousness and the Brain: A Scientific and Philosophical Inquiry*, Plenum Press, New York.

Putnam, H.: 1960, 'The Analytic and the Synthetic', in H. Feigl and G. Maxwell (eds.), *Minnesota Studies in the Philosophy of Science*, Vol. III, University of Minnesota Press, Minneapolis.

Quine, W. V. O.: 1953, 'Two Dogmas of Empiricism', in *From A Logical Point of View*, Harvard University Press, Cambridge, Mass.

Russell, B.: 1927, *The Analysis of Matter*, Allen and Unwin, London.

Russell, B.: 1948, *Human Knowledge: Its Scope and Limits*, Simon and Schuster, New York.

Salmon, W. C.: 1965, *The Foundations of Scientific Inference*, University of Pittsburgh Press, Pittsburgh.

Salmon, W. C.: 1975, 'Confirmation and Relevance', in G. Maxwell and R. M. Anderson, Jr. (eds.), *Minnesota Studies in the Philosophy of Science*, Vol. VI, University of Minnesota Press, Minneapolis.

ERNAN MCMULLIN

HISTORY AND PHILOSOPHY OF SCIENCE: A MARRIAGE OF CONVENIENCE?

In a recent article, Ronald Giere has argued that the currently fashionable union of history and philosophy of science is no more than a marriage of convenience, justified like so many youthful marriages by unhappiness with the parental homes – history and philosophy – rather than by compatibility and mutual need. He allows that it *is*, for the moment at least, a convenience from the institutional standpoint to encourage the two to pair off together, but insists that this pairing has no "strong conceptual rationale."[1] In a session devoted to analyzing the directions in which philosophy of science is developing, this topic would seem to merit an important place, for it can hardly be gainsaid that one of the two most obvious shifts in the philosophy of science over the past fifteen years has been its growing involvement with the history of science; the other shift is, of course, the shift away from the empiricist orthodoxies inherited from the Vienna Circle. And the two are not unconnected. If the former ought (as Dr. Giere suggests) be construed as a growing sensitivity to the realities of actual scientific procedure (whether contemporary or in the past) rather than a call upon the history of science as such, the consequences for the directing of our future efforts would be quite different from what many historicist philosophers of science have been urging upon us.

I want to argue, as I have done before,[2] that the relation between the two fields, is an intimate and a complex one, one that must inevitably affect the future development of both. This is not to say that all (or even most) problems in the philosophy of science require explicit reference to the history of science for their solution; nor is it to say that history of science and philosophy of science in the university context are best carried on in a single department or program. Philosophy of science is, first and foremost, part of philosophy. And what it is studying is science. It sounds as though I am agreeing with Giere's assertion that "the primary relationships for philosophy of science are with philosophy and science,"[3] but I can accept this only if "science" be enlarged beyond the consideration of present theories and practice (i.e. what would be taught in a "science" course) to include a knowledge of their historical roots as well. To put this even more concretely, my position is that some quite central issues in philosophy of science could not be properly handled solely on the basis of the sort of knowledge that a good science course would give. Among these issues, I would judge the principal ones to be: (1) theory-assessment; (2) scientific growth; (3) the ontology of theoretical entities.

In this paper, I will restrict myself, since time is short, to the first of these.[4] Underlying all three of them, of course, is the question of scientific rationality and how it is to be determined. If it can be shown that an adequate treatment even of the most "logical" of the three questions, i.e. that of theory-assessment, cannot avoid reference to the history of science, it seems safe to conclude that the afore-mentioned marriage of the history and philosophy of science is not just one of convenience. My strategy will be to examine the different strands in theory-assessment and show why it is that the history of science has assumed an increasing importance in their determination.

I. LOGICIST VERSUS HISTORICIST

But first, let me try to separate two quite different sorts of consideration that might be brought in support of a theory of science. It might rest on a series of logical intuitions in regard to the epistemic terms used, terms like 'confirm', 'explain', 'falsify', and so forth. Thus for example, someone might say that it is intuitively evident that if E "confirms" H (as the term 'confirms' is normally understood), E also "confirms" H' where H' is logically equivalent to H. It seems reasonable, they will add, that if a piece of evidence confirms an hypothesis, the same evidence will confirm any other hypothesis which is, to all intents and purposes, identical with the first. Notice the pattern here. One appeals to an intuition based on a shared language: "anyone who knows what 'confirms' means would say that..." The only experience involved is the unspecific experience required to allow one to come to use the term, 'confirms', correctly. This need not be an experience with the technicalities of scientific method, since the term has a much wider scope. To construct a "logic of confirmation", one has to articulate a set of intuitive principles of this kind and weave them into a single deductive system, paying close attention to the choice of logical connectives to represent such apparently innocuous terms as 'equivalent' and 'implies'. The "logicist" approach (as it is often called) thus relies on our ability to articulate with certainty some very general principles governing what would count for us as, for example, confirmation. It does not require any references to instances, any induction over (say) various types of confirming case. It is assumed that these would bear out theorems derived from the intuitive first principles; if they did not, we would simply conclude that they were being incorrectly applied to the instances in question, not that the principles were wrong.

Contrasted with this is the descriptive approach to the theory of science: one begins from a survey of the actual practice of the scientist, the ways in which he tests his theories, for instance, or the degree of importance he attaches to crucial experiments. One seeks to formulate "laws", just as one does in inductive inference at first level, except that these will now be laws of method rather than laws of nature. There is, of course, one crucial difference: scientists can make mistakes, so that some instances of method may have to be discounted. Practice cannot therefore, be taken as *altogether* determinative, an immediate problem for the descriptivist program. One may be content to scrutinize contemporary practice to test one's generalizations. Or one may turn to the history of science, either as a richer source of the sort of information the contemporary scene still could in principle reveal, or perhaps as a source of information that a limited epoch (whether the present or any other moment) could never yield. This last is the "historicist" emphasis; not only must we look to the practice of science, but we have to look to it in the detail of its historical manifestation. Historicism and logicism are clearly at the opposite poles in epistemological inquiry such as ours: the historicist will be content with an account which is tentative, descriptive, subject to change; whereas the logicist will hope to enunciate principles that are exact, normative and immutable. The historicist's work is empirical; it is controlled by what he finds in the records. The logicist can lay down norms in what almost seems an *a priori* way; I say "almost" because the logicist's explication *does* depend upon experience, albeit the very general experience that may be required in order to allow him to use terms like 'confirm' and 'falsify' with assurance and exactness.

I have delayed on this distinction, familiar though it is, in order to make sure that my use later on of the terms 'logicist' and 'historicist' will be unambiguous. They refer to two different, and apparently complementary, ways of warranting claims about scientific method. When they are taken to extremes, that is when the all-sufficiency of the intuitive explication or of the historical review are over-emphasized, I will speak of "logicism" or "historicism". It is clear that the logicist must make sure that his intuitions are anchored in the experience which the term he is explicating is intended to articulate; the more complex that experience is, the more he will have to rely on the help of descriptivist techniques. It is also clear that the historicist must accept the general principles of logic, and must admit too that scientists make mistakes of method on occasion in their work. "Pure" logicism or "pure" historicism in the context of the theory of science, are thus plainly untenable and of no practical concern. What we *are* interested in is the spread of the spectrum in between.

II. THE ASSESSMENT OF SCIENTIFIC THEORY

How are scientific theories validated? How are the relative worths of two competing theories to be assessed? When scientists say of a theory that it is "confirmed" by all the available evidence, what does this amount to? There have been a variety of answers to these questions. Let me recall the three most obviously "logicist" types of response that have been given, each based on a perfectly sound logical intuition but one which in each case is manifestly far too simple to serve as sole guide in the search for a theory of theory-assessment.

1. *Deductivist.* In the classic Greek account, the warrant of a scientific statement consisted in its deductive derivability from first principles, themselves intuitive in nature and not strictly part of the science. This is the most secure of all warrants, the simplest of all modes of assessment. One finds it still glimmering like a will o' the wisp in the seventeenth century, beckoning scientists like Descartes on to fresh but unavailing efforts. The ideal it held up was unattainable, but it took centuries of unsuccessful attempts to discover the intuitive first principles, on which the entire program depended, to convince philosophers – and scientists – of this. Aristotle's analysis of an axiomatic science was not at fault; what proved wrong was his (empirical) assumption that human intuition was sufficiently powerful and nature sufficiently transparent to allow the discovery of the principles his science needed.

2. *Inductivist.* The second attempt fared a little better. Since science is the most general and best-warranted knowledge of nature, it clearly has to work from the singulars of experience to universals or laws, by means of a process of induction or generalization. From a finite set of points we get a smooth curve linking two variables in a lawlike manner. Later experiment may show a fine-structure in the curve, or may extend the curve in either direction, but within its original limits of accuracy and specification of variables, the law is a permanent acquisition, and its assessment a straightforward affair. Or so it was thought.

The main problem, of course, with this is that the element of *theory* is overlooked. If science really *did* consist of empirical laws, making use of unproblematic operationally-defined concepts, the inductivist account might suffice, although it would still have to face Hume's challenge to the logicist attempt to infer from the particular to the general. But what of the explanatory dimension of science? its hypothetical structures? its theoretical concepts? Some other mode of validation must be found for these. Once again, a single-strand logicist theory of assessment has failed.

3. *Falsificationist.* Another attractive possibility suggests itself when we reflect on the logical relationship of hypothesis and evidence. Because of its finite character, the evidence can never be sufficiently strong to prove the hypothesis. But surely it can *disprove* it? And if disproof operates in a definitive way then science can progress by successive attempted refutations and theory can be assessed in terms of its potential refutability and survival after successive severe tests. Crucial experiment then becomes the central feature of scientific method.

Of course, we know that it is nothing of the sort. But even without consulting the historian, it is possible to see why falsificationism cannot work as a general account of theory-assessment. One reason, as we all know, is that theories are not single indivisible entities; they are networks of assumptions, postulates, generalizations, many of which could be separately modified while still leaving the main theory intact. Theory refutation is not the simple *modus tollens* of the logician, then; it involves decisions about whether the modifications that would save the theory are *ad hoc* or not, whether the rival theory does not do better in setting up verified predictions, and so forth.

Furthermore, the proposed falsifying fact is never in practice quite as "hard", as unproblematic, as the logicist model demands. One has to take seriously the possibility that it too may have to be modified or even abandoned. But these modifications take us far away from the original insight of falsificationism;[5] to retain the old label, as Lakatos does, may indeed be filial but it runs the risk of making his program look an "epicyclic" one (as a recent critic has charged) where one complication is piled upon another in an effort to maintain an older view.[6] Or to use his own language, does this steadfast adherence to the falsificationist label indicate that we have here a "degenerating problem-shift"? I think not, but the way in which his methodology of research programs was originally proposed might easily lead one to suppose so.

Here then are three "pure" strategies, the deductivist, the inductivist and the falsificationist. To show that they are of themselves separately inadequate for the assessment of theories may seem a simple matter today, since it requires only the smallest familiarity with contemporary science, but it is worth recalling how hard-earned an insight it was in the seventeenth century. Indeed, one still sometimes finds scientists proposing with undiminished enthusiasm one or other of these strategies as an attractively simple account of what they are doing and what all other scientists ought to be doing.[7] And there is no gainsaying that scientists *do* make use of crucial experiments on occasion, and *do* accumulate thousands of detailed empirical laws. There is nothing wrong with the inductivist or the falsificationist accounts, provided that they are seen as part of a wider and much more complex scheme.

There are two other elements in this scheme that have not been mentioned so far, the use of criteria *internal* to a theory to assess its worth, and the so-called HD mode of validation. Each can be considered as a general strategy of assessment in its own right; each has, in fact, been proposed at some time or another as the main strategy for use in the assessment of scientific theories.

4. *Coherentist.* One might look at the *intrinsic* properties of the theory, such properties as coherence, elegance, simplicity, convenience. Scientists are influenced by such considerations. But why *should* they be? One can see why an inductivist or a falsificationist strategy might lead one to the truth about the world. But why should a "coherentist" one, if I may use the term, do so? Is there any reason why a simple theory is more likely

to be correct than a more complex one? Can one use words like 'correct' in the context of this sort of strategy at all? It is this sort of question that leads this sort of strategy to be called a "conventionalist" one, though, be it noted, it need not be so. If one can give reasons why a particular intrinsic feature of a theory is likely to be an index of its adequacy as an account of the real world, then one is not a conventionalist. If of two competing theories, one is more coherent than the other, in some definable sense of that term, one might well argue that the more coherent is to be preferred as more likely to prove out in the long run.

Conventionalism is not, then, a strategy of assessment, strictly speaking. Rather, it is an evaluation of all the available strategies and an assertion that in the end a partially arbitrary choice will have to be made. The conventionalist is not saying that the adoption of a scientific theory is *entirely* arbitrary; he recognizes the constraints of evidence. But he will argue that the decision as to which part of a theory a damaging fact ought lead us to modify, or which curve among the infinitely many available to us one ought to choose to represent an inductively-based law, will ultimately have to rest on some criterion internal to the theory itself, like simplicity or convenience. And this criterion cannot in turn be justified in truth-terms; it is no more than a convention, arbitrarily chosen only because *some* convention is needed in order that a theory may be formulated and defended.

I shall come back briefly to conventionalism later; as a sceptical attitude in regard to the effectiveness of the available strategies of theory-assessment, it can be overcome only by showing that these strategies are in fact able to accomplish what they purport to do. As a theory of theory-truth, it is allied with instrumentalism and positivism, and opposed to realism. I have introduced it here, let me repeat again, only because it so often crops up, in a rather confusing way, in the discussion of internal criteria of theory-assessment. A conventionalist will always fall back on such criteria, interpreted as conventions. But the use of such criteria by a scientist or theorist of science in no way implies a commitment to conventionalism.

Can these criteria be illuminated by the history of science? Yes, in several ways. First, notions like simplicity or elegance are notoriously difficult to analyze; logicist attempts to explicate them have met with very little success. To get a grip on them, there seems to be no other resource than case-studies drawn not just from the present but from the past of science; the widest possible variety of contexts would have to be scrutinized to discover, what, for instance, the criterion of "simplicity" has amounted to in the work of men like Newton or Darwin or Einstein.

There are further questions that could also be asked. In what sorts of situation did scientists fall back on this criterion? To what extent was it a successful one? This last is the historicist question, par excellence. That it has not been particularly emphasized either by historians or philosophers suggests that neither party feel that the answer is likely to be affirmative. Or perhaps, the general lack of historical interest in internal criteria of theory-choice may derive from the difficulty in determining just what internal-criteria *did* affect a particular scientific research program; it will not be enough to discover the scientist using words like 'simplicity' or 'beauty'. And there will be a suspicion that these criteria are in any event secondary to those involved in successful prediction, and are justified only insofar as they reduce to the latter. So to these we now finally turn, having first investigated the deductivist, the inductivist, the falsificationist, and the coherentist strategies of assessment.

5. *Hypothetico-deductivist.* If what we are assessing are hypotheses, surely the best way to do so is to see how successful we are in deriving verified predictions from them. This advice is common in scientific literature from the time of Descartes onwards. And recent attempts to construct a logic of confirmation have taken it as their starting-point. But it is not as simple as it seems. I would like to focus on one particular difficulty, one that has come in for a lot of attention of late.[8] It amounts to this, that deduction and prediction are *not* equivalent. Successful prediction is a deduction which when tested observationally proves to be correct. But the notion of *test* here demands that the result not be known in advance. There is a world of difference, from the point of view of confirmation, between a theory which simply accounts for all the data in the light of which it was originally formulated, and one which predicts novel results, which when tested prove to be the case. We are much more impressed by the latter than the former. Yet from the purely logical point of view, there is no way of distinguishing them: in both cases the data are deduced from the theory, and that is all there is to be said. In his theory of confirmation, Carnap asked: to what degree is the hypothesis, H, confirmed by

the evidence, *e*, but he made no distinction between "evidence" known in advance (i.e. the data the theory was intended to explain) and verified predictions of novel facts, which clearly serve as "evidence" in a quite different and much stronger sense. The logical relation of deducibility cannot of itself convey this difference; a temporal analysis of some kind is needed.

III. THEORY-ASSESSMENT AND THE HISTORY OF SCIENCE

Is this where history of science makes its appearance?[9] Yes and no. Once again, no particular expertise in history of science is needed to make the general point I have just made about the manner in which the verification of a prediction of something new differs from the mere deduction of something already utilized in the construction of the theory. The fact that a *temporal* relation now enters in does not, of itself, mean that history of science need be called on in some special way. Part of the problem lies with the notorious double meaning of the word, 'history'. The history of science is a body of information about the past of science; it is also that past itself. The reference to before and after which we have seen to be essential to an understanding of confirmation forces us to take confirmation to be a temporal process, one where we have to wait for the results to come in. It is thus a *historical* process too, though ordinarily not one which is spread out over a long period. But it may be sufficiently simple to allow us to formulate some more or less intuitive principles governing it; in that event, even though it is *historical*, we would not need to have recourse to the *history of science* in order to formalize it.

Would the formalization constitute a formal logic in such a case? It would be a formal system but its principles would be of a more material sort than those of deductive logic. That is, they would be limited to a very special sort of epistemic context; they would have a lesser degree of assurance than the ordinary rules of logic in the sense that it might not be quite certain how well they actually applied to the empirical process we call theory-assessment, or confirmation, in empirical science. The fact that the relation of confirmation has a temporal dimension is, of itself, no barrier to formalization or to logicality (unless one restricts the term 'logicality' to general deductive logic). Logicality and temporality are not incompatible, as Giere quite rightly points out.[10]

However, the issue is perhaps, not quite as open and shut as it may seem. After all, Toulmin rests much of the argument of his book, *Human Understanding*, on the claim that the temporal dimension of science (which he takes to define the rationality peculiar to science) cannot be analyzed in terms of formal structures at all; one must fall back on a sort of ecological and developmental analysis. One may well ask whether this analysis will not involve some formal structures akin to a logic; the selection among alternative conceptual-changes will certainly require a systematizable notion of assessment. One has to be careful not to fall into a category error as old as Plato and assume that formal structures cannot in principle be applied exactly to changing realities.

But suppose (as Toulmin appears to be assuming) the formal structures *themselves* are changing, i.e. suppose that the notion of theory-assessment itself has changed between Descartes' time and ours, what then? Presumably, one would no longer be able to formulate an intuitive set of principles governing such assessment, and one would have to rely not only on a descriptive account of how assessments are actually made in contemporary science but also on at least some historical detail regarding the kinds of change that have occurred in scientific modes of hypothesis-assessment and why these *did* occur. To the logicist such a suggestion is, of course, anathema. His view is that there is an unchanging set of epistemological principles underlying proof in science, as elsewhere; it is possible that these principles are not well-understood at a given time and are incorrectly applied on occasion. But insofar as they are accepted *as* principles, it is not (in his view) because they have been successfully used in the past, but because they carry intuitive weight in their own right, once they are properly understood. One is right back to Aristotle, and the theory of science of the *Posterior Analytics*. And the issue is clearly joined between logicist and historicist; to the one, the testimony of history is irrelevant, to the other, it is indispensable. If the issue be taken at *this* level, i.e. at the level of changing theories of science and not just at the level of changes in science itself, then Toulmin may have a point in his opposing of logicality and temporality. But this is not because of any inherent tension between them, only that *his* theory of science is going to come perforce from an empirical inspection of history, not from an intuitively-warranted logical system. The contrast thus reduces to the accepted one between historicism and logicism as two basically different modes of justifying theories of science.

But have we any reason to believe that notions of HD-validation *have* altered? And even if they have, might it not be that they have gradually approximated to a norm that can be more directly revealed by careful analysis of intuitive principles? Do we really want to say that our theory of theory-assessment rests on the contingency of the historical record? Let me leave these questions aside for the moment to focus on a more tractable one.

Is our intuition, informed presumably by some rather general experience with the validation of hypotheses, able to come up with a set of principles that everyone could agree on? The main problem is going to be with the word 'novel'; what constitutes a *novel* fact and how much more should its prediction count than the assimilation of previously-known facts? A novel fact is not necessarily one in the future; it is one unknown to the person formulating the theory. But suppose it is already known to others, though not to him? Or suppose he did know about it, but did not take it into account? And surely what we need is a novel *kind* of fact, not just a novel fact? Musgrave formulates a whole series of puzzles of this sort that forcefully bring home the fallibility of our intuitions in this complex area of theory-assessment. Of course, one can easily set up a postulational system of plausible principles of evidence. But will it apply to what actually goes on in science? And if it does not, which is to yield? Are our intuitions, or even the intuitions of the skilled scientist, sufficiently articulable in this domain to allow us to set forth a system that in no way needs to rely on the testimony of actual practice, contemporary or historical? I think not. One has only to see the wide divergences in the published accounts of method over the last century, ranging from Mill and Keynes, who see no virtue whatever in the claim that the prediction of a novel fact should count for more than the deduction of a fact already known on the one hand, to Popper and Lakatos at the other extreme who would hold that a theory that can claim no novel facts to its credit has no sort of confirmation (corroboration) at all.[11]

My conclusion is now clear: logicism will not work unaided in theories of HD-assessment. Our intuitions are simply not secure enough in this domain. They have to be instructed, and supported, and challenged by the testimony of history. It will not be enough to draw on contemporary practice; a more diverse context is needed, because theories can be of widely-different sorts and the ways in which they can come to be accepted are quite various. This does not mean that the philosopher will have to be making constant explicit reference to history of science in his work of formal explication. Only when a question arises about the articulation of a particular axiom or principle might such a reference be needed. To the extent that clear principles can be isolated, the work of formal construction can then go on much as the rational mechanics of the eighteenth century went on, without much reference to the messy empirical world of practice.

IV. THE UNIT TO BE ASSESSED

So far I have assumed that the main type of appraisal that goes on in science is of *theories*; that it is in terms of theory-change that science grows; that it is in the assessment of rival theories that scientific rationality most clearly manifests itself. But there have been other proposals as to what the unit of assessment should be. One, the inductivist claim that the unit should be an empirical law of nature, we have already rejected. Another more recent one, that of Toulmin, is that variation and assessment should be considered at the level of the individual concept. He needs this assumption for his evolutionary model of scientific change to work; he rejects systematic for what he calls "populational" analysis, where the "populations" are historically-developing populations of concepts in which "mutations" occur, the more advantageous of which are selected for survival.[12] The main problem with this metaphor is that it makes each concept an individual competing entity; in fact, a concept would ordinarily only "compete" with those concepts proposed to fill roles similar to its own. More seriously, it assumes that conceptual variations are more or less independent of one another and can be assessed separately. But this is assuredly not the case. Scientists do not evaluate concepts in isolation from one another; they weigh up models or theories involving interconnected sets of concepts. Furthermore, when one of these concepts is altered, it frequently involves a shift across the entire network.

In short, then, Toulmin's unit is much too narrow. Though we can and should study individual concept-changes in science, it is dangerous to isolate these from the wider theoretical context in which they occur. Assessment and change proceed alike on a broader basis than the populational metaphor allows.

At the other end of the scale of generality is Kuhn's "paradigm", so broad and enduring that when it changes, we can talk of revolution. The ambiguity of this concept is well-known, and Kuhn himself (among others) has done something to remedy it.[13] But clearly the work of assessment in science goes on at a finer level than that of paradigm; indeed, the message most people took from the first edition of his *Structure of Scientific Revolutions* was that the change from one paradigm to another is not the work of assessment at all but rather more akin in its immediacy and inarticulable quality to conversion. There are some few things one can say about pressures generated by anomalies, about demands for coherence, and so forth, but, on the whole, paradigm-shift is presented as almost opaque to the methodologist's scrutiny. I am not concerned here with the merits of Kuhn's thesis, nor with the various modifications of it that have been proposed, I merely want to remark that the units to be considered in a theory of assessment in science are clearly not paradigms.

What, then, of the unit we used above, i.e. *theory*? Will not this suffice? It would, except for one thing, and this brings us to one last strand in the tangle of assessment strategies, a sixth if you have been keeping count. What counts, perhaps, most of all in favor of a theory is not just its success in prediction but what might be called its *resilience*, its ability to meet anomaly in a creative and fruitful way. This is *not* a matter of prediction, let it be stressed. It is, rather, a quality of metaphor in the theory which suggests to the scientist how its conceptual structures can be further developed to derive new results or to meet new challenges. Obviously, this is something which manifests itself only gradually over the course of time; one cannot attest to it until the theory has survived many tests and been extended in illuminating new ways. If one looks at the best-established theories of science, the kinetic theory of gases for instance, or the nuclear theory of the atom, one immediately realizes that the confidence we place in them results not merely from their successful predictions of novel facts, but at least as much from their behavior as lead-metophors in the process of conceptual and model change over a considerable period.

Thus, if this last type of assessment strategy be accepted, our unit cannot simply be the theory, it has to be the theory as traced over the creative and competitive phase of its career. This is close to what Lakatos means by a "research program", and it is his merit to have stressed the all-important sort of theory-assessment which makes necessary the extension of our scrutiny from the theory considered as a conceptual structure at a moment of time to the theory considered in terms of its entire historical career. I do not think the term 'research program' quite captures this, though it comes closer than any other proposed so far.

My quarrel with Lakatos' theory of assessment is not over terms, however; it is over the unfortunate falsificationist residues still found in it, which impair its plausibility in numerous ways. He assumes, for instance, that the typical case of assessment occurs when two theories are in competition with one another. But what of the far commoner case when a single theory is in possession: how is *it* to be assessed? If it is not under test, must it be dismissed as *ad hoc*, as he sometimes seems to suggest? What of the well-established theories that no longer produce novel facts, that have been completely explored? Are they now in a "degenerating" stage? So great is the prejudice against inductivism of any sort, against positive confirmation, against the notion of an *established* theory, that the "research program" is still viewed rather more in terms of the rough and tumble of the arena than the creative and convincing development of a single unified metaphor through a historical series of challenges, most of them coming from anomaly or unexplained fact rather than from the pressure of competing full-fledged theories.

In particular, it leads Lakatos to an extraordinary procedure that he terms the "rational reconstruction" of history. This is one way of using the history of science I do *not* recommend.[14] He takes "scientific research programs" of the past and tells us how they *ought* to have been conducted. What results from this technique is not history, nor is it the use of history to warrant a theory of science. In fact, it has very little to do with history, which is being exploited only as a source of problems which can serve to illustrate how his theory of assessment works.

There is a serious ambiguity about this procedure, for the reader who works through the wealth of apparent historical detail in a Lakatos scenario might easily suppose that this detail somehow supports the epistemology he is proposing. And indeed, it often does. But Lakatos cannot acknowledge this without the spectre of justificationism threatening him. If a methodologist can learn from the experience of history, then presumably a scientist can too; and Lakatos remains enough of a Popperian to regard this inference with a shudder. The entire thrust of Lakatos' acute analysis ought lead him to take history more seriously, and embrace at least a quasi-historicist view. But historicism for him is just one more form of inductivism, so that as well as the usual logicist suspicion of appeals to history, he has the additional burden of a falsificationist faith to live with.

V. CONCLUSION

The focus of this paper has been a dual one: the nature of theory-assessment in science, and the degree to which we are dependent on the history of science in such an investigation. Let me now summarize my conclusions. The individual strategies of theory-assessment are so complex that the logicist technique of isolating intuitively acceptable principles frequently fails. Disagreement is common, and there is no alternative to invoking the historical record. What is even more serious is that, in practice, the various strategies discussed above have to be combined with one another, and the ensuing problems of what relative weight to give to each are almost intractable, not only to the logicist epistemologist but quite frequently to the working scientist as well. How should one compare a gain in coherence to a loss in predictive power, or the elimination of an anomaly with a successful prediction of some novel facts? There are no hard-and-fast answers here; one can appeal to the intuitive skills of the working scientist or to the testimony of history but it should not be thought that a completely consistent formalism is going to emerge. In that respect, the historical record is likely to retain its importance. No one bothers to chronicle the times when *Modus Ponens* has worked; it would be a rather dull story. But historians spend prodigious energies in determining the occasions when new theories have arisen in science, and the detail of how they were appraised, confirmed, modified, rejected. Their efforts are not likely to lose their point, if what I have been arguing here is correct.

It is worth comparing first-level and second-level options here. At the first level, which is that of science itself, there have been three broadly different views as to how scientific claims about the world are warranted: they are the intuitive-deductive view sometimes called rationalism, the inductive view and the conventionalist view. Likewise, on the second level when we ask how theories of science themselves are warranted, there are three main alternatives: logicism, historicism, and conventionalism. The balance between them has, however, shifted since intuition obviously works much better in the determination of principles of method than it does in discovering the natures of physical things; logicism is thus much more powerful in the metatheory of science than is rationalism in science.

Despite the efforts of Duhem, Poincaré and many other talented people, conventionalism ultimately fails to convince, both at the first level and especially at the second level of the scientific enterprise. It simply does not account for the history of science; it fails to make sense of it. There are really only two contenders at the second level, therefore, logicism and historicism. Neither one (if we are right) can claim to be all-sufficient. Where a valid logicist explication is available, it is preferred by the philosopher because of its intuitive and (hopefully) coercive character. But where intuitions diverge, there is no option but to turn to the historical record, not only to discover what scientists in fact *have* done, but also to sharpen the philosopher's own intuitions which (let it be said again) are formed not in a vacuum but in the context of very definite experiences of the world and of ways of dealing with the world.

The dialectic between logicist and historicist modes is of especial importance in determining whether and to what degree, the theory of assessment proposed ought be taken as normative. A logicist theory is ordinarily normative, but to the extent that history has served specifically as warrant, the claim may have to be softened. It is perhaps here above all that the issue between logicist and historicist is joined. If the case I have presented is correct, the epistemology of science is likely to continue to require the best efforts of both.

NOTES

[1] 'History and Philosophy of Science: Intimate Relationship or Marriage of Convenience', *British Journal for the Philosophy of Science* **24** (1973), 282–297; see p. 296.

[2] 'The History and Philosophy of Science: A Taxonomy', *Minnesota Studies in the Philosophy of Science*, Vol. 4 (ed. by R. Stuewer), pp. 12–67. Giere's essay is a detailed review of this volume, and specifically of this article.

[3] *Loc. cit.*

[4] I have touched briefly on (3) in 'History and Philosophy of Science: A Taxonomy', pp. 63–67.

⁵ In a well-known exegesis of the Popperian canon, Lakatos distinguished between three different sorts of falsificationism, "dogmatic" (the straightforward logicist model defined above), "naive methodological" (a strongly conventionalist view), and "sophisticated methodological" (his own view, and one which has justificationist elements). See 'The Methodology of Scientific Research Programmes', *Criticism and the Growth of Knowledge* (ed. by I. Lakatos and A. Musgrave), Cambridge, 1970, pp. 91–195.

⁶ Errol Harris, 'Epicyclic Popperism', *BJPS* **23** (1972), 55–67.

⁷ See, for instance, J. R. Platt, 'Strong Inference: The New Baconians', *Science* **146** (1964), 347–353; P. Medawar, *The Art of the Soluble*, London 1967.

⁸ See A Musgrave, 'Logical versus Historical Theories of Confirmation', *BJPS* **25** (1974), 1–23; E. G. Zahar, 'Why did Einstein's Programme Supersede Lorentz's?', *BJPS* **24** (1973), 95–123; 223–262.

⁹ By calling theories of confirmation that take this temporal distinction into account "historical," Musgrave (*op. cit.*) would seem to be suggesting this, but a closer look at his well-reasoned essay shows that the principles of confirmation he discusses do not depend especially heavily on a review of historical instances of confirmation. It would have been better for him to retain his first label, 'logico-historical', because his analysis does not come down more on one side than the other.

¹⁰ One section of my paper: 'Logicality and Rationality: a Comment on Toulmin's Theory of Science' (*Boston Studies* **XI** (1974), pp. 415–430) is devoted to this issue.

¹¹ Musgrave, *op. cit.*

¹² Toulmin makes the concept correspond to the individual organism (the entity in which mutations may occur and which competes with other broadly similar entities for survival). He also tends to take a "discipline" ("characterized by its own body of concepts, methods, and fundamental aims") to correspond to a species: "if intellectual disciplines comprise historically-developing populations of concepts, as organic species do of organisms, we may then consider how the interplay of innovative and selective factors maintains their characteristic unity and continuity." (*Human Understanding*, Princeton, 1972, pp. 139–141). Competition occurs both between species, and between individual members of species, thus presumably between disciplines as a whole and between individual concepts in each discipline.

¹³ In the *Postscript* to the second edition of the *Structure of Scientific Revolutions*, Chicago, 1970.

¹⁴ See §5 'History of science and some philosophers' of my 'History and Philosophy of Science: A Taxonomy.'

HILARY PUTNAM

PHILOSOPHY OF LANGUAGE AND PHILOSOPHY OF SCIENCE*

For over a hundred years, one of the dominant tendencies in the philosophy of science has been verificationism: that is, the doctrine that to know the meaning of a scientific proposition (or of any proposition, according to most verificationists) is to know what would be evidence for that proposition. Historically, verificationism has been closely connected with positivism: that is, at least originally, the view that all that science really does is to describe regularities in human experience. Taken together, these two views seem close to idealism. However, many twentieth century verificationists have wanted to replace the reference to experience in the older formulations of these doctrines with a reference to "observable things" and "observable properties". According to this more recent view, scientific statements about the color of flowers or the eating habits of bears are to be taken at face value as referring to flowers and bears; but scientific statements about such "unobservables" as electrons are not to be taken as referring to electrons, but rather as referring to meter readings and the observable results of cloud chamber experiments. It is not surprising that philosophers who took this tack found themselves in a certain degree of sympathy with psychological behaviorism. Just as they wanted to "reduce" statements about such unobservables as electrons to statements about "public observables" such as meter readings, so they wanted to reduce statements about phenomena which, whatever their private status, were publicly unobservable, such as a person's sensations or emotions, to statements about such public observables as bodily behaviors.

At this point, they found themselves in a certain bind. On the one hand, the doctrine that talk about sensations or emotions is simply talk about a person's behavior is so implausible that almost no philosopher has been able to maintain it, or at least to maintain it for long. On the other hand, if the intuition behind recent verificationism is right, and to know the meaning of a statement is to know what would be *public* evidence for it, then it seems as if there has to be something right about behaviorism.

R. S. Cohen et al. (eds.), PSA 1974, 603–610. All Rights Reserved.
Copyright © 1976 by D. Reidel Publishing Company, Dordrecht-Holland.

And so philosophers tried to develop a philosophy to this effect – a philosophy that would say that "naive behaviorism" was false but that nevertheless there was *some* kind of semantical or logical relation between statements about emotions and feelings and statements about behavior.

In my opinion, verificationism and behaviorism are fundamentally misguided doctrines. While it is certainly possible to do a certain amount of philosophy of science from a non-verificationist and non-positivist point of view, but without developing in detail a theory of meaning alternative to the positivists', one of the major tasks in the present period is the development of such a theory of meaning, a non-verificationist theory of meaning, and the critique of verificationist philosophy of mind.

I. THE DEFECTS OF VERIFICATIONISM

One of the defects of verificationism that was early noticed by the more sophisticated verificationists themselves, and especially by Hans Reichenbach, was a certain distortion of the character of actual scientific methodology and inference. Naive verificationism would say that the statement "There is current flowing in this wire" means "The voltmeter needle is displaced," or something of that kind. That is, the relation between the so-called theoretical statement that current is flowing in the wire and the evidence for it is assimilated to the relation between "John is a bachelor" and "John is a man who has never been married". Now the latter relation is itself not as simple a thing as it may seem at first blush, but it is roughly right that the relation is a *conventional*[1] one: "John is a bachelor" is equated by some kind of conventional agreement with "John is a man who has never been married". But, as Reichenbach pointed out in *Experience and Prediction*, the relation between the theoretical statement and the evidence for it (say, "There is current flowing in the wire" and "The voltmeter needle is displaced") is a probabilistic inference within a theory. It is not that we equate the *sound-sequence* "There is current flowing in this wire" with "The voltmeter needle is displaced" by an act of conventional stipulation; it is rather that we accept a theory of electricity and of the structure of voltmeters from which it follows that *with a high probability*, the voltmeter needle will be displaced if there is current flowing in the wire, and vice versa. To represent what are in fact probabilistic inferences within theories as logical equivalences is a serious distortion. To represent these inferences as purely conventional meaning equivalences is an even more serious distortion.

But sophisticated verificationism found that it had escaped from one difficulty to land in another. If *meaning* is conflated or confounded with *evidence*, and what is evidence for a statement is a function of the total theory in which the statement occurs, then every significant change in theory becomes a change in the meaning of all the constituent words and statements of the theory. One of the early verificationists, Charles Peirce, anticipated this difficulty in the last century when he came to the conclusion that every change in a person's "information" is a change in the meaning of his words. But the distinction between the meaning of a man's words and what he believes about the facts, the distinction between disagreement in the meanings of words and disagreement about the facts, is precisely central to any concept of linguistic *meaning*. If we come to the conclusion that that distinction is untenable then, as Quine has long urged, we should abandon the notion of meaning altogether. With the exception of Quine, most verificationists have found this course unattractive. Thus they were caught in a serious dilemma – caught between their desire to continue talking about linguistic meaning and their adherence to the network theory of meaning which taken seriously implies that nothing can be made of the notion of linguistic meaning.

For a realist, the situation is quite different. No matter how much our theory of electrical charge may change, there is one element in the meaning of the term "electrical charge" that has not changed in the last two hundred years, according to a realist, and that is the reference. "Electrical charge" *refers to the same magnitude* even if our theory of that magnitude has changed drastically. And we can identify that magnitude in a way that is independent of all but the most violent theory change by, for example, singling it out as the magnitude which is casually responsible for certain effects.

But the realist has his problems too. Traditionally realists thought that reference was determined by mental or Platonic entities, intensions. This doctrine of fixed "meanings", either in the head or in the realm of abstract entities (and somehow connected to the head), determining reference once and for all, is open, interestingly enough, to some of the same objections that can be brought to bear against verificationism.

Thus, very recently certain realists like Kripke and myself have begun to redevelop our theory of meaning. Instead of seeing meanings as entities which determine reference, we now are trying to see meanings as largely determined by reference, and reference as largely determined by casual connections.

I would not wish to give the impression that the only problem with verificationism is its inability to give a correct account of our notion of linguistic meaning, however. Truth and falsity are the most fundamental terms of rational criticism, and any adequate philosophy must give some account of these, or failing that, show that they can be dispensed with. In my opinion, verificationism has not succeeded in doing either. There is a sense in which Tarski's technical work in mathematical logic enables one to explicate the notion of truth in the context of a language with fixed meanings and as long as there is no doubt that the terms of that language have clear reference. (Even in that context, as Hartry Field has pointed out, one may question whether we have been given an account of what "true" means, or simply a substitute for the word "true" designed for that specific context.) But if the meaning of words is a function of the theory in which they occur, and changes as that theory changes, then if we limit ourselves to Tarski's methods, "true" and "false" can only be defined in the context of a particular theory. In particular, Tarskian semantics gives no explanation of the meanings of "true" and "false" when they are used to compare and criticize different theories, if meaning is really theory-dependent. But it is just the extra-theoretic notions of truth and falsity which are indispensable for rational criticism[2], which is why they have always been taken as fundamental in the science of logic. In particular, a verificationist cannot explain why, if even the commonest scientific terms (e.g., "voltage", "density", "pressure") have different meanings in the context of different theories, it should ever be justified to *conjoin* a proposition verified by one group of scientists and a proposition verified by a different group of scientists.[3] The simple fact that the conjunction of true statements is true becomes replaced by the mysterious fact that scientists are in the habit of conjoining statements which use words with different meanings and somehow, nevertheless, manage to get successful results. In an insightful unpublished essay titled 'Realism and Scientific Epistemology', Richard Boyd has recently argued that these defects of verificationism and positivism are symptomatic of a deeper defect; that even if verificationism could give a correct *description* of the practice of scientists, it lacks any ideas which would enable one to explain or understand why scientific practice succeeds.)

II. THE A PRIORI AND THE ANALYTIC-SYNTHETIC DISTRACTION

In 1951, Quine caused a commotion in the community of professional philosophers by publishing an attack on the venerable distinction between analytic and synthetic propositions. In their reply to Quine, Grice and Strawson advanced two arguments:

(1) When there is so much agreement among the relevant speakers (in this case, professional philosophers) upon how to use a pair of terms with respect to an open class of sentences, then that pair of terms must mark *some* distinction;

(2) Grice and Strawson argued that the cases in which it appears that an analytic proposition was falsified can be explained away by contending that in each case the meaning of the words changed, and so the proposition that was at one time genuinely analytic was not the same proposition that was later falsified, although it was expressed by the very same sentence.

I agree with the first argument. There is an *obvious* difference (even if we have difficulty stating it) between, say, "all bachelors are unmarried", as a representative analytic sentence, and "my hat is on the table", as a representative synthetic sentence. It seems impossible to say that so obvious a distinction does not really have *any* basis. But Grice and Strawson's second argument seemed to me to be far less successful. Consider the statement that one cannot return to the place from which one started by traveling in a straight line in space in a constant direction. If this statement was once analytic or *a priori* (in 1951, few philosophers of an analytic persuasion would have troubled to distinguish the two notions), and was later falsified by the discovery (let us say) that our world is Riemannian in the large, then the Grice-Strawson rescue move would consist in saying that some term, say, "straight line" has changed its meaning in the course of the change from Euclidean to Riemannian cosmology. But even if "straight line" has changed its "connotations" – even if the theoretical aura surrounding the term is different – still this would not effect the *truth value* of the sentence unless the very reference of the term "straight line" has changed, unless we are now referring to different paths in space as straight lines. But, having studied philosophy of physics and philosophy of geometry with Hans Reichenbach, I was not satisfied with this story at all. Whatever the nature of the conceptual revolution involved in the shift from Newtonian to relativistic cosmology may have been, it was not simply a matter of attaching the old labels, e.g. "straight line", to new curves. What seemed *a priori* before the conceptual revolution was precisely that there *are* paths in space which behave in a Euclidean fashion; or, to drop reference to "paths", what seemed *a priori* was precisely that there were infinitely many non-overlapping places (of, say, the size of an ordinary room) to get to. What turned out to be the case (or, rather, what will turn out to be the case if the universe in the large has compact spatial cross-sections), is precisely that there are only finetely many disjoint places (of the size of an ordinary room) in space to get to, travel as one will. Something literally *inconceivable* has turned out to be true; and it is not just a matter of attaching the old labels ("place", "straight line") to different things.

To state the same point more abstractly: it often happens in a scientific revolution that something that was once taken to be an *a priori* truth is given up; and one cannot say that what has happened is simply that the words have been assigned to new referents, because, from the standpoint of the new theory, there are not and never were any objects which could plausibly have been the referents of the words in question. Nor can we say that the proposition in question used to mean that certain entities ("Euclidean straight lines", "Euclidean places") *would* have certain properties if they existed, and that what has happened is that words ("straight line", "place") which used to have no referents at all have now been assigned referents; for in the geometrical case there certainly were such entities as *places the size of a room*, and what seemed necessary was that these *places* had the property of being infinite in number.

To put it another way, it seemed *a priori* that the terms "path in space" and "place the size of an ordinary room" had referents. To say that the existence propositions, "There are places the size of an ordinary room" and "There are paths in space", were *a posteriori* (in the old sense of the words), whereas if the if-then proposition "If anything is a place the size of a room, then there are infinitely many such places" is *a priori* is utterly unmotivated, since these propositions did not differ in epistemological or methodological status prior to the conceptual revolution under discussion.

I am driven to the conclusion that there is such a thing as the overthrow of a proposition that was once *a priori* (or that once had the status of what we *call* an "*a priori*" truth). If it could be rational to give up truths as self evident as the geometrical propositions just mentioned, then, it seems to me that there is no basis for maintaining that there are any *absolutely a priori* truths, any truths that a rational man is *forbidden* to even doubt. Grice and Strawson were wrong; the overthrow of "*a priori*" propositions is not a mere illusion that can be explained away as change in the meaning of words. Quine's attack on the analytic-synthetic distinction seems to me to be correct. At the same time, if by an analytic truth, one means a statement which is reducible to something like principles of elementary logic via meaning relations that are in some sense conventional, then it still seems to me that there are analytic truths. Empiricist philosophers had bloated the analytic-synthetic distinction by making it coextensive with the *a priori-a posteriori* distinction; the question of the existence of analytic truths, in the sense just mentioned, has to be separated from the question whether any truths, even truths of elementary logic, are *a priori*.

III. CONVENTIONALISM

An issue which is closely connected to the issues surrounding the analytic-synthetic distinction, and its misuse by philosophers, is the issue of conventionalism. Just as some philosophers try to clear up some philosophical puzzles by contending that certain statements which appear to be statements of fact are really "analytic", so some philosophers contend that certain statements which appear to be statements of fact are really "up for grabs", i.e. their truth-value is a matter of convention. It is of interest that conventionalism in the philosophy of space and time was originally motivated by a desire to give an account of the reference of scientific terms. Thus the critique of conventionalism naturally involves one in the very questions about reference that are in the forefront nowadays.

I have not attempted in this talk to put forward any grand view of the nature of philosophy; nor do I have any such grand view to put forward if I would. It will be obvious that I do not agree with those who see philosophy as the history of "howlers", and progress in philosophy as the debunking of howlers. It will also be obvious that I do not agree with those who see philosophy as the enterprise of putting forward *a priori* truths about the real world (since, for one thing, there are no *a priori* truths on my view). I see philosophy as a field which has certain central questions, for example, the relation between thought and reality, and, to mention some questions about which I have *not* written, the relation between freedom and responsibility, and the nature of the good life. It seems obvious that in dealing with these questions philosophers have formulated rival research programs, that they have put forward general hypotheses, and that philosphers within each major research program have modified their hypotheses by trial and error, even if they sometimes refuse to admit that that is what they are doing. To that extent philosophy is a "science". To argue about whether philosophy is a science in any more serious sense seems to me to be hardly a useful occupation. The important thing is that in spite of the stereotypes of science and philosophy that have become blinkers inhibiting the view of laymen, scientists, and philosophers, science and philosophy are interdependent activities; philosophers have always found it essential to draw upon the scientific knowledge of the time, and scientists have always found it essential to do a certain amount of philosophy in their very scientific work, even if they denied that that was what they were doing. It does not seem to me important to decide whether science is philosophy or philosophy is science as long as one has a conception of both that makes both essential to a responsible view of the real world and of man's place in it.

Harvard University

NOTES

* A slightly altered version of this paper appears as forward to my book *Mind, Language, and Reality* (*Philosophical Papers*, vol. 2), Cambridge, 1975.
[1] The sense in which this is so is discussed in 'The Analytic and the Synthetic', chapter 2 of my *Mind, Language, and Reality* (Cambridge 1975), and in David Lewis' *Convention*, (Harvard 1969).
[2] I say this even though these terms are not often used "neat". Very often we say a theory is *probably false*, or *likely to be close to the truth*, not "false" or "true" *simpliciter*. For probable truth and approximate truth presuppose the notions of truth and falsity themselves.
[3] This difficulty cannot be avoided by saying that every group of scientists should be thought of as sharing a "formalization of total science". For then it becomes a miracle that I can say anything true, if the very meaning of my terms presupposes a theory *most of which I don't know*.

SYMPOSIUM

HISTORY AND PHILOSOPHY OF BIOLOGY

KENNETH F. SCHAFFNER

REDUCTIONISM IN BIOLOGY: PROSPECTS AND PROBLEMS

I. INTRODUCTION

Explications of what it means for one science to be reduced to another have been the focus of considerable interest in recent years, and when the reducing science is physics (and chemistry) and the reduced science is biology, additional concerns seem to arise. In this paper I wish to represent a model for theory reduction which has occupied my attention for some years now, and consider its applicability in the area of the biological sciences.[1]

I should state in advance that the model to be discussed represents an ideal standard for accomplished reductions, and does not characterize the research programmes of molecular biologists. Recently I have argued, in point of fact, that a research programme which might be generated from the tenets of the reduction model to be outlined in this paper would not represent some of the most significant advances in molecular biology, such as the development of the Watson-Crick model of DNA or the *genesis* of the Jacob-Monod operon theory.

However, even though I would subscribe to what I have termed a 'peripherality thesis' as regards reductionism as the primary *aim* of molecular biology, I believe that molecular biology is *resulting in* at least *partial* reductions of biology to physics and chemistry, and that even these partial instances are in accord with the ideals proposed as part of the general reduction model to be considered below. In any unfinished and rapidly advancing science which is resulting in reductions, however, it is possible to find aspects which can be interpreted as difficulties for such a model such as I wish to present and defend. I shall mention some of these difficulties in this paper; the other symposiasts develop these and related problems in considerably more detail.

II. A GENERAL MODEL OF THEORY REDUCTION

The model to be given in this paper is but a slight modification of a model which I sketched several years ago, and which is indebted to various earlier forms which have existed in the literature of the philosophy of science since the pioneering work of Ernest Nagel in 1949.[2] Recently, the model I shall outline has been criticized, I believe mistakenly, by several philosophers including the present symposiasts.[3] It will be appropriate to respond to some of those criticisms as I think that an answer to those attempted refutations both clarifies and deepens the original model.

Following standard practice, I shall term a theory to be reduced the reduced theory, and the theory which does the reducing the reducing theory. Theories can be constructed as attempts to capture the essentials of the subject areas, or domains to use Dudley Shapere's term,[4] which they explain, and thus a reduction of a theory (*assuming its adequacy*) is a reduction of the area of application of a theory. For the sake of precision as well as to insure the adequacy of a reduction, I shall assume that the basic principles of the reducing and reduced theories have been codified and that the primitive terms of both theories have been determined.[5] We can thus assume that we know the fundamental entities, processes, and generalizations (or laws) that constitute the reducing and reduced theories and their subject areas.

The conditions for the adequate reduction of a theory have both formal and informal aspects.[6] On the formal side, terms, e.g., 'gene' and 'phenotype' which appear in the reduced theory but not in the reducing theory, must be somehow associated with terms in the reducing theory. The simplest and least question begging way to do this seems to be to require (i) that the entity terms (e.g., 'gene' or 'DNA') to be associated, be construed as extensionally referring to the same entity, even though that entity is described in different ways and (ii) that predicates and relations (e.g., 'dominant') in the two theories, be interpreted as referring to the same states of affairs, characterized with the help of the entity relations spoken above. (This point will be developed more extensively later.) Thus, for example, the term 'gene' can be understood to refer to the same entity which is named by a sequence of nucleotides of DNA (or RNA in some special cases involving viruses). Sentences which formulate such extensional references are best construed as *synthetic* identities, i.e., identities which at least initially require empirical support for their warrant.[7] (Genes were not discovered to be DNA via the analysis of *meaning*;

important and difficult empirical research was required to make such an identification.) I am going to term such sentences 'reduction functions' since they have values which exhaust the universe of the reduced theory, e.g., gene, for arguments in the universe of the reducing theory, e.g., DNA.[8] Structural relations between chemical constituent parts, e.g., DNA *sequences*, appear in the chemical side of the reduction functions. The reduction functions thus contain part of the 'organization' which characterizes biological systems. This last point is relevant to some of Michael Polanyi's claims concerning the impossibility of the reduction of biology.[9]

Again, continuing to comment on the formal aspects of theory reduction, in order to insure that a reduced theory is *in principle eliminable*, i.e., that the reducing theory can do all that the reduced theory accomplished in terms of explanation and systematization of data, we stipulate that the reduced theory must de *derivable* from the reducing theory when the reducing theory is supplemented with the reduction functions mentioned above. The reduced theory thus stands to the reducing theory somewhat as a set of *theorems* in geometry stands to the basic *axioms* of geometry. The reduced theory is derivative, and its ontology and empirical content are more limited aspects of the reducing theories ontology and basic assertions about the world.

Formalizable relations between the reducing and reduced theory do not, however, exhaust the relationship between these theories. As has been stressed over the last decade by Paul Feyerabend,[10] reducing theories often contradict or are incommensurable with aspects of the purportedly reduced theory. The reduction functions of which I spoke above are thus often very difficult to formulate in such a way that they are not fundamentally misleading, confusing, and incoherent. Feyerabend has offered a number of examples from the physical sciences to support his claim. Consider the concept of temperature in classical phenomenological thermodynamics.[11] This notion of temperature, definable in terms of a reversible Carnot cycle and heat exchanges, can be shown to be inextricably associated with the strict, i.e., nonstatistical, form of the second law of thermodynamics. Kinetic theory, Feyerabend notes, cannot "give us such a concept". The terms in both theories cannot be associated by a reduction function without contradiction (if the function is conceived to be analytic) or without falsity (if the function is thought to be a physical hypothesis). The term 'temperature' is thus only a homonym for Feyerabend, and reduction of the macro-theory, thermodynamics, to the micro-theory, statistical mechanics, is impossible. The theories are incommensurable. At best, statistical mechanics replaces phenomenological thermodynamics. (Feyerabend's views are quite similar to Thomas Kuhn's, which are probably more widely known.)[12]

Without attempting to analyze Feyerabend's most interesting arguments in this paper, what I wish to do is to consider the *possibility* that something similar is at work in the relationship of classical genetics to molecular genetics, and to consider what its import may be for the reduction of classical genetics by molecular biology.[13]

It is clear that some modification of the concepts of classical genetics has occurred since 1950 both as a consequence of more sophisticated genetic techniques, e.g., the use of bacteria and phage, and the development of chemical models of genetic processes. Biological dictionaries in the early 1950's used to characterize the gene as the unit of mutation, recombination, *and* function (in the sense of being a necessary condition for a phenotypic characteristic). With the advent of fine structure genetics, Benzer and other geneticists saw the necessity to redefine the notion of the classical gene and to distinguish between these heretofore extensionally equivalent concepts. The early 1960's saw the introduction of new classes of genes, e.g., regulator genes such as repressor-making genes and operator genes. A more careful analysis of phenotypic characteristics via biochemical analysis allowed for a much more precise characterization of genetic effects.[14]

Classical genetics of the type understood by Morgan and Muller in the 1920's, say, has evolved and become much richer even when presented without reference to underlying biochemical details. (Fine structure genetics and regulatory genetics can, though they usually are not, be presented in a roughly phenomenological manner without much stress on the biochemical foundations.)[15] Let us indicate that an evolution of classical genetics has taken place by saying that classical genetics has been modified or corrected to a more adequate form of classical genetics, say neo-classical genetics or classical genetics*. This correction has in part been caused by the successes of research into the chemical nature of the gene and into chemical characterizations of phenotypes. One can, nonetheless, speak in *biological* terminology and use expressions which are characteristically *nonchemical* in presenting classical genetics*, e.g., we could talk about genetic interaction or the blocking effect of the inductibility gene (i^+) on the lactose metabolizing gene (z^+) in the *lac* region of *E. coli*, without knowing that the i^+ *DNA region* makes a special *protein* which prohibits transcription of the z^+ *DNA region*, thus ultimately preventing the synthesis of the *enzyme* β-galactosidase.[16] (It should be) noted here however that because the partial reductions of genetics thus far achieved have resulted in a kind of 'intertwining' of chemistry and genetics,[17] such a separation as is analytically desired for our purposes is not explicitly found in textbooks and research papers.)

The point of the foregoing discussion is as follows: I believe that we can take a modification of classical genetics and explain that *modification* by physics and chemistry, i.e., by molecular genetics, even if reduction functions associating an *unmodified* classical genetics with physics and chemistry cannot be formulated. In this way I hope to be able to outflank the Feyerabend type of objection. In such a way we can, assuming the validity of the model sketched in this paper, obtain a formally precise characterization for the reduction of classical genetics* by physics and chemistry, but only an *informal* relationship between either (i) the physics and chemistry and the *uncorrected* form of classical genetics, or (ii) the corrected and the uncorrected forms of classical genetics. To say, however, that we have reduced classical genetics and not some *totally* different theory to physics and chemistry, we require that the corrected reduced theory bear a strong analogy with the *uncorrected* reduced theory, and, that the reducing theory explain why the uncorrected reduced theory worked as well as it did historically, e.g., by pointing out that the theories lead to approximately equal laws, and that the experimental results will agree very closely except in special extreme circumstances. These relations of approximate equality, close agreement, and strong analogy have yet to find formally precise characterizations, and to date represent informal aspects of a reduction. These elements in the reduction should not, however, be taken as implying that the relation between the reducing theory and reduced theory, in its corrected form, is vague or imprecise.[18] The vagueness lies in the historical relation of strong similarity between the uncorrected and the corrected reduced theory. Such vagueness is in fact logically demanded since the corrected theory is though to be more adequate, and perhaps even completely true, whereas the uncorrected theory is inadequate and in part false, and it would be most unusual if a formal relation that could be codified in logical terms existed between such earlier and later versions of the reduced theory.

A summary of the conditions for an adequate reduction of theory T_2^* by the reducing theory T_1 is given in Figure 1.

General Reduction Model
T_1 – the reducing theory.
T_2 – the original reduced theory.
T_2^* – the 'corrected' reduced theory.

Reduction occurs if and only if:

(1) All primitive terms of T_2^* are associated with one or more of the terms of T_1 such that:
   (a) T_2^* (entities) = function $[T_1$ (entities)$]$.
   (b) T_2^* (predicates) = function $[T_1$ (predicates)$]$.
   (This is the condition of referential identity.)

(2) Given fulfillment of condition (1), that T_2^* be derivable from T_1 supplemented with 1(a) and 1(b) functions. (Condition of derivability.)

(3) T_2^* corrects T_2, i.e., T_2^* makes more accurate predictions.

(4) T_2 is explained by T_1 in that T_2 and T_2^* are strongly analogous, and T_1 indicated why T_2 worked as well as it did historically.

Fig. 1. The general reduction model.

III. CRITICISMS OF THE MODEL

Criticisms of the model sketched above have tended to focus on three issues: first, the nature of the reduction functions and whether these can be specified. David Hull in particular has questioned the possibility of being able to state these functions and has also suggested the conditions that in point of fact exist between classical and molecular genetics are so complex that we would encounter a bizarre many-many set of relations if we attempted to articulate the reduction functions.[19] Secondly, both Hull and Michael Ruse have queried whether the distinction between reduction and *replacement* is not blurred to the point where what I term a reduction is not in fact a replacement of classical genetics by molecular genetics.[20] Finally, Tom Nickles has recently suggested that the model presented here characterizes but one of several different types of reduction, and that emphasis on the derivational or deductive relation between T_1 and T_2* diverts our attention from construing intertheoretic relations under alternative useful non-deductive rubrics.[21] I believe that Bill Wimsatt is in agreement with Nickles suggestions here.[22]

In the scope of this paper I cannot develop a full set of replies to each of these criticisms. It will, however, be useful to examine David Hull's criticisms in some detail and touch very briefly on the replacement and non-deductive relation issues.

Let us begin by looking at Hull's claims, reiterated in the present group of papers, that argues for non-connectability of classical and molecular genetics. The essence of Hull's view is that connections between classical genetics and molecular genetics are many-many, i.e., that to one Mendelian term (say the predicate 'dominant') there will correspond many diverse molecular mechanism which result in dominance. Further, Hull adds, any one molecular mechanism, say an enzyme-synthesizing system, will be responsible for many different types of Mendelian predicates: dominance in some cases, recessiveness in others. With the increase in precision which molecular methods allow, furthermore, major reclassifications of genetic traits are likely, and when all of these difficulties are sorted out, any corrected form of genetics that can be reduced is accordingly likely to be very different from the uncorrected form of genetics. To Hull, "given our pre-analytic intuitions about reduction [the transition from Mendelian to molecular genetics] is a case of reduction, a paradigm case. [However on] ... the logical empiricist analysis of reduction [i.e., on the basis of a type of reduction model introduced earlier] ... Mendelian genetics cannot be reduced to molecular genetics. The long awaited reduction of a biological theory to physics and chemistry turns out not to be a case of 'reduction' ..., but an example of replacement".

It will be useful, in an attempt to present Hull's position as fairly as possible, to allow him to give his arguments in his own terms. Hull contends:

Numerous phenomena are now explained in Mendelian genetics in terms of recessive epistasis [i.e., a certain form of gene interaction]; e.g., color coat in mice, feather color in Plymouth Rock and Leghorn chickens, a certain type of deaf-mutism in man, feathered shanks in chickens, and so on. There is little likelihood that all of these phenomena are produced by a single molecular mechanism. At best, several alternative mechanisms are involved. Conversely, there is little likelihood that these various molecular mechanisms always produce phenomena which are appropriately characterized by the Mendelian predicate term 'recessive epistasis'.

... For instance, one might expect dominant and recessive epistasis to be produced by a combination of those molecular mechanisms that produce epistasis with those that produce dominance and recessiveness.

One does not have to look very deeply into the relation between Mendelian and molecular genetics to discover how naive the preceding expectations actually are. Even if all gross phenotypic traits are translated into molecularly characterized traits, the relation between Mendelian and molecular predicate terms express prohibitively complex, many-many relations. Phenomena characterized by a single Mendelian predicate term can be produced by several different types of molecular mechanisms. Hence, any possible reduction will be complex. Conversely, the same types of molecular mechanism can produce phenomena that must be characterized by different Mendelian predicate terms. Hence, reduction is impossible. Perhaps these latter ambiguities can be eliminated by further 'correcting' Mendelian genetics but then we are presented with the problem of justifying the claim that the end result of all this reformulation is reduction and not replacement. Perhaps something is being reduced to molecular genetics once all these changes have been made in Mendelian genetics, but it is not Mendelian genetics.

When one combines all of the complexities mentioned thus far with the requirement that the relations between the original Mendelian predicate terms be retained, one begins to appreciate the scope of the task confronting a serious proponent of reduction in genetics. Not only must he list all the mechanisms which eventuate in recessive epistasis as well as those that eventuate in dominant epistasis, but also these two lists of molecular mechanisms must be symmetrical in the same way in which traits which are simple dominant of recessive are symmetrical. According to the current analysis of reduction supplied by the logical empiricists and their contemporary descendents, a certain amount of reshuffling and re-classification of the phenomena of the reduced theory is to be expected, but in this instance it seems as if the vast scaffolding of Mendelian genetics must either be ignored or else dismantled and reassembled in a drastically different form.[23]

I believe that Hull has misconstrued the application of the general model of reduction which I outlined earlier in several ways, and that he thus sees logical problems where they do not exist. Empirical problems do exist, namely, exactly what mechanisms are involved in the cytoplasmic interactions that result in genetic epistasis of the dominant and recessive varieties, but these ought not be confused with the logical problems of the type of connections which have to be established and with how a corrected form of classical genetics is to be derived from molecular genetics. Let me argue for this position in some detail.

Hull's primary concern is with the reduction functions for Mendelian predicates. Let us briefly recall what predicates are prior to sketching the manner in which they can be replaced in a reduction. In earlier characterizations of the general model of theory reduction as outlined above, I have followed modern logicians such as Quine in construing predicates in an extensional manner. From this perpective, a predicate is simply a class of those things or entities possessing the predicate. This nominalistic position can be put another way by using standard logical terminology and understanding predicates to be open sentences which become true sentences when the free variable(s) in the open sentence are replaced by entities which possess the predicate characterized by the open sentence. Thus the open sentence 'x is odd' where x ranges over positive integers, say, introduces the predicate 'odd' (or 'is-odd'). When the free variable x is replaced by a named object, then the sentence becomes closed and can be said to be true or false depending on whether the named objects in fact possesses the predicate. Relations can be similarly introduced by open sentences containing two free variables, e.g., the relation of being a brother, by 'x is the brother of y'. Following Quine, we may then say that "the extension of a closed one place predicate is the class of all things of which the predicate is true; the extension of the closed two place predicate is the class of all the pairs of which the predicate is true; and so on".[24] The extensional characterization of predicates and relations is important because reduction functions relate entities and predicates of reduced and reducing theories extensionally, via an imputed relation of synthetic indentity.

Let us examine the predicate 'dominant' or 'is-dominant' from this point of view.

First it will be essential to unpack the notion. Though the predicate is one which is ascribed to genes and might prima facie appear to be a simple one-place predicate,[25] in fact the manner in which it is used in both traditional classical genetics and molecular genetics indicates that the predicate is a rather complex relation. The predicate is defined in classical genetics as a relation between the phenotype produced by an allele of a gene in conjunction with another different allele of the same locus (a heterozygous condition), compared with the phenotype produced by each of these alleles in a double dose of the same allele (a homozygous condition). The predicate 'epistatic', which Hull believes to be a particularly troublesome one for the general reduction model, is similarly unpackable as a rather complex relation between different genes and not simple between different alleles of the same gene.

More specifically, suppose that the phenotype of a gene a when the gene is in a homozygous situation, i.e., aa is **a**.[26] Similarly suppose that genotype AA biologically produces phenotype **A**. (The causal relation of biological production will be represented in the following pages by a single arrow: $\rightarrow$.) If a and A are alleles of the same genetic locus, and the phenotype produced by genotype aA is **a**, then we say that gene a is dominant (with respect to gene A).[27] The predicate 'dominant' therefore is in reality a relation and is extensionally characterized by those pairs of alleles of which the relationship $(aa \rightarrow \mathbf{a})$ & $(AA \rightarrow \mathbf{A})$ & $(aA \rightarrow \mathbf{a}) =_{df} a$ is dominant (with respect to A), is true.

Epistasis can be similarly unpacked. If two non-allelic genes, a and c say, are such that $aa \rightarrow \mathbf{a}$ and $cc \rightarrow \mathbf{c}$, when each genotype is functioning in isolation from the other, but that the genotype $aacc$ biologically produces **a**, we say that a 'hides' c or 'is epistatic to' c. The gene a may so relate to c that a in its dominant allelic form obscures c. (This is 'dominant epistasis'.) The gene a may also relate to c such that a's absence (or recessive allele of a) prevents the appearance of c's phenotype. (This is 'recessive epistasis'). **a** and **c** here are contrasting traits.

The various interactions between genes, of which the two forms of epistasis mentioned above are but a subset, can be operationally characterized since they result in a modification of the noninteractive independence assortment ratio of $9:3:3:1$ for two genes affecting the phenotypes of the organisms. (For a survey of these interactions and the specific modifications of the ratios which they produce the reader must be referred to other sources.)[28]

I have now sufficiently analyzed some of the predicates which appear in classical genetics so as to return to the problem of specifying reduction functions for such predicates.

Specification of reduction functions, which are like dictionary entries that aid a translator, can be looked at in two ways analogous to the two directions of translation: (1) From the context of the theory to be reduced one must be able to take any of the terms (entity terms or predicate – including relations – terms) and univocally replace these by terms or combinations of terms drawn from the reducing theory. (2) From the context of the reducing theory one must be able to pick out combinations of terms which will unambiguously yield equivalents of those terms in the reduced theory which do not appear in the reducing theory. Such sentences characterizing replacements and combination selections must, together with the axioms and initial conditions of the reducing theory, entail the axioms of the reduced theory if the reduction is to be effected.

To obtain reduction functions for Mendelian predicates we proceed in the following manner. (1) The predicate is characterized extensionally by indicating what class of biological entities or pairs of entities, etc., represent its extension. This was done for dominance and epistasis above in terms of genotypes and phenotypes. (2) In place of each occurrence of a biological entity term, a chemical term or combination of terms is inserted in accord with the reduction functions for entity terms. (3) Finally, in place of the biological production arrow which represents the looser or less detailed generalizations of biological causation, we write a double-lined arrow ($\Rightarrow$), which represents a promissory note of a chemically causal account, i.e., a specification of various chemical mechanisms which would yield the chemically characterized consequent on the basis of the chemically characterized antecedent. Thus we could do the following: Let the reduction function for gene a, say be 'gene a = DNA sequence α' and 'gene A = DNA sequence β'. We also suppose phenotype **a** possesses a reduction function 'phenotype **a** = amino acid sequence א', and phenotype **A** = amino sequence ב. Then if we represented the Mendelian predicate for dominance as $(aa \rightarrow \mathbf{a})$ & $(AA \rightarrow \mathbf{A})$ & $(aA \rightarrow \mathbf{a})$ we would obtain:

Allele a is dominant (with respect to A) =
(DNA sequence α, DNA sequence α $\Rightarrow$ amino acid sequence א) &
(DNA sequence β, DNA sequence β $\Rightarrow$ amino acid sequence ב) &
(DNA sequence α, DNA sequence β $\Rightarrow$ amino acid sequence א).

This reduction function for dominance is, as was the characterization of dominance introduced earlier, somewhat oversimplified and for complete adequacy ought to be broadened to include Muller's more sensitive classification and quantitative account of dominance.[29] This need not be done, however, for the purposes of illustrating the logic of reduction functions for predicates. I have deliberately left the interpretation of the double-lined arrow in the right-hand side of the above reduction function unspecified. As noted, it essentially represents a telescoped chemical mechanism of production of the amino acid sequence by the DNA. Such a mechanism must be specified if a chemically adequate account of biological production or causation is to be given. Such a complete account need not function explicitly in the reduction function per se, even though it is needed in any complete account of reduction. Such a reduction function, and similar though more complex ones, could be constructed in accordance with the above-mentioned steps for epistasis in its various forms, allows the translation from the biological language into the chemical and vice versa. Notice that the relations among biological entities which yield the various forms of dominance and epistasis, for example, can easily be preserved via such a construal of reduction functions. This is the case simply because the interesting predicates can be unpacked as relations among biological entities, and these selfsame entities identified with chemical entities. The extensional characterization of predicates and relations as classes and (ordered) pairs of entities accordingly allows an extensional interpretation of reduction functions for predicates.[30]

Assuming that we can identify any gene with a specific DNA sequence and that, as Francis Crick says, "the amino acid sequences of the proteins of an organism ... are the most delicate expression possible of the phenotype of an organism",[31] we have a program for constructing reduction functions for the primitive entities of a corrected classical genetics. (A detailed treatment would require that different types of genes in the corrected genetic theory, e.g., regulator genes, operator genes, structural genes, and the recon and muton subconstituents of all these, be identified with specific DNA sequences).

Substitution of the identified entity terms in the relations which constitute the set of selected primitive predicates, which yields reduction functions for predicates, then, allows for the two-way translation, i.e., from classical genetics to molecular genetics and vice versa. Vis-à-vis David Hull, molecular mechanisms do not appear in connection with these reduction functions *per se*, rather they constitute part of the reducing theory. Chemically characterized relations of dominance can be and are explained by a variety of mechanisms. The only condition for adequacy which is relevant here is that chemical mechanisms yield predictions which are unequivocal when the results of the mechanism(s) working on the available DNA and yielding chemically characterized phenotypes are translated back into the terms of neoclassical genetics.

What Hull has done is to misconstrue the locus where molecular mechanisms function, logically, in a reduction, and he has thus unnecessarily complicated the logical aspects of reduction. Different molecular mechanisms can appropriately be appealed to account for the same genetically characterized relation, as the genetics is less sensitive. The same molecular mechanism can also be appealed to account for different genetic relations, but only if there are further differences at the molecular level. It would be appropriate to label these differences as being associated with different initial conditions in the reducing theory.[32]

There are several appeals which I believe can be made to contemporary genetics which will support my position over Hull's. First, traditionally discovered gene interactions, not only epistasis but also suppression and modification, are in fact currently being treated from a molecular point of view. Several textbooks and research papers can be cited in support of this.[33] The logic of their treatment is not to drastically reclassify the traditionally discovered gene interactions, but rather to identify the genes with DNA, and the phenotype results of the genes with the chemically characterized biosynthetic pathways, and to regard gene interactions as due to modifications of the chemical environments within the cells. Traditional genetics is not drastically reanalyzed, it is corrected and enriched and then explained or reduced, at least in part.

Second, I would note that it is sometimes the practice of molecular geneticists to consider what types of genetic effects would be the result of different types of molecular mechanisms. A most interesting case in point is the predictions by Jacob and Monod of what types of genetic interaction effects in mutants would be observable on the basis of different hypotheses concerning the operator region in the lac system of *E. coli*.[34] In brief, they were able to reason from the chemical interactions of the repressor and the different conjectures about the operator, and infer whether the operator gene would be dominant or recessive, and pleiotropic or not. The fact that in such a case molecular mechanisms yielded unequivocal translations into a corrected genetic vocabulary is a strong argument against Hull's concerns.

Finally, I should add that if the relation between modified classical genetics and molecular genetics were one of replacement, then it is very likely that considerably more controversy about the relations of classical to molecular genetics would have infused the literature. Further, it is also likely that those persons trained in classical genetics would not have made the transition as easily as they clearly have, nor would they comfortably be able to continue to utilize classical methods without developing, in a conscious manner, an elaborate philosophical rationale about the pragmatic character of classical genetics. Clearly such was the case with the type of historical transitions which are termed replacements and not reductions, e.g., the replacement of the theories of Aristotle by Newton's or of Ptolemy's by Copernicus' and Kepler's.

I believe that these three considerations count heavily against a replacement interpretation of the relationship between classical genetics and classical genetics* (or between classical genetics and molecular genetics). I am gratified to see that Ruse, who originally seemed disposed to accept a replacement interpretation, has recently come to accept more of a close-analogy or a reduction relationship.[35] He has also in his recent paper provided an additional example from a contemporary dispute in population genetics to support the reduction interpretation.

Finally I would briefly like to consider Nickels' views concerning non-deductive accounts of reduction. Nickels believes that a deductive relationship between reduced and reducing theories is too narrow and that the Nagel type of relation, which he terms reduction$_1$, has to be supplemented with an additional concept of reduction, a reduction$_2$. This latter concept arises out of the senses of reduction which refer to "being led back from one thing to another ... and to the related notion of transforming something into a different form by performing an operation on it" ... [such as] the "reduction of ores to their metals, the reduction of wood to pulp by pounding it...., and indeed, the reduction of $m_0 v/\sqrt{(1-v^2/c^2)}$ to $m_0 v$ by taking the limit as v goes to zero."[36]

Reduction$_1$ is 'essentially *derivational reduction*' or a deductive explanation, whereas reduction$_2$ is neither a theoretical explanation nor an ontological reduction.[37] Reduction$_2$ is rather 'justificatory and heuristic'. The general idea is that a later reducing$_2$ theory T_L yields a limiting consequence C which is apparently prima facie equivalent to some important constituent of an earlier theory T_E (if not prima facie equivalent to the entire T_E). Reduction$_2$, Nickels contends, characterizes the relation in which T_L reduces in *different ways to* T_E as different limits are taken, whereas in reduction$_1$, there is only one deductive relation between T_L and T_E. Finally, Nickels suggests that reduction$_2$ is not very sensitive to the problem of meaning change and logical compatability since its function is heuristic and justificatory.

With this distinction between two concepts of reduction in view, Nickels explicitly raises some problems with the general reduction model sketched above.

Nickels points out that the analogy condition (condition (4) in Figure 1 above) is problematical in that the notion of analogy has, as I had previously pointed out,[38] not yet received an adequate philosophical treatment. In place of this, Nickels suggests:

> ... the concept of reduction$_2$ may be of some help. Presumably limit relationships and the like will sometimes be useful in spelling out the analogy of T_2 and T_2* in particular cases. That is, in cases of approximative reduction, we can regard T_2* as a (fictitious) successor theory to the historical T_2, and in many of these cases reduction$_2$ techniques should be useful in spelling out the analogy. There is no reason to think that reductions$_2$ can in all cases take over the work of Schaffner's analogy relation, but when it can we shall have T_1 reducing T_2 approximately by reducing T_2*, which in turn reduces$_2$ to T_2.[39]

Reduction$_2$ can be stated in more general terms by employing the notion of a set of 'intertheoretic operations' performed on T_1 which yield T_2. Nickels has proposed the following schema:

> Instead of a single reduction relation (the logical-consequence relation), we now recognize a set of intertheoretic operations O_1 (or corresponding relations) which we might write $O_2[O_1(T_1)] \to T_2$ or simply $O_2 O_1 (T_1) \to T_2$, signifying that by performing operations this notational scheme into a general account of reduction$_2$.[40]

Nickels goes on to argue that one of the problems is that the approximate nature of the relation between T_2 and T_2* characterized in the general reduction model is still basically *derivational*, and that instead of attempting to conceive of approximative reduction as a *failure* to achieve a cleaner non-approximative reduction, approximate reductions ought to be construed under an entirely different rubric, as instances of reduction$_2$.

Let us briefly consider some of the issues arising out of the reduction of classical genetics (T_2) and classical genetics* (T_2*) by molecular genetics (T_1).[41] I have argued that the relation between T_1 (with added reduction functions) and T_2* is deductive or 'derivational' in Nickels terminology. Presumably the issue on which Nickels would favor his alternative construal is whether there is a relation of reduction$_2$ between T_1 (with reduction functions) and T_2.[42]

The problem that I see with Nickel's interesting suggestion is applied to the genetics case is that the intertheoretic operations which are not limit notions are difficult to specify and are deficient compared with the more intuitively appealing notion of a strong analogy. Indeed, as Nickels is willing to admit, it is not difficult to trivialize the notion of reduction$_2$ unless constraints on the types of operations allowed are articulated, and Nickels does not see any general way to do this. Though there may be disagreements about the details of the strength of analogies involved in the genetics case, I think that it can be said that there is general agreement among the participants in the present symposium (with the possible exception of Hull) that there is a very close analogy between T_2 and T_2^*. I would predict also that most members of the community of geneticists would also agree with this position. We are thus dealing with an unanalyzed or primitive relation, but with one which is testable.

Using a set of unspecified operations with no general restrictions would not, in my view, clarify the relations involved in the genetics case as much as the analogy condition does. As I perceive the situation, it would be unwise to dispense with the comparatively clearer analogy condition and with the paradigmatically clear relation of 'deductive consequence', to replace it with a set of unspecified intertheoretic relations characterizing (possibly vacuously) a concept of reduction$_2$.

University of Pittsburgh

NOTES

[1] See Schaffner (1967a) and (1969a), and (1974b). The present paper relies extensively on some of the arguments developed in this latter paper.
[2] Nagel (1949); also see Nagel's more developed ideas in his (1961).
[3] See M. Ruse (1971a) and (1971b) and D. Hull (1972), (1973) and (1974).
[4] D. Shapere (1974).
[5] A formalization, say in first order logic or in set theoretical language, is not necessary on this view, though an axiomatization (in English, for example) is very desirable to insure this 'codification' and the determination of the primitive terms and relations. The requirement is a logical one and requires fulfillment for a determination of the logical 'effectiveness' of a reduction. Such an axiomatization may not be a necessary condition of a de facto reduction since molecular biologists themselves probably can proceed informally, identifying, for example, a particular problem area which has been recalcitrant to physicochemical characterization and explanation. They thus demarcate areas which have been reduced from those which have not. In their research, their description of the area in terms of 'biological' terminology would amount to a provisional codification of the basic principles characterizing the subject area.
[6] This distinction between formal and informal aspects of a reduction is not the same as

Nagel's similar distinction between formal and nonformal conditions made in his (1961), pp. 358ff.
[7] See H. Feigl (1958), p. 370, and L. Sklar, (1964) for a discussion of synthetic identities in reduction. For an argument similar in spirit to Feigl's point that only synthetic identities are satisfactory in that they are not 'nomological danglers' see R. Causey, (1972).
[8] The locution 'reduction function' is patterned on Quine's (1964) 'proxy function' in his article on ontological reduction.
[9] Other aspects of the organization of chemical systems will appear in the section explaining chemical mechanisms (see text below). M. Polanyi has offered an intriguing 'vitalistic' interpretation of the need to accept organization in chemical systems in his (1968). His arguments are not successful, however, inasmuch as (1) structural relations embodying chemical organization are statable in chemical terms and explicable in their workings by normal chemical theories, and (2) the existence of these structures or 'boundary conditions' as Polanyi terms them, is explicable in principle by a chemical evolutionary theory. See my (1967b), (1969b), for more specific details. For a recent lucid introduction to the issues of the chemical origin of life and chemical evolution, see L. E. Orgel, (1973). For a completely sequenced replicating molecule which tells us much about Darwinian evolution in chemical terms, see D. R. Mills, F. R. Kramer, and S. Spiegelman (1973).
[10] Feyerabend (1962) and (1965).
[11] The example is from Feyerabend's (1962). Also see Feyerabend's (1965), p. 223, for a later discussion of his views, and also p. 226 for a presentation of a modified thermodynamics with fluctuation (due to Leo Szilard) which is relevant for the argument given in the text below concerning a 'corrected' T_2^*.
[12] T. S. Kuhn (1962).
[13] The need for redefinitions of basic terms in the reduced science in order to insure that adequate reduction functions are formulatable was discussed in my (1967a). The prospect that the reducing theory might have to be altered to accomplish a reduction was considered in my (1969a) esp. pp. 331–332. Recently David Hull has discussed a number of problems which are associated with the issue of formulating specific reduction functions between classical genetics and molecular genetics, sharply raising the Feyerabend type of problem for this example of reduction. See text below and notes 19 and 20.
[14] The changing definition of a gene is not simply a 'philosophical' problem; the need for a more precise conceptual analysis of the notion of a 'gene' occasioned by new discoveries in molecular genetics has been argued for in the editorial pages of the distinguished British scientific journal *Nature New Biology* **230** (1971), 194.
[15] To obtain a very rough example of what genetics without the underlying biochemical foundations might look like it is instructive to compare the first edition of R. P. Wagner's and H. K. Mitchell's *Genetics and Metabolism* (New York: Wiley, 1955), with the 1964 second edition, which contains more molecular material, and to compare both of these with the manner of presenting genetics in J. D. Watson's *Molecular Biology of the Gene*, 2nd ed. (New York: Benjamin, 1970).
[16] This case of genetic interaction effects in the lac operon is discussed extensively in my (1974a) and (1974b), section 2. In point of historical fact, the genetic interactions were discovered prior to a determination of much of the underlying biochemistry.
[17] The felicitious term 'intertwining' as used to characterize what happens in actual reductions is, I believe, due to Sidney Morgenbesser.
[18] See for example G. Massey's comment in his (1973) p.208.
[19] See Hull (1972), (1973), and (1974).
[20] Hull (1974), (1975) and Ruse (1971a and b).

[21] Nickels (1973).
[22] Wimsatt (1975).
[23] Hull (1974) pp. 40, 39, 41.
[24] Quine (1959), p. 136.
[25] David Hull has suggested, in his comments on a version of the present paper (delivered at a meeting of a Chicago seminar on reduction on 17 July 1973) that the cellular environment ought to be explicitly included in a characterization of dominance, since the environment can affect the dominance relation. In reply I noted that (i) in the case considered the environment is assumed, for reason of simplicity and logical clarity, to be constant and (ii) expanding the relation provided from a two-to a three-place predicate does not, in any event, interestingly alter the logic of the predicate reduction functions.
[26] This notation is a slight change from that used in my (1974a) and is due to suggestions by David Hull and William Wimsatt.
[27] In my (1967a), which outlined the tenets of the general reduction model (or 'paradigm'), I suggested that the model would be better comprehended if a simple example clarifying the dual (entity and predicate) nature of the requisite reduction functions was provided to illustrate the rather abstract form of the model. The example I chose was from genetics, with the entity in the reduced science being a 'gene' and the predicate being 'dominant'. I suggested that the predicate dominant could be interpreted molecularly as "the ability to make an active enzyme". This is a useful first approximation and, in point of historical fact, is not too different from the early and oversimplified 'presence and absence' theory of dominance proposed by Bateson in 1906. (See E. A. Carlson's (1966), Ch. 8, for a discussion of this theory and its weaknesses.) Hull has rightly criticized this interpretation as an inadequate explication at the molecular level of the property of dominance, though he has misinterpreted the function of the example in the 1967 article. In my (1969a), which was explicitly on reduction in biology, I indicated more precisely the complexity of predicate reduction functions. Essentially the same logic proposed there is offered in the present paper, along with a more detailed but still schematic explication of dominance and together with some suggestions for reducing still more complex predicates, such as 'epistatic', which Hull feels are particularly troublesome in reduction.
[28] For survey of the different types of ratios produced by genetic interaction, see R. C. King (1965), pp. 89ff, and R. P. Wagner and H. K. Mitchell (1964), pp. 392–438.
[29] Muller's original article suggesting his analysis (or, better, reanalysis) of dominant and recessive genes into amorph, hypomorph, and hypermorph classes appears in his (1932), p. 213. A more recent account of this quantitative theory of dominance appears in Wagner and Mitchell (1964).
[30] The synthetic identity interpretation of entity reduction functions and the extensional construal of n-ary predicates as ordered n-tuples of entities suggests, although we provisionally distinguish between reduction functions for entities and predicates, that ultimately all reduction functions connect reduced entities and reducing entities. It may also be useful to point out here that we can, for similar heuristic reasons, relax the requirement that the reduction functions for entities only range over (ordered n-tuples of) the reducing entities and allow predicates of the reducing theory to appear on the right-hand side of the entity reduction functions. Such a relaxation would very probably allow for an initially easier statement of the reduction functions, e.g., one could not incorporate properties of the DNA in reduction functions for genes. Again, though, if we ultimately construe predicates as ordered n-tuples of entities, the distinction formally collapses.
[31] See F. H. C. Crick (1958). In addition to primary structure, specification of secondary, tertiary and quatenary structures are probably also needed, given the current state of folding theories.

[32] This point is important and is one of the main areas of continued disagreement between Hull and myself. Hull seems to believe that the list of 'initial conditions' is likely to be impossibly long. I, on the other hand, believe that molecular biologists are able to identify a small set of crucial parameters as initial conditions in their elucidation of fundamental biochemical mechanisms. These initial conditions are usually discussed explicitly in the body of their research articles (sometimes in the 'materials and methods' sections). The specification of the initial conditions, e.g., the temperature at which a specific strain of organisms was incubated and the specific length of time, is absolutely necessary to allow repetition of the experiment in other laboratories whose scientists know the experiment only through the research article. The fact that mechanisms such as the Krebs cycle and the *lac* operon are formulatable in a relatively small set of sentences, with a finite list of relevant initial conditions specifiable, indicates to me that Hull is too pessimistic.

[33] See Wagner and Mitchell (1964) pp. 394–401, for a discussion of a biochemical explanation of gene interaction.

[34] See my (1974c) for a discussion of this strategy in connection with the development of the operon theory of Jacob and Monod.

[35] See Ruse, this symposium, p. 633.

[36] Nickels (1973), p. 184.

[37] Nickels (1973), p. 185.

[38] See my (1967a), p. 146.

[39] Nickels (1973), p. 195.

[40] Nickels (1973), p. 197.

[41] Nickels does not specifically address this case so I am perhaps unfairly characterizing his views without sufficient reason.

[42] Nickels has also raised certain problems about identification, but I do not think these come up forcefully in the context of the present case. See his comments in Nickels (1973), p. 194.

REFERENCES

Carlson, E. A., 1966, *The Gene: A Critical History*, Saunders, Philadelphia.
Causey, R., 1972, *J. Phil.* **69**, 407.
Crick, F. H. C., 1958, *Symp. Soc. Exp. Biol.* **12**, 138.
Feigl, H., 1958, in *Minnesota Studies in the Philosophy of Science, II*, (ed. by H. Feigl, M. Scriven, and G. Maxwell), University of Minnesota Press, Minneapolis.
Feyerabend, P. K., 1962, in *Minnesota Studies in the Philosophy of Science, III*, (ed. by H. Feigl and G. Maxwell), University of Minnesota Press, Minneapolis.
Feyerabend, P. K., 1965, in *Boston Studies in the Philosophy of Science, II*, (ed. by R. S. Cohen and M. W. Wartofsky), Humanities Press, New York.
Hull, D., 1972, *Phil. Sci.* **39**, 491.
Hull, D., 1973, in *Logic, Methodology, and Philosophy of Science, IV: Methodology and Philosophy of Biological Science* (ed. by P. Suppes *et al.*), North Holland, Amsterdam.
Hull, D., 1974, *Philosophy of Biological Science*, Prentice-Hall, Englewood Cliffs, N.J.
Hull, D., 1976, 'Informal Aspects of Theory Reduction', this symposium, p. 653.
King, R. C., 1965, *Genetics*, Oxford University Press, New York.
Kuhn, T. S., 1962, *The Structure of Scientific Revolutions*, University of Chicago Press, Chicago.
Massey, G., 1973, *Ann. Japan Assn. Phil. Sci.* **4**, 203.
Mills, D. R., Kramer, F. R., and Spiegelman, S., 1973, *Science* **180**, 916.

Muller, H. J., 1932, *Proc. Intern. Congr. Genet.*, Ithaca, N.Y., I, 213.
Nagel, E., 1949, in *Science and Civilization* (ed. by R. C. Stauffer), University of Wisconsin Press, Madison.
Nagel, E., 1961, *The Structure of Science*, Harcourt, Brace, and World, New York.
Nickels, T., 1973, *J. Phil.* **70**, 181.
Orgel, L. E., 1973, *The Origins of Life: Molecules and Natural Selection*, Wiley, New York.
Polanyi, M., 1968, *Science* **160**, 1308.
Quine, W. V. O., 1959, *Methods of Logic*, Holt, Rinehart, and Winston, New York.
Quine, W. V. O., 1964, *J. Phil.* **51**, 209.
Ruse, M., 1971a, *Dialectica* **25**, 17.
Ruse, M., 1971b, *Dialectica* **25**, 39.
Schaffner, K., 1967a, *Phil. Sci.* **34**, 137.
Schaffner, K., 1967b, *Science* **157**, 644.
Schaffner, K., 1969a, *Brit. J. Phil. Sci.* **20**, 325.
Schaffner, K., 1969b, *Amer. Sci.* **57**, 410.
Schaffner, K., 1974a, in R. Cohen and R. J. Seeger (ed.), *Boston Studies in the Philosophy of Science, II*, Reidel, Dordrecht, p. 207.
Schaffner, K., 1974b, *J. Hist. Biol.* **7**, 111.
Schaffner, K., 1974c, *Stud. Hist. Phil. Sci.* **4**, 349.
Shapere, D., 1974, in *The Structure of Scientific Theories*, (ed. by F. Suppe) University of Illinois Press, Urbana, p. 518.
Sklar, L., 1964, 'Intertheoretic Reduction in the Natural Sciences', unpubl. diss., Princeton University.
Wagner, R. P. and Mitchell, H. K., 1955, *Genetics and Metabolism*, Wiley, New York.
Wagner, R. P. and Mitchell, H. K., 1964, *Genetics and Metabolism*, 2nd ed., Wiley, New York.
Watson, J. D., 1970, *Molecular Biology of the Gene*, 2nd. ed., Benjamin, New York.
Wimsatt, W., 1976, this symposium, p. 671.

MICHAEL RUSE

REDUCTION IN GENETICS

There is a disagreement between Kenneth Schaffner and David Hull about the relationship between the biological theory of Mendelian genetics and the physico-chemical theory of molecular genetics.[1] Schaffner believes that the logical-empiricist thesis about theory-reduction is, in important respects, applicable to this relationship and illuminating when so applied. Hull denies its applicability and its illumination – indeed, he has gone as far as to say that "I find the logical empiricist analysis of reduction inadequate at best, wrong-headed at worst." (Hull, 1974a, 12) And he adds that "the conclusion seems inescapable that the logical empiricist analysis of reduction is not very instructive in the case of genetics. For my own part, I found that it hindered rather than facilitated understanding the relationship between Mendelian and molecular genetics." (Hull, 1974a, 44)

More precisely, the disagreement between Schaffner and Hull seems to be this. Schaffner follows Ernest Nagel (1961) in arguing that a theory-reduction has occurred when all the claims [2] of one theory T_1 (the reduced theory) can be shown to be deductive consequences of the claims of another theory T_2 (the reducing theory), or, when the two theories talk in different languages, of T_2 together with the principles which allow one to translate from the language of one theory to the language of the other. However, Schaffner goes beyond Nagel in arguing that we can legitimately weaken the above criteria of theory-reduction and still meaningfully talk of reduction (of T_1 to T_2) if, although a theory T_1 cannot be deduced from T_2, a corrected version of T_1, T_1^*, can be deduced from T_2. How much correction is permissible Schaffner leaves open, but he does specify that there must be 'strong analogy' between T_1 and T_1^*.

Applying this thesis about theory-reduction to biology, Schaffner agrees that traditional Mendelian genetics can not in fact be deduced from molecular genetics (even when the latter is supplemented with appropriate translation principles), if for no other reason than that the two theories make conflicting claims. Mendelian genetics denies that the basic heritable unit of biological function (the Mendelian gene) is divisible by crossing-over – molecular genetics allows for such a division in its unit of function (the molecular gene, which is in fact a strip of DNA). However, argues Schaffner, Mendelian genetics can easily be corrected, and this corrected version can be deduced from molecular genetics. Indeed, this correction has taken place – we now have the biological theory of transmission genetics, where the unit of function, the cistron, is divisible by crossing-over. Moreover, argues Schaffner, there is strong analogy between traditional Mendelian genetics and transmission genetics. Hence, he concludes that we can meaningfully speak of theory-reduction in the case of genetics – 'reduction' in the weaker, modified sense outlined above.

Hull's disagreements seem to be three. In the first place, he does not think that transmission genetics has in fact been deduced from molecular genetics and he is not convinced, at least not convinced *a priori*, that this can in fact be done. In the second place, even if such a deduction were done, he feels that there would be so great a difference between the deduced transmission genetics and traditional Mendelian genetics that one could not properly talk of reduction, even in Schaffner's weaker sense. In the third place, perhaps most importantly, Hull feels that the demand for a reduction is misleading – it does not tell us much about present real science, and it leads to little of new scientific or philosophical interest.

Since, after some initial dithering, I have committed myself fairly firmly to Schaffner's side of the argument,[3] it behoves me to try to answer Hull's objections. This I shall now do.

1. Hull points out rightly that as things stand at the moment, transmission genetics has not been deduced from molecular genetics – apart from anything else such deduction requires that the two theories be in hypothetico-deductive form and both are in a far looser state than that. Hull has also done great service, to me at least, in pointing out what an immense amount of work such a deduction would still require. However, Hull wants to go farther than this, and it is here that I start to feel uncomfortable. As Hull notes correctly, there are four possible relations between the concepts of transmission and molecular genetics – one to one (i.e. the entities[4] referred to by a concept in one theory are the very same entities referred to by a concept in the other theory), one to

many (a concept in one theory refers to entities referred to by two or more concepts in the other theory), many to one, and many to many. Now, if we have a one to one relationship, a deduction is at least possible. Letting small Greek letters stand for concepts in molecular genetics, and capital Arabic letters stand for concepts in transmission genetics, then if we have two laws $\alpha \to \beta$, and $A \to B$, and one to one correspondences $A = \alpha$ and $\beta = B$, then we can deduce $A \to B$ from $\alpha \to \beta$.[5] If we have one to many correspondences from transmission genetics to molecular genetics, reduction is still possible. Suppose we have transmission genetical law $A \to B$, molecular genetical laws $\alpha_1 \to \beta_1$, $\alpha_2 \to \beta_2$, and correspondences $A = (\alpha_1 \vee \alpha_2)$, $B = (\beta_1 \vee \beta_2)$, then we can deduce $A \to B$ from the molecular laws. But if we have many-one or many-many correspondences, deduction is impossible. Thus, for instance, from $\alpha \to \beta$, $(A_1 \vee A_2) = \alpha$, and $(B_1 \vee B_2) = \beta$ we can get neither $A_1 \to B_1$, nor $A_2 \to B_2$. All we can get is the weaker $(A_1 \vee A_2) \to (B_1 \vee B_2)$.

Hull seems to feel that in an important sense we are stuck with many-one or many-many correspondences of this kind just described. Having pointed out (truly) that we often get one-many correspondences (from transmission to molecular genetics), he writes:

Conversely, the same molecular mechanisms can produce different phenotypic effects. For example, in a heterozygote, one allele may be completely operative, the other completely inoperative. Depending on the nature of the alleles and the biosynthetic pathways in which they are functioning the effect can vary. It may be the case that the single allele can produce all the product necessary to maintain the reaction at full capacity. If so, the phenomenon would be termed "dominant." Or it might be the case that the presence of the single operative allele decreases the rate of the reaction only slightly or perhaps cuts it in half. In such cases, the phenomenon would be termed "incomplete dominance." Hence, it would seem that even at the molecular level the relation between Mendelian predicate terms and molecular mechanisms is many-many. (Hull, 1974a, 40–41)

Clearly, if this is all that can be said on the matter, a deduction (and hence a reduction) is impossible. But there is a fairly obvious reply to this objection. Water is H_2O whether it be at $20\,°C$ or $80\,°C$; but being of the same chemical constitution at the two temperatures hardly rules out a molecular explanation of the phenomenological differences of water at the two temperatures. More information must be fed in at the molecular level – for example, information about the differences in energy of the molecules at the two temperatures. Similarly, what seems to follow from the genetical case is not the conclusion that a deduction is in principle impossible, but that more information is required at the molecular level showing why the various phenotypic effects occur. Then, we can get away from such correspondences as $\beta = (B_1 \vee B_2)$, replace them by such as $\beta_1 = B_1$, and $\beta_2 = B_2$, and hence deduction is once again possible.

Hull is not unaware of this counter-move.

At this point, one might object that perhaps the same molecular mechanism can result in different phenotypic effects, but that is because relevant factors are being left out – the temperature, hydrogen-ion concentration, and so on. Once all these relevant factors have been included, the one-many relation from the molecular to the phenotypic level will have been converted to a one-one relation. (Hull, 1974a, 42)

However, Hull seems unimpressed by this move. It is, we are told, "a covert restatement of the principle of deterministic causality. The same cause always produces the same effect. If the effects are different, then the causes must be different." (Hull, 1974a, 42). It is certainly not Hull's intention to deny the principle of causality – that would be a bit like attacking motherhood – but he feels that appeals to causality (perhaps like appeals to motherhood) are really so sweeping as to be practically uninformative. We have been offered some rather strong claims by the logical empiricist about the relationship between molecular and biological genetics, and suddenly these collapse into truisms about causality.

Although there is certainly truth in this position, I feel Hull's conclusion is nevertheless an overstatement. Had the logical empiricist nothing more to go on than a belief in the principle of causality, then it could hardly be claimed that there is a reduction – even of the weaker kind suggested by Schaffner – between molecular and Mendelian genetics. However, there is more than this – there is evidence which I think makes the logical empiricist's position far more plausible. In particular, there are cases where, given a general molecular mechanism which could lead to various Mendelian effects, by specifying the peculiarities of a particular case, biologists have actually followed through to a uniquely determined effect. In other words, logical empiricists have more to rely on than general beliefs in causality – they can point to actual cases where the kinds of connexions they believe ought in principle to hold have in practise been shown to hold.

Take for example a molecular mechanism which Hull cites as implying a many-many relationship between Mendelian and molecular genetics, namely molecular heterozygosity. This arises when one has different allelic molecular genes, and Hull's claim is that all kinds of biological phenotypes are possible – hence a reduction is impossible (or one has to resort to general claims about causality). An instance of this heterozygosity occurs in the case of sickle-cell anaemia in humans.[6] This disease, endemic in certain African tribes (and thus to be found in American negroes), normally leads to death in early childhood. The disease is known to be caused by the mutant allele of a human hemoglobin gene – homozygotes for the so-called 'sickle-cell' gene suffer from the disease. Heterozygotes however do not suffer in any appreciable way from anaemia, although the 'wild-type' allele is not entirely dominant over the sickle-cell allele. Where the heterozygote differs from the normal homozygote is that, apparently because of the heterozygosity, it has an increased resistance to malaria. Thus the sickle-cell gene is maintained in malarially infected districts in a balanced situation.

Now, given the way Hull presents the case, one would hardly expect that the geneticist could go from the molecular heterozygosity for the sickle-cell allele to increased malarial resistance in the heterozygote. We have a one-many relationship, and if we assume more then all we are doing is committing ourselves to a "covert restatement of the principle of deterministic causality." However, in fact, a great deal is known about the molecular mechanics of the sickle-cell case, so much so in fact that geneticists can practically give an entire molecular explanation of the (biological) phenotypic effects, without need for blind faith in principles of causality. In particular, hemoglobin which consists essentially of two pairs of polypeptide chains (α and β) is thought to be the product of two (non-allelic) genes. The sickle-cell mutant causes the substitution of one amino acid in the β-chain for another amino acid (valine for glutamic acid). This substitution leads to 'stacking' of the molecules in a uniform alignment, which increases the viscosity of the hemoglobin, which leads in turn to the characteristic sickle-shaped distortion of the cell. In the heterozygote, because of the existence of some normal hemoglobin one gets less stacking in a cell, which can thus function, but still come increased viscosity. Parasitical infection by the falciparum malarial organism of such a heterozygote cell increases sickling because the parasite uses oxygen, thus further increasing viscosity. It is then believed that the infected, much sickled cell is removed by the body by phagocytosis. Thus the malarial parasite sows the seeds of its own destruction.

It seems to me that the conclusion that all that one has in this case is a general molecular mechanism and a belief in the principle of causality is unwarranted. Knowledge of the general molecular situation and of the specific molecular peculiarities of the sickle-cell case points in a fairly unambiguous way to the biological genetical consequences – as the logical empiricist suggests they should. In short, given that this example is but one of many that could be cited, whilst I agree entirely with Hull that an actual reduction might be a lot more complex than some of us have supposed – and much credit is due to Hull for underlining this fact – I do not find troublesome his doubts about the possibility of a reduction even in principle.

In his most recent writings on this subject, Hull has a variant on the argument being considered. He suggests that even if one's assumption of the principle of causality bore fruit and one were to link up molecular and biological mechanisms entirely, one might still have no reduction. So detailed an analysis may be required, that at best one could link individual instances with individual instances – something falling short of a reduction which is between theories, which (by definition) go beyond the individual to the general. Hull writes:

> It may well be true that a particular molecular mechanism of some kind or other can be found for each instance of a particular pattern of Mendelian genetics on a case by case basis, but more than that is required in reduction. Reduction functions relate *kinds* of entities and predicates, not particulars. The problem is to discover natural kinds in the two theories which can be related systematically in reduction functions. (Hull, 1974b, 20)

Again I find Hull's objection not to be devistating in theory (although again I think he points to an important practical question). Clearly Hull is right in thinking that we are going to need a terrific number of connexions between molecular mechanisms and Mendelian (or transmission) genetical effects, so many so in fact that one might well question whether the whole enterprise of articulating a fully worked-out reduction is worthwhile. But is the result really going to be that *every* instance of a Mendelian effect will have to be treated separately? Surely, for example, one is going to have similar instances between members of the same species. Take for example the M and N blood groups in man, which in the Mendelian (or transmission) world are believed a function of the segregation of two alleles. Is Hull really suggesting that every case of a human homozygous for the M-gene (and thus with blood which works in a particular way in agglutination tests) is going to call for a different molecular mechanism? If he is, then more argument by him is needed. Or take for example the sickle-cell phenomenon just discussed. Is it Hull's claim that geneticists are wrong in dealing with the heterozygotes (and homozygotes) on a collective basis, rather than on an individual by individual basis? If so, then yet more argument is needed – argument which I think Hull might find rather embarrassing for he must show that scientists ought not be behaving as they do, whereas the thrust of much of his criticism of the logical empiricists is that logical empiricists are too ready to prescribe the ideal course of science, when they should be describing its actual course.

In short, it seems clear that when dealing with members of the same species (or sub-groups within a species) generalities can be made between molecular and biological levels. *A priori* it seems plausible also to suggest that in some cases such generalities might go beyond one species – particularly to members of closely related species. Thus, required generalities do seem available. Moreover, although one cannot deny that the available generalities might be a lot more restricted than most reductionists have supposed, Hull has really given no arguments ruling out general similarities between widely different organisms linking molecular mechanisms and biological effects – general similarities which can be applied to specific cases. Thus take for example the biological phenomenon of epistasis, where the phenotypic effects of alleles at one locus seem to be a function in part of alleles at another locus. In such a situation, although different combinations of alleles a_1 and a_2 at one locus might cause phenotypes A_1 and A_2 under normal circumstances, it may be that neither phenotype will occur unless one has at least one instance of allele b_1 at another locus. Hull rightly points out that epistasis may have several molecular mechanisms – nevertheless there are some general models which seem applicable to widely differing cases. Thus, for example, the most obvious model supposes that we have a cellular product which is being produced sequentially by enzymes caused by the two sets of alleles.

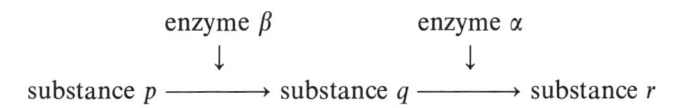

$$\begin{array}{ccc} \text{enzyme } \beta & & \text{enzyme } \alpha \\ \downarrow & & \downarrow \\ \text{substance } p \longrightarrow \text{substance } q & \longrightarrow & \text{substance } r \end{array}$$

Using the above terminology, we might suppose that enzyme β is produced when and only when allele b_1 is present, that enzyme α is produced when and only when allele a_1 is present, that phenotype A_1 occurs when and only when substance r is produced, that phenotype A_2 occurs when and only when substance r is absent but substance q is present, and that when neither substances q nor r are present a different phenotype (say A_3) occurs. This model is known to fit many cases of epistasis. One instance is epistatic expression of hydrocyanic acid in white clover, something which leads to vigorous growth (Burns, 1972, 81–2). Taken alone, one pair of alleles in clover segregate 3:1 for HCN:cyanogenic glucoside (*i.e.* the HCN-producing allele is dominant). However, the segregation of this pair of genes is in turn controlled epistatically by another pair which segregate 3:1 for the production of cyanogenic glucoside. And it is known moreover that this case of epistasis fits the above model.

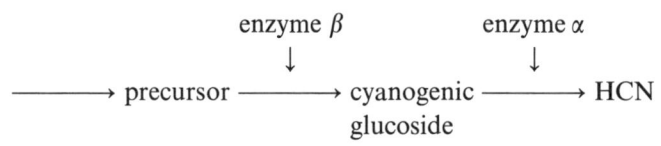

$$\begin{array}{ccc} & \text{enzyme } \beta & \text{enzyme } \alpha \\ & \downarrow & \downarrow \\ \longrightarrow \text{precursor} & \longrightarrow \text{cyanogenic} & \longrightarrow \text{HCN} \\ & \text{glucoside} & \end{array}$$

There are many other cases of epistasis known to fit this model, particularly those affecting colour. (Strickberger, 1968, gives examples of epistasis involving colour, and indeed, uses the above model to explain them.) Hence, although Hull is no doubt right in doubting that one molecular mechanism will fit all cases of epistasis, it seems wrong to conclude that something completely new must be found for every case. Some generalities do apply.

2. Hull's second criticism of Schaffner's position is that were one to deduce a biological genetics from molecular genetics, the deduced biological genetics would be so different from Mendelian genetics that talk of "reduction" would still be inappropriate. There would be no strong analogy between traditional Mendelian genetics and the corrected biological genetics ("transmission genetics"). "The amount of reconstruction necessary to permit the deduction of transmission genetics from molecular genetics would seem to preclude the existence of any strong analogy between classical Mendelian genetics and reconstructed transmission genetics." (Hull, 1974a, 43)

Of course, as Hull rightly points out, part of the problem here is that to date no one has been prepared to give a thorough explication of what might be meant by "strong analogy," and it is clear that if the debate about reduction in genetics is to continue, someone soon must turn to this task. At present one must work by example and rough intuition – Kepler's astronomy and that which can be deduced from Newtonian axioms do seem strongly analogous, Darwin's theory of evolution through natural selection and Richard Owen's Platonic theory of organic origins seem to be miles apart. But even though we have to work with such crude guidelines a these, I would suggest that Hull's conclusion is not well taken.

It is agreed by all that Mendelian genetics and transmission genetics are not identical – if they were, then there would be no argument. Mendelian genetics rolls together the units of function, mutation, and crossing-over, whereas transmission genetics separates these out, allowing mutation of just part of the unit of function and crossing-over within the unit of function. But where else are there essential differences? It seems to me that just about everything else is retained from the old biological genetics in the new biological genetics, and what differences there are do not make themselves much felt in a lot of cases. Take for example the work of someone like Th. Dobzhansky. The third and final edition of his classic *Genetics and the Origin of Species* appeared in 1951, before the work of Watson, Crick, or Benzer. In it he relied on the Mendelian law of segregation, its generalization to large groups (the Hardy-Weinberg law), and so on. The revised and retitled edition of his work *Genetics of the Evolutionary Process* appeared in 1970. The early chapters discuss, as accepted scientific fact, the major findings of molecular biology, for instance the Watson-Crick model of DNA. But then what do we find? All the old favourites like the Hardy-Weinberg law make their reappearance and play just as great a role in Dobzhansky's theorizing as they ever did. Hence, not much change between the old and new seems to have occurred.

Of course one might argue that Dobzhansky is holding to and working with two contradictory theories – traditional Mendelian genetics and modern molecular genetics. But I see no reason why one should assume this. He identifies explicitly his unit of function (which he normally calls the "gene") with the unit of function of transmission genetics, the cistron. But, having made this change, he can immediately make use of just about all his old theory, because for the kinds of problems he tackles the modern sophisticated analysis (replacing the old Mendelian gene) is not needed. Basically, Dobzhansky is untroubled by crossing-over within the unit of function, because he works at a rather cruder level analysis. Hence, it seems to me true to say that the modern *biological* geneticist incorporates into his new biological genetics much of the old Mendelian genetics.

To conclude this section let me look briefly at an argument Hull puts forward in support of his position that there can be no strong analogy between traditional Mendelian genetics and the corrected biological genetics derivable from molecular genetics (what I have been calling 'transmission genetics'). Hull's argument is that any biological genetics derived from molecular genetics must reflect differences at the molecular level, even though there may be no such differences at the biological level. But if there are no differences at the biological level, they will not presently be mentioned in biological genetics (either Mendelian or transmission). Hence, argues Hull, a great deal of correction in this respect will be required before Mendelian genetics can be converted to a derivable biological genetics – therefore, strong analogy seems out of the question. Hull writes:

Transmission geneticists claim no interest in the variety of biochemical mechanisms that eventuate in the characters whose transmission they follow, unless these differences are reflected in different Mendelian ratios. If not, this additional information about biochemical pathways is of no use to them. However, if reduction functions are to be established that associate the terms of transmission and molecular genetics, the important biochemical differences must be read back into transmission genetics. If the same molecular structure is produced in two different ways, this difference must be reflected in the relevant reduction functions and Mendelian genetics changed accordingly. (Hull, 1974a, 43)

Put simply, this argument seems fallacious. As Hull himself points out, a many-one relationship from the molecular to the biological does not prevent a reduction. But this kind of many-one relationship is what is being supposed here. Suppose in a species a molecular gene product α which leads to phenotype A can be produced in two ways, as follows. (a is the biological gene, G the molecular gene, α, β, γ, δ are molecular cell products, the alleles B_1, B_2, C_1, C_2 are producing enzymes driving the changes, alleles at B and C are supposed always linked, perhaps through suppression of crossing-over). We have here the kind of situation envisioned by Hull – biological theory could be quite ignorant of the

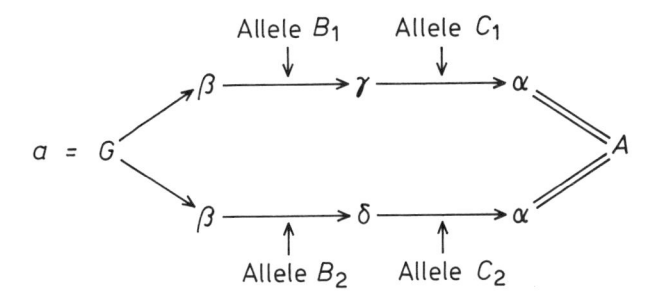

$$\begin{array}{c}\text{Allele } B_1 \quad \text{Allele } C_1 \\ \downarrow \quad\quad \downarrow \\ a = G \begin{array}{c} \nearrow \beta \longrightarrow \gamma \longrightarrow \alpha \searrow \\ \searrow \beta \longrightarrow \delta \longrightarrow \alpha \nearrow \end{array} A \\ \uparrow \quad\quad \uparrow \\ \text{Allele } B_2 \quad \text{Allele } C_2 \end{array}$$

different molecular pathways – but a reduction is possible. From, $G \to \beta$, $\beta \to \gamma$, $\beta \to \delta$, $\gamma \to \alpha$, $\delta \to \alpha$, $a = G$, $\alpha = A$, we can get $a \to A$. We can also get $a \to A$ if the two pathways lead to slightly different molecular end products (α_1 and α_2), if, as supposed, the molecular difference makes no difference at the gross phenotypic level. There is certainly no need, as Hull suggests, to reconstruct our biological theory (say from A to A_1 and A_2) to reflect the differences at the molecular level.

3. Hull's third criticism is probably his most important. This is that there is something really rather irrelevant about the logical empiricist thesis of reduction. Somehow it directs us away from an understanding of what is happening in genetics, rather than towards it. Hull writes:

> To my knowledge, no biologist is currently engaged in the attempt to reconstruct Mendelian and molecular genetics so that Mendelian genetics can be derived from molecular genetics. What is more, I can think of no reason to encourage a biologist to do so. This state of affairs strikes me as strange.... Either this reduction is a peculiar case or else whatever it is that makes the pre-analytic notion of reduction seem important has dropped out of the logical empiricist analysis of reduction. (Hull, 1973, 622)

Somewhat paradoxically Schaffner (and Ruse) agree fully with Hull about the actual place reduction seems to play in geneticists' research programmes. Schaffner concedes that the really exciting advances in molecular genetics have not come through the attempt to spell out a reduction, and he writes:

> I wish to propose that the aim of reductionism is *peripheral* to molecular biology, and that an attempt to construe the development of molecular biology as exemplifying a research program or set of research programs whose conscious and constant *intent* is reductionism is both inappropriate and historically misleading. (Schaffner, 1974a, 111)

In a similar vein, Ruse thought it prudent to add the following footnote to his most recent discussion of reduction: "It is perhaps important to emphasize that no actual deduction from a purely molecular theory to a purely biological theory seems yet in existence. Indeed, reading the works of biologists one gets the feeling that the whole question of reduction is more of a philosopher's problem than a scientist's." (Ruse, 1973, 207n) What more damning comment could one make than that! Perhaps Hull is right and we should find better things to do with our time.

In defence of the logical empiricist, I offer the following two arguments. First, although some of us may secretly have a hope that our most recent attack on the covering-law model will some day win us a Nobel prize in science, we are at present trying to do philosophy not science. Our job is not to make brilliant new scientific discoveries, but to analyse and understand the way science works. In John Locke's great words "it is ambition enough to be employed as an under-labourer in clearing the ground a little, and removing some of the rubbish that lies in the way to knowledge..." (Locke, 1959, 14) Hence, that scientists are not pushing full-speed ahead on a reduction does not mean that philosophers ought not use the formal, idealized analyses of logical empiricism to see if they can throw light on the relationship between molecular and biological genetics. Scientists have their job to do, we have ours, and ours is certainly not to follow slavishly in the footsteps of science – nor is it necessarily a mark that we have done our job badly if scientists do not immediately rush to put flesh on our formal analyses.

Since this point I am trying to make is rather important, it may perhaps be worthwhile to consider for a moment a closely related example. No one could deny that in its present form evolutionary theory taken as a whole is not a rigorously formulated hypothetico-deductive system. But it does not necessarily follow that the hypothetico-deductive ideal is irrelevant to evolutionary studies – that the philosopher ought not use the hypothetico-deductive model in his attempt to understand what the evolutionist is doing. Certain parts of evolutionary science, particularly those concerned with the spread of genes in populations, do approximate to a hypothetico-deductive theory and the philosopher can therefore use his model directly to throw light on these parts and on how these parts relate to other parts of evolutionary studies (like organic geographical distribution). Moreover, those places where the theory fails to fit the model can guide the philosopher to an understanding of the problems facing the evolutionist – essential data irretrievably lost, vast timespans, and so on. And although it is certainly true that logical empiricists would (or at least should) feel most uncomfortable were there no parts of evolutionary theory in any sense hypothetico-deductive or were there no sense in which evolutionary studies had progressed towards a greater manifestation of the hypothetico-deductive ideal, there is nothing in their epousal of this ideal which insists that scientists must put this ideal before all else. It is fully recognized that things like missing data might make a significant complete hypothetico-deductive evolutionary theory a practical impossibility, and that scientists might feel the technical details of filling out such a theory with full rigour to be rather boring and not something which would lead to dramatic new insights – as in mathematical proofs we often see how a problem can be solved but might not wish to fill in every last step.

An exactly analogous situation seems to me to prevail with respect to theory reduction and genetics. Because of their understanding of DNA, the genetic code, and so on, biologists can see in broad outline how transmission genetics follows from molecular genetics (in the manner suggested by logical empiricists), and as we have seen in the sickle-cell case, when it is in their interests they can often trace through the details of some particular case with precision. But for various reasons, particularly because it does not seem very exciting and would seem only to involve filling out details of a general picture already broadly understood, biologists are not trying to set up a fully articulated deduction of transmission genetics from molecular genetics. The logical empiricist, however, can and does recognize this fact. It is the broad outline which interests him – the general relationship between molecular and non-molecular genetics – and the way in which particular details are filled in when there is a specific need or interest. Hence, I would suggest that Hull is wrong when he writes that:

> The crucial observation is that no geneticists to my knowledge are attempting to derive the principles of transmission genetics from those of molecular genetics. But according to the logical empiricist analysis of reduction, this is precisely what they should be doing. (Hull, 1974a, 44)

But, in a sense this is all rather tangential. Hull's main claim is not so much that the logical empiricist thesis on reduction is scientifically irrelevant (although he thinks it is), but that it is philosophically harmful. It "hindered" his understanding of the true situation in genetics, rather than facilitated it (Hull, 1974a, 44). It is here that I want to make my second argument, for I think Hull is wrong in claiming this. I think rather that the logical empiricist analysis does help us towards an understanding of what is really happening in genetics.

In order to defend the logical empiricist analysis, let me first ask of Hull the question – What, if the logical empiricist account is incorrect, is indeed the true situation? At times Hull seems to have no answer to this question, and seems almost in despair to give up hope of finding such an analysis. Rather, somewhat wearily, even he seems prepared to go along with the logical empiricist account.

> If my estimation of the situation in genetics turns out to be accurate, then even the latest versions of the logical empiricist analysis of reduction are inadequate and must be improved. This conclusion, however, should not be taken as being too damning of the logical empiricist analysis of reduction. In spite of its shortcomings, it is currently the best analysis which we have of reduction. In fact, it is the only analysis which we have. (Hull, 1973, 634)

However, at other times he is more positive in his opposition. The logical empiricist sees the relationship between the two theories as one of logical, deductive connexion. The one theory, in some sense, 'contains' the other. Hull suggests however that the two theories belong in some important way to different worlds – at least, to incommesurable ways of viewing this world. He writes:

> Most contemporary geneticists know both theories. They can operate successfully within the conceptual framework of each and even leap nimbly back and forth between the two disciplines, but they cannot specify how they accomplish this feat of conceptual gymnastics. Whatever connections there might be, they are subliminal. In a word, those geneticists who work both in Mendelian and molecular genetics are schizophrenic. The transitions which they make from one conceptual schema to the other are not so much inferences as *gestalt* shifts... (Hull, 1973, 626)

We have then two sharply different analyses of the situation in genetics. One sees the reduced theory T_1 in some sense contained in the reducing theory T_2, the other sees T_1 and T_2 as being logically quite distinct. To use N. R. Hanson's example of a drawing which looks, in one way, like a rabbit, and, in another way, like a duck, T_1 is a rabbit and T_2 is a duck. This is a somewhat extreme position I am ascribing to Hull, and it should in fairness be added that he does qualify the passage just quoted by saying "To be sure, these observations are psychological in nature, and reduction as set out by philosophers of science is a logical relation between rational reconstructions of s jentific theories..." On the other hand he adds "but these psychological facts provide indirect evidence for the position I am about to urge concerning the logical relation." What is clear is that even if Hull would not accept all the implications that Hanson might draw from the situation, he believes that from a formal viewpoint molecular genetics is a new theory replacing Mendelian genetics, not one absorbing Mendelian genetics.

If the logical empiricist analysis of reduction is correct, then Mendelian genetics cannot be reduced to molecular genetics. The long-awaited reduction of a biological theory to physics and chemistry turns out not to be a case of "reduction" after all, but an example of replacement. (Hull, 1974a, 44)

Hull is certainly right in pointing to the way in which geneticists switch blithely to and fro between the molecular and biological levels, without giving too much thought to the relationship between them. But the extreme position to which he pushes this observation clouds rather than clears the view to correct understanding. Consider for a moment a situation in genetics where geneticists do work at both molecular and biological levels, namely over the problem of the amount of genetically caused variation in natural populations.[7] There are, or rather were, two hypotheses about variation when the problem was considered entirely at the biological level. On the one hand, followers of H. J. Muller argued for the so-called 'classical' hypothesis, namely that most organisms in a population are genetically similar, with just the occasional mutant allele, usually recessive, usually deleterious, deviating from the 'wild-type.' On the other hand, followers of Th. Dobzhansky supported the 'balance' hypothesis – they saw selection primarily maintaining genetic variability in a population, for example through heterosis (this refers to a situation like the sickle-cell case, where the heterozygote for two alleles is fitter than either homozygote – at equilibrium both alleles are retained in a balance). Hence, for balance hypothesis supporters there is in an important sense no such thing as a wild-type – almost every genotype in a population is different. Diagrammatically we can represent the differences between the two populations as follows.

$$
\begin{array}{ll}
& \text{Two randomly chosen individuals} \\
\text{Classical} & \dfrac{+++m+\cdots++}{+++++\cdots m+} \quad \dfrac{+++++\cdots m+}{+++++\cdots ++} \\
\text{hypothesis} & \\
& (+\ = \text{wild-type allele};\ m = \text{deleterious mutant})
\end{array}
$$

Balance hypothesis $\dfrac{A_3B_2C_2DE_5\cdots Z_2}{A_1B_7C_2DE_2\cdots Z_3}$ $\dfrac{A_2B_4C_1DE_2\cdots Z_1}{A_3B_5C_2DE_3\cdots Z_1}$

(A_1, A_2, = different alleles)

Now, at the biological level it is all but impossible to decide between these hypotheses. Balance hypothesis supporters argued that the phenotypic differences caused by allelic changes would normally be very slight (although such differences could have significant evolutionary implications). But the measurement of slight differences due to genetic change is incredibly difficult, mainly because environmental changes can also cause phenotypic differences and one can never be absolutely sure that one is keeping the environment constant (remember, the environment of an individual includes the other members of its population).

What geneticists have done in the past ten years, however, is switch the debate from the biological to the molecular level. Functional strips of DNA lead, via RNA, to polypeptide chains – strings of linked amino acids. Changes in the DNA lead (taking note of redundancies) to changes in the polypeptide chains. Using a technique known as 'gel electrophoresis' (which is based on the fact that some amino acids have different electrostatic charges and that hence changes in polypeptide chains lead to changes in charge) geneticists can detect changes in polypeptide chains, which, in turn, they interpret as changes in the DNA structure. Both classical and balance hypothesis supporters think that the findings about polypeptide chains are relevant to their debate. In particular it is found that polypeptide chains (caused by DNA strips at the same locus) do vary greatly. Balance hypothesis supporters take this to be confirmation of their position, namely that selection maintains genetic variability in populations. Classical supporters, whilst they cannot deny this variation, argue that the variation is non-adaptive, that it is due to drift not selection, and they can give both theoretical and empirical arguments to support their position.

Clearly the resort to the molecular level has not brought debate to an end – if anything it rages more fiercely than ever before. But what I want to ask is what is the most reasonable way to interpret geneticists' resort to the molecular level at all? Why did geneticists get so excited about gel electrophoresis and so on? On Hull's interpretation of the situation it is difficult to see why geneticists should have turned (in the instance we are considering) to the molecular level, thinking (as they obviously did) that the molecular level was going to break down some barriers. Given Hull's analysis, geneticists had a debate at the biological level. It was not getting untangled. So they gave up, made a gesalt switch, and turned to a different problem, a problem at the molecular level. Two levels, two problems, and never the twain shall meet. On the logical empiricist analysis however, all is readily explicable, including geneticists' enthusiasm for the molecular level. Mendelian genes (or cistrons) can be identified with strips of DNA, DNA causes polypeptide chains, and these either lead to or can be identified with gross phenotypic characteristics. Hence, inasmuch as one gathers information about genetically caused variation amongst polypeptide chains, one is throwing light on genetically caused variation at the gross phenotypic level – the matter at the heart of the debate in the first place.

Thus, what we find in the case of balance hypothesis supporters is the belief that much of the difference in the polypeptide chains reflects as difference at the gross phenotypic level– difference which they believe must be maintained by selection. And in support of their case they point to several facts which they believe explicable only on their hypothesis (at least, certainly not on the classical hypothesis). One thing in particular they believe proves their case, namely that polypeptide ratios between closely related groups are often very similar, pointing to a systematic cause, which they believe can only be selection and is certainly not drift.

But the position of classical hypothesis supporters is also readily explicable given the logical empiricist analysis. They cannot deny the differences in polypeptide chains, however they argue that either different chains lead to the same gross phenotypic effects, or that different chains lead to different effects but that these effects are selectively neutral between themselves. In other words, they argue either for a many-one relationship (from molecular to biological) or for a one-one relationship, either of which, as we have seen, is permissible in a reduction. But either way, we see that they do believe that polypeptides have direct implications for gross phenotypic differences, which is what the logical empiricist account leads us to expect, and indeed demands.

In short, I suggest that the logical empiricist account throws much light on the current debate in population genetics, something which Hull's alternative does not. Thus I suggest that the logical empiricist account of reduction as applied to genetics is not as irrelevant or misleading as Hull claims.

University of Guelph,
Ontario, Canada

NOTES

[1] Pertinent references are Schaffner (1967), (1969), (1974a), (1974b); Hull (1972), (1973), (1974a), (1974b).
[2] Particularly the axioms.
[3] See Ruse (1973); but see also Ruse (1971).
[4] I mention here only entities – there is also the question of properties, but they introduce no new principles into the discussion.
[5] '→' stands for some kind of causal connexion; '=' stands for some kind of sameness – Schaffner suggests synthetic identity. The concepts in these laws and identities might well refer to combinations of entities and properties.
[6] A general discussion of this phenomenon can be found in Ruse (1973). The specific details I am about to give now can be found in Winchester (1972), Strickberger (1968), Luzzatto *et al.* (1970), and references.
[7] Full details can be found in Lewontin (1974).

Burns, G. W.: 1972, *The Science of Genetics*, 2nd ed., Macmillan, New York.
Dobzhansky, Th.: 1951, *Genetics and the Origin of Species*, 3rd ed. 1951, Columbia University Press, New York.
Dobzhansky, Th.: 1970, *Genetics of the Evolutionary Process*, Columbia, New York.
Hull, D. L.: 1972, 'Reduction in Genetics – Biology or Philosophy?', *Phil. Sci.* **39**, 491–99.
Hull, D. L.: 1973, 'Reduction in Genetics – Doing the Impossible', in P. Suppes *et al.* (eds.), *Logic, Methodology and Philosophy of Science* **IV**, 619–35.
Hull, D. L.: 1974a, *Philosophy of Biological Science*, Prentice-Hall, Englewood Cliffs.
Hull, D. L.: 1974b, 'Informal Aspects of Theory Reduction'. Read at P.S.A. conference 1974.
Lewontin, R. C.: 1974, *The Genetic Basis of Evolutionary Change*, Columbia University Press, New York.
Locke, J.: 1959, A. C. Fraser (ed.), *An Essay Concerning Human Understanding*, Dover, New York.
Luzzatto, L., Nwachuku-Jarrett, E. S., and Reddy, S.: 1970, 'Increased Sickling of Parasitised Erythrocytes as Mechanism of Resistance Against Malaria in the Sickle-Cell Trait', *Lancet* **I**, 319–22.
Nagel, E.: 1961, *The Structure of Science*, Routledge and Kegan Paul, London.
Ruse, M.: 1971, 'Reduction, Replacement, and Molecular Biology', *Dialectica* **25**, 39–72.
Ruse, M.: 1973, *The Philosophy of Biology*, Hutchinson University Library, London.
Schaffner, K. F.: 1967, 'Approaches to Reduction', *Phil. Sci.* **34**, 137–47.
Schaffner, K. F.: 1969, 'The Watson-Crick Model and Reductionism', *Brit. J. Phil. Sci.* **20**, 325–48.

Schaffner, K. F.: 1974a, 'The Peripherality of Reductionism in the Development of Molecular Biology', *J. Hist. Biol.* **7**, 111–39.
Schaffner, K. F.: 1974b, 'Reductionism in Biology: Prospects and Problems', Read at P.S.A. conference 1974.
Strickberger, M. W.: 1968, *Genetics*, Macmillan, New York.

DAVID L. HULL

INFORMAL ASPECTS OF THEORY REDUCTION

The issues which separate the members of this symposium concern the nature of reduction and scientific theories. However, these issues involve primarily informal aspects of theory reduction. But more than this, they concern the nature of philosophy of science itself.

Kenneth Schaffner (1967) has set out what he terms a general reduction paradigm, a development of earlier efforts by such philosophers as Ernest Nagel, J. H. Woodger, and Carl Hempel. This model, according to Schaffner, "represents an ideal standard for *accomplished* reductions, and does not characterize the research programmes of molecular biologists." Following the lead of his predecessors, Schaffner does not intend for his model of theory reduction to characterize the ongoing process of science (Wimsatt's rational$_1$) but an abstract, formal relation between atemporal rational reconstructions of scientific theories (Wimsatt's rational$_2$).

Numerous objections have been raised to this way of doing philosophy of science. In this paper, I will deal with three. First, very little has been said about the process of rational reconstruction. How does one go from scientific theories as scientists set them out (what might be termed "raw science") to the rational reconstructions of philosophers? By what criteria are alternative analyses to be judged? Second, there is little agreement about how explicit and complete these analyses have to be. Must philosophers actually axiomatize the relevant theories, set out the appropriate reduction functions, and then carry out the derivation, or are vague gestures about in principle possibilities good enough? Finally, even if a Nagel-type analysis of theory reduction adequately captures one aspect of theory reduction, mightn't there be more to it than that? For example, one might wish to provide an analysis of reduction as a temporal process. Why must philosophers studiously exclude all temporal considerations from their analyses?

For all the importance which philosophers of science have placed on the process of extracting a scientific theory from the scientific literature and reconstructing it rationally, they have said precious little about how to do it. At times they seem to act as if scientific theories exist right out there in nature, as if there were some one thing called a "Mendelian theory of inheritance" and some one thing called a "molecular theory of inheritance." But such is clearly not the case. At any one time in the development of a particular theory, numerous partially incompatible versions of this theory can be found in the primary literature of science. When these clusters of theories are traced through time, the multiplicity only increases. As David M. Knight (1967:2) has observed with respect to atomic theory, "There was no one classical, received atomic theory but rather a number of theories overlapping in their explanatory ranges." At the turn of the century, when Mendelian genetics was rediscovered, a half dozen geneticists set out differing versions of it. During the first ten years of the development of Mendelian genetics, these versions changed radically. About the only important substantive claim that remained unchanged was the "law of segregation" or, as it was also called, the principle of the purety of the gamotes. According to this principle, genes do not contaminate each other in the heterozygote. During the next half century, additional changes were made in Mendelian genetics, including the discovery that genes as physical entities *do* contaminate each other. As both Ruse and Schaffner have pointed out, crossover occurs within genes as well as between them, resulting in the physical mixing of alleles.

Traditional logical empiricist philosophers of science would surely reply, and several have, that all the preceding is irrelevant. An historian of science might be interested in this diversity but not a philosopher. The subject matter of philosophy of science is not raw science but the philosopher's own rational reconstructions of science. There are two problems with the preceding response. First, philosophers of science must get their rational reconstructions of science from somewhere. If it is not from the multifarious primary literature of science, then where? I suspect the answer to this question is, from college textbooks. Textbooks present just the sort of ahistorical, after the fact simplifications required by philosophers of science. Second, if the subject matter of philosophy of science is the rational reconstructions of philosophers of science and these reconstructions are produced in accordance with their own analyses of science, how then are these analyses to be evaluated? For example, the analyses of theory and theory reduction produced by the logical empiricists are interdependent. Theories are exactly the sorts of things that can be reduced, and reduction is exactly the sort of relation which can exist between theories. Obviously, reconstructing Mendelian and molecular genetics as theories on the logical empiricist analysis of "theory" stacks the deck in favor of reduction. The issue is whether or not the bias introduced at this step is sufficient to preclude independent evaluation of the resulting claims of a successful reduction.

Philosophers have exerted considerable effort on investigating the relationship of scientific theories, laws, and observation statements to the empirical world. "Snow is white" is true, if and only if, snow is white. But very little time has been spent discussing the relation of philosophical analyses of science to science. Karl Popper's doctrine of falsificationism is true, if and only if, ...? Because so little has been said about this general issue, the participants in this symposium have been placed in the uncomfortable position of not having any generally acknowledged criteria to use in evaluating the particular example of reduction under investigation. The subject matter of philosophy of science may well be the philosopher's own rational reconstructions, but some external constraints must be placed on his freedom to reconstruct science, or else philosophy of science runs the risk of being as much a self-justifying activity as theology. The most likely candidate for these external constraints is the actual practice and productions of scientists. For this purpose, textbook expositions will not do.

Scientists do not produce theories in precisely the form advocated by logical empiricist philosophers. These philosophers reply that they should. However, if too many of the things which scientists produce as scientific theories depart too radically from the philosopher's notion of a scientific theory, and if scientists insist on producing their deviant theories even after they are made aware of philosophers' views on the subject, then philosophers just might be led to reconsider their original analyses. But how true to raw science must these rational reconstructions be? Only Imre Lakatos (1971:107) has had the courage (or foolhardiness) to claim that the connection is as free as one might wish to make it. He authorizes philosophers to reconstruct history of science the way it should have happened and to "indicate *in the footnotes* how actual history 'misbehaved' in the light of its rational reconstruction."[1] Philosophy of science may be to some extent normative, but until the source of these normative powers are more fully understood, it would be wise for philosophers of science to exercise some restraint lest we find ourselves once again arguing that space is necessarily Euclidean and species necessarily immutable.

Ruse objects to my claim that the logical empiricist analysis of theory and theory reduction have been philosophically harmful, that they have hindered rather than facilitated understanding of the true situation in genetics (Hull, 1972, 1973, 1974). He himself cannot see what harm it has done. On the logical empiricist analysis, scientific theories are individuated on the basis of substantive content. Two theories are two different theories because they differ with respect to one or more substantive claims. From this perspective, the molecular theory of Watson and Crick is different from the biological theory of Mendel, but Watson and Crick's theory is also different from the molecular theories of Jacob and Monod, Lederberg, and Kornberg as is Mendel's theory from those of de Vries, Correns, Morgan, Dobzhansky and Muller. But all these theories are not equally different. One would expect them to form two clusters, one roughly molecular, the other Mendelian. But they do not if these clusters are to be formed on the basis of the substantive claims made in these theories. As Wimsatt has pointed out, theories evolve in a process which he terms "successional reduction." According to Wimsatt, successional reduction, like Schaffner's criterion of strong analogy, is an intransitive similarity relation. An early version of a theory can be similar to a later version of that same theory (that is why they are versions of the *same* theory) and this version similar to an even later version of that theory, without the earliest and latest versions being similar at all. They are all versions of the same theory because of continuous and unitary development.

I tend to agree with Wimsatt's analysis of successional reduction as far as it goes, but similarity in substantive content, even interpreted as a serial relation, is not sufficient for individuating scientific theories. The substantive content of science changes too rapidly and sporadically for that. Instead the continuing commitment on the part of scientists to certain procedures, goals, problems, and metaphysical presuppositions supply most of the continuity to be found in science. Early Mendelian geneticists abandoned or modified nearly all of the basic tenets of Mendelian genetics, but they were still Mendelian geneticists improving

Mendelian theory, not opponents producing competing theories. Similar stories can be told for other episodes in the history of science. Jacques Loeb's disciples rejected all of Loeb's substantive claims about animal tropisms, yet were still disciples, not rebels. They were merely "developing" his mechanistic conception of life (Fleming, 1964). T. H. Huxley was a Darwinian, though he thought that evolution was usually saltative and that natural selection might well play a subsidiary role in evolution.

Previously I have characterized Mendelian genetics as a theory of hereditary transmission in which the inheritance of various phenotypic traits is followed from generation to generation, ratios discerned in the distribution of these traits, and the appropriate number and kind of genes postulated to account for the distributions. Although Mendelian geneticists assumed that some sort of causal chains connected genes with the resultant traits, Mendelian genetics was not a developmental theory. So far no one has objected to this characterization, yet the substantive content of Mendelian genetics plays almost no role in it. Ruse cannot see how the logical empiricist analysis of science could impede an adequate understanding of science. Here are but two ways. It ignores the temporal dimension to science and directs attention away from the chief means by which various stages of a scientific theory can be integrated to form a single theory.[2]

The final criticism I wish to make of the philosophical literature on reduction in genetics is that it is too sketchy and programatic. None of the philosophers who have discussed the problem of reduction in genetics have actually axiomatized the two theories, set out the necessary reduction functions, and performed the derivation. Everyone comments that Mendelian genes are to be identified with some segment of DNA, possibly those segments delineated by the *cis-trans* test, and predicate terms of Mendelian genetics, like dominant, with various kinds of molecular mechanisms such as the production of an active enzyme. To date, this has been the sum total of all the efforts of philosophers to show how Nagel's formal conditions of theory reduction can be fulfilled in the case of genetics. Perhaps I expect too much of the philosophical notion of reduction. Perhaps my standards are unrealistically high. But on any standards, these efforts do not seem impressive. The problem is whether or not there is any point to fulfilling the requirements of the logical empiricists analysis of theory reduction. If not, why not? If so, who should be doing it?

Ruse says that scientists have their jobs to do, philosophers theirs, and among neither is the task of setting out reductions in any sort of detail. Schaffner subscribes to what he has termed a peripherality thesis to the effect that reductionism is not a primary aim of molecular biology. Even so, it "is *resulting in* at least *partial* reduction of biology to physics and chemistry." I find these views puzzling. If crucial experiments are as important in science as philosophers once claimed they were, then one has every right to expect scientists to be performing them. If not, then a philosopher is within his rights to urge them to do so. Similar observations seem to be warranted for other philosophical theses about science. To claim simultaneously that falsification is the chief distinguishing feature of science and that no scientist need ever attempt to falsify a scientific law or theory seems to me to be more than a little paradoxical.

In the middle of the 19th century, physicists such as Bernoulli, Joule, and Kronig suggested that the behavior of gases could be accounted for by considering the motion of their constituent molecules. Was thermodynamics thereby reduced to statistical mechanics? Later Clausius, Maxwell, Boltzman and others developed a detailed theory of gases. With a few modifications, several thermodynamic laws were actually derived from these statistical theories.[3] Boyle's law was explained in terms of the gas molecules striking the walls of the container, balancing the external pressure applied to the gas. If the volume of the gas is cut in half, each of the molecules strikes the walls twice as often, thereby doubling the pressure. In a similar manner the law of Charles and Gay-Lussac can also be explained. If the absolute temperature of a gas is doubled, the speed of the molecules is increased by a factor $\sqrt{2}$. This causes the molecules to make $\sqrt{2}$ times as many collisions as before, and each collision is increased in force by $\sqrt{2}$, so that the pressure itself is doubled by doubling the absolute temperature. Avogadro's law is also explained by the fact that the average kinetic energy is the same at a given temperature for all gases.

Where are comparable derivations for genetics? At the turn of the century, Garrod suggested that either genes were enzymes or else controlled enzymatic action. Was Mendelian genetics thereby reduced to molecular genetics? By the middle of the 20th century, we knew that genes were made primarily out of DNA and proteins out of twenty or so amino acids. Was it then that Mendelian genetics was reduced to molecular genetics? I suggest not. What is needed is the derivation of the basic principles of Mendelian genetics from molecular biology. Ruse (1973:203) concludes that the derivation of Mendel's first law from molecular premises "looks very promising." According to this law, alleles do not contaminate each other while residing on homologous chromosomes in the heterozygote. If a Mendelian gene is defined as a "piece of DNA which is just enough to serve as the cause of a cellular product (i.e. a polypeptide chain)," as Ruse suggests, then Mendel's first law is false, because crossover takes place as commonly within such functional units as between them. Only single nucleotides remain pure in a physical sense. One must correct Mendelian genetics to accommodate these findings. As Ruse (1973:206) says:

Now, if we consider not the older classical Mendelian genetics but this new yet still entirely biological, fine-structure genetics, then the task of showing that one can have a Nagelian-type reductive relationship between molecular and non-molecular genetics seems once again to be within the realm of possibility. One links the (biological) cistron with the (non-biological) molecular gene and the (biological) muton and the (biological) recon with a very few non-biological nucleotide pairs of the DNA molecule.

Ruse claims that the Mendelian gene is the basic unit of biological function and that these units of function at the level of DNA are distinguished by the *cis-trans* test. Yet Mendel's first law must be "corrected" to apply to the recon, the smallest unit of recombination. I have several objections to this suggestion. First, like any operationally-defined unit, the cistron cannot fulfill the functions of a theoretical entity. There *is*, for example, intra-cistronic complementation because two or more cistrons quite commonly cooperate to produce single polypeptide chains. The cistron is hardly *the* unit of biological function. Second, on Ruse's interpretation, Mendel's first law becomes a tautology. If a recon is defined as the smallest unit of recombination and Mendel's first law is reformulated to state that recons do not contaminate each other at meiosis by recombination, then Mendel's first law is converted from a false synthetic claim to one that is true by definition. Of course, given Mendel's original intention, his first law is not all that false. Mendel was concerned with explaining how recessive traits could reemerge from the heterozygote in as pure a form as when they went in. Finally, we still are left with no justification for calling this new, corrected biological theory "Mendelian genetics." *And* I'd like to see a few sample derivations.

Who is going to do all this? Schaffner seems to think that molecular biologists are laying the groundwork for some such derivation, but only as a peripheral part of their endeavors. Ruse admits that scientists are not rushing to flesh out the formal analyses of theory reduction presented by philosophers. I do not see them even strolling in that general direction. Perhaps, then, philosophers might attempt to flesh out their own analyses. By this, I do not mean that they should proceed to carry out empirical investigations but that on the basis of the primary scientific literature, they should attempt to reconstruct the two theories, set out representative reduction functions, and perform representative derivations – just to show that it can be done. In the past, some philosophers of science *have* occupied themselves with tasks such as these and without any intent or hope of winning a Nobel prize. Woodger (1937), for example, actually tried to axiomatize part of Mendelian genetics. (Unfortunately, his axiomatization is extremely deficient; see Kyburg, 1968.) In doing so, such formalistically-inclined philosophers are performing one of the traditional tasks of philosophy of science.

Given the turn which the controversy over reduction in genetics has taken recently, a formal treatment would seem to be more important than ever. Unless one goes off in a completely new direction, as Wimsatt suggests, the current dispute can be resolved only if the reductionist thesis is presented in a more coherent and unified manner than it has been in the past. Piecemeal observations are no longer good enough. Everyone now seems to agree that Mendelian genetics is not just a special case of molecular genetics. Both theories must be modified if reduction in any formal sense is to take place. How similar must the modified and unmodified versions of these two theories be in order to term the relation between the modified versions "reduction"?

In his early writings, Ruse (1971) found the differences between Mendelian and molecular genetics too extensive to permit reduction in any strict, formal sense. His current intuitions are that the differences are not all that great. From the first, Schaffner has felt that a strong analogy existed between corrected and uncorrected Mendelian genetics, even though no one had presented these theories explicitly or had explicated the notion of strong analogy. My intuitive impression continues to be that the differences between the corrected and uncorrected versions of these theories are too numerous and too fundamental to consider the relationship between the two corrected theories reduction in the formal sense of the term. Pre-analytically, the relation between Mendelian and molecular genetics is a paradigm case of theory reduction, but from the point of view of the logical empiricist analysis of theory reduction, it looks more like replacement. However, I do not see how our continuing to trade intuitions is going to get us anywhere. If this dispute is to be resolved, the two theories will have to be set out in greater detail and the necessary changes specified with greater precision than they have been in the past.

The absence of any explicit formulation of the two theories also serves to blunt a second objection which has been raised to the straightforward application of the logical empiricist analysis of theory reduction to the situation in genetics. Schaffner and I do not disagree about the discrepancies which exist between historically accurate rational reconstructions of Mendelian and molecular genetics. Our major point of contention is how much slippage exists between the two and where to put it. Should it be reflected in the reduction functions, incorporated into the theories, or buried in the boundary conditions, auxiliary hypotheses, background knowledge, and the like? Until a reasonably complete and explicit analysis of Mendelian and molecular genetics has been presented, including a specification of which elements are to be put into each of these categories, any objection can be evaded by shunting it somewhere else. If a reduction function becomes too complicated, then get rid of the complexity by some vague reference to differences in the environment. I am not claiming that such evasive maneuvers are inherent in the position currently espoused by Schaffner and Ruse, only that in the absence of a more formal analysis, they are all but impossible to avoid.[4]

Before turning to issues more intimately connected to the specific example of reduction under investigation, let me summarize in diagrammatic form the message of the preceding pages. Let $\{T_1\}$ stand for all the various versions of molecular genetics to be found in the primary literature of that discipline and $\{T_2\}$ stand for all the various versions of Mendelian genetics to be found in the literature of that field. Let T_1 and T_2 stand for historically accurate rational reconstructions of $\{T_1\}$ and $\{T_2\}$ respectively, corrected only to the extent necessary to fulfill the logical empiricist analysis of a scientific theory. I doubt that anyone has ever supposed that $\{T_2\}$ could be derived from $\{T_1\}$, but someone might be tempted to think that T_2 is derivable from T_1. But as Schaffner (1967 & 1969) admits, both theories need further modification, some of it for the sole purpose of fulfilling the logical empiricist analysis of theory reduction. Following the symbols introduced by Schaffner, let T_1^* and T_2^* stand for rational reconstructions of molecular and Mendelian genetics respectively, further modified to take such corrections into account. Thus, when a logical empiricist says that Mendelian genetics is reducible to molecular genetics, all he is claiming is that a modified, corrected rational reconstruction of Mendelian genetics, T_2^*, is derivable from a modified, corrected rational reconstruction of molecular genetics, T_1^*.

Molecular Genetics   *Mendelian Genetics*

$\{T_1\}$ (raw theories) $\{T_2\}$

↓   (Theories reconstructed to fulfill the   ↓

           logical empiricist analysis of theory)

T_1 T_2

↓   (Theories reconstructed to fulfill the   ↓

           logical empiricist analysis of reduction)

T_1^* --→ T_2^*

(derivation)

Fig. 1. Diagram of the logical empiricist analysis of theory reduction.

In Figure 1, the broken arrow represents derivation. If derivation is taken to be deduction, then a highly formal analysis of it is available. The solid arrows represent rational reconstruction, modification, and correction. So far very little has been said about these processes. In the absence of any explicit analysis of these currently informal aspects of theory reduction, the claim that Mendelian genetics can or cannot be reduced to molecular genetics seems to be premature. With enough ingenuity (and lack of intellectual integrity), the phlogiston theory could probably be reduced to statistical mechanics, once the two theories had been sufficiently "modified" and "corrected."

Previously, I (Hull, 1972, 1973, 1974) have argued that the reduction functions necessary for the reduction of anything which might legitimately be called Mendelian genetics to molecular genetics are prohibitively complex many-many relations. Many-one relations from molecular to Mendelian genetics make the reduction complicated, so complicated that specific derivations would rarely be carried out in practise. Like three-body problems in physics, it is nice to know that they can be solved, but no one is tempted to do so very often. One-many relations from molecular to Mendelian genetics make reduction impossible. Assume for the moment that the goal of reduction functions is to identify the thing terms of Mendelian genetics (e.g., genes and traits) with some sort of entity terms at the molecular level (e.g., DNA and proteins) and the predicate terms of Mendelian genetics (e.g., dominant and epistatic) with various molecular mechanisms (e.g., production of an active enzyme or a necessary substrate). This is what all of the philosophers involved in the dispute seem to have had in mind in their early writings (Schaffner, 1967, 1969; Hull, 1972, 1973, 1974; Ruse, 1971, 1973; Simon, 1971). Thus, the one situation which is precluded, if reduction is to be possible, is a single molecular entity or mechanism being identified with two or more Mendelian entities, properties, or relations.

In order to decide whether or not Mendelian genetics is reducible to molecular genetics, it is not enough to correlate textbook versions of these two theories. The actual complexities of both types of phenomena must be recognized and taken into account. For example, in an early publication, Schaffner (1967:144) set out a sketch of what a reduction function for "dominant" should look like. According to Schaffner, $gene_1$ is dominant $\equiv$ DNA $segment_1$ is capable of directing the synthesis of an active enzyme. But like the definition of "gene" in terms of the production of polypeptide chains, this reduction function for "dominant" required considerable refinement. Not all dominant genes function in the production of enzymes, active or otherwise. But of greater significance, "dominant" is a relational term. One allele can be dominant to another of its alleles but recessive to a third. In addition, one allele can be dominant to another with respect to one trait which it controls but recessive with respect to a second. The age of the organism must also be taken into account. One allele can be dominant to another allele with respect to a particular trait early on in the ontogenetic development of an organism; recessive later. For example, red shell color in a particular species of snail is dominant to yellow early on but then gradually changes to recessive.

Both the genetic and external environments also matter. A change in a gene in one place in the genome can affect the functioning of genes elsewhere. Thresholds also exist such that below a particular level one trait develops; above that level another trait. Sometimes such switches are only stochastic, determining only the probability of certain phenotypes. For example, the castes in some social insects are genetically similar and are determined by the quality and quantity of food fed to the larvae. Flowering in many plants and diapause in many insects are induced by the length of daylight and darkness. In the marine worm Bonellia sex is determined by the environment: larvae growing on a female's proboscis become males; anywhere else females. And I'm sure no one will object to my adding at this point, "And so on."

A comparable story can be told for molecular genetics. There seems to be a reasonably close correlation between molecular and Mendelian genes, especially if the genome is divided into genes on the basis of those segments of DNA which produce single molecules of RNA. The correlation between specific DNA segments and specific molecules of RNA is pretty much one-one, though some exceptions do exist. But once these molecules of RNA begin to function in the production of proteins, the situation becomes a good deal more complicated. As Schaffner (1969:411) has observed:

The requirements for protein synthesis include about 60 transfer RNA molecules, at least 20 activating enzymes which attach the amino acids to the t-RNAs, special enzymes for initiation, propagation, and termination of peptide synthesis, ribosomes containing three different structural RNAs, and as many as 50 different structural proteins, messenger RNA, ATP, GTP, and Mg^{+2}.

And other factors matter as well, including temperature, pH, ionic strength, and so on.

I have not listed all the preceding complexities just for the sake of obscurantism. Perhaps a few, many, or all of them can be taken into account. The point is that some hint must be given as to which of these factors are to be included in the relevant theory, which in the environment and in what sense "environment," and how these theories and environments are to be correlated. As Robert Causey (1972:194) observes in his discussion of uniform microreductions, the reducing theory and its primitive nonlogical predicates will have to make reference to environmental conditions and to state some relations between the elements of the domain of the reducing theory and the relevant aspects of the environment. That these factors be stated *explicitly* and correlated *systematically* does not seem to be asking too much. Neither vague promises nor a case by case treatment is good enough.

Ruse has responded to my claim that the relation between particular *kinds* of molecular mechanisms and *patterns* of Mendelian inheritance is one-many by arguing that "given a general mechanism which could lead to various Mendelian effects, by specifying the peculiarities of a particular case, one can follow through to a uniquely determined effect." He even toys with the idea that I might be claiming that "*every* instance of a Mendelian effect will have to be treated separately." Instead, Ruse counters, "there are some general models which seem applicable to widely differing cases." As I have noted previously (Hull, 1974:41), one likely mechanism for epistasis is the production by the epistatic gene of a necessary precursor for the reaction controlled by the hypostatic gene. As Ruse observes, this "model is known to fit many cases." True, but which ones? After all, reduction functions are supposed to be synthetic *identity* statements, not statements of loose correlations. But Ruse replies, if one supplements the description of the general model with a specification of the *particular circumstances*, then one can infer unerringly which pattern of Mendelian inheritance should result. I take this statement to be fairly safe, because the only way that it could be false is by miraculous intervention of a supernatural power. The issue is the relative importance of the general model and the particular circumstances in such inferences. As it turns out in the case of genetics, knowledge of the general model does not contribute much to the inference. Such models are too ubiquitous for that. Instead it is the peculiarities of the particular case which are decisive. For example, sometimes the total inactivation of a particular allele which codes for an enzyme will result in no enzyme being produced, sometimes in the continued production of the enzyme at full capacity, and sometimes any of the intermediary states. Given knowledge of the general mechanism, no inference can be made to any sort of general Mendelian pattern of inheritance.

I realize that from a general law one can infer to the particular case only by introducing statements of particular circumstances. In order to infer the path of the earth around the sun, one must know the masses, positions, and velocities of these two bodies. Newton's laws alone will not suffice. What is needed in the case of the reduction functions in genetics is a specification *in advance* of the molecular mechanisms (what factors are to be included, which excluded, and how these factors are related) and what *sorts* of things are part of the particular circumstances. To be told that in each case some molecular mechanism or other combined with whatever particular circumstances turn out to be relevant are responsible for the observed Mendelian pattern of inheritance is just a covert statement of determinism. I have no quarrel with macroscopic determinism, but the issue is reduction, not determinism. The logical empiricist analysis of theory reduction concerns the inferential relationship between *theories*. In order for the thesis to be significant, a *systematic* relationship must exist between the key elements in the two theories, in this case between various *kinds* of molecular mechanisms and various *patterns* of Mendelian inheritance.

Although Schaffner's original suggestion for a reduction function for the Mendelian notion of dominance initiated the recent dispute about the actual relationship between molecular mechanisms and Mendelian patterns of inheritance, Schaffner in the interim has taken a completely different tack. On Schaffner's current analysis, all one needs to know in order to reduce Mendelian genetics to molecular genetics is that genes are made out of DNA and proteins out of amino acids. This is the total empirical content of the necessary reduction functions. If we let A and a stand for two alleles, Ⓐ and ⓐ stand for the resulting phenotypic traits, and a single arrow stand for "Mendelianly produce," then the Mendelian claim that allele A is dominant to allele a with respect to character states Ⓐ and ⓐ can be diagrammed as follows:

$AA \rightarrow$ Ⓐ, and
$Aa \rightarrow$ Ⓐ, and
$aa \rightarrow$ ⓐ.

Now if we identify Mendelian alleles A and a with DNA sequences α and β respectively and the phenotypic traits Ⓐ and ⓐ with amino acid sequences א and ב respectively, and let a double arrow stand for "molecularly produce," then the preceding Mendelian claim can be rewritten as follows:

$\alpha\alpha \Rightarrow$ א, and
$\alpha\beta \Rightarrow$ א, and
$\beta\beta \Rightarrow$ ב.

As the reader can see simply by comparing these two formulations, all Schaffner has done is substitute particular DNA sequences for particular Mendelian genes, an arrow standing for "molecularly produce" for an arrow standing for "Mendelianly produce," and amino acid sequences for phenotypic traits. At first one might have expected the formulation of reduction functions for Mendelian predicate terms to be quite difficult and require considerable knowledge of molecular genetics. On Schaffner's current analysis, the sum total of the empirical content of his *synthetic* identity statements is the identification of Mendelian genes with segments of DNA and phenotypic traits with amino acids. All the rest of the empirical content of molecular genetics is buried in the double arrow. Schaffner justifies this move by noting that the double arrow "essentially represents a telescoped chemical mechanism of production of the amino acid sequence by the DNA. Such a mechanism must be specified if a *chemically adequate* account of biological production or causation is to be given. Such a complete account need not function explicity in the *reduction function per se*, even though it is needed in any complete account of reduction."

Schaffner has chosen one possible way to represent the reduction of Mendelian to molecular genetics. In his characterization, the reduction functions for "gene" and "phenotypic trait" are empirically significant. Those for the large array of Mendelian predicate terms follow automatically from these initial two reduction functions and contribute nothing empirically significant to the reduction. I must admit that I find Schaffner's move disheartening. I had thought that Nagel, Woodger, Hempel *et al.*, were attempting to analyze an important notion of reduction. Whatever it was that made it important either philosophically or scientifically seems to have dropped out along the way. Defenders of the logical empiricist analysis of theory reduction seem to be willing to trivialize their analysis in order to salvage it. As with similar disputes over the existence of God, I personally much prefer to retain a significant notion of the concept at issue and admit that it does not apply rather than trivialize it to the extent necessary to insure its application. "Does God exist?" Certainly! God is the universe, and only lunatics doubt the existence of the universe. "Ah yes, then physicists are theologians."

In summary, both Ruse and Schaffner have defended the possibility of providing the reduction functions necessary to derive Mendelian from molecular genetics. They both agree that these reduction functions are *synthetic identity* statements. To put my objections to this thesis succinctly, I have argued that Ruse's reduction functions are not *identities* and Schaffner's reduction functions for the predicate terms of Mendelian genetics are not *synthetic*. In both of their accounts, all empirical difficulties are shunted into some area whose elements happily never need to be stated explicitly. Schaffner's leaving the causal relation between Mendelian genes and the phenotypic traits which they control unspecified is perfectly legitimate. Although most Mendelian geneticists assumed that some sort of causal chains connected genes with phenotypic traits, the specification of these causal chains played no role in transmission genetics. The same cannot be said for molecular genetics. Molecular genetics sets out in detail the molecular structure of genes and proteins and the mechanisms by which DNA replicates itself and produces molecules of RNA which cooperate in the synthesis of proteins. Leaving the double arrow unexplicated as a "promissory note for a *chemically* causal account" to be specified later is to leave the reduction itself largely promissory, to be supplied later. Nor is the burying of all empircal difficulties in the theory, particular circumstances, auxiliary hypotheses, and what have you just to keep the reduction functions simple a very propitious maneuver. The complexity of the relationship between molecular and Mendelian genetics must be measured in terms of *all* the operative elements, not just in terms of the simplicity of the reduction functions. The decision to keep the reduction functions simple merely requires that the complexities which I have mentioned be accommodated elsewhere. They lay in wait for anyone attempting to set out a complete analysis of theory reduction in genetics.

One of the major impediments to the straightforward application of the logical empiricist analysis of theory reduction to the case in genetics is that Mendelian genetics is almost exclusively a transmission theory. Mechanisms are irrelevant. Whereas molecular genetics is concerned first and foremost with mechanisms. In fact, it is difficult to set out molecular genetics in the form of a "theory," in the usual logical empiricist sense of the term. Wimsatt has questioned whether reduction is best construed as a relation between theories at all:

> At least in biology, most scientists see their work as explaining types of phenomena by discovering mechanisms, rather than explaining theories by deriving them from or reducing them to other theories, and that *this* is seen by them as reduction, or as integrally tied to it.

I suggest that the explanation of phenomena at one level of analysis by setting out the mechanisms which produce it at a lower level is the important sense of reduction which has gradually been eliminated as the logical empiricist analysis of theory reduction has been reduced to its current sad state. Little by little the notion of reduction has been so trivialized that it is difficult to imagine a situation which could not be made to fit it.

The University of Wisconsin-Milwaukee

NOTES

[1] Following Lakatos's suggestion, I have stated in the text what Lakatos should have said and now, in a note, indicate how he misbehaved. Lakatos distinguishes between internal history (history as it should have happened given a particular set of philosophical views about science) and external history (history as it actually happened regardless of any such philosophical views). The more internal history coincides with external history, Lakatos argues, the better the philosophy of science which gave rise to the internal history. However, Lakatos never tells us how one can write a factually accurate external history independent of any views about the nature of science.

[2] Numerous philosophers have argued for an evolutionary interpretation of science, most recently Stephen Toulmin, *Human Understanding* (Princeton University Press, 1972). On this analysis, scientific theories are historical entities developing continuously in time; see my 'Central Subjects and Historical Narratives', *History and Theory* (forthcoming). With respect to the particular example of reduction at issue, Michael Simon, *The Matter of Life* (Yale University Press, 1971) assumes an evolutionary viewpoint. Even though radical changes took place in the transition from Mendelian to molecular genetics, Simon argues that the process is one of successive correction and refinement, not replacement, because none of the changes took place abruptly.

[3] At least, so the story goes. L. Sklar argues for a position with respect to the reduction of classical thermodynamics to statistical mechanics similar to the one that I am urging with respect to genetics; see his contribution in this volume.

[4] Schaffner at least should not complain of my request for a more explicit and complete explication of the two theories and the necessary reduction functions because he raised a similar objection to an argument set out by Bentley Glass ('The Relation of the Physical Sciences to Biology – Indeterminacy and Causality,' in B. Baumrin (ed.), *Philosophy of Science: The Delaware Seminar*, Interscience, New York, 1963). In this paper, Glass argued that the statistical laws found for phenomena at one level of organization are not derivable from the statistical laws at a lower level of organization. Schaffner responded that in order to eliminate the possibility of interlevel reduction between statistical laws, "Glass would have to present an appropriate axiomatization of a true probabilistic theory in biology and demonstrate that the identification of biological entities with physiochemical entities and explanation of the biological entities' behavior on the basis of either causal or statistical laws involving physiochemical terms would entail a contradiction" ('Antireductionism and Molecular Biology', *Science* **157** (1967), 645). If Glass must do all Schaffner claims to substantiate his antireductionist thesis, there is no good reason to expect Schaffner to do less if he is to substantiate his reductionist thesis.

BIBLIOGRAPHY

Causey, R. L.: 1972, 'Uniform Microreductions', *Synthese* **25**, 176–218.
Fleming, Donald (ed.): 1964, *The Mechanistic Conception of Life*, The Belknap Press, Cambridge, Mass.
Hull, D. L.: 1972, 'Reduction in Genetics – Biology or Philosophy?', *Philosophy of Science* **39**, 491–499.
Hull, D. L.: 1973, 'Reduction in Genetics – Doing the Impossible,' in P. Suppes (ed.), *Logic, Methodology and Philosophy of Science*, vol. IV, North-Holland Publishing Co., pp. 619–635.
Hull, D. L.: 1974, *Philosophy of the Biological Sciences*, Prentice-Hall, Inc. Englewood Cliffs.
Knight, David M.: 1967, *Atoms and Elements*, Hutchinson University Library, London.
Kyburg, H.: 1968, *Philosophy of Science*, Macmillan, New York.
Lakatos, I.: 1971, 'History of Science and Its Rational Reconstruction,' *Boston Studies in the Philosophy of Science*, vol. **8**, 91–136.
Ruse, Michael: 1971, 'Reduction, Replacement and Molecular Biology,' *Dialectica* **25**, 39–72.
Ruse, Michael: 1973, *The Philosophy of Biology*, Hutchinson University Library, London.
Schaffner, Kenneth: 1967, 'Approaches to Reduction,' *Philosophy of Science* **34**, 137–147.
Schaffner, Kenneth: 1969, 'The Watson-Crick Model and Reductionism,' *The British Journal for the Philosophy of Science* **20**, 325–348.
Simon, Michael: 1971, *The Matter of Life*, Yale University Press, New Haven.
Woodger, J.: 1937, *The Axiomatic Method in Biology*, Cambridge University Press, Cambridge.

WILLIAM C. WIMSATT

REDUCTIVE EXPLANATION:
A FUNCTIONAL ACCOUNT*

I

Philosophical discussions of reduction seem at odds or unsettled on a number of questions:

(i) Is it a relation between real or between reconstructed theories, and if the latter, how much reconstruction is appropriate?[1] Or is reduction best construed as a relation between theories at all?[2]

(ii) Is it primarily connected with theory succession, with theoretical explanation, or with both?[3]

(iii) Is translatability *in principle* sufficient, or must we have the translations in hand, and if the former, how do we judge the possibility of translation when we don't have one?[4]

(iv) What is the point of defending the formal model of reduction if it doesn't actually happen (Hull[5], Ruse[6]), or if the defense has the consequence that if reductions occur, they are trivial and uninformative (Hull[7]), or merely incidental consequences of the purposeful activity of the scientist *qua* scientist in devising explanations (Schaffner[8])?

Furthermore:

(v) At least in biology, most scientists see their work as explaining types of phenomena by discovering mechanisms, rather than explaining theories by deriving them from or reducing them to other theories, and *this* is seen by them as reduction, or as integrally tied to it.[9]

(vi) None of the symposiasts present are suggesting inadequacies in the kinds[10] of mechanisms postulated by molecular geneticists for the explanation of more macroscopic genetic phenomena.

(vii) Nonetheless, two of them (Ruse in his earlier work, though no longer, and Hull) seem to suggest that there is no reduction (only a replacement), and the third (Schaffner) suggests that a reduction is occurring, but is a merely incidental consequence of the activity of these scientists.

What possibly can explain this wide disagreement between scientists who appear to take reductive explanation seriously and to regard it as an – indeed, as perhaps *the* important consequence of their work, and philosophers who are attempting to faithfully characterize their activity and its rationale? Can reduction be as unimportant (or nonexistent) in science as these philosophers seem to suggest? I think the answer must be 'no', and that there are four main factors which are responsible for the present philosophical confusion on this point:

(1) Philosophers have taken the 'linguistic turn' and talk about relations between linguistic entities, whereas biologists are more frequently unabashed (or sometimes abashed) realists, and talk about mechanisms, causal relations and phenomena. Though not necessarily vicious, I think that the linguistic move has lead philosophers astray. I will here defend a realistic account of reduction.[11]

(2) While virtually everyone agrees that a philosopher by the nature of his task must be interested in doing some rational reconstruction, doing so serves different ends in different contexts. A failure to distinguish these ends and how they may be served contributes to the apparent defensibility of the formal model of reduction.

(3) No real competitor to the formalistic (or more generally structuralist) account of reduction has been forthcoming. Therefore there has been a tendency to regard 'informal' reductions (Ruse)[12] as either non-reductions or as *deficient* reductions, which can be remedied by becoming formalized. I will outline some aspects of a functional account of reduction which suggests that *'informal reductions'* are the proper end of scientific analyses aiming at reductive explanations.

(4) An emphasis on structural (deductive, formal, logical) similarities has led to a lumping of cases of theory succession with cases of theoretical explanation, with the result that discussions of reduction, replacement, identification and explanation (which have radically different significances in the two contexts) have become thoroughly muddled.[13] A functional account of these activities yields important clarifications of their nature.

I wish to say something about (2), before turning to my analysis of reduction, which concerns primarily (3) and (4). The first point enters mainly by implication.

II. TWO KINDS OF RATIONAL RECONSTRUCTION

There are at least two (and probably more) contexts where talk of rational reconstruction seems appropriate in connection with plausible and useful activities of philosophers of science:

Rational$_1$ – An Optimal Strategy

One might want to abstract from the often irrelevant details and sometimes mistaken moves of the actual practice of science to reconstruct the significant patterns of scientific activity.[14] Insofar as these patterns can be claimed to be a relatively efficient, or even an *optimal* way of achieving or trying to achieve the ends of such activity, the reconstruction could claim to be a rational reconstruction in the sense of rational decision theory – that it represented the way one ought to do that activity. As such the philosopher of science is a *therapist with respect to scientific strategy*.

Rational$_2$ – A Canon of Logical Rigor

A physicist (and nowadays with increasing frequency, a biologist) might ask a mathematician for 'formal' help. He might wish to prove a mathematical conjecture whose truth or falsity he is uncertain of and which has important implications for his work. Or he may have an argument which he can formulate more informally, but desires more rigor either to buttress the argument or to determine more precisely the conditions under which it holds. As such a mathematician is a *therapist with respect to formal argument*, logic, and 'critical thinking', and these are also roles which could legitimately and usefully be played by a philosopher of science.

In either case the philosopher of science would be analyzing or criticizing an activity in terms of how well it served the ends of the scientist, and in each case, the activity itself and the analysis of it further these ends.

Note that the functions of the philosopher of science in these two cases are, at least *prima facie* not equivalent. It is not at all clear that improvements in rigor, *per se*, are a rational *qua* efficient way to do science – say, for finding explanations – nor even that the ultimate end state of science will be to improve the rigor of theories *which are otherwise adequate* – i.e., after their other problems have been solved. Improvements in rigor are sometimes useful, but not always. Philosophers of science have sometimes talked as if improvement in rigor is a scientific-end-in-itself, but no one here is doing so. I believe that the sort of confirmation and troubleshooting suggested above is the main function of rigorous argument in science, and that rigor is not a scientific-end-in-itself.

One effect of logical empiricism (with its emphasis on the 'logical structure' of laws, theories, explanations, predictions, and experiments) has been to blur – even to obliterate – the distinction between these two senses of 'rational reconstruction'. This conflation has had a disastrous effect upon the analysis of reduction – proceeding as it has in terms of the formal model. Schaffner's thesis of the 'peripherality' of reduction suggests that any successful defense of the formal model would win a pyrrhic victory. In terms of the above distinctions, I would describe this 'peripherality' of the formal model as follows: It is not rational$_1$ to view formal (i.e., rational$_2$) reduction as a scientific-end-in-itself because science then becomes an inefficient and ineffective way of pursuing known scientific ends (such as explanation). And although the formal model of reduction is by definition a rational$_2$ model, it is not even an effective *means* to some end because it is not the answer to a request for formal (i.e. rational$_2$) assistance which anyone has made or would be likely to make! Thus, although early discussions of formal reduction seemed to hold out the hope that it would perform the functions of both kinds of rational criticism, it is my impression that more recent sophisticated discussions (such as Schaffner's) have given up on both claims. But these claims are not peripheral and readily dispensable. They represent one of the major motivations for pursuing either a formalistic or a reductionistic strategy in science. If they must be given up, one's claim to be analyzing reduction as that concept is *used* in science must be suspect.

Paradoxically, if a non-formal (or perhaps 'partially formal') account of reduction is allowed, it can be seen to be a rational activity in both senses: It is an efficient (rational$_1$) way in which to proceed, and it proceeds by using logical instruments for the critical (rational$_2$) evaluation of theoretical and observational claims. Because it is a partially formal model, the use of formal methods (as discussed by Schaffner and Ruse) is to be expected on this model also, and it derives confirmation from the cases they adduce to support the formal model. It does not require total systematization, however, which has *not* been exemplified in any of the cases they discuss and which formal reduction requires (See, e.g., Schaffner, 1976, p. 614).

How do we get such an alternative to the formal model of reduction? Just as a characterization of logical *structure* (a rational$_2$ reconstruction) suggests and is suggested by a formal model of reduction, the view of scientific activity as purposive suggests a *functional* analysis and characterization – a rational$_1$ reconstruction – of reduction. Such an analysis distinguishes activities which may, in some respects have similar structure,[15] and may point to and explain further structural differences which have been ignored on the formal approach. Most importantly, I believe that a functionalist approach shows why the research aims of the scientist *contribute to* (in the sense of moving in the direction of) fulfilling the aims of the formal model, but are in fact *different from* and even, *inconsistent with*, actually getting there. Then a stronger version of Schaffner's (1974b) 'peripherality' thesis is justified:

(P1) Not only is progress toward formal reduction incidental, but

(P2) It also seems to be epiphenomenal, since this progress towards formal reduction appears to have no *further* consequences.

(P3) Finally, if (as I believe) getting there is inconsistent with the real aims of science, this 'progress' is bound to remain incomplete.

III. SUCCESSIONAL VS. INTER-LEVEL REDUCTIONS

The functional viewpoint is perhaps best explicated by expanding upon and modifying Schaffner's model, which has many useful features, although the end result will be quite different. (See Figures 1 and 3.) Most importantly, Schaffner distinguishes between and includes both a derivability condition between the reducing theory (T_1), and a corrected version of the reduced theory (T_2^*), and a condition of strong analogy between T_2^* and its uncorrected predecessor, T_2. These two relations are prototypic of two distinct relationships, each of which has been called 'reduction'.

Schaffner's condition of strong analogy is closely related to Nickles' 'reduction$_2$' (Nickles, 1973, p. 194ff.) and to what I elsewhere (Wimsatt, 1975) and below call 'successional' or 'intra-level' reduction. Nickles' account, emphasizing transformational and possibly non-deductive

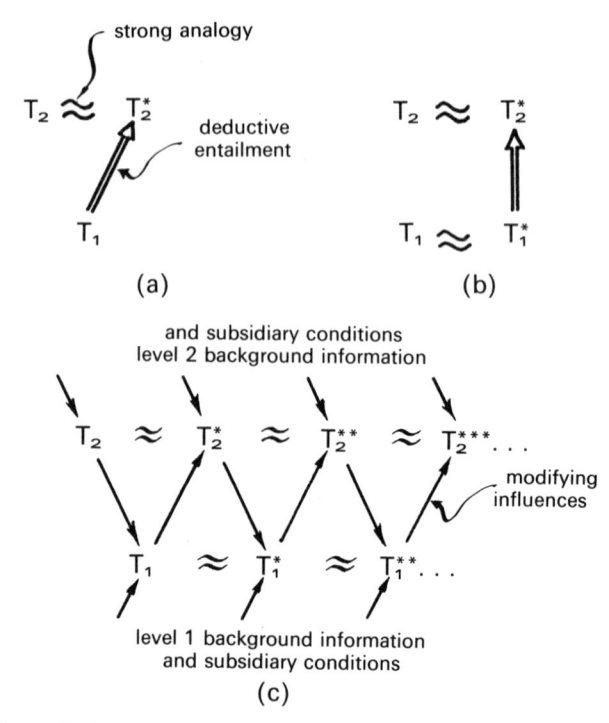

Fig. 1. (a) Theory Reduction: Schaffner (1967). T_2: reduced theory; T_2^*: corrected reduced theory; T_1: reducing theory. (b) Theory Reduction: Schaffner (1969). T_1^*: modified (corrected?) reducing theory. (c) 'Coevolution of theories at Different Levels': Wimsatt (1973) [an early draft of (1975)].

relations between successive competing theories affords an important partial explication of 'strong analogy'. A functional account of this activity explains many of the structural features Nickles proposes, and others which he does not mention.

What is not clear on Schaffner's model, but implicit in Nickles' is that 'reduction$_2$' (which is a kind of 'pattern matching' problem and could also be regarded as *demonstrating and analyzing* the 'strong analogy' between T_2 and T_2^*)[16] is neither automatic nor self evident. It has a point, involves work, and is performed for reasons separate from the functions of the 'other' reductive relation. Nickles suggests that reduction$_2$ performs heuristic and justificatory functions *vis-a-vis* the uncorrected older T_2.[17]

I believe that reduction$_2$ is fundamentally connected with theory succession (of T_2 by $T_2{}^*$) and performs rather more functions than Nickles makes out. *It is most immediately a transformational operation whose function is to localize and analyze the similarities and differences between T_2 and $T_2{}^*$* which in turn serve a variety of further functions. Most interestingly, because none of these functions are served by making comparisons other than between $T_2{}^*$ and its immediate predecessor, T_2, and in any case, similarities and differences become *less* localizeable as changes accumulate, successional reduction would be expected to be *intransitive*, and to behave as a similarity relation.[18] *Thus the intransitivity of successional reduction is an explicable feature, not a given, on the functional account of this activity.*

For further analysis of the specific uses made of these localized similarities and differences between T_2 and $T_2{}^*$ and diagrammed in Figure 2, I refer you to part II of Wimsatt (1975). The following contrasts between 'successional' and 'explanatory' reductions should be noted here however:

(1) *Successional reduction is* and must be *a relation between theories* (since it it is these which exhibit the similarities and differences), unlike *explanatory reduction* which *is not*, in any but degenerately simple cases.

(2) *Replacement* occurs only with the *failure* of successional reduction – failure to localize similarities and differences among successive competing theories. Replacement and successional reduction are opposites. But for explanatory reductions, replace*ability* is closer to and is by many treated as a *synonym* for reduce*ability*. A failure of T_1 to reduce T_2 (perhaps derivatively, by reducing $T_2{}^*$) would make T_2 and its successors *emergent* and *irreplaceable* relative to T_1. *Replacement obviously has two different meanings here.*

(3) *Successional reductions are intransitive.* A number of them 'add up' to a replacement. *Explanatory reductions are transitive.* (It is this last fact which raised the hopes among advocates of 'unity of science' for great ontological economies through reduction, about which I have more to say (1975 and) below.)

(4) Talk about elimination might be appropriate for the posited entities of corrected and replaced theories if the new theory is sufficiently different that there is no significant continuity between old and new entities. But such talk is frequently illegitimately extended to contexts of

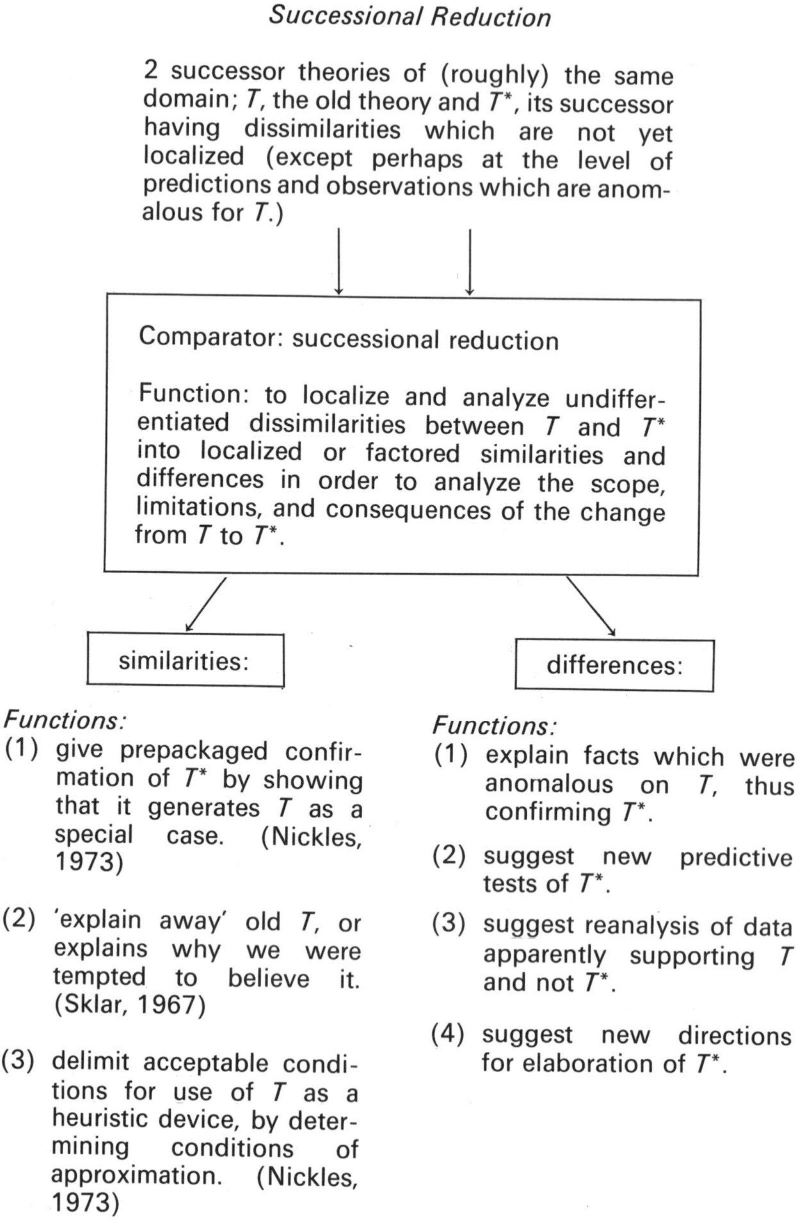

Successional Reduction

2 successor theories of (roughly) the same domain; T, the old theory and T^*, its successor having dissimilarities which are not yet localized (except perhaps at the level of predictions and observations which are anomalous for T.)

⬇ ⬇

Comparator: successional reduction

Function: to localize and analyze undifferentiated dissimilarities between T and T^* into localized or factored similarities and differences in order to analyze the scope, limitations, and consequences of the change from T to T^*.

↙ ↘

similarities: differences:

Functions:
(1) give prepackaged confirmation of T^* by showing that it generates T as a special case. (Nickles, 1973)

(2) 'explain away' old T, or explains why we were tempted to believe it. (Sklar, 1967)

(3) delimit acceptable conditions for use of T as a heuristic device, by determining conditions of approximation. (Nickles, 1973)

Functions:
(1) explain facts which were anomalous on T, thus confirming T^*.

(2) suggest new predictive tests of T^*.

(3) suggest reanalysis of data apparently supporting T and not T^*.

(4) suggest new directions for elaboration of T^*.

Fig. 2.

explanatory reduction. This is often motivated by talk of ontological or postulational simplicity in the light of supposed translateability and deduceability, (discussed further below), but in at least some cases looks suspiciously like treating reduction and replacement as opposites. Thus, in arguing that the formal model of reduction doesn't fit the relation of Mendelian to molecular genetics, Hull and Ruse[19] each suggest that it looks more like a case of replacement. As I suggested in (2) above, the opposition between reduction and replacement is appropriate for successional reduction, but *not* for interlevel or explanatory reduction. Their claim is thus misplaced if it concerns the relation between T_1 and T_2. Though intelligible if construed as concerning the relation between T_2 and T_2^*, I would disagree on the facts of the case, and agree with Schaffner (1976) and Ruse's (1976) most recent view that there is no replacement, but a reduction. To explain why, I must say a great deal more about explanatory reductions. In what follows, I will be talking about them unless otherwise indicated.

IV. LEVELS OF ORGANIZATION AND THE CO-EVOLUTION AND DEVELOPMENT OF INTER-LEVEL THEORIES

Rather than talking directly about reductive relations between theories, the approach I have taken (Wimsatt, 1975) is the realistic one of regarding levels of organization – features of the world – as primary, and defined in such a way that it is natural that theories should be about entities at these levels of organization. The notion of a level implies a partial ordering, such that higher level entities are composed of lower level

---

Notes:
(1) Nickles (1973) also suggests that 'reductions$_2$' may be done in a variety of ways. This is understandable if the point of the transformation is how best to factor out similarities and differences.

(2) Successional reductions may be possible 'locally' (for parts of theories) even when not possible globally (for the whole theory.).

(3) Differences in meanings of the key terms may be regarded as irrelevant as long as they are localizable in a way that allows fixing praise and blame on specific components of T and T^* in comparatively evaluating them. (see also Glymour, 1975). Thus the 'meaning change' objection is avoidable.

entities, and, in a universe where reductionism is a good research strategy, the properties of higher level entities are predominantly best explained in terms of the properties and interrelations of lower level entities.

But I argue further that levels of organization are primarily characterized as local maxima of regularity and predictability in the phase space of different modes of organization of matter. Given this, selection forces (and at lower levels, the stability considerations into which these shade) suggest that the majority of readily defineable entities will be found in the (phase space) neighborhood of levels of organization, and that the simplest and most powerful theories will be about entities at these levels.[20]

Nothing in this approach entails that levels defined as local maxima of regularity and predictability must always be well-defined and delineated, or strictly linearly orderable, (although they usually are for simpler systems) and in fact certain conditions can be suggested (in *this* world) where these assumptions are false (see Wimsatt, 1974 and 1975, part III). These are conditions where neat composition relations cannot be specified for all (or perhaps even for any) of the entities in these different 'perspectives'. (Level talk *requires* the possibility of specifying composition relations, so I talk about 'perspectives' when this condition is not met.) This failure of orderability leads to the 'intertwining' of theories mentioned by Schaffner (1974b) in discussing the operon model, (see also his 1974a), in support of his thesis of the 'peripherality of reduction', and to the much more extreme situation suggested by Roth (1974) in her penetrating analysis of the same case – which she sees as the development of an inter-(multiple) level theory rather than is the tying or merging together of preexisting theories.

These sorts of complexities have been ignored in discussions of the standard model of reduction, and Hull's discussions of the difficulties of translation just begin to characterize one of their major effects. Nor is this problem limited to genetics. Fodor's recent (1974) discussion supports the view that it is of substantially greater scope and provides a careful analysis of problems that arise for the standard ('type reduction') account of reduction in these areas. But the standard model just looks so right that it is hard to see how it *could* be wrong. In this light, claims like those of Hull and Fodor look almost counterintuitive, and it becomes easy to give them short schrift. There are several sources of bias in favor of the 'standard model' which contribute to this appearance:

(1) There is a general tendency to characterize the lower level theory (T_1) as 'more general' and 'more explanatory' than the upper level theories (T_2 and T_2^*), trading on our general reductionistic prejudices in favor of using compositional information (rather than, e.g., contextual information) in an explanation. This has complex sources which I have discussed elsewhere (Wimsatt, 1975), and has as one of its effects the tendency to assume that lower-level theories correct upper-level theories, but not conversely.[21]

(2) Another important source of bias leading to this error is the distinction between contexts of justification and contexts of discovery, and the attention paid to the former at the expense of the latter. We primarily worry about justifying edifices – theoretical structures that have already undergone substantial revision and selection, and that we have begun to presuppose in a variety of other areas and are thus loath to revise in any substantial way. We discover and propose models – tentatively and usually without much commitment. We give them up or modify them easily because little else depends upon it. For reductions (or at least for those which look much like they will come close to satisfying the formal model) the lower level theory is already well into the edifice stage, and it is thus not surprising that lower level corrections are less visible, having for the most part already occurred.

(3) A bias towards the 'standard model' is introduced *via* the view that explanations involve giving laws, rather than citing causal factors or giving causal mechanisms. How this is introduced (laws suggest greater systematization than do causal factors) and avoided (by accepted Salmon's account (1971) of statistical explanation) is discussed below in part V.

(4) Discussions of translateability tend to revolve around those cases where it looks easiest to give a translation. It is often easier for properties than for objects (which are characterized by a variety of theoretically relevant properties if they are important objects). It is easier for objects if they are not functionally defined (or are fallaciously *treated* as if they were not) since function makes features of the *context* highly relevant. (As linguists know, a context-dependent translation is an incomplete translation.) Functionally defined processes can be the most difficult,

since they will often be associated with a number of objects which will also be involved in *other* functional processes (see Wimsatt, 1974), and can be realized in a variety of different ways.

Discussions of reduction in genetics have not even approached the translation of some of these terms. Terms from population genetics like 'heterosis', 'additive (multiplicative, non-additive, non-multiplicative) interactions in fitness', (see Lewontin, 1974) and Lewontin's 'coupling coefficient' (*Ibid*, p. 294), represent things we look for and find mechanisms for, but general or context-independent translations at a molecular level seem absurd – both impossible and pointless. 'Context-dependent translations' are easy to come by, of course. Discovering the mechanisms in specific cases *gives* us that. But that won't do for the formal model: for those purposes a *'context-dependent translation' is not a translation*.

What would a new view of inter-level reduction look like? Schaffner's later move (1969) in allowing modifications to T_1 in order to affect the reduction (Figure 1b) is a step towards the picture I would draw: *Theoretical conceptions of entities at different levels coevolve and are mutually elaborated* (particularly at places where they 'touch' – where we come closest to having inter-level translations)[22] *under the pressure of one another and 'outside' influences*. (See Figure 1c.) In this picture, both successional reductions (or replacements) and explanatory reductions are occurring in an intricately interwoven fashion. Very roughly, all corrections in theory get packed into a 'successional' component, (because Leibniz's Law applied to inter-level identities ferrets them out of the other component) and all unfalsified explanatory and compositional statements get packed into the 'explanatory reduction' component. Theory at different levels progresses by piecemeal modification, in a manner paradigmatically exemplified by Roth's discussion of the operon theory (1974, ch. II). (See Figure 3.)

Three things should be noticed about these modifications:

(1) Their form may well be deductive or quasi-deductive in character, but if so, the arguments are usually both enthymematic and riddled with *Ceteris paribus* assumptions. Typically, it is decided that a T_1-level mechanism cannot accomodate a T_2-level phenomenon without modification to T_1^*, in which case inferential failure of T_1 is the source of the change; or from T_1 and appropriate boundary conditions, we infer, predict, or deduce that a phenomenon which is incompatible with T_2,

REDUCTIVE EXPLANATION: A FUNCTIONAL ACCOUNT

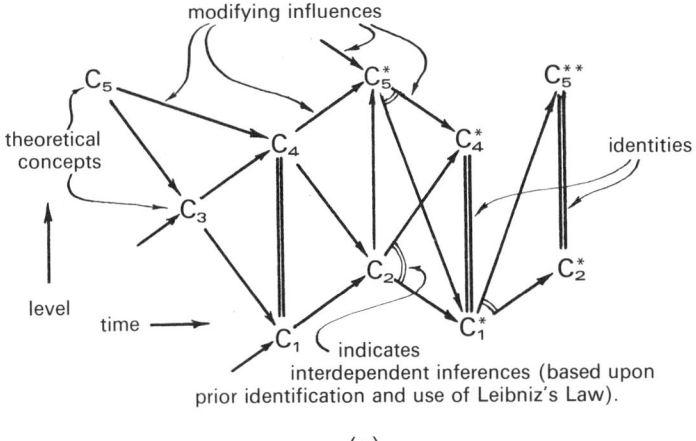

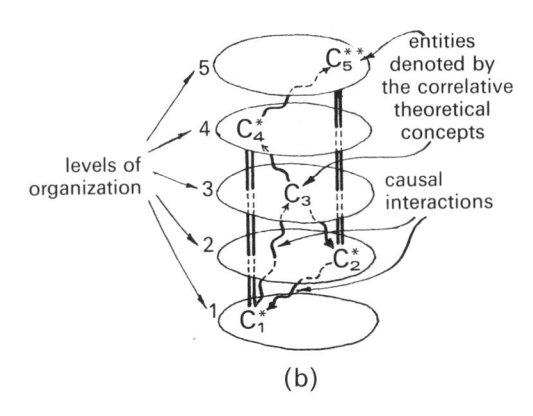

Fig. 3. (a) 'Inference Structure of the Development of the Theory'; (b) 'Resultant Causal Structure of the Mechanism According to the Theory'; 'Development of an Inter-level Theory'.

An extension of the model of Roth (1974) involving the use of identities as proposed in Wimsatt (1975) in the coevolution of concepts in the development of an inter-level theory of the operation of a causal mechanism. Strong analogy between concepts and their descendants (C_m^{n*}, C_m^{n+*}) is assumed generally (but not necessarily universally) to hold, but is not represented here to simplify the diagram.

but not with a T_2^* and observed results should occur, in which case an inferential success of T_1 and its associated mechanisms is the source of the change.

(2) The modification occurs without a total deductive systematization, or often even an informal recodification of the theories. The new theories are characterized in terms of the changes from the preceding theories, but since they were similarly characterized, there is hardly ever a thorough systematization.

(3) The important difference of this picture from Schaffner's is that it is primarily the *changes* in theories which result from deductive arguments. Seldom if ever is any even sizeable fragment of a theory deduced wholesale from another, and seldom if ever is even a single theory sufficiently systematized to meet the conditions for applying the formal model. Furthermore, it is so clearly unnecessary and irrelevant to the search for explanations to do so.

Schaffner's own accounts (1974a, 1974b) and that of Roth (1974) are beautiful confirmation of this highly efficient, but formally, highly confusing strategy of theory evolution. These suggest that the vertical arrows *not* be interpreted as total entailments between theories (or reductions, where upwards arrows are concerned), but as single rough deductions or inferences from attempts to match the structure of causal mechanisms as described at different levels resulting in changes in *parts* of theories. There is, to be sure, use of deductive argument, and lower-level explanation of upper level phenomena. The examples of Ruse (1976), (hemoglobin and sickle cell anemia), Roth (1974), and Schaffner (1974a, 1974b) are marvelous. But as Hull points out, they do *not* touch the issue of whether a total deductive systematization is occurring since such cases would also be expected on the view of reduction advanced here. But if this is all that happens, why should one bother to attempt to characterize reduction along the lines of the formal model? There just seems to be too big a gap between principle and practice for the principle to be very interesting.

Aside from philosophical predelictions of an 'eliminative' sort, there seem to be two reasons for holding onto the formal model of reduction:

(a) The belief that as the 'fit' gets better between upper and lower level theories, their relationship asymtotically approaches the conditions of the formal model of reduction.

(b) The belief that even if the 'fit' never asymptotes, or if it does, doesn't converge on the formal model, the latter represents an aim of scientists.

While Schaffner (1974b) has questioned whether trying to accomplish the reductionistic program *per se* is a good scientific strategy, I suspect that he (and perhaps many scientists) believe that it is at least a secret hope or end. I want to examine the grounds for this latter belief, and suggest an alternative interpretation which is more consistent with scientists' actual behavior. This interpretation also raises serious questions about the first assumption.

Finally, the formal model would not be nearly as tempting if there were not, for each philosopher talking about 'translating away' upper-level vocabulary, a scientist talking about 'analyzing away' upper-level entities. It thus looks as if a claim about words can be 'cashed in' for a claim about entities, and a claim about entities which many scientists appear to accept. The formal model thus appears to have direct support in the talk of many scientists of the 'nothing more than' persuasion. But of what are they persuaded? Are the translations or analyses like those promised by Schaffner *immediately* forthcoming? Usually not. No one actually ground them all out, but that's said to be just a *practical* difficulty. It is *in principle* possible. But 'in principle' claims have been failing, only to be replaced by new ones, since the time of Democritus. Given their history, such *in principle* claims could not plausibly be treated as self-warranting. But then what else warrants them? How can we evaluate these *in principle* claims, to distinguish good ones from bad ones? Or perhaps these *in principle* claims are not the claims they seem to be to knowledge the claimant cannot have: I suggest rather that they are important tools in the task of looking for explanations. Before discussing this (in Section VIII) I must talk about explanation.

V: TWO VIEWS OF EXPLANATION: MAJOR FACTORS AND MECHANISMS VS. LAWS AND DEDUCTIVE COMPLETENESS

I accept Salmon's (1971) account of explanation as a successful search for 'statistically relevant' partitions of the reference class of the event being explained, with two provisos: First, I will make some modifications (to be explained below and in the appendix) to bring it into line with a view of science as an activity conducted according to cost-benefit considerations. Secondly, I assume that in finding 'statistically relevant' partitions, we are doing so with the aim of partitioning the reference class into *kinds of mechanisms*, or kinds of cases involving a given mechanism. (I am thus giving a realist interpretation to his model). In a *reductive* explanation, these mechanisms or factors are at a lower level of organization than that of the phenomenon being explained.

One of the intriguing features of Salmon's account is his move from constructing (statistical) *laws* to a search for statistically relevant *factors*. Laws suggest the need for a complete account of the conditions under which they apply and are correct, and the connection of explanation with laws thus naturally suggests the sort of exhaustive search for factors and conditions that would go along with a complete translation of terms or a complete deductive reduction. By contrast, a search for factors (especially a search for the *major* factors – enter cost-benefit considerations!) ties in naturally with a view of explanation as a search for the mechanisms which produce a given phenomenon, and as an account of how they do it. This search stops short of an exhaustive deductive account by sticking much of the initial and boundary conditions and many background assumptions into a *Ceteris paribus* qualifier on the explanation *because they are too unimportant or insufficiently general to be accounted part of the 'mechanism'*.

The deductivist or formal account *can* give superficial recognition to such differences of importance by different labelling (laws, boundary conditions, initial conditions, etc.) of different parts of the deductive basis. However, in looking first for a valid deduction, the formal account treats all such information as if it were fundamentally alike because it is all equally necessary for the deduction to go through. It thus rides roughshod over realistic intuitions as to differences in the roles and importance of these different kinds of information. Hull is sensitive to this in arguing that a single molecular mechanism can lead to different Mendelian traits, for which he has been criticized by Ruse (1976) and Schaffner (1976). Neither Hull nor I nor the scientists who would agree with us are anti-reductionists or anti-determinists. We are simply responding to widespread and reproduceable intuitions as to when a change in the total state-description is counted as a change in the 'mechanism', and when it is not.

This judgment and its reproducibility are explicable on a combination of realistic, evolutionary, and cost-benefit considerations about the nature of scientific theorizing: A mechanism is a 'kind', and cost-benefit considerations on the complexity of the theory introduce a 'crossover point' beyond which a phenomenon or state is too infrequent or unimportant in a theory to be reified as a kind. There will thus be cases involving the same 'mechanism' with different outcomes which will be attributed to differences in the (more variable and less central) initial or boundary conditions, or to violation of the nebulous *Ceteris Paribus* clause.

The deductivist also makes and must make such judgments of relative importance, but the baggage of having to construct a valid deduction and of having to treat the correspondences between lower and upper levels as 'translations' leads to dangerous misdescriptions of what is going on in several respects:

(1) It is only too easy to assume that variations in the boundary conditions are predictively negligible because they are treated as of negligible or lesser general explanatory importance. A failure to include them as part of the 'mechanism' as Hull has done indicates the latter, but in no way implies either that the same mechanism always produces the same output, or that this failure indicates that the same total state of the system is on different occasions yielding different outcomes. These are mistaken interpretations which become tempting when Hull's discussion of mechanisms is read as if it were about state-descriptions, and when the only differences of importance are assumed to be differences of deducibility or predictability.

(2) Schaffner's claim (1976, pp. 624–25) that Hull's discussion of mechanisms misconstrues the logic of the formal model is double-edged. He would in effect substitute talk about state-descriptions. But if the scientists are interested in *mechanisms* and Hull's point is defensible in terms of the way we investigate and reason about mechanisms, (as I think are so), of what relevance is Schaffner's probably correct claim that the formal model is defensible if we translate from talk about mechanisms to talk about state-descriptions? If scientists aren't interested in state-descriptions, Schaffner has apparently defended the formal correctness of his model at the cost of showing its irrelevance to how scientists talk and reason about reduction. Schaffner's claim about the peripherality

of reduction begins to look more and more as if it applies more modestly and correctly *to the formal model of reduction.*

(3) An equally dangerous move acompanies Schaffner's account of the relation between micro- and macro-descriptions as 'translation'. Schaffner (1975, note 25) *assumes* the constancy of the environment and unstated initial and boundary conditions over a range of different cases in constructing his 'translation' for the dominance relation. This is done 'for reasons of simplicity and logical clarity' (*Ibid.*). But while this is an appropriate defense of simplifying assumptions in a model or idealization, it is not an appropriate move in defense of a 'translation' which is to be used in the way that his are. Thus *one thing his assumption does is to mask the real context-dependence of his 'translation' by artificially assuming that the context in constant!* But if one is trying to establish that context-independent translations can be given (a necessary move if one is to use these translations as general premises in a deduction over range of cases in which the context changes), this move is to beg the question. It is to hide deductive incompleteness by trading it for translational incorrectness or equivocation. Schaffner *cannot* do so. (See his (1976), ms., pp. 622–23.)

Schaffner would not assume this constancy if it were admitted or discovered that there were an important variable (or 'part of the mechanism') contained in that set of things assumed constant. He would then attempt to delineate that variable, and include it in the 'translation'. Thus the boundary between what is in the 'translation' and what is 'assumed constant' is fixed by the same judgments of importance used in delineating 'mechanism' from 'background' on the model which I (and I believe Hull) would defend. But what is not in the translation (or mechanism) is not thereby 'constant'. It is quite variable in fact, and *its very variability is one of the reasons for not including a detailed specification for it in the general theoretical account.* Its variability makes it unimportant for theory construction, and often for selection as well [23], though it can often produce divergent predictive results, and frustrate attempts at 'translation'.

Although Salmon is probably not be considered a scientific realist, his account of scientific explanation is thus a natural ally of realistic accounts of science because of its natural structural affinities for such explanations in terms of major factors and mechanisms, in general, and lower level mechanisms in the case of reductive explanations. (See Shimony, 1971; Boyd, 1973, 1974; Campbell, 1974a, 1974b; and Wimsatt, 1975.)[24]

VI. LEVELS OF ORGANIZATION AND EXPLANATORY COSTS AND BENEFITS

Suppose that the primary aim of science and of inter-level reduction is explanation. We wish to be able to explain every phenomenon under every informative description by showing, first if possible, how it is a product of causal interactions at its own level, but barring that, how it is a product of causal interactions at lower levels (a micro-level or reductive explanation), or least probably and desireably in our reductionistic conceptual scheme, (but absolutely unavoidably in a world of evolution driven by selection processes), how it is a product of causal interactions at higher levels, (most commonly, a functional explanation).

This order of priorities in the search for an explanation follows naturally from the account of levels as local maxima of regularity and predictability, together with acceptance of a weakly but generically reductionistic world view, and the assumption that the *search* for explanatory factors is also conducted according to some sort of efficiency optimizing or cost-benefit considerations. The rationale for this is discussed more fully in (Wimsatt, 1975) and is roughly as follows:

(1) The characterization of levels of organization as local maxima of regularity and predictability implies that most entities will most probably interact most strongly with (and most phenomena will be most probably explained in terms of) other entities and phenomena at the same level.

(2) A reductionist conceptual scheme (or world) is at least one in which when explanations are not forthcoming in terms of other same-level entities and phenomena, one is more likely to look for (or find) an explanation in terms of lower-level phenomena and entities than in terms of higher-level phenomena and entities.

(3) If a search for explanatory factors is conducted along some such principle as 'Look in the most likely place first, and then in other places in the order of their likelihoods of yielding an explanation', then the above order of priorities is established.[25]

Salmon's account of explanation will be generally presupposed here, but with a 'cost-benefit' clause added to it: not only are 'statistically irrelevant' partitions products of a choice of explanatorily irrelevant variables, (as he points out), but 'statistically negligible' partitions are similarly products of explanatory negligible variables. This change is consonant with the remarks of the preceding section on recognizing the different roles and importance of mechanisms, boundary conditions, and the like in an explanation, but also has some extremely important further ramifications. The most important of these is that the intuitive sense of what it is for one variable to 'screen off' another changes (in a manner described in the appendix), from the account of Salmon with consequences to be explored below.

The idea that there can be explanatorily negligible partitions of the reference class of the event or phenomenon being explained suggests an asymmetry of explanatory strategy for cases which do and cases which do not meet macroscopic regularities or laws. When a macro-regularity has relatively few exceptions, redescribing a phenomenon that *meets* the macro-regularity in terms of an *exact* micro-regularity provides no (or negligibly) further explanation. All (or most) of the explanatory power of the lower level description is 'screened off' (Salmon, 1971, p. 55 but see Appendix below) by the succes of the macro-regularity. The situation is different however for cases which are anomalies for or exceptions to the upper level regularities. Since an anomaly does not meet the macro-regularity, the macro-regularity *cannot* 'screen off' the micro-level variables. If the class of macro-level cases within which exceptions occur is significantly non-homogeneous when described in micro-level terms, *then* going to a lower-level description can be significantly explanatory, in that it may be possible to find a micro-level description partitioning the cases into exceptional and non-exceptional ones at the macro-level. We would then have a micro-explanation for the deviant phenomenon.

Thus, for example, the ideal gas law (or its corrected phenomenological successor), as a relationship between macroscopic causal factors, is explanation enough for occasions when gases obey it. Going to the micro-level in such a case is not (or negligibly) more explanatory. Of course, if all of the molecules go to one corner of the container, the micro-level must be invoked since the macro-level law does *not* apply, and in *that* case partitions in terms of micro-variables will be statistically relevant.

I have discussed one main reason to look for information at lower levels: to explain exceptional cases at the upper level. The other main reason is to explain upper-level regularities. But part of explaining exceptional cases involves explaining why they are exceptional in a way that is consistent with the patterns found in the motley of cases explained by the upper level law (*qua* set of interrelated causal factors.) This usually involves explaining exceptional and motley cases in terms of a single class of mechanisms or micro-variables. This requires that the relevant kinds of micro-descriptions necessary to explain the exceptional cases *also* be usable in generating the upper law as a 'special case' or 'limiting' or 'approximate' description. It thus leads to an explanation of a revised version of the upper level law.[26]

But what is a law, and why bother to explain one if, as I have argued, mechanisms and major factors bear the primary role in explanations of events that laws have been thought to do? The answer that suggests itself in the cases I have looked at where laws are being explained in terms of lower-level factors and mechanisms is that *laws are regularities involving distributions of cases characterized at the macro-level*. They are explained as the product of the interaction of the mechanisms and major factors invoked at the micro-level with the micro-level distributions of initial and boundary conditions. They are not *mere* regularities (or 'accidental generalizations' as Nagel (1961) characterizes the infirm statement of lawlike form) because they are exhibited as the product of *causal* interactions of micro-level mechanisms, factors, and initial and boundary conditions. Such law-statements thus support the appropriate counterfactual and subjunctive conditionals. Indeed, when a macro-regularity is explained in this manner, an understanding of the micro-level mechanisms and conditions which generate the macro-level distribution and how they do so give a much richer structure of counterfactuals expressable in terms of micro-descriptions than before.

I am not sure whether this characterization of a law is generalizable. It might seem limited to cases where the phenomena of a law admit of meaningful redescription at a lower level. But, at least in those cases where this characterization applies, (and this would appear to cover all cases of (inter-level) reductive explanation) a law should be explicable in the same general way as an event. The only difference would be that instead of talking about individual constellations of mechanisms, factors and conditions, we are talking about assumed *distributions* of the above.

The reduction of thermodynamics to statistical mechanics would provide useful examples of explanations of this sort. (See, e.g., the much discussed explanations of the second law of thermodynamics.) But so also would the history of the assumption of the 'purity of the gametes in the heterozygote' which Hull (1974, 1976) makes much of in arguing that molecular genetics replaces, rather than reduces Mendelian genetics. I believe that Hull is incorrect in his conclusion, and that an illustration of how this 'law' is explained reductively helps us to see how much real continuity there is between Mendelian and molecular genetics.

VII. AN EXAMPLE: THE ASSUMPTION OF 'THE PURITY OF THE GAMETES' IN THE HETEROZYGOTE

This assumption began life as Mendel's 'law of segregation' – to explain the fact that some apparently lost characters ('recessives') reappeared apparently unchanged in successive generations. Mendel's explanation was that in the company of certain alleles ('dominants') the factors did not express themselves as characters, *but that they were transmitted to offspring unchanged (by their allelic factors or anything else)* to express themselves in future genotypes in which they were homozygous or dominant.[27]

In the Mendelism that Castle attacked, with his belief that the allelic genes 'contaminated' one another in the heterozygous state, it was accepted that genes affecting a given character came in pairs (were alleles), but Mendel's other law – of 'independent assortment' (that non-allelic genes assorted independently of one another in the offspring) was being challenged, both experimentally and theoretically, by Bateson and others, including Morgan and his students.

The 'linear linkage' model of the Morgan school explained some of Castle's results (gradual changes in coat color conformation in rats) by the gradual accumulation through selection of so-called 'modifier' genes at *other* loci (presumably linked on the same chromosome) which modified the *effect* of the genes identified as producing coat color, *without modifying the allelic genes themselves*. There was thus no need to suppose (in this case) that allelic genes 'contaminated' one another in the heterozygous state. Castle's supporting claim that these modifications were irreversible were successfully contested experimentally.

The Morgan model supposed that the genes were linearly arranged on chromosomes, with allelic genes on corresponding places on the homologous paired chromosomes. According to this model, homologous chromosomes would, at a certain part of the cell cycle, wind around one another forming 'chiasmata', break, and exchange segments. This was called crossing-over and recombination. A central feature of the model was that genes on the same chromosome would tend to assort together, constituting linkage groups. This was in contradiction to Mendel's law of independent assortment. A prediction of the linear model and the mechanisms of recombination was that the probability of recombination between two points along the chromosome was a monotonic increasing function of the distance between points, (being approximately linear for small distances and approaching 50% (or random assortment) for large distances.)[28] These also were experimentally confirmed. Furthermore, *in the absence of any 'atomistic' assumptions* (placing a lower bound on minimum distance between recombinations) *this model would predict a finite frequency for crossing-over within genes of any finite size.*

A gene has a size, and this was recognized by members of the Morgan school, though different ways of estimating it[29] produced different results. Although it was usually assumed that the genes behaved like 'beads-on-a-string' (or independent atoms) as far as recombination was concerned, Muller, a Morgan student, questioned whether these 'atoms' were the same for recombinational and for mutational events. Other observed phenomena (like 'position effect') also raised questions about the 'beads-on-a-string' model. It also was generally supposed that genes had an underlying molecular nature, though it was unknown what this was, and how it produced the properties manifested by genes, so the idea that genes had a molecular infrastructure was not new. Indeed, the 'atomicity' of the genes was clearly believed, to the extent that it was only with respect to the genetic or biological properties of the genes.

The details of how the molecular account of the gene explain 'position effect' and the possibility of differences between recombinational, functional and mutational criteria for individuating genes are well known (see, e.g., Hull, 1974), or any modern genetics text) and uncontroversial here. All of these have the effect of compromising the view of genes as monolithic, monadic 'atoms' with respect to some of their biological properties. If there are any 'atomic' units of DNA, it is the individual base pair – again not because smaller changes are impossible, but because if they occur, they are not counted as *genetic* changes. But while this would show that there were no 'atomic' genes of the size Morgan and his school had assumed, and that their different criteria of individuation picked out *different* larger compound assemblages of bases as genes, it is not necessarily a disproof of their genetic 'atomism'. It could just as well be taken as a demonstration that their 'atoms' were smaller than they had thought, (see Note 32) and (being of at that time unknown constitution) had some unexpected properties which explained others that the genes had been thought to have.

How does the assumption of the 'purity of the genes in the heterozygote' fare? This becomes a question of the possibility of intra-genic recombination – but not a simple question: we must ask not only what happens, but also, what an experiment detects.

We can now explain in terms of the design (I nearly said 'logic'!) of the recombination experiment why, even if they should occur readily, it was very difficult to find intra-genic cross-overs and recombinations. We can do this in terms of the molecularly characterized gene, *but there is no need to do so*. Morgan could have done so himself, as it is an obvious consequence of the 'classical model' of the genome.

(1) On this model, there were a large number of genes on each chromosome. Muller estimated in 1919 that there were at least 500 genes on the X chromosome in *Drosophila*, and we now know that to have been at least a four-fold underestimate.[30]

(2) It was taken as a given then as now that any individual gene has a very high stability, which would have applied either to intra-genic recombination or to any other mutational event.

(3) The design of a recombination experiment involved looking at a small number of 'marker' genes spaced along the chromosome in order to see how frequently they (or more accurately, the traits which signal their presence) stay together in offspring. The usual number of marker genes was 2, though 3 and 4 were occasionally used to detect multiple crossing-over by Sturtevant.[31] Supposing even that one could detect any intra-genic recombination occurring in any of the marker genes

(see (4)), the very small fraction of the genome being used as marker genes renders it very probable that recombinational events will not occur in any of the markers, but will occur elsewhere along the chromosome, separating whole the marker genes on either side of the break.

(4) We now know that intra-genic recombination would produce a non-functioning gene. This would have been scored by the Morgan school as a 'loss' or 'mutation' of a gene, rather than as an intra-genic recombination, so they probably did *not* detect any such events that did occur. (Only with the later work on intra-cistronic complementation were the classical techniques sufficiently refined to detect such intra-genic events. But it is worth emphasizing that the problem was a technical one, and not a conceptual one for the classical approach.)

The net effect of this is twofold:

(a) The classical model itself predicts that if genes are as small and as numerous as they had to be (and they were smaller and more numerous), intragenic recombination would be hard or impossible to detect, even if virtually all recombinational events were intra-genic.

(b) What was *seen* in recombination experiments was whole (marker) genes separating from one another untouched.

The first fact might have produced caution. It did not. The second observation led to an extrapolated assumption that recombination occurred *between genes, generally*, rather than just *between the observed genes*. But the first fact means that the new molecular picture is *not* that different from the old model. By analogy with the old model:

(1) Crossing-over should be a monotonic increasing function of the length of the DNA involved.

(2) The probability of crossing-over should be very near 0 for lengths of DNA of the order of functional genes – e.g. cistrons.

(3) Individual base pairs, *at least*, still have the 'atomistic' status of the bead-like genes of the old model, since crossing-over cannot meaningfully be said to occur within a base.

(4) The linear arrangement of the genes on chromosomes (preserved in the linearity of the primary structure of the DNA molecule) is unchanged in the modern account, and plays a central role in accounting for the high stability of the genes, the high reliability of the segregation mechanisms (without which genetics would be impossible), and the low frequency of 'contamination' in the heterozygote.

But intra-genic recombination *is* assumed to be possible on the molecular account,[32] and not on the beads-on-a-string model. Does this make the molecular theory a 'neo-contaminationist' theory rather than a neo-classical one?

Castle had no well worked out mechanism, but only a set of experiments which purported to show that classical (pre-Morganian) Mendelism did not work. There was little in Castle's work from which 'neo-contaminationists' could claim descent. The purported phenomena of Castle's experiments for 'contamination' turned out to be non-existent or to admit of Morganian explanations. His explanations had no important connections with the explanations a molecular 'neo-contaminationist' would give for his 'neo-contamination' phenomena, but Morgan's did. Thus, without a theory, a mechanism, or a set of phenomena persisting through time to call their own, there is no 'Castlian genetics', and there are no molecular 'neo-contaminationists'.

The kinds of connections between the two accounts clearly support the claim that the mechanism of the Morganian and molecular theories (especially when looked at with the time and size scale appropriate to the Morganian account – a move appropriate to showing that T_2 and T_2^* are strongly analogous) are indeed strongly analogous. I thus agree with Schaffner and Ruse on this issue.

Indeed, there has been so little change, and what has changed has done so with such continuity that it is tempting not to describe this as a case of successional reduction at all. It is very tempting to say that Morgan's gene *is* the molecular gene, at a different level of description, and conversely. But to make this identification in the same breath with a claim of strong analogy is to invite confusion of identity by descent of concepts in successive theories (which is a similarity relation) with referential identity of different level descriptions of the same object (which is an identity relation.) The former notion requires no further attention now, but the latter concept and its role in reductive explanations and analyses is radically different on this account from that suggested by the formal model. Furthermore, the much better fit of this account of the role and uses of identity hypotheses with actual scientific practice is one of the strongest arguments for this account and against that of the formal model.

VIII. IDENTIFICATORY HYPOTHESES AS TOOLS IN THE SEARCH FOR EXPLANATIONS

In its earlier formulations, the classical model of reduction had nothing to say about the role of identifications in reduction. Thus, Nagel (1961) suggests that bridge laws or correspondence rules might be grounded in definitions, conventions, or empirically discovered correlations or hypothesized identifications, as if one was as good as another. The widespread instrumentalism and mistrust of identifications as metaphysical, and as going beyond the evidence, has perhaps led many writers away from asking why scientists might prefer to make one claim rather than another. In the one area where this has been hotly debated (and where postulating identities or postulating correspondences is seen as making a metaphysical difference which bears immediately on matters of importance) philosophers of mind appear to almost universally believe that identity claims are a solely metaphysical and evidentially unsupportable extension beyond the evidence of observable correspondences. (See Kim (1966) for a representative and influential view.) Only recently (see e.g., Causey, (1972)) have philosophers of science found a necessary role for identities in reduction. I wish to suggest a heretofore unexplored and absolutely central role for hypothesized identifications as tools in the search for explanations which, among other things explains a number of features concerning their use which have been considered to be unjustified, unjustifiable, or otherwise anomalous. (I have discussed some aspects of this analysis more fully in Wimsatt, 1975, 1976).

I will assume that we are faced with some upper-level explanatory problem: some phenomenon for which we have no micro-level explanation, or perhaps something which lower level accounts would lead us to expect at the upper level, but which has not been observed. Such an explanatory failure suggests inaccurate compositional information, or none. How do we discover the source of these inaccuracies, of the locus of our incomplete information? An identity claim, with its subsequent application of Leibniz's Law provides the most rigorous detector of possible error or of a failure of fit of applicable descriptions at different levels: *Two things are identical if and only if any property of either is a property of the other*. If there are properties apparently had by one but not by the other, then either the identity claim is false (as many are) or else *there are as yet undiscovered translations between descriptions at the different levels* which show that the relevant properties are indeed shared.

Thus, *in principle* translateability (or analyzability) is a corollary to and the cutting edge of an identity claim. The identity claim is in turn a tool to ferret out the source of explanatory failures which, by its transitivity, allows one to delve an arbitrary number of levels lower if need be to pinpoint the mismatch, or by its scope, to any properties – however diffuse and relational – to detect a relevant but ignored interaction. (For this reason, I do not share the view of some writers that Leibniz's law should be weakened in all sorts of ways for intensional contexts, and the like.)

Several interesting features follow from this account:

(1) It would be expected that identity claims and claims of translateability should be honored more in the breach than in the observance. They function primarily as templates, which help us to locate and to focus upon *relevant* differences – differences which can help us to solve explanatory problems – in order to remove these differences and thereby to make more accurate identity claims. Thus the warrant for claims of *in principle* translateability, which was questioned above in Section IV, is the same as that for making the identity claim from which if flows.

(2) The warrant for this claim is in part the warrant for using a good tool appropriately: that its employment at this time and in this place may help us to discover a description or suggest a redescription which will allow us to explain some heretofore unexplained phenomenon. There is *no* warrant for using the claim if it is *known* to be false. The strength of the claim, which makes it such a sensitive template, renders it easily falsified, and like any strong claim, its negation carries no or little significant information. Thus, if one of the standard defeating conditions for identification, such as causal relation or failure of spatio-temporal coindence is known to obtain, the claim is dropped – though perhaps in favor of a correspondence claim (see Wimsatt, (1975, part II).)

(3) This kind of warrant can however apply early in the stages of an investigation, and explains behavior which seems irrational and unjustifiable on a more inductivist account of the making of identity claims. Identity claims are often made on the basis of correspondences between

or explanations of only 2 or 3 properties, often together with some subsidiary background information of a non-correlational nature. I have argued (in Wimsatt, (1976)) that this was in fact true for the early identifications, by Boveri and by Sutton, of Mendel's 'factors' with the chromosomes. To the inductivist, this would look like a wildly irresponsible claim: a projection from 2 or 3 properties of a pair of entities to *all* properties of those entities. Moreover, to add insult to injury, the burden of proof after the making of such a claim is not upon its maker (as one would expect on an inductivist account), but upon those who *doubt* the claim to come up with a counterinstance. Only then is the maker obligated to respond to the putative counterinstance, either by elaborating and defending his claim, or by giving it up, as the case seems to demand. Sutton and Boveri proposed a number of new correspondences on the basis of their identifications, and these were later observed, though subsequent conceptual modifications and clarifications led to an elaboration of the identification claims by Morgan and his students, and the generation of many new predicted correspondences (Wimsatt, 1976). The early stages at which identities are proposed; the fact that they seem to provide the basis for, rather than be made on the basis of claims of correspondence; and the location of the burden of proof after the making of an identity claim all support this account of the role of identity claims against the inductivist, who should expect the opposite in each case.

(4) The fragility and falsifiability of identity claims are hidden by the 'open texture' of our concepts (Waismann, (1951)), and in more severe cases, by the same tendency to claim identity by descent of our concepts that makes successional reduction possible. With successional reduction, the similarities *and* differences in the successive theories are analyzed critically and used. Only afterwards is the similarity implied by the possibility of performing a successional reduction invoked to maximize the apparent continuity in this identity-by-descent of theoretical concepts. Similarly, with inter-level identifications, the similarities are used critically to ferret out the differences, and only afterwards are the newly assimilated differences reified after the fact into the original identification. The fact that it has become more specific, more detailed, and sometimes has undergone outright changes is hidden from us, so that we see only the continuity of 'identity by descent' in our concept of the specific identifications we have made.

(5) This analysis suggests that scientists should prefer indentity claims to claims of correspondence when there is no specific reason (such as the violation of one of the identity conditions mentioned in (2) above) to prefer correspondence. They should do so because they prefer the stronger tool, and not for reasons of 'ontological simplicity' (or whatever) as suggested by Kim (1966). From a specific identification, after all, one can generate all necessary correspondences, including new ones which might arise as new properties and relationships are discovered at one level or another. But from the set of correspondences one might derive from an identification given what is known at a given time, one could *not* (without covert reintroduction of the identification) know how to generate new correspondences to fit the new information as it comes in. Identifications are an effective guide to theory elaboration. Correspondences are not. Thus one can understand not only why identity claims might be made early in the course of an investigation, but also why the metaphysically more conservative strategy of making correspondence claims instead will not work. In a static view of science, identity claims and corresponding claims of correspondence only may be empirically indistinguishable. But in a dynamic view of science, only identity claims can effectively move science forward.

The analysis of reduction and of correlative activities proposed here has differed from most extant analyses in two important respects: it has been primarily functional, with the aim of deriving and explaining salient structural features (including some not explained by the standard model) in terms of their functioning in efficiently promoting the aims of science; most notably, explanation. Secondly, it has aimed at a dynamical account of science, in which optimally efficient change and elaboration are the primary process, and in which stasis is either an artificial construct, a temporary blockage which must be explained, or an end state which we are not likely to reach in the forseeable future. I believe further that it supports realistic conceptions of the nature of theoretical entities, and of the functions and roles of scientific theory, and does so while being truer to the ways in which scientists *actually* behave than the extant analyses of these activities deriving from the structuralist, static, and often instrumentalist logical empiricist tradition. Finally, it fits into a broader generically evolutionary account of man and his activities, and encourages me to believe that biology may soon be a source for paradigms and analyses which will inform philosophy and philosophy of science generally, rather than being little more than the backwards field for the brushfire skirmish in which philosophical imperialists moving out from the 'hard' sciences stop to try their weapons. The latter time is now fast receding into the past, but it is not yet so far that almost all of us cannot remember it.

APPENDIX I. MODIFICATIONS APPROPRIATE TO A COST-BENEFIT VERSION OF SALMON'S ACCOUNT OF EXPLANATION

Salmon (1971, p. 55) defines what it is for one variable to 'screen off' another as follows:

...D screens off C from B in reference class A if and only if:
(i) $P(B/A.C.D) = P(B/A.D)$ [C adds nothing to D.]
(ii) $P(B/A.C.D) \neq P(B/A.C)$ [D adds something to C.]

Thus, on this interpretation, microstate description D in statistical thermodynamics *screens off* the macrostate description C from B (a macrostate in accordance with a phenomenological macro-law) in A (a macroscopically characterized assumed-ideal gas). This is so because of those fluctuations from the equilibrium state predictable from D, but not predictable from C, which generates the inequality in (ii).

Note how this definition handles an upper-level anomaly (say, a macroscopically unpredictable fluctuation). Since it would be true that:

(1) $P(B^*/A.C.D) = P(B^*/A.D)$
(2) $P(B^*/A.C.D) \neq P(B^*/A.C)$

where all is as before except that B^* is a macrostate violating phenomenological macro-laws, it is clear that according to the above definition, D screens off B^* from C in A.

It is the consequence and intent of Salmon's definition that any strict improvement in information requires saying that the variables generating the improvement screen off any other set of variables which they represent this sort of improvement upon. *This is so no matter how small the improvement and how great the cost resulting from adopting the new set of variables.* It is another consequence of accepting a view of scientific method appropriate to Laplacean demons.

I think that scientific practice and good sense suggest the value of a different notion of 'screening off', which, because of its obvious connections with cost-benefit analysis might be called the 'effectively screens off' relation:

C effectively screens off D from *B* in reference class *A* if (and perhaps not only if):

(a) $P(B/A.C.D) = P(B/A.D)$
(b) $P(B/A.C.D) \simeq P(B/A.C)$ [*D* improves the characterization only a little.]
(c) $C(D) \gg C(C)$ [*D* is enormously more expensive information to get than *C*.]
(c′)   *D* is a *compositional redescription* of *C*.

Some comments are in order about conditions (c) and (c′), which are probably alternatives, or nearly so. The second condition comes closer to capturing the intended application of the effective screening off relationship in the present context, since I am here considering inter-level explanatory reductions, where the lower level is a compositional redescription of the upper level. Furthermore, at least empirically, the truth of (c′) appears to guarantee the truth of (c), at least for those kinds of cases we are likely to regard as interesting compositional redescriptions, and thus for all of those cases where we are likely to find any room for debate in the matter of inter-level reduction. Indeed, I am inclined to feel that the proposed 'upper level' is not at a distinct level unless at least most of the compositional redescriptions of upper level phenomena in terms of lower-level entities meet condition (c), which would, in turn, guarantee that any inter-level reduction would be non-trivial.

Condition (c) gives explicitly the cost part of the cost-benefit condition, whereas the approximate equality in (b) guarantees that the benefits, if any, of using redescription *D* are small. Obviously, the deviation from strict equality in (b) and the cost-ratio in (c) required for the effective screening off relation to hold are interdependent, and are in turn both dependent upon outside factors which determine the importance of additional information and level of acceptable costs. These may vary with the purposes for which the theory is being used, and with any other factors (such as the current explosion in the development of computers and computational facilities) that may radically affect these costs or importances.

The situation where the approximate equality in (b) is in fact an inequality is by far the most interesting one, for *under these circumstances, D screens off C* (according to Salmon's definition) *but C effectively screens off D* (according to my characterization.) Thus, in this case, the two criteria would pick out different factors to include in an explanation of phenomenon B.

Condition (a) was also included for the same reason: it is the same as condition (i) in Salmon's definition of the screening off relation, and thus points directly to a class of cases in which X screens off Y but Y effectively screens off X. Condition (a) would presumably be met in any case in which a successful and total theory reduction (along deductivist lines outlined by Nagel and Schaffner) holds between two theories, such that D is a description imbedded in the reducing theory and C is a description imbedded in the reduced theory. (I would guess that this should be provable as a theorem in the probability calculus from the characteristics of their model of reduction.)

I am not sure however, how or even whether this result would be provable for reduction as I have characterized that relation. I rather suspect that it is not. Furthermore, in cases where no reduction or only a partial reduction has been accomplished, it would at least be true that condition (a) would not be known to be met for at least some descriptions C in the upper-level theory (and further, that on a subjectivist notion of probability, condition (a) would almost certainly *not* be met for these cases.)

In fact, I see no reason why condition (a) should not be dropped for the effective screening off relation, since conditions (b) and (c) (or (c′)) seem to include all that is necessary – namely, the cost-benefit conditions. I have included it for the time being because it heightens the contrast between the screening off and effective screening off relations, and because I think that substantial further work is necessary to see what if any other modifications and applications seem desirable in developing a cost-benefit model of explanation. The need for at least one further clarification should be immediately obvious: since Salmon (1971, p. 105) points out that his screening off rule follows from his characterization of explanation, if I believe that the effective screening off relation says something fundamental about the notion of explanation (as I do), it is necessary for me to produce an appropriately modified concept of explanation. This is better left to some future date.

An important consequence of adopting the effective screening off relation rather than the screening off relation was assumed in the text. This was that although upper level descriptions meeting upper level laws would effectively screen off lower level redescriptions, upper level anomalies – upper level descriptions which failed to meet upper level laws – would *fail* to effectively screen off lower level redescriptions. This introduced an important asymmetry between cases which met upper-level laws (and thus which were acceptably explained at the upper level) and cases which were upper-level anomalies (and which thus had to be explained at the lower level.) On Salmon's screening off relation, there is no asymmetry of course, since both cases which meet and cases which fail to meet upper level laws are explained at the lower level, because lower level variables screen off upper level variables in either case.

This assymmetry arises in the following way for the effective screening off relation. Suppose as before that B^* represents an upper level description which is anomalous for upper level theory. Presumably then:

(a) $P(B^*/A.C.D) = P(B^*/A.D)$
(b) $P(B^*/A.C.D) \neq P(B^*/A.C)$

The failure of condition (b) occurs because if B^* is an anomaly, then $P(B^*/A.C)$ must either equal zero, or be very low, and much lower, for example than the probability of states which are held to be explained by the upper level theory under similar circumstances. On the other hand, if B^* is to be explicable by an account in terms of lower level variables, it must be that there exists an appropriate description of B^* such that $P(B^*/A.D)$ is appreciably greater than zero – and in general of the order that similar phenomena held to be explicable on the lower level theory would exhibit. Thus the failure of condition (b) means that the benefits of redescribing B^* at a lower level are not neglegible, and in general justify the greater costs implied by conditions (c) or (c').

Dept. of Philosophy,
The University of Chicago

NOTES

* The major portion of this paper was written while I was a visiting research fellow in Humanities, Science, and Technology at Cornell University. I wish to thank the program and especially Max Black and Stuart Brown for their support.
[1] See Ruse (1971), and Hull (1974).
[2] See Roth (1974), and Wimsatt (1975).
[3] See Nickles (1973), and Wimsatt (1975).
[4] On the point of *in principle* translateability, see Boyd (1972) for a masterful discussion and doubts of a more general and pervasive nature.
[5] Hull (1974).
[6] Ruse (1971).
[7] Hull (1974).
[8] Schaffner (1974b).
[9] Boyd (1973), (1974, unpublished manuscript), and especially Kauffman (1970).
[10] It is naturally important to distinguish between disputes over details of particular mechanisms from objections (e.g., like those of Haldane (1914), or Elsasser (1965)) which challenge the adequacy of an entire approach.
[11] See Boyd (1974), and Wimsatt (1975), parts II and III. Boyd tends to locate the primary difficulty in verificationism and in acceptance of the Humean account of causation, but as a realist, would also agree with the views advanced here and in my (1975).
[12] Ruse has (in this symposium) retreated from his earlier attack on the formal model and attempt to characterize 'informal reduction'. I am more in sympathy with his earlier views.
[13] This line of criticism was initiated by Nickles (1973). See also Wimsatt (1975), part II and below.
[14] All of us believe that some reconstruction is necessary. Hull and I appear to believe that less is necessary (me) or appropriate (both of us) than Schaffner or Ruse. See Hull's discussion in this symposium (1976), which however does *not* mention the specific alternative discussed here: reconstruction of reduction as an efficient end-directed activity.
[15] This approach is explicit in Kim's (1964) analysis of the deductive-nomological (or D-N) model for explanation and prediction, though Kim advances this as a defense of the D-N model, (by suggesting that the differences are pragmatic and epistemological rather than structural), and I am using it as an attack on the formal model (by suggesting that the structural similarities are more superficial than the functional differences).

Nickles (1973) individuates two types of reduction on both functional and structural grounds, but concentrates on what I call 'successional' or 'intra-level' reductions, largely accepts the formal model for the other kind (which is most relevant here) and does not draw the close links between functional and structural characterizations that can be made for each of the two types. Schaffner's (1974b) argument for the peripherality of (formal) reduction in the development of molecular biology invokes Bayesian arguments for choosing scientific research strategies, which presupposes a purposive account of scientific activity, but he has not attempted a functional analysis of reduction or of other related activities.

[16] Schaffner (1976, pp. 626–28) appears to regard Nickles' 'reduction $_2$' and the correlative notion of a transformation as a competitor to his condition of strong analogy, and criticizes it, uncharitably, I think, for being too open ended in that there seems (says Nickles) to be no general way to characterize what kinds of transformations should be

allowable and what should not. He claims, by contrast that the notion of 'strong analogy' can be applied with general agreement (3 out of 4, at least – *pace* Hull!) in the case of genetics and thus, though it is unanalyzed and primitive, it is at least testable. But surely Nickles could claim as much for the notion of an allowable transformation. I suspect that there would be general agreement in any given case on what transformations would be allowable in constructing a 'reduction$_2$'. I believe that Nickles despaired of finding something which I don't think exists: a general theory-independent criterion which would determine the allowable transformations. This is impossible for the same reason that a theory independent notion of 'strong analogy' would be impossible: what transformations are allowable (or even interesting) and what features of an analogy are salient depend upon usually quite general and important features of theory in that area. And on these, there would usually be general agreement. Further, the notion of a transformation is mathematically an extremely powerful and suggestive one, and is less tied down to intuitive notions of similarity than analogy. For three relevant examples which are very different in terms of allowable transformations, but for each of which there would be agreement on what transformations would be allowable, see Minsky and Papert's applications of linear transformations to the analysis of the data-manipulating capabilities of certain classes of neural networks (1969); the 'law of similitude' and its use in building scale models of ships and aircraft for testing in wind tunnels and towing tanks; and the continuous deformations allowable in the applications of conformal mapping to 2-dimensional airfoil theory (see Prandtl and Tietjens (1957)) and in D'Arcy Thompson's application of his (1961) theory of transformations to problems of development and allometric growth. Indeed, none of these has been seen as involving anything like reduction, and it is one of the more provocative aspects of Nickles' analysis that it suggests the possibility of seeing them in a new light.

[17] Nickles gives a more complete account of theory succession and elaboration in his paper in this volume (Nickles, 1975). His paper suggests and may require modifications to the account of successional reduction adumbrated here, but seems to lend further support to the general functionalist approach. His more recent account seems to show some of the features of both intra- and inter-level reduction, but this is to be expected in the analysis of any multi-level historical case, which should involve both components of change. Further, those ways in which his new account differs from his earlier one, or from the view advanced here should be of no comfort to advocates of the 'standard model'.

[18] Ruse (1971) suggests that reducibility is a similarity relation, but gives different reasons (which I do not accept) for saying so.

[19] In his symposium paper (Ruse, 1976), Ruse gives up this view and attacks Hull for holding it. In this matter, I agree with Ruse (and Schaffner) though in virtually all other respects, I agree with Hull.

[20] For the earliest statement of a closely related view, see Simon (1969, chapter 4). See also Bronowski (1970). For my general approval of and some dissatisfactions with Simon's view, see my (1974), and for a thorough discussion of levels, see my (1975), part III.

[21] Thus, e.g., in his first (1967) presentation of his general reduction paradigm, Schaffner made provision for upper-level modifications or corrections, but not for lower-level ones, a matter which he corrected later (in his (1969)).

[22] This picture is in this respect very close to that drawn by Friedrich Waismann in his penetrating essays, 'Verifiability' and 'Language Strata' (Waismann, 1951, 1953), though he put more weight on the language and less on the underlying structure of the world than I would. In particular, Waismann suggests that different language strata might not fit exactly, but would permit nearly exact translations at some points and none, or only very

rough and partial ones at others. This is roughly what I believe to be true for the languages which best describe phenomena and entities at different levels of organization.

[23] Traits which are highly variable in irregular ways are unusually difficult to select for in most cases, so one might argue that it would be highly unlikely that they would be included as part of a *functional* mechanism. But all or virtually all mechanisms which are of interest in biological organisms are functional. Thus highly variable things would not likely be included as parts of biological mechanisms. No less an ecologist than G. E. Hutchinson used this elegantly to argue (Hutchinson, 1964) that certain trace materials probably could not be utilized by organisms to perform any characteristic functions because they were present in amounts of less than about 10^4 atoms per cell, which Hutchinson suggests as a rough stochastic threshold below which fluctuation phenomena rendered their presence too unreliable to be used by selection in any biological processes. Unfortunately, this reasoning does not apply symmetrically to allow one to assume (as Dinman, 1972 does) that lower concentrations of trace elements could not *disrupt* functional processes.

[24] This realism may look superficially very much like a kind of instrumentalism, because our perceptual apparatus, senses, cognitive apparatus, and theories are all treated as instruments designed by biological psychological and social selection processes according to cost-benefit constraints which naturally introduce biases. But the biases are taken seriously as deviations from a correct portrayal of the real world. We regard the biases of the senses, theories, etc. as leading to *false* judgements which we try to correct when appropriate. That a good theory is a useful *instrument* for getting around in the world is a product of the fact that it contains a good deal of *truth*. This is no form of instrumentalism.

[25] I do not think that this is the *best* way to argue for this conclusion, primarily because I believe that judgements as to where one should look for an explanation of a phenomenon are made on other grounds which determine whether a standard causal, micro-level, or functional explanation is appropriate, and that the judgements of relative likelihood follow from these in any given case. Nonetheless, at least globally (not in specific cases), I think that the likelihoods are assumed to be as they are in the argument in the text, and the matter is clearly worth further study.

[26] If it were to turn out that there were a single micro-variable which partitioned the macroscopic reference class into exceptions and non-exceptions to the macro-law, this micro-variable would give the relevant lower-level type-descriptions for a reduction. The force of Hull's complaint concerning the complexity of reduction functions is that there isn't even a small number of such variables. The force of ergodic theory is to suggest that the same problem affects statistical mechanics, but that the number of 'pathological' states involved is so small (of measure 0) that we nonetheless treat it as a reduction. (See Sklar, 1974). The number of 'pathological' states in the case of genetics is *not* likely to be of measure 0 however.

[27] I am not in all respects using Mendel's terminology (or even his assumptions) in this description, but the respects in which it is thereby distorted do not affect the present argument.

[28] I am here talking about the possibility of a single break, so the complications of 'interference' and multiple crossing over do not arise. But even this ignores the complication that breakage strength may vary along the chromosome. All of these factors were recognized and discussed by the Morgan school.

[29] See Carlson, 1966, pp. 83, 85, 158ff.

[30] The underestimates in the number of genes was a crucial factor in overestimating their size. This was one area in which further progress raised questions about the classical

model, such as Muller's doubts that the unit of mutation was the same as the unit of recombination.

[31] There are a variety of reasons why it becomes experimentally more difficult to handle a large number of markers in a given experiment, and the largest number ever followed at once to my knowledge was 6, by Muller, and that for a very special kind of test of the linearity hypothesis. (See Muller, 1920, especially Table II, for discussion of why smaller numbers of marker genes were usually followed.)

[32] Indeed, this may exaggerate the difference. Evidence is accumulating in *Neurospora* (a bread mold widely used in genetic experiments) that there is a strong or even an absolute bias against intra-genic recombination at a molecular level. This is a product of site specificities in where the 'nickases' (enzymes which nick open the DNA to allow recombination) will act. If this phenomenon is veridical and generalizable, then the 'beads on a string' view of the genome is inappropriate only for suggesting a macro-mechanical metaphor rather than a chemical or a micro-mechanical one. (See Whitehouse, 1973, pp. 367–369, for relevant discussion.) I thank Thomas Kass for helpful discussion of this and other related points.

BIBLIOGRAPHY

Boyd, Richard: 1972, 'Determinism, Laws, and Predictability in Principle', *Philosophy of Science* **39**, No. 4 (December), pp. 431–450.
Boyd, Richard: 1973, 'Realism, Underdetermination, and a Causal Theory of Evidence', *Nous* **7**, No. 1 (March), pp. 1–12.
Boyd, Richard: 1974, 'Materialism Without Reductionism: Non-Humean Causation and the Evidence for Physicalism', mimeographed draft, 140 pp.
Bronowski, Jakob: 1970, 'New Concepts in the Evolution of Complexity: Stratified Stability and Unbounded Plans', *Synthese* **21**, pp. 228–246.
Campbell, Donald T.: 1974a, 'Evolutionary Epistemology', in P. A. Schilpp, ed., *The Philosophy of Karl Popper*, v. 1, (LaSalle Illinois: Open Court), pp. 413–463.
Campbell, Donald T.: 1974b, "Downwards Causation' in Hierarchically Organized Biological Systems', in F. J. Ayala and T. Dobzhnazky, eds., *Studies in the Philosophy of Biology*, (University of California Press: Berkeley), pp. 179–186.
Carlson, Elof A.: 1966, *The Gene: A Critical History*, (Philadelphia: Saunders).
Causey, R. W.: 1972, 'Attribute-Identities in Micro-Reductions', *Journal of Philosophy* **69**, No. 14, (August 3), pp. 407–422.
Dinman, Bertram D.: 1972, "Non-Concept' of 'No-Threshold' Chemicals in the Environment', *Science* **175**, (February 4), pp. 495–497.
Elsasser, Walter M.: 1965, *Atom and Organism*, (Princeton: Princeton University Press).
Fodor, Jerry A.: 1974, 'Special Sciences (Or: The Disunity of Science as a Working Hypothesis)', *Synthese* **28**, pp. 97–115.
Glymour, Clark: 1975, 'Relevant Evidence', *Journal of Philosophy* **72**, (August 14), pp. 403–425.
Haldane, J. S.: 1914, *Mechanism, Life and Personality*, (New York: Dutton).
Hull, David L.: 1972, 'Reduction in Genetics Biology or Philosophy?', *Philosophy of Science* **39**, (December), pp. 491–499.
Hull, David L.: 1974, *Philosophy of Biological Science*, (Prentice-Hall: Englewood Cliffs).
Hull, David L.: 1976, 'Informal Aspects of Theory Reduction', this volume, p. 653.
Hutchinson, G. E.: 1964, 'The Influence of the Environment', *Proceedings of the National Academy of Sciences*, v. 51, pp. 930–934.

Kauffman, Stuart A.: 1972, 'Articulation of Parts Explanation in Biology and the Rational Search for Them', in *PSA-1970*, R. C. Buck and R. S. Cohen, eds., *Boston Studies in the Philosophy of Science*, v. 8, pp. 257–272.
Kim, Jaegwon: 1964, 'Inference, Explanation and Prediction', *Journal of Philosophy* 61, No. 12, (July 11), pp. 360–368.
Kim, Jaegwon: 1966, 'On the Psycho-Physical Identity Thesis', *American Philosophical Quarterly* 3, pp. 227–235.
Lewontin, Richard C.: 1974, *The Genetic Basis of Evolutionary Change*, (New York: Columbia University Press).
Maull, 1974; see Roth, 1974.
Minsky, Marvin, and Papert, Seymour: 1969, *Perceptrons: A Study in Computational Geometry*, (Cambridge: M.I.T. University Press).
Muller, Herman J.: 1920, 'Are the Factors of Heredity Arranged in a Line?', *American Naturalist* 54, (March-April), pp. 97–121.
Nagel, Ernest: 1961, *The Structure of Science*, (New York: Harcourt).
Nickles, Thomas: 1973, 'Two Concepts of Inter-theoretic Reduction', *Journal of Philosophy* 70, No. 7, (April 12), pp. 181–201.
Nickles, Thomas: 1976, 'Theory Generalization, Problem Reduction, and the Unity of Science', this volume, p. 33.
Prandtl, Ludwig, and Tietjens, O. G.: 1957, *Fundamentals of Aero- and Hydro-mechanics*, (New York: Dover) (reprint of original volume published in 1934 by McGraw-Hill).
Roth, Nancy Maull: 1974, 'Progress in Modern Biology: An Alternative to Reduction', Ph.D. dissertation, Committee on Conceptual Foundations of Science, University of Chicago.
Ruse, Michael: 1971, 'Reduction, Replacement, and Molecular Biology', *Dialectica* 25, pp. 39–72.
Ruse, Michael: 1973, *The Philosophy of Biology*, (London: Hutchinson University Library).
Ruse, Michael: 1976, 'Reduction in Genetics', this volume, p. 633.
Salmon, Wesley C.: 1971, *Statistical Explanation and Statistical Relevance*, (Pittsburgh: University of Pittsburgh Press).
Schaffner, Kenneth F.: 1967, 'Approaches to Reduction', *Philosophy of Science* 34, (June), pp. 137–147.
Schaffner, K. F.: 1969, 'The Watson-Crick Model and Reductionism', *British Journal for the Philosophy of Science* 20, pp. 325–348.
Schaffner, K. F.: 1974a, 'Logic of Discovery and Justification in Regulatory Genetics', *Studies in History and Philosophy of Science* 4, No. 4, pp. 349–385.
Schaffner, K. F.: 1974b, 'The Peripherality of Reductionism in the Development of Molecular Biology', *Journal of the History of Biology* 7, No. 1 (Spring), pp. 111–139.
Schaffner, K. F.: 1976, 'Reductionism in Biology: Prospects and Problems', this volume, p. 613.
Shimony, Abner: 1971, 'Perception from an Evolutionary Point of View', *Journal of Philosophy* 68, No. 19, (October 7), pp. 571–583.
Simon, Herbert A.: 1969, *The Sciences of the Artificial*, (Cambridge: M.I.T. University Press).
Sklar, Lawrence: 1967, 'Types of Inter-Theoretic Reduction', *British Journal for the Philosophy of Science* 18, No. 2, (August), pp. 106–124.
Sklar Lawrence: 1973, 'Statistical Explanation and Ergodic Theory', *Philosophy of Science* 40, (June), pp. 194–212.
Thompson, D'Arcy W.: 1961, *On Growth and Form*, abridged edition, edited with commentary by J. T. Bonner, (London: Cambridge University Press).

Waismann, Friedrich: 1951, 'Verifiability', in A. G. N. Flew, ed., *Logic and Language*, (first series), (London: Blackwell), pp. 117–144.
Waismann, F.: 1953, 'Language Strata', in A. G. N. Flew, ed., *Logic and Language*, (second series), (London: Blackwell), pp. 11–31.
Whitehouse, H. L. K.: 1973, *Towards an Understanding of the Mechanisms of Heredity*, third revised edition, (New York: St. Martin's Press).
Wimsatt, William C.: 1974, 'Complexity and Organization', in K. F. Schaffner and R. S. Cohen, eds., *PSA-1972, Boston Studies in the Philosophy of Science*, v. 20, (Dordrecht: Reidel), pp. 67–86.
Wimsatt, W. C.: 1975, 'Reductionism, Levels of Organization, and the Mind-Body Problem', in *Conseiousness and the Brain*, edited by G. G. Globus, G. Maxwell, and I. Savodnik, (New York: Plenum, 1976), pp. 205–267.
Wimsatt, W. C.: 1976, 'Correspondence versus Identity and the Problem of Spatiality in the Localization of the Genome and Determining the Configuration of the Mental Realm', invited address, Section VIII (Foundations of Biology), *5th International Congress on Logic, Methodology and Philosophy of Science*, London, Ontario, August 31, 1975. To be published in the proceedings, edited by Jaako Hintikka, by D. Reidel (Dordrecht).

CONTRIBUTED PAPERS

SESSION V

NANCY DELANEY CARTWRIGHT

HOW DO WE APPLY SCIENCE?

Until recently, philosophers have either denied that there is such a thing as philosophy of technology or have held it in contempt. A philosophy of technology would be as absurd as a philosophy of sport, they scoffed. Nowadays philosophers pay more attention to this crucial area. We recognize that technology raises economic, moral, and social problems that demand philosophical attention. But we still tend to overlook that applied science has conceptual problems – problems for the philosopher of science – independent of the moral, social, and aesthetic problems of technology. It is the need for philosophical study of these conceptual problems that I want to stress.

To make clear that these problems are scientific problems separate from the familiar problems of decision theory, costbenefit analysis, and social responsibility, I shall choose two examples at a fairly high level in the application of physics – one quantum, one classical. Both use well developed theories to generate low-level phenomenological laws. In both cases the laws describe concrete irreversible processes like the flow of electricity, the diffusion of a component in a mixture, or the transfer of heat.

When in traditional philosophy of science we have studied the relation between theories and low-level laws our attention has been directed not at the generation of new phenomenological laws from theories, but rather at the testing of theories by their ability to reproduce known phenomenological laws. It is commonly assumed that once we have found an adequate model for the second activity we can apply it, *mutatis mutandis*, to the first. My two examples will show that this assumption is mistaken.

Some of the reasons for this are almost trivial. In testing we look for new consequences of a theory on those occasions when we can bring about the boundary conditions needed for the use of a covering law. In applying the theory, however, the boundary conditions are set, and we must discover what the theory predicts under those conditions. The task has seldom been performed already and in general it is not possible to

R. S. Cohen et al. (eds.), PSA 1974, 713–719. All Rights Reserved.
Copyright © 1976 by D. Reidel Publishing Company, Dordrecht-Holland.

determine directly from the theories bearing on a domain what follows from a given, natural, set of boundary conditions.

This difference between testing and application matters because most of our laws are *ceteris paribus* laws. They describe processes under the fiction that the only causally relevant phenomena are those in the antecedent of the law. This is fine for testing because we can often fabricate situations in which this fiction is almost correct. In dynamics we guard against 'external' forces such as electromagnetism, sometimes literally, by erecting shields. Other forces like those of gravity can be acknowledged in the experimental design to minimize their effect. In application, however, we cannot usually concern ourselves with the tiny closed systems of which our theories speak. We want to work in the heterogenous domains of different theories and we lack a practicable hypertheory that tells us the upshot of interacting elements, elements that are treated only within disparate and isolated theories. The classical study of irreversible processes provides an instructive example.

Flow processes like diffusion, heat transfer, or electric current ought to be studied by the transport equations of statistical mechanics. Usually, however, the model for the distribution functions and the details of the transport equations are too complex: the method is unusable. A colleague of mine in engineering estimates that 90% of all engineering systems cannot be treated by the currently available methods of statistical mechanics. "We analyze them by whatever means seem appropriate for the problem at hand," he adds. (Stephen J. Kline, Stanford University, who studies methods of approximation in Kline, 1965).

In practice engineers frequently handle irreversible processes with old fashioned phenomenological laws describing the flow (or flux) of the quantity under study. Most of these laws have been known for quite a long time. For example there is Fick's law, dating from 1855, which relates the diffusion velocity of a component in a mixture to the gradient of its density ($J_m = - D\, \partial c/\partial x$). Equally simple laws describe other processes: Fourier's law for heat flow, Newton's law for sheering force (momentum flux) and Ohm's law for electric current. Each of these is a linear differential equation in t (e.g. the J_m in Fick's law cited above is dm/dt), giving the time rate of change of the desired quantity (in the case of Fick's law, the mass). Hence a solution at one time completely determines the quantity at any other time. Given that the quantity can be controlled at some point in a process, these equations should be perfect for determining the future evolution of the process. They are not.

The trouble is that each equation is what I have called a *ceteris paribus* law. It describes the flux only so long as just one kind of cause is operating. More realistic examples set different forces at play simultaneously. In a mixture of liquids, for example, if both the temperatures and the concentrations are non-uniform, there may be a flow of liquid due not only to the concentration gradients but also to the temperature gradients. This is called the Soret effect.

The situation is this. For the several fluxes J we have laws of the form,

$$J_m = f_1(\alpha_m)$$
$$\vdots$$
$$J_q = f_n(\alpha_q).$$

Each of these is appropriate only when its α is the only relevant variable. For cross-effects we require laws of the form,

$$J_m = g_1(\alpha_m, \ldots, \alpha_q)$$
$$\vdots$$
$$J_q = g_n(\alpha_m, \ldots, \alpha_q).$$

We have such laws only in a few special cases. There is no uniform procedure for obtaining them from the laws that govern the individual variables in isolation. Yet the laws and theories for the separate phenomena are certainly relevant. The cross-effect laws are not generated by naive induction nor by explicit high level theorizing. Instead, they are obtained by adapting existing laws. We can illustrate with the Soret effect. In this case we assume simple linear additivity of effects, and obtain a new law by adding a "thermal diffusion factor" into Fick's law.

Are there any principles to be followed in modifying to allow for cross-effects? There is only one systematic account of cross-effect modification for flow processes. It originated with Onsager in 1931, but was not developed until the 1950's. Onsager theory defines force-flux pairs, and prescribes a method for writing cross-effect equations involving different forces. As C. A. Truesdell describes it, "Onsagerism claims to unify and correlate much existing knowledge of irreversible processes" (Truesdell, 1969, p. 140). Unfortunately, it does not succeed. Truesdell continues:

As far as concerns heat conduction, viscosity, and diffusion... this is not so. Not only does Onsagerism not apply to any of these phenomena without a Procrustean force-fit, but even in the generous interpretation of its sectaries it does not yield as much reduction for the theory of viscosity as was known a century earlier and follows from fundamental principles....

Truesdell claims that the principles used in Onsager theory are vacuous. The principles must sometimes be applied in one way, sometimes in another, in an *ad hoc* fashion demanded by each new situation. The prescription for constructing laws, for example, depends on the proper choice of conjugate flux-force pairs. Onsager theory offers a general principle for making this choice but if the principle were followed literally, we would not make the proper choice in even the most simple situations. In practice on any given occasion the choice is left to the physicist's imagination.

It seems that after its first glimmer of generality the Onsager approach turns out to be a collection of *ad hoc* techniques. But we should not in philosophical purity turn our backs at the word "*ad hoc*." Instead, we should try to discover standards for what constitutes a good correction or modification of an existing law. Ultimately or course any preferred law must be tested. But the problem of applied science is not the testing of laws (for which we have ample methodology). Rather the problem of applied science is to articulate and evaluate techniques for modifying and applying laws. To insist that such standards are merely rules of thumb that cannot be described, or, if described, cannot be evaluated or justified or improved, is to abdicate a central task for the philosophy of science. Since applied science *is* successfully conducted at this level, and yet fits no current methodological model, we must suppose not that the applied science is irrational but that our models of rationality are inadequate.

Broadly, our current models are either inductive or deductive. A rule of application is justified if it leads to formulas that work, or if its approximate success can be deduced from a general theory. The first method does not distinguish among untried rules; and even when evidence of success is available, the second method is preferred. It explains why the rules work and what their limitations will be. In our example, however, the deductive method is unavailable. The implications of the general theory are too complicated. Admittedly, it is possible that any cross-effect laws which are justified are *in principle* deducible within an elaborated version of statistical mechanics. Reasons now given for a particular modification cannot refer to this theory. Nevertheless if the expanded theory were at hand, the success of the reasoning would be understood, or else the reasons would be seen as the lucky guesses they are.

This viewpoint seems to me ingenuous, wishfully putting off the hard work that needs to be done. A second example illustrates why. The example is a quantum mechanical one, concerning the use of joint distribution functions for non-commuting observables. The functions are used to calculate quantum effects in transport processes. Here the implications of the general theory are clear: the theory explicitly rejects any such functions. Yet calculations using joint distributions are reasonably successful. Justification for their use is not merely inductive, however. Only certain functions are used in these calculations, and it is possible to give reasons for choosing some, rather than others, among equally prohibited functions. We can give good reasons for the choice; but we cannot explain *why* the reasons are good. The reasoning does not fit into any of our current models.

Let us look at the example. Quantum theory predicts the distribution of momentum values p for an ensemble of identically prepared systems. It also predicts the distribution of position values q in the same ensemble. But it does not predict a joint distribution for position and momentum, and it is possible to prove that no such distribution exists. Yet so called "joint distributions" of position and momentum are regularly used in both quantum optics and quantum thermodynamics. They are used as methods of approximation in the all too common situation in which the Schroedinger equation is impossible to solve. Although incompatible with formal quantum mechanics, the joint distributions are used to obtain quantum corrections to phenomenological laws.

The study of quantum effects in the thermodynamics of irreversible processes provides illustrations. Joint distributions have been used in both of the two systematic approaches to irreversible processes that we have discussed. (1) Joint distributions have been used to develop quantum corrections to classical transport equations, and (2) for a quantum version of Onsager theory. (For a more detailed discussion of these cases, see Cartwright, 1974.) Two facts argue in favor of using the functions. (1) Some of the calculations made via joint distributions have since been carried out using methods of pure quantum theory and the results have agreed closely. (2) Non classical predictions made using the joint distributions are generally in good agreement with experiment.

When these functions are used, usually some minimal requirements are stated for any function which is to serve as a joint distribution. Indefinitely many functions satisfy these explicit requirements, but only a handful are used. A starting point, then, is to ask, "Are there any further constraints on these functions which are not made explicit?"

It appears there are. For one, in both quantum optics and quantum thermodynamics, all the commonly used functions are ones which yield the same covariance in position and momentum. It is possible to characterize the family of distributions which give the designated covariance. We can then add to the explicit requirements another, adopted in practice, that any admissable function must be a member of this family. This is a good example of the kind of rule of thumb, used but not stated, that ought to be made explicit.

This is only part of the job, however. The question of justification remains. We have a (partial) characterization of functions used. Is there any reason in future applications to continue to use distributions that satisfy this characterization? I think so. Many of the cases in which joint distributions figure essentially involve the development of a quantum Liouville equation. Classically, the Liouville equation describes situations in which there are fixed correlations between position and momentum. Taking covariance as a direct measure of correlation, if we restrict the joint distribution functions to ones in our covariance-constant family, this classically important feature of Liouville equations will carry over into quantum mechanics.

What kind of justification is this? What relevance is the classical analogy for making a choice *within* quantum theory of a function *prohibited by* quantum theory? I do not know. It seems that it is not an accident that out of the infinitely many possible pseudo joint distributions, only a handful have been used. Similarly, going back to our classical example, it is not an accident that out of the indefinitely many ways of modifying the laws of irreversible processes only a few are deemed to be obvious candidates. We, as philosophers of science, should be explaining why it is not an accident. Out of the disorganized mass of experimental practice in testing hypotheses, philosophers of science have evolved an elaborate normative code – or rather, several competing opinions about what the norms are. Applied science deserves similar scrutiny.

Stanford University

BIBLIOGRAPHY

Cartwright, N.: 1974, 'Correlations without Joint Distributions in Quantum Mechanics', *Foundations of Physics* **4**, 127–135.
Kline, S. J.: 1965, *Similitude and Approximation Theory*, McGraw-Hill, New York.
Truesdell, C. A.: 1969, *Rational Thermodynamics*, McGraw-Hill, New York.

WILLIAM DEMOPOULOS

WHAT IS THE LOGICAL INTERPRETATION OF QUANTUM MECHANICS?

Let me begin by briefly explaining the concept of 'interpretation' relevant to this discussion.

Certain physical theories postulate abstract structural constraints which events are held to satisfy. Such theories are termed 'principle theories'. Interpretations of principle theories aim to explain their relation to the theories they replace. Interpretations are therefore concerned with the nature of the transitions between theories.

Theories of space-time structure provide the most accessible illustration of principle theories. For example, Newtonian mechanics in the absence of gravitation represents the 4-dimensional geometry of space-time by the inhomogeneous Galilean group, which acts transitively in the class of free motions, i.e. the inhomogeneous Galilean group is the symmetry group of the free motions: it is a subgroup of the symmetry group of every mechanical system, and the largest such subgroup. Einstein's special principle of relativity is the hypothesis that the symmetry group of the free motions is the Poincaré group. The transition from the Galilean group to the Poincaré group is associated with a corresponding modification in space-time structure. The absolute time and Euclidean metric of Newtonian mechanics are dropped altogether, and the metrical relations of space-time are determined by the Minkowski tensor.

The special theory of relativity represents the transition from Newtonian mechanics to Maxwell's electrodynamics as involving a modification of the structure of space-time. In this sense, the spectral theory may be regarded as an *interpretation* of classical electrodynamics.

I turn now to the quantum theory. To motivate the later discussion I want first to consider why the transition to an essentially statistical theory is problematic.

Let us suppose that a certain class K of physical systems may be known with complete precision. For any system of this class it is possible to completely specify its type and its state. Both pieces of information – the possible types of system, and their possible states – are theory relative.

What is assumed is that for any S in K there is no extra theoretical limit on the amount of information obtainable concerning S. I shall say that a statistical theory is fundamental if it is based on a maximal amount of information concerning the systems of K. That is to say, for a fundamental theory, the degree of imprecision of our knowledge may be ignored, since the theory is supposed to hold even when this is made arbitrarily small. By contrast, a statistical theory which is not fundamental is explicitly designed to take account of the case where, for whatever reasons, a maximal amount of information is not available.

For example, in classical statistical mechanics, the theoretically important states are characterized by some positive dispersion. In this case the dispersion is easily explained in terms of the incompleteness of our knowledge of the exact phase point of the system. The fundamental theory is given by classical mechanics which represents the time evolution of the phase point of the system, and the possible phase points are in one-to-one correspondence with dispersion-free states; these are the pure statistical states of classical mechanics.

The pure statistical states of the quantum theory are not dispersion-free. In this sense, the theory is *significantly* statistical. This suggests the question, Under what conditions is a significantly statistical theory correctly regarded as fundamental?

In the case of atomic systems, the response favored by many physicists consists in denying that any theory can be fundamental in the sense just outlined. (Cf. e.g. Pauli's letters 115 and 116 to Born, as well as the subsequent commentary by Born.) Knowledge of the systems dealt with by the quantum theory is essentially incomplete in the sense that any predictively adequate theory must accept the existence of a significant restriction on what can be known concerning this class of systems.

Beginning with Heisenberg's γ-ray microscope thought experiment, there is a long series of quasi-physical arguments aimed at making this view plausible. All of these arguments appeal to the operational incompatibility of direct measurements of certain pairs of physical magnitudes. This is quite irrelevant as Einstein showed. His argument may be briefly reconstructed as follows.

Two systems S and S' are *coupled* if there exist magnitudes A and A' of S and S' (respectively) such that the probability that $A = \lambda$ is 1(0) if and only if the probability that $A' = \lambda'$ is 0(1). It is a theorem that there exist interactions which lead to a coupling in this sense. A classic example is given by a pair of spin-½ particles in the singlet spin state. In any theory admitting the existence of coupled systems, it is unnecessary to interact directly with S, say, in order to determine the value of the magnitude A; it suffices to measure the magnitude A' of S'. Since the systems are spacially separated, this cannot possibly affect S.

The fact that quantum mechanics admits the existence of coupled systems means that the theory does not support the usual (operationist) interpretation of the statistical character of the theory. The idea that our knowledge is essentially incomplete assumes that a direct measurement of all magnitudes is not possible. This of course may well be true. The difficulty is that direct measurements are not necessary for determining the values of the A_i; moreover, this fact is a consequence of the quantum theory. Einstein made a definitive contribution to this phase of the problem by showing that the rejection of verificationism removes any methodological objection to fundamental theories of atomic systems.

For the logical interpretation, a statistical theory is fundamental only if it is complete; moreover, the quantum theory is complete. This removes the only serious obstacle to regarding the theory as fundamental. The analysis therefore depends on the concept of completeness. This is explicated within the framework of the following analysis of statistical theories. We restrict our attention to the statistical theory of a fixed system S. Such a theory consists of an algebra **M** of physical magnitudes (it is assumed that **M** is at least a partial algebra) together with a set **S** of statistical states. Elements $\psi \in \mathbf{S}$ assign probabilities to ranges of values of magnitudes in M: for each $A \in M$, and $\psi \in \mathbf{S}$, $P_{A,\psi}(U)$ denotes the probability that the value of the magnitude A lies in the (measurable) subset U of Real numbers. $(P_{A,\psi}: \mathbf{F}(R) \to [0, 1]$ is the distribution function of the magnitude A determined by ψ.)

The two algebraic structures of relevance to this discussion are: **M** is the algebra R^{Ω} of real valued functions on a classical phase space. **M** is the partial algebra $\mathbf{N}(H)$ of self-adjoint operators on a separable Hilbert space. These correspond, respectively, to classical and quantum mechanics.

The subalgebra of idempotent elements of **M** form a partial Boolean algebra **L**. In the case of R^{Ω} this is the subalgebra B of characteristic functions on Ω. This is isomorphic to the field $\mathbf{F}(\Omega)$ of measurable subsets of Ω. For $\mathbf{N}(H)$, this is the partial Boolean algebra N of projection operators. N is isomorphic to the partial Boolean algebra $B(H)$ of closed linear subspaces of H. A partial Boolean algebra $\mathbf{L}$ may be pictured as a collection of Boolean algebras such that the intersection of two algebras of the collection is well-defined and is a Boolean sub-algebra of $\mathbf{L}$. Moreover if every pair of any n elements of $\mathbf{L}$ belong to a common Boolean algebra, there is a Boolean algebra (in the collection) containing all of them.

The notion of a partial Boolean algebra is a generalization of the concept of a Boolean algebra since not every pair of elements of $\mathbf{L}$ may belong to a common Boolean subalgebra. Elements which belong to a common subalgebra are said to be compatible; the Boolean operations are restricted to compatible elements. For example, let $\cup$, $\cap$, $'$ denote the span, intersection, and orthocomplement of the subspaces in H. Then for a, b, in H, a is compatible with b ($a \leftrightarrow b$) if there exist mutually orthogonal subspaces a_1, b_1, c such that $a = a_1 \cup c$ and $b = b_1 \cup c$. The operations $\cup$, $\cap$ are restricted to $\leftrightarrow$. H is the unit, and $\{0\}$ is the 0 of the partial Boolean algebra $B(H)$.

Physical properties are introduced in terms of the magnitudes A in $\mathbf{M}$ as follows. Take a real number λ in the range of A. Then $A = \lambda$ (this is read 'the value of A is λ') is a property of S. More generally, given a subset U of R and magnitude A, $A \in U$ – the value of A lies in U – represents a property of the system S.

Von Neumann (Ch. III, Sect. 5) observed that every property of S is represented by an idempotent magnitude in $\mathbf{L}$. This is simply seen in the case of classical mechanics. Let Ω be a subset of n-dimensional Euclidean space. An elementary event in the history of S is represented by a point in Ω. As is well-known, an event ω is associated with a pure statistical state of classical mechanics: the 2-valued measure on $\mathbf{F}(\Omega)$ determined by ω. Now let f_A be a real valued function in R^Ω representing the magnitude A. A property $A \in U$ holds for S if and only if S is in a state ω such that $f_A(\omega) \in U$. Let Γ denote the subset $f_A^{-1}(U)$ of Ω. It is clear that S has the property $A \in U$ if and only if the state ω of S lies in Γ. The property $A \in U$ is said to be associated with Γ. In general there are many properties $A_i \in U_j$ such that $f_A^{-1}(U_j) = \Gamma$ for some $U_j \in R$. Now let a be the characteristic function of Γ, and let P be a property associated with Γ. By the correspondence between properties of S and subsets of Ω, it follows that every property P is represented by the characteristic function a, in the sense that P holds if and only if S is in a state ω such that $a(\omega) = 1$. Since a is two-valued this is equivalent to $a(\omega) \neq 0$.

The situation in quantum mechanics is exactly analogous. Elementary events, represented by rays **K** in H, are associated with pure statistical states. In quantum mechanics statistical states are given by measures on the closed linear subspaces of H. The pure state associated with **K** is determined by taking the square of the norm of the projection of a unit vector lying in **K** onto each subspace of H. Since there is a one-to-one correspondence which associates each projection operator with the subspace which is its range, this determines a probability measure on N. Recall atoms in B are characteristic functions of singleton subsets $\{\omega\}$ of $\mathbf{F}(\Omega)$. N is also atomic. An atom in N is a projection operator onto a one-dimensional subspace of H. Thus in each theory, there is a one-to-one correspondence between elementary events and atoms in **L**.

To summarize, every magnitude may be replaced by a set of properties, and every property corresponds to a two-valued quantity, i.e. to an idempotent magnitude. (For this reason it suffices to consider the algebra **L** of idempotent magnitudes.) The correspondence is not one-to-one since very many properties are associated with the same subset of Ω (or subspace of H), and therefore, represented by the same idempotent magnitude. There is a one-one correspondence between idempotent magnitudes and equivalence classes of properties represented by the same idempotent in **L**. An idempotent may therefore be thought of as the equivalence class of properties it represents.

Within this framework the mathematical investigations of Gleason and Kochen and Specker suggest a general concept of completeness applicable to statistical theories. The explication depends on the notion of a proper extension of a statistical theory. Extensions are defined relative to the category $\mathscr{C}$ of algebraic structures representing the idempotent magnitudes of the theory and the set S of statistical states. Take **L**, **L'** in $\mathscr{C}$. Suppose there is an imbedding ϕ carrying **L** into **L'**. $\psi' \in S'$ is an extension of $\psi \in S$ if $\psi = \psi' \circ \phi$. (**L'**, S') is an extension of (L, S) if every $\psi \in S$ has an extension $\psi' \in S'$. The extension is proper if $\psi \neq \psi' \mid \phi[\mathbf{L}]$ for some $\psi' \in S'$. Complete statistical theories have no proper extensions.

Gleason's theorem excludes certain extensions of the quantum theory. For the logical structures $B(H)$ Gleason's theorem is equivalent to the general result that the quantum theory has no proper extensions in the category of partial Boolean algebras. A Boolean extension may be defined as one for which L' is a Boolean algebra. Theorems 0 and 1 of Kochen and Specker imply that the quantum theory has no Boolean extensions. This is a similar (but weaker) result concerning a sub-category of the category of partial Boolean algebras. The completeness of classical mechanics, on this analysis, follows from the fact that $F(\Omega)$ is perfect and reduced so that no new measures result from extensions in the category of Boolean algebras.

Hence a consequence of this analysis is that classical mechanics and quantum mechanics are complete in exactly the same sense. In neither theory do there exist proper extensions in the category of algebraic structures of their idempotent magnitudes. As principle theories classical mechanics, on this analysis, follows from the fact that $bf(\Omega)$ is perfect and the possible events open to a physical system, i.e. they determine different *possibility structures* of events, and each theory is complete relative to the category of algebraic structures defined.

Quantum mechanics is indeterministic in the sense that the pure statistical states of the theory are not degenerate measures concentrated on 0 and 1 as in classical mechanics; hence the maximal amount of information concerning a physical system is significantly probabilistic. This arises from the fact that certain properties are stochastically independent, as are the idempotents which represent them. More exactly, the properties are *strongly independent* in the sense that they are stochastically related to other properties in a way which excludes their being stochastically related to each other. Thus the significance of incompatibility is that it leads to indeterminism.

Indeterminism in this sense must not be confused with the very different concept of indeterminateness. The theory is indeterministic or significantly statistical in the sense that the pure states take values in the open interval $(0, 1)$. The thesis that the theory is indeterminate holds that there are properties P such that P neither holds nor fails to hold of S. This is not implied by indeterminism nor is it in any way required by the view that the theory provides a maximal amount of information concerning the system. Rather, indeterminateness is suggested by the idea that incompatible propositions (idempotents) are somehow inconsistent, i.e., basically the operationist interpretation of the statistical character of the theory. But just the opposite is the case: since a pair of propositions are incompatible, they *cannot* be inconsistent.

Classically – i.e. when **L** is a Boolean algebra – a proposition (idempotent) a in **L** may be replaced by the sequence of its truth values under all possible realizations (homomorphisms onto the two element Boolean algebra, Z_2). This is a consequence of the semi-simplicity of Boolean algebras i.e. the fact that every Boolean algebra is imbeddable into a direct union of Z_2. Theorem 1 of Kochen and Specker establishes that a finite partial Boolean subalgebra of $B(H^3)$ has no homomorphisms onto Z_2. Therefore the possibility of such a Boolean representation of the truth of the propositions of N is in general excluded, since this requires the existence of a homomorphism onto Z_2. But this in no way implies that a proposition in N is sometimes neither true nor false.

To conclude this part of the discussion: There are two different accounts of indeterminism which are historically important. The first, which apparently goes back to Aristotle, rejects bivalence: A theory is indeterministic if it assumes that there are propositions whose truth value is indeterminate. The second represented by the quantum theory, retains bivalence while rejecting semi-simplicity. An indeterministic theory is then characterized by the absence of two-valued homomorphisms, and therefore, of two-valued measures. The coherence of indeterminateness seems to rest on the Aristotelian metaphysic of act and potency. But nothing of this sort is required by the indeterminism of quantum mechanics. This form of indeterminism implies that there is no Boolean representation of the properties obtaining at a given time; but for any property P, it is completely determinate whether or not P holds.

BIBLIOGRAPHY

Birkhoff, G. and Von Neumann, J.: 'The Logic of Quantum Mechanics', *Annals of Mathematics* **37** (1936), 823–843.

Born, M.: *The Born-Einstein Letters: The Correspondence Between Albert Einstein and Max and Hedwig Born: 1916–1955*, Walker and Company, 1971.

Bub, J.: *The Interpretation of Quantum Mechanics*, Reidel, 1974.

Demopoulos, W.: 'Contributions to the Interpretation of Quantum Mechanics' in C. Hooker (ed.), *The Logico-Algebraic Approach to Quantum Mechanics*, Reidel, 1975.

Einstein, A.: 'Quantum Mechanics and Reality' (1948), in *The Born-Einstein Letters: The Correspondence Between Albert Einstein and Max and Hedwig Born: 1916–1955*, Walker and Company, (1971) pp. 168–173.

Finkelstein, D.: 'Logic of Quantum Physics', *Transactions of the New York Academy of Science* **25** (1963), 621–635.
Gleason, A.: 'Measures on the Closed Subspaces of Hilbert Space', *Journal of Mathematics and Mechanics* **6** (1957), 885–893.
Kochen, S. and Specker, E. P.: 'The Problem of Hidden Variables in Quantum Mechanics', *Journal of Mathematics and Mechanics* **17** (1967), 59–87.
von Neumann, J.: *Mathematical Foundations of Quantum Mechanics*, Princeton University Press, 1955.
Putnam, H.: 'Is Logic Empirical?', *Boston Studies in the Philosophy of Science*, Vol. V (ed. by R. S. Cohen and M. W. Wartofsky), Reidel, 1969.

INDEX OF NAMES

In the following index italic numbers denote a reference in the Notes, and bold numbers the first page of a contribution.

Achinstein, P. *265*, *288*, 289, *299*
Ackermann, R. 359
Ackoff, R. L. 547
Acland, H. 188
Adams, H. *118*, 122
Addis, L. **361**, **369**, *376*
Agassi, J. *71*, 477, **449**
Alexander, T. 179
Anastasi, A. *140*, *141*, 141
Anderson, A. R. xi-xii
Anderson, R. M. Jr. **549**, 560
Aquina, T. *100*
Arp, H. 320
Arrow, K. J. 447
Ayer, A. J. *390*

Bache, R. M. 179
Bahcall, J. 320
Bane, M. J. 188
Barnes, S. B. 547
Bass, R. W. *476*, 476
Bateson *630*
Belnap, N. D. Jr. xi
Benjamin, A. C. 497
Bereiter, C. *179*, 179
Bergh, G. van den *476*, 476
Berlin, B. M. *299*
Berofsky, B. *358*, 359
Berscheid, E. 179
Berzelius, J. *119*
Binet *178*
Biakhoff, G. 513, 727
Blake, R. M. 102
Block, N. J. *13*, 13, **127**, *140*, *141*, 141, *178*, 179, *320*, 320
Bloom, B. 188
Bohr, N. *72*, *73*, 74

Boltzmann 72
Born, M. 727
Bowles, S. 179
Bowman, P. A. **423**, *432*, *433*, 433, 434
Boyd, R. *140*, *705*, 708
Braithwaite, R. B. *390*
Brand, M. *359*, 360
Bridgman, P. W. *118*, *119*, *120*, *121*, 122, *433*, 433, 497
Britton, D. 486
Brock, W. H. *119*, 122
Brodbeck, M. *299*
Bronfenbrenner, U. 179
Bronowski, J. *706*, 708
Brown, W. C. *178*, 179
Bruner, J. S. *299*
Bryan, M. M. 188
Bub, J. *420*, 421, 727
Buchdahl, G. *119*, *120*, 122
Bunge, M. 13, 421
Burgers, J. M. 74
Burns, G. W. 650
Burt *178*
Burtt, E. A. *497*, 497

Cahn, S. M. 421
Campbell, D. T. *213*, 708
Campbell, N. R. *233*, 234, 498
Carlson, E. A. *630*, 631, *707*, 708
Carnap, R. *233*, 234, *265*, *266*, *376*, *390*, *560*
Cartwright, N. D. *713*, 719
Carugo, A. *100*, *101*
Cattell, J. M. 179
Causey, R. L. **3**, 13, *629*, 631, *670*, 708
Churchman, C. W. 547
Clausius *73*
Cohen, D. 188

INDEX OF NAMES

Cohen, I. B. 103
Cohen, R. S. *525*
Coleman, J. S. 179
Cornman *288, 289*
Cournot, A. *121*, 123
Crick, F. H. C. *630*, 631
Crombie, A. C. *100, 101,* 102
Cronbach, L. J. *140, 141,* 141, 188
Cronin, J. *178*, 179

Dalton *121*
Dampier-Whetham, W. C. D. 498
Daniels, N. *140,* 141, **143,** *177, 178,* 179
Davidson, D. 360, *391*
Davis, A. 188
Debye, P. *73*, 74
Demopoulos, W. **721,** 727
Derden, J. K. *266*
Descartes *465*
Dinman, B. D. *707*, 708
Divers, E. *119*
Dobzhansky, T. *178,* 179, 650
Dolby, R. G. 547
Drake, S. 102, 103, *119*, 123
Dray, W. *368*
Dretske, F. 289
Drobisch *121*
Ducasse, C. J. 102
Duhem, P. 103
Dworkin, G. *140*, 141, *178,* 179, *320,* 320

Earman, J. 434
Easton, L. *119, 120,* 123
Eccles, J. C. *560,* 560
Edson, L. 179
Eells, K. 188
Ehrenfest, P. *72, 73, 74,* 74
Eiduson, B. T. 547
Einstein, A. *71, 72, 74, 320,* 727
Ellis, B. D. *432, 433,* 434
Elsasser, W. M. *705,* 708
Emery, F. 547
Evans, H. M. 498

Farber, M. 498
Favaro, A. 103
Feigl, H. *560,* 560, *629,* 631
Feller, W. *419, 420,* 421
Fetzer, J. H. **377,** *390, 391*

Feyerabend, P. K. XII, *288,* 289, *311, 312,* 547, *629,* 631
Field, H. 32, *140,* 141
Field, G. 320
Findley, W. G. 188
Finkelstein, D. 728
Finnochiaro, M. A. 103
Fitzgerald, P. **235**
Fleming, D. 670
Fodor, J. A. 13, *140, 141,* 141, 708
Fraassen, B. C. van **215,** *233,* 234, 421
Fredette, R. 103
Friedlander, E. 498
Friedlander, M. W. **477,** 486
Friedman, M. *319,* 320, 434

Gaines, A. M. *299*
Galileo, G. **79,** *100, 101, 102*
Garber, H. 179
Gardner, M. *299*
Gendron, B. 13
Gibbons, M. 447
Gibbs, J. 31, 32
Giere, *600*
Gintis, H. 179, 188
Glansdorff, P. 32
Glass, B. *669*
Gleason, A. M. 513, 728
Glymour, C. 708
Gold, T. 486
Goldman, A. *358,* 360
Goodman, N. *376, 390, 391,* 421
Grad, H. 31, 32
Grünbaum *390, 432, 433,* 434
Guilford, J. *140,* 141

Haldane, J. B. S. 498, *705,* 708
Hanson, N. R. *311,* 547
Harré, R. 103
Harris, E. *601,*
Havighurst, R. 188
Hawkins, D. *420, 421,* 421
Hawkins, G. S. *476,* 476
Haynes, F. *311, 312*
Heber, R. *179,* 179
Hegel *120*
Heilbron, J. L. 74
Heisenberg, W. 421
Helmholtz, H. von 498

INDEX OF NAMES

Hempel, C. G. *233, 299, 368,* **369,** *375, 376, 390, 391*
Herrick, R. 188
Herrnstein, R. J. *177,* 180
Heyns, B. 188
Hilpinen, R. *359,* 360
Hirsch, J. *178,* 180
Hölling, J. 421
Hooker, C. A. *420, 421,* 421
Hoyt, H. P. 486
Hubel, D. H. *560,* 560
Hull, D. L. 13, *628, 629, 630, 631,* 631, *650,* 650, **653,** 670, *705, 706, 707,* 708
Hullet, J. *299*
Humphreys, W. C. 498
Hunt, I. E. 434
Hutchinson, G. E. *707,* 708

Jacob 631
Jacobs, G. H. 560
Jammer, M. *74,* 74
Jarvie, J. 447
Jauch, J. M. 513
Jaynes, E. 31, 32
Jeans, J. H. *72,* 74
Jeffrey, R. C. *376*
Jencks, C. 188
Jensen, A. R. *140, 141,* 141, *179,* 180, 188
Joffe, M. *72*
Jordan, P. 498

Kabrisky, M. *560,* 560
Kaila *376*
Kalke, W. 13
Kamber, F. 513
Kamin, L. *178,* 180
Kamlah, W. 513
Kampen, N. van 32
Kant *120*
Kardos, J. L. 486
Kass, T. *708*
Kauffman, S. A. *705,* 709
Kaye, K. **181,** 188
Keisler, C. A. 547
Kekulé *119*
Kemeny, J. G. 498
Kennard, E. H. 75
Kim, J. 360, *705,* 709
Kimball, M. E. *299*

King, M. D. 547
King, R. C. *630,* 631
Kirschenmann, P. **393,** 421
Klein, M. J. *72, 73,* 74, 75
Kleinpeter, H. *119, 120,* 123
Kline, S. J. 719
Knezo, G. J. *465*
Knight, D. M. *119,* 122, 123, 670
Kochen, S. 513, 728
Koertge, N. *102,* 103, *300*
Köhler, W. 560
Koyré, A. 103
Kramer, F. R. *629,* 631
Kreisler, M. 321
Kron, A. *359,* 360
Kuhn, T. S. 74, *213, 311, 312, 320,* 321, 547, *629,* 631
Kyburg, H. *390,* 670

Lakatos, I. XII–XIII, 547, *601, 669,* 670
Landé, A. 421
Landsberg, P. 32
Lauritsen, T. 75
Layzer, D. 180
Levenstein, P. 180
Lewis, D. *359,* 360, *610,*
Lewontin, R. C. *178,* 180, *650,* 650, 709
Locke, J. 650
Lockyer, J. N. *476,* 476
Lorentz, H. A. *433,* 434
Lorenz, K. 514
Lorenzen, P. 513, 514
Luzzatto, L. *650,* 650
Lyon, A. 360
Lyttleton, R. A. 486

McClelland, D. C. 547
Mach, E. *119, 121, 122,* 123
Mackey, G. W. 421
Mackie, J. *359,* 360
Maclachlan, J. 103
McMullin, E. **585**
McTighe, T. P. 103
Madden, E. H. 102
Mandelstam, S. 75
Marshall, W. H. 561
Martin, E. Jr. 289
Martin, J. 321
Martin, M. **293,** *299, 311, 319,* 321

Marstrand, P. K. 447
Mason, R. O. 547
Maslow, A. H. 498
Massey, G. *629*, 631
Maxwell, G. *140*, 141, 289, *560*, 560, **565**, *582*, *583*, 583, 584
Maxwell, J. C. *122*, 123
Medawar, P. *601*
Meehl, P. E. 560
Mellor, D. H. *391*, 421
Mendel *707*
Mercer, J. R. *178*, 180
Merton, R. K. 547
Meyerson, E. *119*, 123
Michelson, S. 188
Mills, D. R. *629*, 631
Minsky, M. *706*, 709
Mises, R. von *390*
Mitchell, H. K. *629*, *630*, *631*, 632
Mitroff, I. I. **529**, *547*, 547
Mittelstaedt, P. **501**, 514, *526*
Miyajima, M. 486
Monod *631*
Morgenbesser, S. 548, *629*
Morton, N. E. 180
Mountcastle, V. B. 560
Mulkay, M. 548
Muller, H. J. 498, *630*, 632, *708*, 709
Münster, A. 31
Musgrave, A. *601*

Nagel, E. *233*, 360, 421, *628*, *629*, 632, 650, 709
Naylor, R. H. 103
Nerlich, G. C. *390*
Neumann, J. von 421, 513, 727, 728
Newton, I. *100*, *121*, 123, *140*, *320*, 421
Nickles, T. **33**, 75, *630*, *631*, 632, *705*, *706*, 709
Niiniluoto, I. 360
Noll, W. *320*, 321
Nwachuku-Jarrett, E. S. 650

Oppenheim, P. 13
Oppolzer, T. R. von *476*, 476
Orgel, L. E. *629*, 632
Ornstein, M. 498
Ortony, A. *311*, *312*
Ostien, P. A. **271**

Ostwald *119*

Palter, R. **313**
Pap, A. *376*, *390*, *391*
Papert, S. *706*, 709
Passmore, J. *213*
Paterson, A. M. *476*, 476, **487**
Pavitt, K. 447
Pearson, K. *119*
Penrose, O. 32
Pepper, S. *560*, 560
Petrie, H. G. **301**
Piron, C. 513, 514
Planck, M. *72*, *73*, 75
Platt, J. R. *601*
Poggio, G. F. 560
Poincaré, H. 421, *433*, 434
Polanyi, M. *311*, *629*, 632
Popper, K. R. *71*, *213*, *312*, *320*, 321, *390*, *391*, *547*, 548
Post, H. R. *71*, 75
Prabhu, N. U. *419*, 421
Prandtl *706*, 709
Pribram, K. H. *560*, 560
Prigogine, I. 32
Purnell, F. 103
Putnam, H. 13, 75, *140*, *141*, 141, *265*, *525*, 584, **603**, 728

Quay, P. 32
Quine, W. V. O. *288*, *289*, 289, *376*, *390*, 584, *629*, *630*, 632

Rao, D. 180
Ratterman, H. A. *119*, 123
Rayleigh, Lord *74*, 75
Reddy, S. 650
Reichenbach, H. *233*, *390*, *433*, 434, 548
Rensch, B. *320*, 321
Richards, I. A. *312*
Richtmyer, F. K. *74*, 75
Riemann *120*, *433*
Rist, R. 180
Robinson, J. T. *299*
Roe, A. 548
Rose, L. E. **469**, *476*, 476
Rosenfeld, L. *74*, 75, *525*
Rosenthal *476*
Roth, N. M. *705*, 709

INDEX OF NAMES

Royce, J. *118*, 123
Ruelle, D. 31
Ruse, M. *628*, *629*, *631*, 632, **633**, *650*, 650, 670, *705*, *706*, 709
Russell, B. *119*, *120*, 123, *390*, 560, *582*, 584
Ryle, G. *376*

Salmon, W. C. *376*, *582*, *583*, 584, 709
Sarvajna, D. K. 486
Schaffner, K. F. *433*, 434, **613**, *628*, 632, *650*, 650, 651, *669*, 670, *705*, *706*, 709
Scheffler, I. *299*, *311*, 498
Schlesinger, G. 13, *421*, 421
Schmitt, C. B. 103
Schumpeter, J. 447
Schwab, J. J. *299*
Sellars, W. *560*, 560
Settle, T. B. *103*, **437,** 447
Shapere, D. 75, 103, 548, *628*, 632
Shea, W. R. 103
Shields *178*
Shimony, A. 709
Shklovsky, I. S. 486
Shockley, W. *179*, 180
Simon, H. 360, *706*, 709
Simon, M. *669*, 670
Sinai, Ya. 31
Skeels, H. M. 180
Sklar, L. *15*, *71*, *73*, *629*, 632, *669*, *707*, 709
Skodak, M. 180
Skolimowski, H. **191,** *213*, **459**
Smart, J. J. *140*, 141
Smith, A. 447
Smith, M. 188
Sommerfeld *73*
Spassky, B. *178*, 179
Specker, E. P. 513, 728
Spector, M. 289
Spiegelman, S. *629*, 631
Stachel, J. **515**
Stachow, E. W. 514
Stallo, J. B. **105,** *118*, *119*, 123
Stenner, A. 359
Stetson, B. R. 180
Strauss, M. *525*, *526*
Strickberger, M. W. *650*, 651
Strong, J. V. **105**
Struik, D. J. 498
Suchting, W. A. *391*, 434

Suppes, P. *418*, 422
Svensson, F. *465*
Swain *359*
Szilard, L. *629*

Tait, P. G. *118*, 123
Talbot, S. A. 561
Taylor, L. W. *497*, 498
Taylor, R. 422
Terman, L. *178*, 180
Thiele, J. *119*, *121*, 123
Thompson, D'Arcy *706*, 709
Thomson, G. H. 141
Tietjens, O. G. *706*, 709
Tisza, L. 32, *71*, 75
Tolman, R. 31, 32
Tomonaga, S. 75
Toulmin, S. *311*, *320*, 321, 548, *601*, *669*
Trautman, A. *320*, 321
Truesdell, C. A. 32, *320*, 321, 719
Tuckerman, B. *476*, 476
Tullock, G. 447
Tuomela, R. **325,** 360
Tyler, L. *140*, 141

Urbach, P. *140*, 141

Valois, R. L. De 560
Velikovsky, I. *476*, 476, **477,** 486, **487**
Vernon, P. E. 141
Voyer, R. 447

Waerden, B. L. van der 75
Wagner, R. P. *629*, *630*, *631*, 632
Waismann, F. *706*, 710
Walker, R. M. 486
Wallace, J. 289
Wallace, W. A. **79,** 103, 104
Wallach, H. 560
Walster, E. 179
Watanabe, S. *419*, 422
Watson, J. D. *629*, 632
Webber, R. *178*
Wechsler, D. *140*, 141
Weisskopf, V. F. *421*, 422
Weller, C. M. *311*
Wessels, L. **215**
Whewell, W. *122*, 123
Whiston, W. *122*

White, L. *465*
Whitehouse, H. L. K. *708*, 710
Whittaker, E. T. 75, *121*, 123
Wien *74*
Wiesel, T. N. *560*, 560
Wiesner, J. B. 498
Wigner, E. P. *421*, 422
Wilkinson, G. D. *119*, 123
Will, F. L. *312*
Williams, H. S. 498
Williams, I. P. 486
Williamson, A. W. *119*
Wimsatt, W. C. *630*, 632, **671**, *705*, 710
Winchester *650*

Winnie, J. 234, *432*, 434
Wisan, W. L. 104
Wolf, N. S. 103
Woodger, J. 670
Wright, G. H. von 360

Yerkes, R. 130
Yourgrau, W. 75

Zahar, E. G. *601*
Ziff, P. *319*, 321
Zímmerman, D. W. 486
Zimmerman, J. 486
Zuckerman, H. *547*, 548

SYNTHESE LIBRARY

Monographs on Epistemology, Logic, Methodology,
Philosophy of Science, Sociology of Science and of Knowledge, and on the
Mathematical Methods of Social and Behavioral Sciences

Managing Editor:

JAAKKO HINTIKKA (Academy of Finland and Stanford University)

Editors:

ROBERT S. COHEN (Boston University)
DONALD DAVIDSON (The Rockefeller University and Princeton University)
GABRIËL NUCHELMANS (University of Leyden)
WESLEY C. SALMON (University of Arizona)

1. J. M. BOCHEŃSKI, *A Precis of Mathematical Logic*. 1959, X + 100 pp.
2. P. L. GUIRAUD, *Problèmes et méthodes de la statistique linguistique*. 1960, VI + 146 pp.
3. HANS FREUDENTHAL (ed.), *The Concept and the Role of the Model in Mathematics and Natural and Social Sciences, Proceedings of a Colloquium held at Utrecht, The Netherlands, January 1960*. 1961, VI + 194 pp.
4. EVERT W. BETH, *Formal Methods. An Introduction to Symbolic Logic and the Study of effective Operations in Arithmetic and Logic*. 1962, XIV + 170 pp.
5. B. H. KAZEMIER and D. VUYSJE (eds.), *Logic and Language. Studies dedicated to Professor Rudolf Carnap on the Occasion of his Seventieth Birthday*. 1962, VI + 256 pp.
6. MARX W. WARTOFSKY (ed.), *Proceedings of the Boston Colloquium for the Philosophy of Science, 1961–1962*, Boston Studies in the Philosophy of Science (ed. by Robert S. Cohen and Marx W. Wartofsky), Volume I. 1973, VIII + 212 pp.
7. A. A. ZINOV'EV, *Philosophical Problems of Many-Valued Logic*. 1963. XIV + 155 pp.
8. GEORGES GURVITCH, *The Spectrum of Social Time*. 1964, XXVI + 152 pp.
9. PAUL LORENZEN, *Formal Logic*. 1965, VIII + 123 pp.
10. ROBERT S. COHEN and MARX W. WARTOFSKY (eds.), *In Honor of Philipp Frank*, Boston Studies in het Philosophy of Science (ed. by Robert S. Cohen and Marx W. Wartofsky), Volume II. 1965, XXXIV + 475 pp.
11. EVERT W. BETH, *Mathematical Thought. An Introduction to the Philosopy of Mathematics*. 1965, XII + 208 pp.
12. EVERT W. BETH and JEAN PIAGET, *Mathematical Epistemology and Psychology*. 1966, XII + 326 pp.
13. GUIDO KÜNG, *Ontology and the Logistic Analysis of Language. An Enquiry into the Contemporary Views on Universals*. 1967, XI + 210 pp.
14. ROBERT S. COHEN and MARX W. WARTOFSKY (eds.), *Proceedings of the Boston Colloquium for the Philosophy of Science 1964–1966, in Memory of Norwood Russell Hanson*, Boston Studies in the Philosophy of Science (ed. by Robert S. Cohen and Marx W. Wartofsky), Volume III. 1967, XLIX + 489 pp.

15. C. D. BROAD, *Induction, Probability, and Causation. Selected Papers*. 1968, XI + 296 pp.
16. GÜNTHER PATZIG, *Aristotle's Theory of the Syllogism. A logical-Philosophical Study of Book A of the Prior Analytics*. 1968, XVII + 215 pp.
17. NICHOLAS RESCHER, *Topics in Philosophical Logic*. 1968, XIV + 347 pp.
18. ROBERT S. COHEN and MARX W. WARTOFSKY (eds.), *Proceedings of the Boston Colloquium for the Philosophy of Science 1966–1968*, Boston Studies in the Philosophy of Science (ed. by Robert S. Cohen and Marx W. Wartofsky), Volume IV. 1969, VIII + 537 pp.
19. ROBERT S. COHEN and MARX W. WARTOFSKY (eds.), *Proceedings of the Boston Colloquium for the Philosophy of Science 1966–1968*, Boston Studies in the Philosophy of Science (ed. by Robert S. Cohen and Marx W. Wartofsky), Volume V. 1969, VIII + 482 pp.
20. J. W. DAVIS, D. J. HOCKNEY, and W. K. WILSON (eds.), *Philosophical Logic*. 1969, VIII + 277 pp.
21. D. DAVIDSON and J. HINTIKKA (eds.), *Words and Objections. Essays on the Work of W. V. Quine*. 1969, VIII + 366 pp.
22. PATRICK SUPPES, *Studies in the Methodology and Foundations of Science. Selected Papers from 1911 to 1969*, XII + 473 pp.
23. JAAKKO HINTIKKA, *Models for Modalities. Selected Essays*. 1969, IX + 220 pp.
24. NICHOLAS RESCHER et al. (eds.), *Essays in Honor of Carl G. Hempel. A Tribute on the Occasion of his Sixty-Fifth Birthday*. 1969, VII + 272 pp.
25. P.V. TAVANEC (ed.), *Problems of the Logic of Scientific Knowledge*. 1969, VII + 429 pp.
26. MARSHALL SWAIN (ed.), *Induction, Acceptance, and Rational Belief*. 1970, VII + 232 pp.
27. ROBERT S. COHEN and RAYMOND J. SEEGER (eds.), *Ernst Mach: Physicist and Philosopher*, Boston Studies in the Philosophy of Science (ed. by Robert S. Cohen and Marx W. Wartofsky), Volume VI. 1970, VIII + 295 pp.
28. JAAKKO HINTIKKA and PATRICK SUPPES, *Information and Inference*. 1970, X + 336 pp.
29. KAREL LAMBERT, *Philosophical Problems in Logic. Some Recent Developments*. 1970, VII + 176 pp.
30. ROLF A. EBERLE, *Nominalistic Systems*. 1970, IX + 217 pp.
31. PAUL WEINGARTNER and GERHARD ZECHA (eds.), *Induction, Physics, and Ethics. Proceedings and Discussions of the 1968 Salzburg Colloquium in the Philosophy of Science*. 1970, X + 382 pp.
32. EVERT W. BETH, *Aspects of Modern Logic*. 1970, XI + 176 pp.
33. RISTO HILPINEN (ed.), *Deontic Logic: Introductory and Systematic Readings*. 1971, VII + 182 pp.
34. JEAN-LOUIS KRIVINE, *Introduction to Axiomatic Set Theory*. 1971, VII + 98 pp.
35. JOSEPH D. SNEED, *The Logical Structure of Mathematical Physics*. 1971, XV + 311 pp.
36. CARL R. KORDIG, *The Justification of Scientific Change*. 1971, XIV + 119 pp.
37. MILIČ ČAPEK, *Bergson and Modern Physics*, Boston Studies in the Philosophy of Science (ed. by Robert S. Cohen and Marx W. Wartofsky), Volume VII. 1971, XV + 414 pp.
38. NORWOOD RUSSELL HANSON, *What I do Not Believe, and Other Essays* (ed. by Stephen Toulmin and Harry Woolf), 1971, XII + 390 pp.
39. ROGER C. BUCK and ROBERT S. COHEN (eds.), *PSA 1970. In Memory of Rudolf Carnap*, Boston Studies in the Philosophy of Science (ed. by Robert S. Cohen and Marx W. Wartofsky), Volume VIII. 1971, LXVI + 615 pp. Also available as paperback.
40. DONALD DAVIDSON and GILBERT HARMAN (eds.), *Semantics of Natural Language*. 1972, X + 769 pp. Also available as paperback.

41. YEHOSHUA BAR-HILLEL (ed.), *Pragmatics of Natural Languages*. 1971, VII + 231 pp.
42. SÖREN STENLUND, *Combinators, λ-Terms and Proof Theory*. 1972, 184 pp.
43. MARTIN STRAUSS, *Modern Physics and Its Philosophy. Selected Papers in the Logic, History, and Philosophy of Science.* 1972, X + 297 pp.
44. MARIO BUNGE, *Method, Model and Matter*. 1973, VII + 196 pp.
45. MARIO BUNGE, *Philosophy of Physics*. 1973, IX + 248 pp.
46. A. A. ZINOV'EV, *Foundations of the Logical Theory of Scientific Knowledge (Complex Logic)*, Boston Studies in the Philosophy of Science (ed. by Robert S. Cohen and Marx W. Wartofsky), Volume IX. Revised and enlarged English edition with an appendix, by G. A. Smirnov, E. A. Sidorenka, A. M. Fedina, and L. A. Bobrova. 1973, XXII + 301 pp. Also available as paperback.
47. LADISLAV TONDL, *Scientific Procedures*, Boston Studies in the Philosophy of Science (ed. by Robert S. Cohen and Marx W. Wartofsky), Volume X. 1973, XII + 268 pp. Also available as paperback.
48. NORWOOD RUSSELL HANSON, *Constellations and Conjectures*, (ed. by Willard C. Humphreys, Jr.), 1973, X + 282 pp.
49. K. J. J. HINTIKKA, J. M. E. MORAVCSIK, and P. SUPPES (eds.), *Approaches to Natural Language. Proceedings of the 1970 Stanford Workshop on Grammar and Semantics.* 1973, VIII + 526 pp. Also available as paperback.
50. MARIO BUNGE (ed.), *Exact Philosophy – Problems, Tools, and Goals*. 1973, X + 214 pp.
51. RADU J. BOGDAN and ILKKA NIINILUOTO (eds.), *Logic, Language, and Probability. A selection of papers contributed to Sections IV, VI, and XI of the Fourth International Congress for Logic, Methodology, and Philosophy of Science, Bucharest, September 1971.* 1973, X + 323 pp.
52. GLENN PEARCE and PATRICK MAYNARD (eds.), *Conceptual Chance*. 1973, XII + 282 pp.
53. ILKKA NIINILUOTO and RAIMO TUOMELA, *Theoretical Concepts and Hypothetico-Inductive Inference*. 1973, VII + 264 pp.
54. ROLAND FRAÏSSÉ, *Course of Mathematical Logic – Volume 1: Relation and Logical Formula*. 1973, XVI + 186 pp. Also available as paperback.
55. ADOLF GRÜNBAUM, *Philosophical Problems of Space and Time*. Second, enlarged edition, Boston Studies in the Philosophy of Science (ed. by Robert S. Cohen and Marx W. Wartofsky), Volume XII. 1973, XXIII + 884 pp. Also available as paperback.
56. PATRICK SUPPES (ed.), *Space, Time, and Geometry*. 1973, XI + 424 pp.
57. HANS KELSEN, *Essays in Legal and Moral Philosophy*, selected and introduced by Ota Weinberger. 1973, XXVIII + 300 pp.
58. R. J. SEEGER and ROBERT S. COHEN (eds.), *Philosophical Foundations of Science. Proceedings of an AAAS Program, 1969.* Boston Studies in the Philosophy of Science (ed. by Robert S. Cohen and Marx W. Wartofsky), Volume XI. 1974, X + 545 pp. Also available as paperback.
59. ROBERT S. COHEN and MARX W. WARTOFSKY (eds.), *Logical and Epistemological Studies in Contemporary Physics*, Boston Studies in the Philosophy of Science (ed. by Robert S. Cohen and Marx W. Wartofsky), Volume XIII. 1973, VIII + 462 pp. Also available as paperback.
60. ROBERT S. COHEN and MARX W. WARTOFSKY (eds.), *Methodological and Historical Essays in the Natural and Social Sciences. Proceedings of the Boston Colloquium for the Philosophy of Science, 1969–1972*, Boston Studies in the Philosophy of Science (ed. by Robert S. Cohen and Marx W. Wartofsky), Volume XIV. 1974, VIII + 405 pp. Also available as paperback.
61. ROBERT S. COHEN, J. J. STACHEL and MARX W. WARTOFSKY (eds.), *For Dirk Struik.*

Scientific, Historical and Polical Essays in Honor of Dirk J. Struik, Boston Studies in the Philosophy of Science (ed. by Robert S. Cohen and Marx W. Wartofsky), Volume XV. 1974, XXVII + 652 pp. Also available as paperback.
62. KAZIMIERZ AJDUKIEWICZ, *Pragmatic Logic*, transl. from the Polish by Olgierd Wojtasiewicz. (1974, XV + 460 pp.
63. SÖREN STENLUND (ed.), *Logical Theory and Semantic Analysis. Essays Dedicated to Stig Kanger on His Fiftieth Birthday*. 1974, V + 217 pp.
64. KENNETH F. SCHAFFNER and ROBERT S. COHEN (eds.), *Proceedings of the 1972 Biennial Meeting, Philosophy of Science Association*, Boston Studies in the Philosophy of Science (ed. by Robert S. Cohen and Marx W. Wartofsky), Volume XX. 1974, IX + 444 pp. Also available as paperback.
65. HENRY E. KYBURG, JR., *The Logical Foundations of Statistical Inference*. 1974, IX + 421 pp.
66. MARJORIE GRENE, *The Understanding of Nature: Essays in the Philosophy of Biology*, Boston Studies in the Philosophy of Science (ed. by Robert S. Cohen and Marx W. Wartofsky), Volume XXIII. 1974, XII + 360 pp. Also available as paperback.
67. JAN M. BROEKMAN, *Stucturalism: Moscow, Prague, Paris*. 1974, IX + 117 pp.
68. NORMAN GESCHWIND, *Selected Papers on Language and the Brain*, Boston Studies in the Philosophy of Science (ed. by Robert S. Cohen and Marx W. Wartofsky), Volume XVI. 1974, XII + 549 pp. Also available as paperback.
69. ROLAND FRAÏSSÉ, *Course of Mathematical Logic* – Volume II: *Model Theory*. 1974, XIX + 192 pp.
70. ANDRZEJ GRZEGORCZYK, *An Outline of Methematical Logic. Fundamental Results and Notions Explained with All Details*. 1974, X + 596 pp.
71. FRANZ VON KUTSCHERA, *Philosophy of Language*. 1975, VII + 305 pp.
72. JUHA MANNINEN and RAIMO TUOMELA (eds.), *Essays on Explanation and Understanding. Studies in the Foundations of Humanities and Social Sciences*. 1976, VII + 440 pp.
73. JAAKKO HINTIKKA (ed.), *Rudolf Carnap, Logical Empiricist. Materials and Perspectives*. 1975, LXVIII + 400 pp.
74. MILIČ ČAPEK (ed.), *The Concepts of Space and Time. Their Structure and Their Development*. Boston Studies in the Philosophy of Science (ed. by Robert S. Cohen and Marx W. Wartofsky), Volume XXII. 1976, LVI + 570 pp. Also available as paperback.
75. JAAKKO HINTIKKA and UNTO REMES, *The Method of Analysis. Its Geometrical Origin and Its General Significance*. Boston Studies in the Philosophy of Science (ed. by Robert S. Cohen and Marx W. Wartofsky), Volume XXV. 1974, XVIII + 144 pp. Also available as paperback.
76. JOHN EMERY MURDOCH and EDITH DUDLEY SYLLA, *The Cultural Context of Medieval Learning. Proceedings of the First International Colloquium on Philosophy, Science, and Theology in the Middle Ages – September 1973*. Boston Studies in the Philosophy of Science (ed. by Robert S. Cohen and Marx W. Wartofsky), Volume XXVI. 1975, X + 566 pp. Also available as paperback.
77. STEFAN AMSTERDAMSKI, *Between Experience and Metaphysics. Philosophical Problems of the Evolution of Science*. Boston Studies in the Philosophy of Science (ed. by Robert S. Cohen and Marx W. Wartofsky), Volume XXXV. 1975, XVIII + 193 pp. Also available as paperback.
78. PATRICK SUPPES (ed.), *Logic and Probability in Quantum Mechanics*. 1976, XV + 541 pp.
80. JOSEPH AGASSI, *Science in Flux*. Boston Studies in the Philosophy of Science (ed. by Robert S. Cohen and Marx W. Wartofsky), Volume XXVIII. 1975, XXVI + 553 pp. Also available as paperback.

81. SANDRA G. HARDING (ed.), *Can Theories Be Refuted? Essays on the Duhem-Quine Thesis*. 1976, XXI + 318 pp. Also available as paperback.
84. MARJORIE GRENE and EVERETT MENDELSOHN (eds.), *Topics in the Philosophy of Biology*. Boston Studies in the Philosophy of Science (ed. by Robert S. Cohen and Marx W. Wartofsky), Volume XXVII. 1976, XIII + 454 pp. Also available as paperback.
85. E. FISCHBEIN, *The Intuitive Sources of Probabilistic Thinking in Children*. 1975, XIII + 204 pp.
86. ERNEST W. ADAMS, *The Logic of Conditionals. An Application of Probability to Deductive Logic*. 1975, XIII + 156 pp.
89. A. KASHER (ed.), *Language in Focus: Foundations, Methods and Systems. Essays Dedicated to Yehoshua Bar-Hillel*. Boston Studies in the Philosophy of Science (ed. by Robert S. Cohen and Marx W. Wartofsky), Volume XLIII. 1976, XXVIII + 679 pp. Also available as paperback.
90. JAAKKO HINTIKKA, *The Intentions of Intentionality and Other New Models for Modalities*. 1975, XVIII + 262 pp. Also available as paperback.
93. RADU J. BOGDAN, *Local Induction*. 1976, XIV + 340 pp.
95. PETER MITTELSTAEDT, *Philosophical Problems of Modern Physics*. Boston Studies in the Philosophy of Science (ed. by Robert S. Cohen and Marx W. Wartofsky), Volume XVIII. 1976, X + 211 pp. Also available as paperback.
96. GERALD HOLTON and WILLIAM BLANPIED (eds.), *Science and Its Public: The Changing Relationship*. Boston Studies in the Philosophy of Science (ed. by Robert S. Cohen and Marx W. Wartofsky), Volume XXXIII. 1976, XXV + 289 pp. Also available as paperback.

SYNTHESE HISTORICAL LIBRARY

Texts and Studies
in the History of Logic and Philosophy

Editors:

N. KRETZMANN (Cornell University)
G. NUCHELMANS (University of Leyden)
L. M. DE RIJK (University of Leyden)

1. M. T. BEONIO-BROCCHIERI FUMAGALLI, *The Logic of Abelard.* Translated from the Italian. 1969, IX + 101 pp.
2. GOTTFRIED WILHELM LEIBNITZ, *Philosophical Papers and Letters.* A selection translated and edited, with an introduction, by Leroy E. Loemker. 1969, XII + 736 pp.
3. ERNST MALLY, *Logische Schriften*, ed. by Karl Wolf and Paul Weingartner. 1971, X + 340 pp.
4. Lewis White Beck (ed.), *Proceedings of the Third International Kant Congress.* 1972, XI + 718 pp.
5. BERNARD BOLZANO, *Theory of Science*, ed. by Jan Berg. 1973, XV + 398 pp.
6. J. M. E. MORAVCSIK (ed.), *Patterns in Plato's Thought. Papers arising out of the 1971 West Coast Greek Philosophy Conference.* 1973, VIII + 212 pp.
7. NABIL SHEHABY, *The Propositional Logic of Avicenna: A Translation from al-Shifā: al-Qiyās*, with Introduction, Commentary and Glossary. 1973, XIII + 296 pp.
8. DESMOND PAUL HENRY, *Commentary on De Grammatico: The Historical-Logical Dimensions of a Dialogue of St. Anselm's.* 1974, IX + 345 pp.
9. JOHN CORCORAN, *Ancient Logic and Its Modern Interpretations.* 1974, X + 208 pp.
10. E. M. BARTH, *The Logic of the Articles in Traditional Philosophy.* 1974, XXVII + 533 pp.
11. JAAKKO HINTIKKA, *Knowledge and the Known. Historical Perspectives in Epistemology.* 1974, XII + 243 pp.
12. E. J. ASHWORTH, *Language and Logic in the Post-Medieval Period.* 1974, XIII + 304 pp.
13. ARISTOTLE, *The Nicomachean Ethics.* Translated with Commentaries and Glossary by Hyppocrates G. Apostle. 1975, XXI + 372 pp.
14. R. M. DANCY, *Sense and Contradiction: A Study in Aristotle.* 1975, XII + 184 pp.
15. WILBUR RICHARD KNORR, *The Evolution of the Euclidean Elements. A Study of the Theory of Incommensurable Magnitudes and Its Significance for Early Greek Geometry.* 1975, IX + 374 pp.
16. AUGUSTINE, *De Dialectica.* Translated with the Introduction and Notes by B. Darrell Jackson. 1975, XI + 151 pp.